P9-CKX-684

METHODS IN CELL BIOLOGY

VOLUME 31

Vesicular Transport

Part A

ASCB

METHODS IN CELL BIOLOGY

Prepared under the Auspices of the American Society for Cell Biology

VOLUME 31

Vesicular Transport
Part A

Edited by

ALAN M. TARTAKOFF
INSTITUTE OF PATHOLOGY
CASE WESTERN RESERVE UNIVERSITY SCHOOL OF MEDICINE
CLEVELAND, OHIO

ACADEMIC PRESS, INC.
Harcourt Brace Jovanovich, Publishers
San Diego New York Berkeley Boston
London Sydney Tokyo Toronto

ACADEMIC PRESS, INC.
San Diego, California 92101

United Kingdom Edition published by
ACADEMIC PRESS LIMITED
24-28 Oval Road, London NW1 7DX

LIBRARY OF CONGRESS CATALOG CARD NUMBER: 64-14220

ISBN 0-12-564131-1 (alk. paper)

PRINTED IN THE UNITED STATES OF AMERICA
89 90 91 92 9 8 7 6 5 4 3 2 1

For Paola Ymayo,
Daniela Helen Elizabeth,
and Joseph Michael

CONTENTS

PART IV. MORPHOLOGICAL PROCEDURES

CONTRIBUTORS

Numbers in parentheses indicate the pages on which the authors' contributions begin.

ALISON E. M. ADAMS, Department of Biology, Massachusetts Institute of Technology, Cambridge, Massachusetts 02139 (357)

GUDRUN AHNERT-HILGER, Universität Ulm, Abteilung Anatomie und Zellbiologie, Oberer Eselsberg, D-7900 Ulm, Federal Republic of Germany (63)

RICHARD G. W. ANDERSON, Department of Cell Biology and Anatomy, The University of Texas Southwestern Medical Center, Dallas, Texas 75235 (463)

CLAUDE ANTONY, European Molecular Biology Laboratory, D-6900, Heidelberg, Federal Republic of Germany (105)

ROBERT BACALLAO, Division of Nephrology, UCLA Medical Center, Los Angeles, California 90024 (437)

DAVID BAKER, Department of Biochemistry, University of California, Berkeley, California 94720 (129)

W. E. BALCH, Department of Molecular Biophysics and Biochemistry, Yale University, New Haven, Connecticut 06405 (91)

C. J. M. BECKERS, Department of Molecular Biophysics and Biochemistry, Yale University, New Haven, Connecticut 06405 (91)

MARK K. BENNETT, European Molecular Biology Laboratory, D-6900 Heidelberg, Federal Republic of Germany (105)

KEITH BROKLEHURST, Laboratory of Cell Biology and Genetics, National Institute of Diabetes, Digestive and Kidney Diseases, National Institutes of Health, Bethesda, Maryland 20892 (207)

WILLIAN J. BROWN, Section of Biochemistry, Molecular, and Cell Biology, Cornell University, Ithaca, New York 14853 (553)

A. LEE BURNS, Laboratory of Cell Biology and Genetics, National Institute of Diabetes, Digestive and Kidney Diseases, National Institutes of Health, Bethesda, Maryland 20892 (207)

IVAN DECURTIS, European Molecular Biology Laboratory, D-6900, Heidelberg, Federal Republic of Germany (105)

RUBEN DIAZ, Department of Cell Biology and Physiology, Washington University School of Medicine, St. Louis, Missouri 63110 (25, 179)

FRANCESCO DI VIRGILIO, C.N.R. Center for the Study of the Physiology of Mitochondria, and Institute of General Pathology, University of Padua, Padua, Italy (453)

STEPHEN DOXSEY, Department of Biochemistry and Biophysics, University of California, San Francisco, California 94143 (157)

DAVID G. DRUBIN,[1] Department of Biology, Massachusetts Institute of Technology, Cambridge, Massachusetts 02139 (357)

HARMA ELLENS, Department of Pharmacology, University of California, San Francisco, California 94143 (157)

MARILYN G. FARQUHAR, Department of Cell Biology, Yale University School of Medicine, New Haven, Connecticut 06510 (553)

SUSAN FERRO-NOVICK, Department of Cell Biology, Yale University School of Medicine, New Haven, Connecticut 06510 (145)

[1] *Present address:* Department of Molecular and Cell Biology, University of California, Berkeley, California 94720.

Karl Josef Föhr, Universität Ulm, Abteilung Anatomie und Zellbiologie, Oberer Eselsberg, D-7900 Ulm, Federal Republic of Germany (63)

Jeffrey S. Glenn, Department of Biochemistry and Biophysics, University of California, San Francisco, California 94143 (157)

Bruno Goud,[2] Department of Cell Biology, Yale University School of Medicine, New Haven, Connecticut 06510 (335)

Manfred Gratzl, Universität Ulm, Abteilung Anatomie und Zellbiologie, Oberer Eselsberg, D-7900 Ulm, Federal Republic of Germany (63)

Jean Gruenberg, European Molecular Biology Laboratory, 6900 Heidelberg, Federal Republic of Germany (265)

Dwijendra Gupta,[3] Institute of Pathology, Case Western Reserve University School of Medicine, Cleveland, Ohio 44106 (247)

Brian K. Haarer, Department of Anatomy and Cell Biology, The University of Michigan, Ann Arbor, Michigan 48109 (357)

Harry Haigler, Department of Physiology and Biophysics, University of California, Irvine, California 92717 (207)

Leah T. Haimo, Department of Biology, University of California, Riverside, California 92521 (3)

Kathryn E. Howell, European Molecular Biology Laboratory, 6900 Heidelberg, Federal Republic of Germany (265)

Elizabeth W. Jones, Department of Biological Sciences, Carnegie Mellon University, Pittsburgh, Pennsylvania 15213 (357)

Alisa Kastan Kabcenell, Department of Cell Biology, Yale University School of Medicine, New Haven, Connecticut 06510 (145)

Jürgen Kartenbeck, Deutsches Krebsforschungszentrum, D-6900, Heidelberg, Federal Republic of Germany (105)

D. S. Keller, Department of Molecular Biophysics and Biochemistry, Yale University, New Haven, Connecticut 06405 (91)

Wolfgang Mach, Universität Ulm, Abteilung Anatomie und Zellbiologie, Oberer Eselsberg, D-7900 Ulm, Federal Republic of Germany (63)

Mark Marsh, Institute of Cancer Research, Chester Beatty Laboratories, London SW3 6JB, England (319)

Luis S. Mayorga, Department of Cell Biology and Physiology, Washington University School of Medicine, St. Louis, Missouri 63110 (179)

Robert F. Murphy, Department of Biological Sciences, and Center for Fluorescence Research in Biomedical Sciences, Carnegie-Mellon University, Pittsburgh, Pennsylvania 15213 (293)

Peter J. Novick, Department of Cell Biology, Yale University School of Medicine, New Haven, Connecticut 06510 (335)

Barbara M. F. Pearse, MRC Laboratory of Molecular Biology, Cambridge CB2 2QH, England (229)

[2] *Present address:* Department of Immunology, Institut Pasteur, Paris, France.
[3] *Present address:* Department of Biochemistry, Faculty of Sciences, University of Allahabad, Allahabad 211 002, U. P., India.

HARVEY B. POLLARD, Laboratory of Cell Biology and Genetics, National Institute of Diabetes, Digestive and Kidney Diseases, National Institutes of Health, Bethesda, Maryland 20892 (207)

ROBERT A. PRESTON, Department of Biological Sciences, Carnegie Mellon University, Pittsburgh, Pennsylvania 15213 (357)

JOHN E. PRINGLE, Department of Biology, The University of Michigan, Ann Arbor, Michigan 48109 (357)

JASPER RINE, Department of Biochemistry, University of California, Berkeley, California 94720 (473)

EDUARDO ROJAS, Laboratory of Cell Biology and Genetics, National Institute of Diabetes, Digestive and Kidney Diseases, National Institutes of Health, Bethesda, Maryland 20892 (207)

J. ROTH, Interdepartmental Electron Microscopy Biocenter, and Department of Cell Biology, University of Basel, CH-4056 Basel, Switzerland (513)

MOSHE M. ROZDZIAL, Department of Biology, University of California, Riverside, California 92521 (3)

HANNELE RUOHOLA,[4] Department of Cell Biology, Yale University School of Medicine, New Haven, Connecticut 06510 (145, 335)

RANDY SCHEKMAN, Department of Biochemistry, University of California, Berkeley, California 94720 (129)

D. D. SCHLAEPFER, Department of Physiology and Biophysics, University of California, Irvine, California 92717 (207)

RUTH SCHMID, SINTEF, 7034 Trondheim, Norway (265)

SAMUEL C. SILVERSTEIN, Department of Physiology and Cellular Biophysics, Columbia University College of Physicians and Surgeons, New York, New York 10032 (45, 453)

KAI SIMONS, European Molecular Biology Laboratory, D-6900, Heidelberg, Federal Republic of Germany (105)

PHILIP D. STAHL, Department of Cell Biology and Physiology, Washington University School of Medicine, St. Louis, Missouri 63110 (25, 179)

TIM STEARNS, Department of Biology, Massachusetts Institute of Technology, Cambridge, Massachusetts 02139 (357)

THOMAS H. STEINBERG, Department of Medicine, Columbia University College of Physicians and Surgeons, New York, New York 10032 (45, 453)

ERNST H. K. STELZER, Confocal Light Microscopy Group, European Molecular Biology Laboratory, D-6900 Heidelberg, Federal Republic of Germany (437)

ALAN M. TARTAKOFF, Institute of Pathology, Case Western Reserve University School of Medicine, Cleveland, Ohio 44106 (247)

JOHN UGELSTAD, Department of Industrial Chemistry, University of Trondheim, 7034 Trondheim, Norway (265)

NANCY C. WALWORTH, Department of Cell Biology, Yale University School of Medicine, New Haven, Connecticut 06510 (335)

ANGELA WANDINGER-NESS, European Molecular Biology Laboratory, D-6900, Heidelberg, Federal Republic of Germany (105)

GRAHAM WARREN, The Imperial Cancer Research Fund, Lincoln's Inn Fields, London WC2A 3PX, England (197)

[4] *Present address:* Department of Biochemistry, University of Helsinki, Finland.

Judith M. White, Departments of Pharmacology and Biochemistry and Biophysics, University of California, San Francisco, California 94143 (157)

Russell B. Wilson, Department of Biological Sciences, and Center for Fluorescence Research in Biomedical Sciences, Carnegie-Mellon University, Pittsburgh, Pennsylvania 15213 (293)

Philip G. Woodman, The Imperial Cancer Research Fund, Lincoln's Inn Fields, London WC2A 3PX, England (197)

Robin Wright, Department of Biochemistry, University of California, Berkeley, California 94720 (473)

PREFACE

The elucidation of the events of vesicular transport along the secretory and endocytic paths has grown out of the combined efforts of electron microscopy, cyto- and immunocytochemistry, autoradiography, genetics, and biochemistry. The traditional cell types investigated exhibited macroscopic (i.e., visible in the light microscope) evidence of their transport activities—for the secretory path, so-called "regulated" secretory cells were frequent objects of study and for the endocytic path an accessible and conspicuous model was provided by phagocytic cells. We now realize that these were both special cases—much as the study of skeletal muscle was a special case which provided essential background for investigating the contractile and cytoskeletal elements of non-muscle cells. Today, the attention of cell biologists has broadened to include cells engaged in constitutive protein transport and cells engaged in pinocytosis and receptor-mediated endocytosis of soluble ligands.

The diversification of choice of cell type has been matched by a diversification of the macromolecules under study since the availability of highly sensitive and specific immune reagents has made it possible to study the transport of essentially any macromolecule. The result has been a major increase in interest in this area of membrane cell biology and the development of an "applied" or "protein-specific" cell biology. The dividends have not, however, been exclusively applied—the insistence of scientists upon investigation of their favorite objects of study has inevitably brought to light basic phenomena of far-reaching importance, such as the numerous observations of general importance for understanding vesicular transport which have resulted from the study of transport of viral envelope glycoproteins or from the study of transport in yeast.

These volumes of *Methods in Cell Biology* highlight procedures of general interest, some of which are of use for investigation of the basic mechanisms operating in intracellular transport and some of which will be most valuable for descriptive studies by investigators monitoring the transport of a particular macromolecule. A limited number of other publications include procedural detail of the sort which is included in these chapters. When possible, these cross-references are given either in the prefatory remarks or in the chapters themselves. Nevertheless, cell biological methods have often not been systematized—they must be witnessed first-hand in order to be exactly reproduced. As motivation for clarity and completeness of exposition, I have urged the contributors to these volumes to consider that their chapters provide an opportunity to reduce their expenditure of time in explaining their methods first-hand to others.

Despite the established importance of nucleic acid-based procedures (transfection, quantitation of mRNA, manipulation of the structure both of the cell's transport apparatus and of the structure of molecules undergoing transport, etc.) for study of vesicular transport, these procedures are not emphasized in these volumes. This is because they are still in a state of development, because they are covered in several up-to-date texts and symposia (Spandidos and Wilkie, 1984; Chirikjian, 1985; Hooper, 1985; Glover, 1985, 1987; Miller and Calos, 1987), and because most of these methods are not designed primarily for the study of vesicular transport.

Methods for the study of vectorial transport to the cisternal space of the RER have also been covered before (Fleischer and Fleischer, 1983), although the newest yeast protocols, which make use of permeabilized cells and post-translational vectorial transport, are presented in Section I,A.

Because of the remarkable genetic accessibility of yeast (Schekman, 1985; Botstein and Fink, 1988), it has come to serve as a minimal model for studies of vesicular transport. I have therefore included several chapters on methods specific to yeast (an overview of the secretory path in the context of SEC mutants, permeabilization procedures, subcellular fractionation, and EM immunocytochemical and fluorescent methods). Two earlier volumes in this series have been entirely devoted to yeast. Their tables of contents are included in the Contents of Recent Volumes.

Apart from the effort to provide a precise description of the covalent and conformational maturation of macromolecules undergoing transport, major issues which I hope these volumes will help address are: (1) the identification of structural determinants which govern the destination and itinerary of individual macromolecules and (2) the description of vesicular carriers and the mechanisms which underlie their functions. It is especially this latter area which is obscure, despite definite progress in development of cell-free models of transport and first analysis of cellular mutants defective in transport.

Outstanding open questions concern: (1) the nature of specific membrane–membrane recognition; (2) the control of membrane fusion; (3) the possible roles of "cytoskeletal" components; (4) the issue of how individual vesicular carriers can participate in cyclic transport; and (5) the issue of the extent to which membrane–membrane interactions are stochastic.

Many well-studied models for specific membrane–membrane interaction concern the approximation and fusion of the ectodomain of membranes, for example, sperm–egg interaction, recognition between *Chlamydomonas* mating types, enveloped virus-cell surface or -endosome fusion. By contrast, events of vesicle fusion along the secretory and endocytic paths must be initiated at the endodomain (e.g., endosome–endosome fusion; secretion granule–plasma membrane fusion). Vesicle departure from a closed surface may, however, involve fusion between the apposed ectodomains of a single continuous surface (e.g., budding from the transitional elements of the RER, endocytosis).

Although fusion of isolated vesicles has been accomplished *in vitro*, and although there are specific proteins on the endodomain of vesicular carriers, it is not known to what extent these components are responsible for the specificity of membrane–membrane interaction. Considering the very large thermodynamic barrier which opposes bilayer fusion, a powerful catalyst must intervene. The extent of resolution of the events of fusion has not yet reached a level which makes it possible to generalize with regard to parameters which regulate fusion, e.g., ATP and/or specific ions. Recent studies of exocytosis argue that although ATP may be important for a priming function, it is not needed for exocytosis itself.

An important consideration in comparing the validity of reconstitution experiments using isolated organelles with experiments on permeabilized cell models is the extent to which the cytoskeleton (conceived in its broadest sense) is involved. Many pharmacologic and anatomic observations have pointed to intimate relations between microtubules and vesicular traffic; nevertheless, much of this information is consistent with tubules playing a passive rather than an active role and the most recent studies of yeast argue against an essential role for tubulin in protein secretion. Since the organization of the cytoplasm is clearly nonrandom, it would nevertheless be surprising if no cytoskeletal proteins were obligatorily involved in membrane traffic.

Ongoing studies of vesicular traffic have identified several examples of membrane recycling (in receptor-mediated endocytosis, in recapture of "exocytic membrane" contributed to the cell surface at the moment of exocytosis, in exit from the RER). These striking examples of membrane economy raise a basic question of vesicle targeting specificity—apparently a single carrier can either fuse with a given partner or separate from it. This situation suggests that the parameters which initially lead to fusion or fission decay, so that the transport options of a given vesicle change through time.

Related to this issue is the question of the extent to which membrane–membrane interactions are stochastic. Studies of the *N*-glycans of glycoproteins which exit from the RER argue strongly that secretory glycoproteins do not undergo exocytosis until they have traversed the Golgi stack; nevertheless, it is by no means clear that they traverse the Golgi in altogether sequential fashion—indeed the considerable kinetic dispersal of a cohort of pulse-labeled newly synthesized secretory proteins suggests that the itinerary may be far from direct. Moreover, along the endocytic path (which appears to have considerable overlap with terminal steps of the constitutive secretory path) growing evidence indicates that a significant portion of endocytic tracers returns to the plasma membrane rather than being uniformly delivered to lysosomes.

A broad overview of the biological issues which are addressed by the procedures described in these two volumes is given in several recent books, reviews, and symposia (Silverstein, 1978; Evered and Collins, 1982; Gething, 1985; Kelly, 1985; Pfeffer and Rothman, 1987; Pastan and Willingham, 1987; Tartakoff, 1987).

I thank Sonya Olsen for help in preparation of these volumes, Michael Lamm and the National Institutes of Health for their support, and Leslie Wilson and the inspiring Mexican countryside for the impetus to edit these volumes.

ALAN M. TARTAKOFF

REFERENCES

Botstein, D., and Fink, G., (1988). Yeast, an experimental organism for modern biology. *Science* **240**, 1439–1444.

Chirikjian, J., ed. (1985). "Gene Amplification and Analysis," Vol. 3. Elsevier, New York.

Evered, D., and Collins, G., eds (1982). "Membrane Recycling" (Ciba Foundation Symposium, New Series #92). Pittman, London.

Fleischer, S., and Fleischer, B., eds (1983). "Methods in Enzymology," Vol. 96. Academic Press, Orlando, Florida.

Gething, M. J., ed. (1985). "Protein Transport and Secretion." Cold Spring Harbor Laboratory, Cold Spring Harbor, New York.

Glover, D., ed. (1985). "DNA Cloning, A Practical Approach," Vol. II. IRL Press, Oxford.

Glover, D., ed. (1987). "DNA Cloning, A Practical Approach," Vol. III. IRL Press, Oxford.

Hooper, M. (1985). "Mammalian Cell Genetics." Wiley and Sons, New York.

Kelly, R. (1985). Pathways of protein secretion in eukaryotes. *Science* **189**, 347–358.

Miller, J., and Calos, M., eds. (1987). "Gene Transfer Vectors for Mammalian Cells." Cold Spring Harbor Laboratory, Cold Spring Harbor, New York.

Pastan, I., and Willingham, M., eds. (1987). "Endocytosis." Plenum, New York.

Pfeffer, S., and Rothman, J. (1987). Biosynthetic protein transport and sorting by the endoplasmic reticulum and Golgi. *Ann. Rev. Biochem.* **56**, 829–852.

Schekman, R. (1985). Protein localization and membrane traffic in yeast. *Ann. Rev. Cell Biol.* **1**, 115–143.

Silverstein, S., ed. (1978). "Transport of Macromolecules in Cellular Systems." Dahlem Conferenzen, Berlin.

Spandidos, D., and Wilkie, N. (1984). *In* "Transcription and Translation" (B. Hames and S. Higgins, eds.). IRL Press, Oxford.

Tartakoff, A. (1987). "The Secretory and Endocytic Paths." Wiley and Sons, New York.

Part I. Gaining Access to the Cytoplasm

Although methods are available for isolating subcellular fractions that make possible significant reconstitution of transport *in vitro* (see Part II), recent efforts have emphasized the need for procedures to introduce macromolecules (e.g., antibodies) into the cytoplasm of intact cells and the validity of permeabilized cell models for analyzing transport. A major input in this area has come from the striking success of transfection of cells in culture.

Many agents have been used to gain access to the cytoplasm of cells in suspension or adhering to Petri dishes. Certain of these are of cell-specific value (e.g., addition of ATP), whereas others are more universally applicable. The latter range from physical rupture (McNeil *et al.*, 1984; Borle *et al.*, 1986; Fechheimer *et al.*, 1987; Tao *et al.*, 1987; McNeil, 1988) to microinjection (Timm *et al.*, 1983; Gomperts and Fernandez, 1985; Celis *et al.*, 1986), osmotic (Okada and Rechsteiner, 1982; Nomura *et al.*, 1986) and electric shock (Baker and Knight, 1983), detergent treatment, and fusion with liposomes (see Leserman *et al.*, Chapter 19, Vol. 32) or with red cells (Schlegel and Rechsteiner, 1978). The universal (optimistic) goal has been to conserve a maximum of subcellular organization. A number of remarkable successes have been reported. Several of these methods are illustrated in the following chapters. Two that have been of particular importance for studies of exocytosis are high-voltage electrical shock and the use of microelectrodes for dialysis of cytoplasmic content.

A broad view of microbial toxins has been published that may be of interest (Harshman, 1988).

References

Baker, P., and Knight, D. (1983). High voltage techniques for gaining access to the interior of cells: Application to the study of exocytosis and membrane turnover. *In* "Methods in Enzymology" (S. Fleischer and B. Fleischer, eds.), Vol. 96, pp. 28–36. Academic Press, New York.

Borle, A., Freudenrich, C., and Snowdowne, K. (1986). A simple method for incorporating aequorin into mammalian cells. *Am. J. Physiol.* **251,** C323–C326.

Celis, J. E., Graessmann, A., and Loyter, A., eds. (1986). "Microinjection and Organelle Transplantation Techniques: Methods and Applications." Academic Press, Orlando, Florida.

Fechheimer, M., Boylan, J., Parker, S., Sisken, J., Patel, G., and Zimmer, S. (1987). Transfection of mammalian cells with plasmid DNA by scrape loading and sonication loading. *Proc. Natl. Acad. Sci. U.S.A.* **84,** 8463–8467.

Furosawa, M., Yamaizumi, M., Nishimura, T., Uchida, T., and Okada, Y. (1976). Use of erythrocyte ghosts for injection of substances into animal cells by cell fusion. *In* "Methods in Cell Biology" (D. M. Prescott, ed.), Vol. 14, pp. 73–80. Academic Press, New York.

Gomperts, B., and Fernandez, J. (1985). Techniques for membrane permeabilization. *Trends Biochem. Sci.* **10,** 414–417.

Harshman, S. ed. (1988). "Methods in Enzymology," Vol. 165. Academic Press, San Diego, California.

McNeil, P. (1988). Incorporation of macromolecules into living cells. *In* "Methods in Cell Biology" (Y.-L. Wang and D. Taylor, eds.), Vol. 29. Academic Press, San Diego, California.

McNeil, P., Murphy, R., Lanni, F., and Taylor, D. (1984). A method for incorporating macromolecules into adherent cells. *J. Cell Biol.* **98,** 1556–1564.

Nomura, S., Kamiya, T., and Oishi, M. (1986). A procedure to introduce protein molecules into living mammalian cells. *Exp. Cell Res.* **163,** 434–444.

Okada, C., and Rechsteiner, M. (1982). Introduction of macromolecules into cultured mammalian cells by osmotic lysis of pinocytic vesicles. *Cell (Cambridge, Mass.)* **29,** 33–41.

Schlegel, R., and Rechsteiner, M. (1978). Red-cell-mediated microinjection of macromolecules into mammalian cells. *In* "Methods in Cell Biology" (D. M. Prescott, ed.), Vol. 20, pp. 341–354. Academic Press, New York.

Tao, W., Wilkinson, J., Stanbridge, E., and Berns, M. (1987). Direct gene transfer into human cultured cells facilitated by laser micropuncture of the cell membrane. *Proc. Natl. Acad. Sci. U.S.A.* **84,** 4180–4184.

Timm, B., Kondor-Koch, C., Lehrach, H., Riedel, H., Edstrom, J.-E., and Garoff, H. (1983). Expression of viral membrane proteins from cloned cDNA by microinjection into eukaryotic cell nuclei. *In* "Methods in Enzymology" (S. Fleischer and B. Fleischer, eds.), Vol. 96, pp. 496–511. Academic Press, New York.

Chapter 1

Lysed Chromatophores: A Model System for the Study of Bidirectional Organelle Transport

LEAH T. HAIMO AND MOSHE M. ROZDZIAL[1]

Department of Biology
University of California
Riverside, California 92521

---

[1] Present address: Department of Molecular, Cellular and Developmental Biology, University of Colorado, Boulder, Colorado 80309.

I. Introduction

A. Properties of Chromatophores

Many animals can rapidly and dramatically alter their coloration, a reflection of the activity of their dermal chromatophores, pigment-containing cells. These chromatophores are particularly prominent among the lower vertebrates, such as fish, in which rapid, short-term color changes provide camouflage or are used for territorial or sexual display (Fig. 1). The dermal chromatophores are of considerable interest as a model system to study organelle transport, as the pigment in these cells, which is contained in thousands of organelles termed pigment granules, is coordinately and cyclically transported to and from the cell center (Fig. 2). When the granules are aggregated (Fig 2a), most of the cell is unpigmented and, as a result, the animal appears lightly colored (Fig. 1). Alternatively, when the granules are dispersed throughout the cytoplasm (Fig. 2b), the entire chromatophore is pigmented, and the animal appears dark (Fig. 1). These color changes occur in a time span of less than a minute to ~1 hour, depending on the type of chromatophores involved. The physiological changes in coloration of the dermal chromatophores, which are mediated by hormonal or neuronal stimulation (for reviews, see Bagnara and Hadley, 1973; Novales, 1983), are in contrast to the morphological changes in epidermal chromatophores, which generally do not transport pigment granules, but rather darken over a time span of hours or days as they synthesize pigment. Epidermal chromatophores will not be discussed further in this chapter; the reader is referred to the excellent treatise on chromatophores by Bagnara and Hadley (1973) for additional information.

B. Chromatophores as a Model System to Study Organelle Transport

The sole function of the dermal chromatophores is to transport pigment and, accordingly, these cells have long been used as an excellent, highly ordered system in which to study organelle movements (for reviews, see Schliwa and Euteneuer, 1983; Schliwa, 1984; Obika, 1986). Because the two directions of transport occur as temporally distinct events, motility in each direction can be examined independently without contributions from one complicating an analysis of the other. Similarly, as transport in both directions can be initiated in cells *in vitro* by various agents, the mechanisms that regulate the direction, which may be distinct from the mechanisms generating movement, can also be examined. In addition to

FIG. 1. The African cichlid, *Tilapia mossambica*, exhibiting two extremes in coloration. (a) Unstressed *Tilapia* are normally mottled white and black. (b) The white fish exhibits camouflage behavior, having been placed in a white container shortly before being photographed. The black fish exhibits male dominant sexual behavior. When several males are placed in the same aquarium, they will all turn black and fight aggressively. One male will dominate and remain black; the others will return to a mottled pattern to avoid attacks from the dominant male. All *Tilapia* are capable of rapidly changing from mottled to white or black. These color changes are a reflection of pigment granule transport occurring in dermal melanophores. Bar = 4 cm (a); 6 cm (b).

mass transport, the dispersed pigment granules undergo saltatory movements in which the granules shuttle to and fro. Saltations of organelles occur in all eukaryotic cells (Rehbun, 1972; Schliwa, 1984), and chromatophores may provide a model system in which to study this phenonemon (Rozdzial and Haimo, 1986a; McNiven and Ward, 1988). Finally, because the organelles are pigmented, they can be readily visualized, unambiguously identified, and their movements easily tracked in the light microscope and quantitated using photometry (Clark and Rosenbaum,

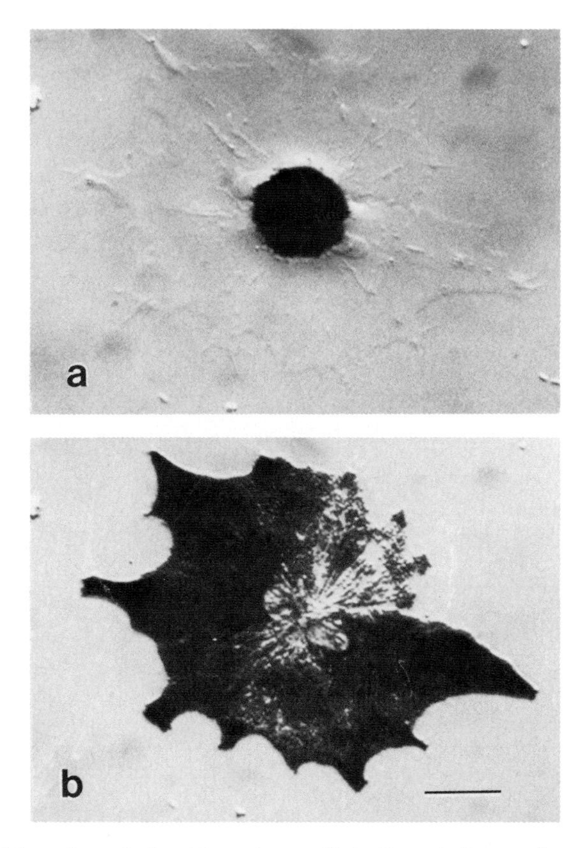

FIG. 2. Melanophores isolated from the angelfish, *Pterophyllum scalare*, demonstrating the changes in distribution of pigment when the granules are (a) aggregated to the cell center or (b) dispersed throughout the cytoplasm. In (a) the boundaries of the cell do not change when the pigment is aggregated. Bar = 25 μm. Reproduced from *Journal of Cell Biology*, 1978, Vol. 76, pp. 229–236 by copyright permission of the Rockefeller University Press (courtesy of M. Schliwa).

1984; Rozdzial and Haimo, 1986a). Moreover, the pigment itself provides an immediate marker for following the granules through various isolation procedures.

C. Types of Dermal Chromatophores

Briefly, the dermal chromatophores are classified according to their color. The most prominent type are melanophores, which contain the black pigment melanin within membrane-bound organelles termed melanosomes. Xanthophores and erythrophores can each contain two different types of pigmented lipid droplets composed of carotenoids or pteridines, which exhibit a range of colors from yellow to red. Chromatophores that are yellow are xanthophores; those that are red are erythrophores. Iridophores do not contain pigment and do not absorb light. Rather, they contain guanine or other purine crystals that reflect light, providing a basis for the iridescent or shimmering appearance of many fish and other animals. In all dermal chromatophores, a rearrangement of the pigment or purine crystals within the cytoplasm changes the amount of light absorbed or reflected by that cell and, thus, the overall appearance of the animal. Most studies have been conducted on the true pigment-bearing chromatophores—the melanophores, xanthophores, and erythrophores—and these will be the focus of this chapter. For additional information on the origins, chemistry of pigment, and physiological behavior of the different chromatophores, the reader is again directed to Bagnara and Hadley (1973).

D. Rates of Pigment Transport and the Role of the Cytoskeleton in Chromatophores

Although it might be expected that a common mechanism generating pigment granule movements will be shared by all chromatophores, there are substantial differences in the characteristics of transport in the various cells. Goldfish xanthophores require about 30 minutes to 1 hour to disperse or aggregate completely (Winchester *et al.*, 1976), transporting pigment at rates <0.1 μm/second. Melanophores transport pigment at a rate about an order of magnitude faster, 1–2 μm/seconds, and complete transit in $\sim$1 minute (Green, 1968; Rozdzial and Haimo, 1986a). These rates of melanosome transport are similar to those of fast axonal transport of organelles (Grafstein and Forman, 1980). Erythrophores from the squirrelfish, *Holocentrus ascensionis,* transport pigment at rates up to 15–20 μm/second (Byers and Porter, 1977), two orders of magnitude faster than that of the xanthophores, and require only several seconds to complete aggregation or dispersion.

In addition to differences in the rate of transport, the cytoskeletal structures responsible for pigment granule transport have not been unambiguously identified (for reviews, see Schliwa and Euteneuer, 1983; Schliwa, 1884; Obika, 1986). The preponderance of data suggest that pigment granule movements in melanophores and erythrophores occur along microtubules (Bilke *et al.*, 1966; Schliwa and Bereiter-Hahn, 1973; Murphy and Tilney, 1974; Beckerle and Porter, 1983; Ip *et al.*, 1984; McNiven *et al.*, 1984; McNiven and Ward, 1988), but substantial controversy on this issue remains (Stearns and Wang, 1987). Furthermore, the pigment granules in xanthophores appear to be intimately associated with intermediate filaments (Lim *et al.*, 1987), rather than microtubules. Thus, the differences in rates of pigment granule transport among the chromatophores may reflect transport along different cytoskeletal stuctures in which distinct force-generating mechanisms are involved. Alternatively, transport in different chromatophores may occur along a common cytoskeletal structure but utilize different force-generating molecules. However, it is possible that a common underlying mechanism generates motility in all these cells, and differences in rates may simply be a reflection of the number of "motor" molecules available or of a superimposed regulatory system. The basis for these differences will be sorted out only when the mechanisms involved in transport are more completely understood.

As a means to study organelle transport, we, as well as several other groups, have been using lysed chromatophores to analyze the mechanisms generating and regulating bidirectional pigment granule movements. The remainder of the chapter will be devoted primarily to the techniques that have been developed to obtain chromatophores and to lyse them and reactivate motility. A brief discussion of the information that these systems have so far yielded is also included.

II. Methods to Obtain, Lyse, and Reactivate Chromatophores

A. Choice of Species

Because the scales of fish are readily obtained and can contain large numbers of chromatophores, they have been the primary source of material for studies of pigment granule movements. Some of the species that have been used are the goldfish (*Carassius auratus*), the marine killifish (*Fundulus heteroclitus*), the swordtail (*Xiphophorus helleri*), the angelfish (*Pterophyllum scalare*), and the marine squirrelfish (*H. ascensionis*). We have chosen to work on melanophores from the freshwater African cichlid, *Tilapia mossambica*. This fish is now being extensively aquacultured in southern California (e.g., Aquafarms International, Mecca, CA), is extremely hardy, and grows to >50 cm in captivity. The scales contain a large number of melanophores in the dermis, and the fish, particularly the males, exhibit dramatic color changes (Fig. 1).

B. Methods for Removal of Scales

For reactivation of pigment granule transport in lysed chromatophores, usually one to four scales will provide sufficient material to conduct a set of experiments. The fish need not be sacrificed in order to obtain small numbers of scales and can grow new scales. In *Tilapia,* new scales are grown to mature size in ~1 month. To remove scales, fish are netted and placed onto a wet surface, usually clean wet paper toweling in a shallow container. To pacify the fish during the procedure, a wet towel is placed over its head and gills. With fine forceps, we grasp a scale behind the region in which the dermis grows, and gently pull it from the body. The scales overlay one another, and the dermis and epidermis grow only onto the posterior, exposed end of each scale. In *Tilapia* and *Fundulus,* the more dorsal and anterior the location of the scale, the higher the density of dermal melanophores. After a scale has been detached from the body, we gently wipe mucus from the surface of the fish over the area of the skin from which the scale was removed. This treatment helps protect the injured skin, allowing a new scale to grow. In order to obtain large numbers of scales for biochemical studies, it is usually necessary to sacrifice the fish. The scales can then be removed rapidly by scraping the skin from the tail toward the head with the flat end of a laboratory weighing spatula.

The scales should be placed into an appropriate Ringer's solution as soon as they have been removed from the fish. For *Tilapia,* a freshwater fish, we use a modified goldfish Ringer's containing 100 m*M* NaCl, 2.5 m*M* KCl, 1m*M* $MgCl_2$, and 10 m*M* $NaHCO_3$, pH 7.4. For *Fundulus,* a saltwater fish, we use a modified teleost marine Ringer's containing 134 m*M* NaCl, 2.5 m*M* KCl, 1 m*M* $MgCl_2$, 0.5 m*M* Na_2HPO_4, and 15 m*M* $NaHCO_3$. For both solutions the sodium bicarbonate is added just before use.

C. Removal of the Epidermis from the Scale

The upper surface of the scale is covered by the dermis, which in turn is covered by the epidermis. The latter must be removed to gain access to the dermal chromatophores. The most commonly used method to remove the epidermis is to incubate the scale in collagenase to digest collagen in the underlying extracellular matrix. We find that this procedure yields excellent results in terms of melanophore viability. However, as the epidermis must be manually dissected, practice and steady hands are required. Alternatively, the epidermis will slough off scales, which are incubated in EDTA (Lo *et al.,* 1982). No finesse is required for this procedure, and it is the only reasonable method to remove epidermis from large numbers of scales. However, melanophores appear to be less viable after this treatment than after collagenase treatment.

1. Collagenase Digestion

Scales are agitated in 5 mg/ml collagenase in Ringer's until the epidermis has loosened so that it can be lifted up. The scales should be well agitated (on a rotary shaker or rocker platform) during this procedure to facilitate entry of the collagenase between the dermal and epidermal layers. One to four scales can be digested in 0.5 ml collagenase, which should be freshly prepared. Several variables contribute to the success of collagenase treatment.

a. Brand of Collagenase. The brand or lot number of the collagenase is the single most important variable in the success or failure of the epidermis loosening. We have used successfully two brands of collagenase, Sigma type II and Worthington type III, although there is variation between lots of these brands.There appears to be no correlation between collagenase purity and its activity, and it is recommended that one buy several different small lots of collagenase and then buy in quantity any that happens to yield good results (Clark *et al.*, 1987).

b. Slicing the Scale into Strips. In addition to problems with collagenase, we have found that when the epidermis near the edges of large scales is ready to lift up, the rest of the epidermis has not yet separated from the dermis. If longer digestion periods are used, the dermis will lift off the scale along with the epidermis. We have circumvented this problem by cutting the scale into several thin longitudinal strips prior to incubation in the collagenase, thereby allowing the enzyme a much greater surface along which to gain entry between the dermis and epidermis. We find the slicing procedure to be essential for removing the epidermis from the very large, and thick scales of *Tilapia,* and helpful with the smaller, thinner scales of *Fundulus*. The scales are sliced using a single-edged razor blade while observing them under a stereo dissecting microscope. Several parallel cuts are made from the posterior end of the scale toward but not completely through to the anterior end. Thus, the slices remain joined at the anterior end of the scale, which can then still be handled as a single unit. The sliced scale is then placed into collagenase for digestion and removal of the epidermis. The slices can later be severed from the scale and used for reactivation studies, so that experimental and control procedures can be conducted on tissue from a single scale.

c. Time of Digestion in Collagenase. It is necessary to determine empirically the length of time that the scales must be incubated in collagenase to remove the epidermis. If the time is too short, the epidermis will shred and tear as one tries to lift it off the scale. If the time is too long, the dermis will also lift up, still attached to the epidermis. Generally, the dermis is then useless for further study. The epidermis, which may be difficult to discern before collagenase treatment, becomes puffy and cloudy during digestion. It is this characteristic that should provide the guidepost for digestion time. The cloudy appearance is most easily visualized using a stereo dissecting microscope equipped with a darkfield stage and transmitted light. Alternatively, the epidermis may be visualized using a black opaque stage and reflected light. With *Tilapia* we find that the scales usually require about 1 to 1.5 hour of digestion, whereas with *Fundulus* about 30 to 45 minutes are usually sufficient. Even within a species, the incubation time varies, seemingly depending on the thickness of the scale.

d. Dissection of the Epidermis. Once digestion has been completed, the scales should be removed from the collagenase and placed into fresh Ringer's in a Petri dish, which is then set on the stage of the dissecting microscope. The scale is held in place with forceps in one hand while the epidermis is grasped using a pair of fine forceps (Dumont No. 5) in the other hand. The epidermis extends more proximally on the scale than does the dermis, and we find that it is generally preferable to take hold of the epidermis at this proximal end and gently, with a smooth motion, pull toward the distal end of the scale. If digestion time is accurately timed, the epidermis will easily peel off each section of scale as an intact sheet. Once the epidermis has been removed, the scale remains viable in Ringer's for several hours and the chromatophores in the dermis can be lysed and reactivated as described in Section II, F and G.

2. EDTA Method for Epidermis Removal

In order to remove the epidermis from large numbers of scales, *Tilapia* scales are agitated for 1–1.5 hour in 100 m*M* NaCl, 81 m*M* EDTA, 53 m*M* $NaHCO_3$, and 2.5 m*M* KCl (modified from Lo *et al.*, 1982). The scales are then washed and agitated with several changes of Ringer's. During this procedure, the epidermis sloughs off the scale in small sheets. The scales have to be checked by differential interference contrast (DIC) microscopy to determine if any epidermis remains on the scale. Small cells, in a focal plane above the chromatophores, will be observed if the epidermis has not been completely removed. Several more washes with Ringer's will usually remove any residual epidermis. We find that melanophores require a recovery period of several hours in either Ringer's or tissue culture media after the EDTA treatment before they again aggregate and disperse pigment in response to external stimulation.

D. Removal of the Dermis from the Scale

Although the dermis can also be removed from the scale, it rolls up and the chromatophores lose their characteristic flattened discoid or stellate shape. Thus, detached dermis is generally not suitable for light-microscope motility studies of pigment granule movements. In order to study pigment granule movements in lysed cells, the chromatophores must be used while still on the scale (Section II,C) or after isolation and spreading on a coverslip, as described in Section II,E. Detached dermis is useful, however, for biochemical analysis such as for assessing changes in protein phosphorylation patterns accompanying pigment granule transport (Rozdzial and Haimo, 1986b).

Following removal of the epidermis, as described in Section II,C, the scale with dermis attached should be returned to collagenase and agitated for a second round of digestion. The dermis will loosen from the scale after ~30 to 60 minutes and can be easily pulled up using fine forceps. The scale, after the second round of digestion but before the dermis has been lifted off, can be used for studies to reactivate pigment granule movement in lysed chromatophores. To analyze phosphorylated proteins during pigment granule transport, we remove the epidermis and then return the scales to collagenase to loosen the dermis. The cells still on the scale are then lysed (Section II,F). Following reactivation of motility using ^{32}P-ATP, the dermis is finally lifted off the scale and prepared for gel electrophoresis followed by autoradiography (Rozdzial and Haimo, 1986b).

E. Isolation of Chromatophores from the Dermis

The major disadvantage of working on chromatophores, from a biochemical standpoint, is that they do not exist on scales as a homogeneous population of cells. The dermis on the scale contains large numbers of nonchromatophore cells, although the chromatophores, because of their size, can constitute the majority of the tissue. In addition, more than one type of chromatophore may be present and it would be necessary to separate melanophores from xanthophores as well as from other cell types. It is possible to isolate chromatophores from the scale, although the procedures to do so are more effective with some species than others.

Xanthophores have been isolated from goldfish scales in large numbers, making them quite useful for biochemical analysis (Lo *et al.*, 1982; Lynch *et al.*, 1986a,b). Erythrophores and melanophores have been isolated from the scales of the squirrelfish, and melanophores isolated from angelfish (Schliwa *et al.*, 1978; Luby and Porter, 1980; McNiven *et al.*, 1984). Because these cells must be pipetted off the tissue, only small numbers have been obtained and no biochemical studies have been conducted on them. Because we have found these procedures do not work well for *Tilapia* melanophores, we recommend consulting the references just cited for details of the methods. Melanophores have been isolated in reasonable amounts from fins of the Black Moor goldfish (Clark *et al.*, 1987). In addition, a melanoma cell line of melanophores has been developed that remain pigmented and aggregate and disperse pigment granules upon hormonal or drug stimulation but that have regained the ability to divide (Ogawa *et al.*, 1987). It should now be possible to obtain biochemical amounts of material from a homogeneous population of melanophores and therefore correlate with greater confidence motile events with biochemical changes.

F. Lysis of Chromatophores

1. Lysis Conditions

a. Detergents. Melanophores, erthyrophores, and xanthophores have all been successfully lysed with detergents, and they retain the ability to undergo partial or complete pigment granule translocations *in vitro*. Melanophores were first lysed by Clark and Rosenbaum (1982) using the detergent Brij 58, which had been previously used in studies to reactivate mitotic movements (Cande, 1982). The Brij-lysed *Fundulus* melanophores were able to undergo pigment granule aggregation but not dispersion *in vitro*. Using the mild glycoside digitonin, which interacts with cholesterol and other β-hydroxysterols, Stearns and Ochs (1982) were able to reactivate pigment granule saltations in erythrophores of *Holocentrus* but were unable to reactivate aggregation or dispersion. It now appears that the ability of these lysed models to undergo only partial pigment granule translocations *in vitro* was a function of the composition of the lysis buffer rather than the choice of detergent. We have used both digitonin (Rozdzial and Haimo, 1986a) and Brij 58 (R. Johnson and L. T. Haimo, unpublished) to lyse melanophores of *Tilapia* and *Fundulus*, and to obtain cells that are capable of undergoing numerous rounds of aggregation and dispersion *in vitro*. Similarly, McNiven and Ward (1988) have used Brij 58 to lyse *Holocentrus* erythrophores, which are also then able to transport pigment bidirectionally. In addition, bidirectional pigment granule transport occurs in melanophores lysed with saponin (Grundstrom *et al.*, 1985). Thus, a variety of mild detergents can be used to lyse chromatophores in which motility can subsequently be reactivated.

Digitonin at low concentrations (0.001%) apparently intercalates into the membrane at cholesterol sites, weakening the membrane and causing vesiculation, which generates small holes (Stearns and Ochs, 1982). The soluble components of the cell are then slowly leaked into the surrounding medium (Sarafian *et al.*, 1987). Higher digitonin concentrations, or rapid flushing, completely lyses the membrane (Stearns and Ochs, 1982). Cholesterol is present at high molar ratios in the plasma membrane but not in organelle membranes; thus digitonin can be used to lyse the plasma membrane selectively without disrupting the internal structures of interest. Thus, a reasonable approach in a new system would be to lyse cells using a low concentration of digitonin and develop conditions in which the intracellular process of interest can be reactivated *in vitro*. Subsequently, the cells can be lysed with higher concentrations of digitonin or with other nonspecific detergents such as Brij 58 if a more thorough lysis of the plasma membrane is desired. PtK1 cells have been lysed and anaphase chromosome movements reactivated using digitonin in the same lysis buffer we developed for melanophores (Spurck and Pickett-Heaps, 1987).

b. Composition of the Lysis Buffer. We developed a lysis buffer in which melanophores are capable of undergoing bidirectional pigment granule transport *in vitro*.The buffer was modified from one used to lyse and reactivate cilia of *Chlamydomonas reinhardii* (Witman *et al.*, 1978) and contains digitonin or Brij 58 in 33 m*M* potassium acetate, 30 m*M* HEPES, 10 m*M* EGTA, 5 m*M* $MgSO_4$, 0.5 m*M* EDTA, 2.5% polyethylene glycol, pH 7.4. A concentration of 0.001% digitonin is used to permeabilize the cells gently without depleting them of endogenous nucleotides, while $\geq$0.0015% digitonin or 0.5% Brij 58 is included to lyse the plasma membrane thoroughly and deplete the cells of soluble components (Rozdzial and Haimo, 1986a).

Digitonin is relatively insoluble in aqueous solutions. We make a 0.4% stock in 50% ethanol, which can be stored at room temperature for about a month. 0.1 m*M* epinephrine and 5 m*M* isobutylmethyl xanthine (IBMX), which respectively induce aggregation and dispersion in live or in gently permeabilized cells, are prepared as concentrated stocks in dimethyl sulfoxide (DMSO) and stored at −20°C. Epinephrine is light-sensitive and should be kept in a dark container.

2. Methods of Lysis

a. Perfusion on a Microscope Slide. The melanophores, on the scale, or isolated and spread on a coverslip, are mounted in Ringer's on a perfusion chamber prepared by elevating a coverslip over a slide using modeling clay or broken pieces of coverslips. The lysis buffer is then gently added at one side of the perfusion chamber and drawn across by placing a filter paper wick at the other side. In order to keep the scales from moving in the flow, the section of scale is immobilized by inserting the anterior, tissue-free end into a dab of silicone vacuum grease placed in the chamber. Thus, it is possible to monitor and record the behavior of individual cells through various procedures. Melanophores from *Tilapia* are usually lysed by gentle perfusion with 0.001% digitonin in ~1–2 minutes. These cells, though lysed, retain or generate sufficient ATP to undergo several rounds of aggregation and dispersion without requiring added ATP (Rozdzial and Haimo, 1986a). By increasing the concentration of digitonin to 0.0015–0.003% and by rapidly perfusing with lysis buffer, the cells can be more extensively lysed so that reactivation of aggregation and dispersion requires addition of ATP. With lysis in the perfusion chamber, we generally find that cells near the posterior end of the scale lyse before those at the anterior end do, possibly because the curvature of the scale affects the flow of lysis buffer over it.

b. Agitation of Scales in Solution. Melanophores can also be lysed by placing pieces of the scale into Microfuge tubes containing lysis buffer and then gently agitating them for 1–2 minutes. The cells are uniformly lysed regardless of location on the scale; thus, we generally find this method to be preferable. The scales are then mounted in a perfusion chamber for reactivation studies.

3. Assay for Lysis

There are a number of ways of determining that chromatophores have been lysed. If large portions of the plasma membrane are solubilized during lysis, the cells will often release pigment granules to the surrounding medium (Stearns and Ochs, 1982; McNiven and Ward, 1988). Even with gentle lysis, pigment granules within the cells sometimes detach from the cytoskeleton and undergo Brownian movements, a behavior not exhibited by pigment granules in live cells. Often the pigment granules translocate in response to lysis. In *Tilapia* melanophores, for example, we observe that the pigment granules initially aggregate in response to treatment with the lysis buffer, and then disperse far into the stellate extensions of the cells, a process we refer to as hyperdispersion (Rozdzial and Haimo, 1986a). We use this feature as a rapid indicator of lysis. In addition, pigment granules in live cells normally undergo saltatory movements, whereas those in lysed cells that have been depleted of endogenous nucleotides become stationary.

There are a variety of methods to verify that the visual signs just outlined are indeed an accurate indicator of cell lysis. Cells can be treated with lysis buffer containing or lacking the detergent and then fixed and reacted with antibodies against an internal cell structure, such as microtubules. Such studies have demonstrated that the microtubules are stained only in the detergent-treated cells (Clark and Rosenbaum, 1982; Stearns and Ochs, 1982; Rozdzial and Haimo, 1986a). Alternatively, thin sections of lysed cells can be examined by transmission electron microscopy (Stearns and Ochs, 1982; Clark and Rosenbaum, 1982) or whole mounts examined by scanning electron microscopy (McNiven and Ward, 1988) to view the integrity of the plasma membrane. These studies reveal that the plasma membrane can be almost completely solubilized, while the microtubules and pigment granules remain intact (Fig. 3). Immunofluorescence and electron microscopy are time-consuming procedures and need be done only initially to define the parameters of lysis. Finally, live cells respond differently than do lysed cells to various treatments. For example, live *Tilapia* melanophores do not respond to exogenous cAMP, while lysed cells disperse pigment in response to it. Similarly, ATP has no effect on live melanophores but will induce aggregation in lysed melanophores that are depleted of endogenous nucleotides (Clark and Rosenbaum, 1984; Rozdzial and Haimo, 1986a).

G. Reactivation of Pigment Transport in Lysed Chromatophores

1. Permeabilized Cells

Melanophores that have been gently lysed with 0.001% digitonin appear to retain functional mitochondria and continue to produce ATP for some period of time after lysis (~30–60 minutes), as these cells can be induced to transport pigment without addition of exogenous ATP. We refer to these as permeabilized cells.

a. Aggregation. Following perfusion of *Tilapia* melanophores (Fig. 4a) with the lysis buffer, the cells hyperdisperse pigment (Fig. 4b). The lysed cells can then be induced to aggregate pigment by addition of 0.1 m*M* epinephrine in lysis buffer (Fig. 4c). ATP is not required if the

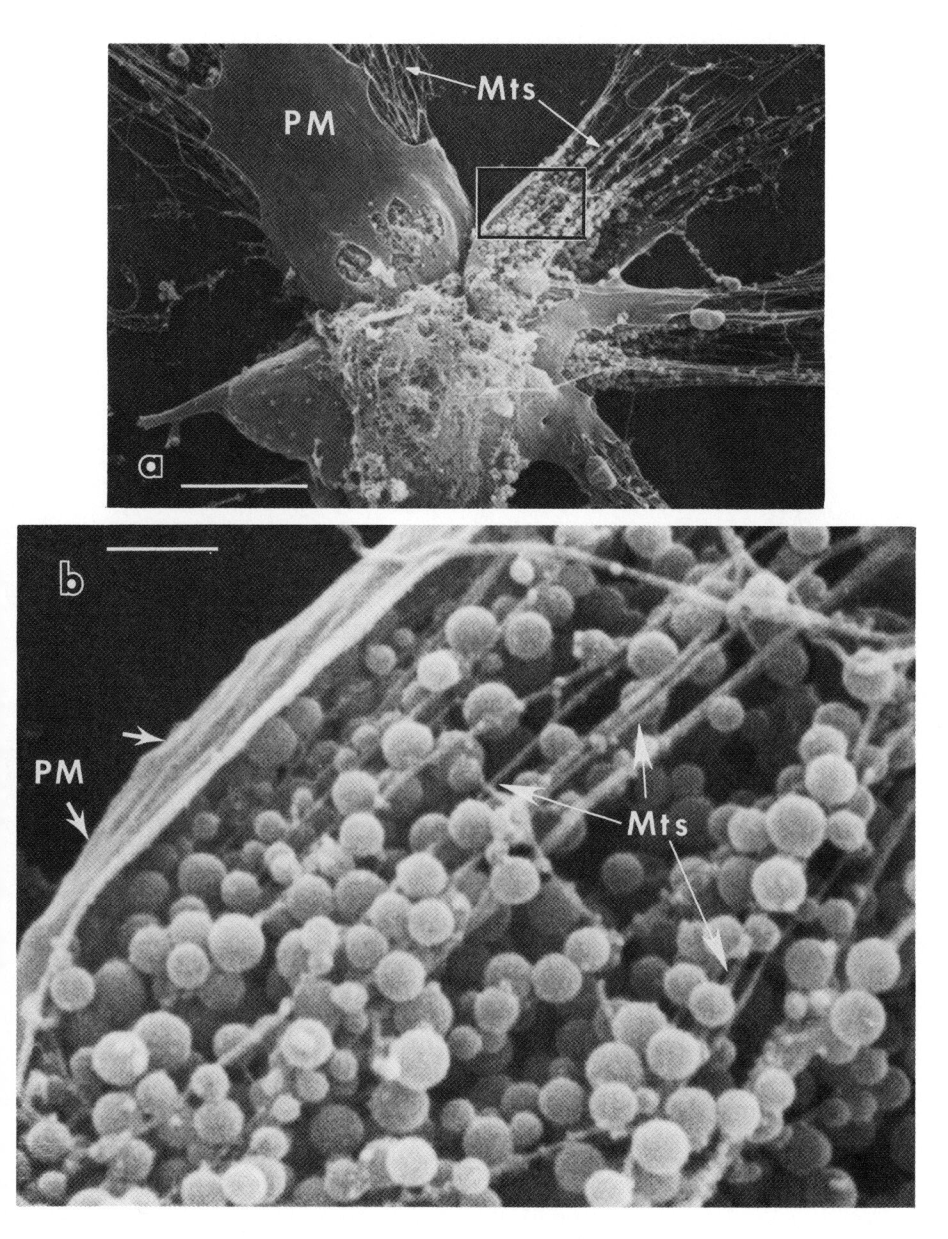

FIG. 3. A detergent-lysed erythrophore from the squirrelfish, *Holocentrus ascensionis*, lacks much of the plasma membrane (PM) but retains microtubules (Mts) with attached pigment granules, as visualized by scanning electron microscopy. The cell has been lysed for 1 minute with 0.5% Brij and is capable of aggregating and dispersing pigment granules *in vitro*. Bar = 5 μm (a); 0.5 μm (b). Reproduced from the *Journal of Cell Biology*, 1988, Vol. 106, pp. 111–125 by copyright permission of the Rockefeller University Press (courtesy of M. A. McNiven).

FIG. 4. Detergent-lysed melanophores aggregate and disperse pigment granules *in vitro*. Live *Tilapia* melanophores (a) respond to lysis buffer containing 0.001% digitonin by hyperdispersing their pigment granules (b). These permeabilized cells will aggregate pigment upon stimulation with 0.1 m*M* epinephrine (c) and disperse it upon stimulation with 1 m*M* cAMP (d). Bar = 100 μm. Reproduced from Rodzial and Haimo (1986b).

cells have not been depleted of endogenous pools by the lysis. Epinephrine will not induce aggregation in depleted cells (see later).

b. Dispersion. Permeabilized cells that have undergone pigment granule aggregation *in vitro* (Fig. 4c) will disperse pigment upon perfusion with lysis buffer containing cAMP (Fig. 4d). Addition of cAMP alone will not induce transport in depleted cells, and it cannot cross the membrane of live cells in sufficient amounts to induce dispersion. Alternatively, IBMX, which inhibits phosphodiesterase activity, will induce dispersion in live or gently permeabilized cells, but not in depleted one.

Multiple rounds of aggregation and dispersion can be induced by alternately perfusing gently permeabilized cells with epinephrine and cAMP. Eventually, the cells will exhaust their endogenous nucleotide pools and cease to respond to these treatments.

2. Depleted Cells

Perfusing the melanophores for ~10 minutes with multiple changes (~1 ml total) of lysis buffer containing higher concentrations of digitonin (generally 0.0015–0.003%) will allow soluble components to be extracted from the melanophores, which will become depleted of endogenous nucleotide pools. Alternatively, the cells rapidly (1–2 minutes) become depleted by agitating pieces of scale with attached dermis in 0.5 ml lysis buffer in a Microfuge tube.

a. Aggregation. Aggregation in depleted cells cannot be induced by epinephrine; rather, it is initiated simply by addition of ATP to the lysis buffer. The rate of pigment transport is a function of the ATP concentration (Rozdzial and Haimo, 1986a), and maximal rates are attained with 1–3 m*M* ATP.

b. Dispersion. In order to examine dispersion in depleted cells, they are incubated in epinephrine during lysis to prevent the spontaneous hyperdispersion of pigment that normally occurs. The cells are depleted by lysis with buffer containing the elevated digitonin concentrations and multiple perfusions or agitation, as already described. Depleted melanophores do not disperse pigment by addition of cAMP alone but require ATP as well.

A depleted cell is capable of undergoing a complete round of pigment transport, aggregating pigment upon perfusion with ATP and dispersing pigment upon perfusion with cAMP and ATP. Extensive lysis does, however, extract the cells of other required components in addition to nucleotides, and depleted cells may then only transport pigment if "rescued" by addition of a particular component (Thaler and Haimo, 1987).

III. Mechanisms of Organelle Transport in Chromatophores

An analysis of pigment granule transport in lysed chromatophores has provided considerable information regarding the mechanisms governing the direction of transport. These studies are also beginning to yield information concerning the mechanisms generating the movements.

A. Role of ATP in Pigment Granule Transport

Studies on live cells that were treated with metabolic inhibitors have revealed that pigment granules aggregate as the cells are depleted of ATP. These studies have led to the proposal that pigment granule aggregation is an ATP-independent process in which dispersed granules, encased in an elastic cytoplasmic lattice, aggregate as the lattice recoils in the absence of ATP (Luby and Porter, 1980). Studies on lysed chromatophores have demonstrated that aggregation does in fact require ATP (Clark and Rosenbaum, 1984; Rozdzial and Haimo, 1986a; McNiven and Ward, 1988). That aggregation occurs during metabolic poisoning is probably a consequence of the fact that dispersion requires an ATP concentration an order of magnitude higher than does aggregation (Rozdzial and Haimo, 1986a), and, further, that dispersion requires elevated cAMP (Rozdzial and Haimo, 1986a) or decreased levels of free Ca^{2+} (McNiven and Ward, 1988), both of which are accomplished at the expense of ATP. Thus, there would be a threshold of ATP in poisoned cells below which only aggregation, and not dispersion, could be supported. As reactivation of pigment granule transport in lysed cells has demonstrated that both directions of transport require ATP, it will be necessary to determine which force-generating molecules use the ATP in these two directions of transport.

B. Regulation of the Direction of Pigment Granule Transport

1. Control by cAMP

Our studies on pigment granule aggregation and dispersion in lysed melanophores have revealed that the direction of transport is controlled by cAMP. Cells depleted of endogenous nucleotides aggregate pigment in ATP and disperse it in ATP and cAMP. Pigment granules in the process of aggregating immediately change direction and disperse if cAMP is added to the lysed cells. Conversely, granules that are dispersing reverse direction and aggregate if cAMP is removed (Rozdzial and Haimo, 1986a).

The effects of epinephrine and caffeine analogs (e.g., IBMX) on pigment granule transport can be explained by the changes they induce in the cAMP levels in the cell. Epinephrine binds to α-adrenergic receptors on fish melanophores and inhibits adenylate cyclase activity (Fujii and Miyashita, 1975). Thus, cAMP levels would become depressed, and the cells could aggregate pigment. Conversely, dispersion requires elevated cAMP levels. Because IBMX inhibits phosphodiesterase activity and allows the cells to build up cAMP levels, IBMX treatment of live or lysed cells with an endogenous nucleotide pool stimulates production of cAMP and promotes dispersion. Lysed, depleted cells, which lack a pool of cAMP and ATP, derive no benefit either from epinephrine addition (because cAMP levels are already depressed) or from IBMX addition (because they have no ATP from which to synthesize cAMP). Accordingly, depleted melanophores aggregate by addition of ATP and require no other stimulation, whereas they disperse only in response to cAMP and ATP.

2. CONTROL BY Ca^{2+}

Studies on pigment granule transport in the erythrophores of the squirrelfish, *H. ascensionis*, have suggested that Ca^{2+}, rather than cAMP, regulates the direction of transport. These cells cannot aggregate pigment when placed in a Ca^{2+}-free buffer. Moreover, addition of a Ca^{2+} ionophore to cells in a Ca^{2+}-containing buffer induces pigment granule aggregation (Luby-Phelps and Porter, 1982). Similarly, in lysed erythrophores aggregation is initiated by elevating free-Ca^{2+} concentrations, whereas dispersion is initiated by lowering free-Ca^{2+} (McNiven and Ward, 1988). We have found that in *Tilapia* Ca^{2+} stimulates aggregation by activating the phosphatase calcineurin (Thaler and Haimo, in preparation).

C. Role of Protein Phosphorylation and Dephosphorylation in Pigment Granule Transport

Pigment granule dispersion in melanophores and xanthophores requires cAMP, suggesting that the process requires the participation of cAMP-dependent protein kinase. Incubation of live xanthophores with $^{32}PO_4$ results in the phosphorylation of a pigment-associated, 57-kDa protein during dispersion and its dephosphorylation during aggregation (Lynch *et al.*, 1986a). Similarly, the same polypeptide is phosphorylated in cell-free extracts incubated in cAMP analogs (Lynch *et al.*, 1986b). Moreover, treatment of lysed melanophores with protein kinase inhibitor, which specifically inhibits cAMP-dependent protein kinase, prevents dispersion and the phosphorylation of the 57-kDa polypeptide (Rozdzial and Haimo, 1986b). Conversely, treatments that interfere with phosphatase activity inhibit both pigment granule aggregation and the dephosphorylation of the 57-kDa polypeptide in lysed melanophores (Rozdzial and Haimo, 1986b). Addition to lysed cells of the catalytic subunit of cAMP-dependent protein kinase can induce dispersion in the absence of cAMP (Thaler and Haimo, 1987; Yu *et al.*, 1987), whereas addition of the Ca^{2+}-stimulated phosphatase calcineurin to extensively depleted melanophore cells can rescue aggregation in cells that are no longer capable of aggregating (Thaler and Haimo, 1987 and in preparation). Together, these studies suggest that pigment granule dispersion in melanophores and xanthophores requires protein phosphorylation, whereas pigment granule aggregation requires protein dephosphorylation.

It is not known if Ca^{2+} regulates the direction of transport in erythrophores by controlling protein phosphorylation and dephosphorylation. Thus, it will be most interesting to determine if changes in the phosphorylation patterns of proteins in lysed erythrophores accompany changes in the direction of transport induced by raising or lowering the free Ca^{2+}.

D. Potential Motors in Pigment Granule Transport

Various inhibitors have been used to identify the motors involved in pigment granule transport. Vanadate, an inhibitor of dynein ATPase (Gibbons *et al.*, 1978), prevents pigment granule aggregation (Clark and Rosenbaum, 1982; Grundstrom *et al.*, 1985; Negishi *et al.*, 1985; Rozdzial and Haimo, 1986b; Ogawa *et al.*, 1987) but not dispersion in lysed cells (Grundstrom *et al.*, 1985; Rozdzial and Haimo, 1986b). However, vanadate is also an inhibitor of phosphatase activity (VanEtten *et al.*, 1974; Lopez *et al.*, 1976), and, as pigment granule aggregation requires phosphatase activity (Rozdzial and Haimo, 1986b), it is not possible to infer from vanadate inhibition alone that dynein is involved in aggregation. However, we have found that melanophores incubated in ATP and vanadate and then irradiated with UV, a diagnostic assay for dynein (Gibbons and Gibbons, 1987), are no longer able to aggregate pigment (Haimo and Fenton, 1988), suggesting that these movements are, in fact, mediated by dynein. Further, erythro-9-[3-(2-hydroxynonyl)] adenine (EHNA), another dynein inhibitor (Bouchard *et al.*, 1981), has some inhibitory effect on aggregation (Beckerle and Porter, 1982; Negishi *et al.*, 1985; Ogawa *et al.*, 1987), and microtubules in melanoma melanophores become stained when incubated in an antibody directed against dynein (Ogawa *et al.*, 1987). Studies on pigment granule dispersion suggest that it may be mediated by protein phosphorylation alone with no involvement of an ATPase (Rozdzial and Haimo, 1986b). Additional studies must be conducted before the motors responsible for bidirectional pigment granule transport can be identified.

IV. Summary

The development of procedures to lyse and reactivate pigment granule movements in chromatophores has provided the only information to date concerning the mechanisms by which cells regulate the direction of organelle transport. Continued analysis of motility in these models as well as in a reconstituted system containing only the pigment granules, the appropriate cytoskeletal structures, and defined soluble cell components should contribute to our understanding of the mechanisms by which protein phosphorylation and dephosphorylation or Ca^{2+} regulate direction of transport and to the identification and characterization of the force-generating proteins responsible for producing bidirectional organelle movements.

Acknowledgments

We are most grateful for the considerable contributions of Ray Fenton, Cathy Thaler, and Ray Johnson. This research has been supported by PHS grant GM-28886, NSF grant DCB 8710428, and the Committee on Research, University of California, Riverside.

References

Bagnara, J. T., and Hadley, M. E. (1973). "Chromatophores and Color Change: The Comparative Physiology of Animal Pigmentation." Prentice-Hall, Englewood Cliffs, New Jersey.

Beckerle, M. C., and Porter, K. E. (1982). *Nature (London)* **295,** 701–703.

Beckerle, M. C., and Porter, K. E. (1983). *J. Cell Biol.* **96,** 354–362.

Bilke, D., Tilney, L. G., and Porter, K. E. (1966). *Protoplasma* **61,** 322–345.

Bouchard, P., Penningroth, S. M., Cheung, A., Gagnon, C., and Bardin, C. W. (1981). *Proc. Natl. Acad. Sci. U.S.A.* **78,** 1033–1036.

Byers, H. R., and Porter, K. E. (1977). *J. Cell Biol.* **75,** 541–558.

Cande, W. Z. (1982). *Cell (Cambridge, Mass.)* **28,** 15–22.

Clark, C. R., Taylor, J. D., and Tchen, T. T. (1987). *In Vitro* **23,** 417–421.

Clark, T., and Rosenbaum, J. L. (1982). *Proc. Natl. Acad. Sci. U.S.A.* **79,** 4655–4659.

Clark, T., and Rosenbaum, J. L. (1984). *Cell Motil.* **4,** 431–441.

Fujii, R., and Miyashita, Y. (1975). *Comp. Biochem. Biophys. C* **51C,** 171–178.

Gibbons, B. H., and Gibbons, I. R. (1987). *J. Biol. Chem.* **262,** 8354–8359.

Gibbons, I. R., Cosson, M. P., Evans, J. A., Gibbons, B. H., Houck, B., Martinson, K. H., Sale, W. S., and Tang, W.-J. Y. (1978). *Proc. Natl. Acad. Sci. U.S.A.* **75,** 2220–2224.

Grafstein, B., and Forman, D. S. (1980). *Physiol. Rev.* **60,** 1167–1283.

Green, L. (1968). *Proc. Natl. Acad. Sci. U.S.A.* **59,** 1179–1186.

Grundstrom, N., Karlsson, J. O. G., and Andersson, R. G. G. (1985). *Acta Physiol. Scand.* **125,** 415–421.

Haimo, L. T., and Fenton, R. D. (1988). *J. Cell Biol.* **107,** 245a.

Ip, W., Murphy, D. B., and Heuser, J. E. (1984). *J. Ultrastruct. Res.* **86,** 162–175.

Lim, S. S., Ris, H., and Schnasse, B. (1987). *J. Cell Biol.* **105,** 37a.
Lo, S. J., Grabowski, S. M., Lynch, T. J., Kern, D. G., Taylor, J. D., and Tchen, T. T. (1982). *In Vitro* **18,** 356–360.
Lopez, V., Stevens, T., and Lindquist, R. N. (1976). *Arch. Biochem. Biophys.* **175,** 31–38.
Luby, K. J., and Porter, K. E. (1980). *Cell (Cambridge, Mass.)* **21,** 13–23.
Luby-Phelps, K. J., and Porter, K. E. (1982). *Cell (Cambridge, Mass.)* **29,** 441–450.
Lynch, T. J., Taylor, J. D., and Tchen, T. T. (1986a). *J. Biol. Chem.* **261,** 4204–4211.
Lynch, T. J., Wu, B., Taylor, J. D., and Tchen, T. T. (1986b). *J. Biol. Chem.* **261,** 4212–2416.
McNiven, M. A., and Ward, J. B. (1988). *J. Cell Biol.* **106,** 111–125.
McNiven, M. A., Wang, M., and Porter, K. E. (1984). *Cell (Cambridge, Mass.)* **37,** 753–765.
Murphy, D. B., and Tilney, L. G. (1974). *J. Cell Biol.* **61,** 757–779.
Negishi, S., Fernandez, H. R. C., and Obika, M. (1985). *Zool. Sci.* **2,** 469–475.
Novales, R. R. (1983). *Am. Zool.* **23,** 559–568.
Obika, M. (1986). *Zool. Sci.* **3,** 1–11.
Ogawa, K., Hosaya, H., Yokota, E., Kobayashi, T., Wakamatsu, Y., Ozata, K., Negisha, S., and Obika, M. (1987). *Eur. J. Cell Biol.* **43,** 3–9.
Rehbun, L. I. (1972). *Int. Rev. Cytol.* **32,** 93–137.
Rozdzial, M. M., and Haimo, L. T. (1986a). *J. Cell Biol.* **103,** 2755–2764.
Rozdzial, M. M., and Haimo, L. T. (1986b). *Cell (Cambridge, Mass.)* **47,** 1061–1070.
Sarafian, T., Aunis, D., and Bader, M. (1987). *J. Biol. Chem.* **262,** 16671–16676.
Schliwa, M. (1984). *Cell Muscle Motil.* **5,** 1–82.
Schliwa, M., and Bereiter-Hahn, J. (1973). *Z. Zellforsch. Mikrosk. Anat.* **147,** 127–148.
Schliwa, M., and Euteneuer, U. (1983). *Am. Zool.* **23,** 479–494.
Schliwa, M., Osborn, M., and Weber, K. (1978). *J. Cell Biol.* **76,** 229–236.
Spurck, T. P., and Pickett-Heaps, J. D. (1987). *J. Cell Biol.* **105,** 1691–1705.
Stearns, M. E., and Ochs, R. L. (1982). *J. Cell Biol.* **94,** 727–739.
Stearns, M. E., and Wang, M. (1987). *J. Cell Sci.* **87,** 565–580.
Thaler, C. D., and Haimo, L. T. (1987). *J. Cell Biol.* **105,** 127a.
VanEtten, R. L., Waymack, P. P., and Rehkop, D. M. (1974). *J. Am. Chem. Soc.* **96,** 6782–6785.
Winchester, J. D., Ngo, F., Tchen, T. T., and Taylor, J. D. (1976). *Endocrinol. Res. Commun.* **3,** 335–342.
Witman, G. B., Plummer, J., and Sander, G. (1978). *J. Cell Biol.* **76,** 729–747.
Yu, F., Wu, B., Taylor, J. D., and Tchen, T. T. (1987). *J. Cell Biol.* **105,** 126a.

Chapter 2

Digitonin Permeabilization Procedures for the Study of Endosome Acidification and Function

RUBEN DIAZ AND PHILIP D. STAHL

Department of Cell Biology and Physiology
Washington University School of Medicine
St. Louis, Missouri 63110

I. Introduction

Over the past few years, attempts have been made to gain access to the cytoplasmic contents of eukaryotic cells without impairing the general organization of the cell. Techniques to permeabilize cells have facilitated the study of complex processes such as organelle function (Yamashiro *et al.*, 1983), vesicle-mediated transport and fusion (Dunn and Holz, 1983; Lelkes and Pollard, 1987; Sarafian *et al.*, 1987), and vesicle movement regulation within the confinements of the cell (Rozdzial and Haimo, 1986).

Because these functions require the interaction of multiple cellular components, they are often difficult to reconstitute in cell-free systems. Permeabilized cell systems also bypass the need for purified components in biological systems. Purified subcellular organelles are difficult to obtain and are often labile and poorly amenable to physiological studies. In a permeabilized cell, on the other hand, the cytosolic contents can be manipulated during experiments while most intracellular components remain functionally intact. Thus, the use of permeabilized cell systems can often provide a rapid and more reliable approach for the study of cell function.

Several methods have been described that selectively permeabilize the plasma membrane without altering the properties of intracellular membranes. High-voltage discharges can be applied to cells to create temporary "holes" in the plasma membrane (Knight and Baker, 1982). Channel-forming bacterial exotoxins (e.g., streptolysin O and staphylococcal α-toxin) assemble in supramolecular amphiphilic polymers that get inserted in the lipid bilayer to form stable pores of various diameters (Ahnert-Hilger *et al.*, 1985; Howell *et al.*, 1987). Their insertion is often limited to the plasma membrane because of their lack of accessibility to other intracellular membranes. These pores, much like those formed with voltage pulses, permit the exchange of the contents of extracellular and intracellular compartments. The exclusion limits of each channel are determined strictly by pore size. Proteins can pass through pores formed by streptolysin O, whereas α-toxin permits the exchange of only small molecules. A report by Steinberg *et al.* (1987) shows that cells exposed to extracellular ATP also become leaky. The effect of ATP is reversible, providing a means to introduce a nondiffusible molecule into the cytoplasm.

Detergents (e.g., digitonin and saponin) have also been used for selective membrane permeabilization (Dunn and Holz, 1983; Brooks and Treml, 1984). Because of the difficulty in restricting detergent permeabilization to the plasma membrane, the conditions for detergent use must be carefully controlled. Digitonin, a steroid glycoside, interacts specifically with 3β-hydroxysterols. The predominance of cholesterol in the plasma membrane of most cells makes this membrane highly vulnerable to digitonin. Digitonin has been successfully used in studies of chromaffin granule exocytosis (Lelkes and Pollard, 1987; Sarafian *et al.*, 1987), measurement of organelle acidification (Yamashiro *et al.*, 1983), and vesicle movement (Rozdzial and Haimo, 1986).

In this chapter, the technique of digitonin permeabilization is used to study receptor-mediated endocytosis. Cells recognize and internalize ligands via cell surface receptors and deliver these ligands to lysosomes or to other intracellular organelles (Stahl and Schwartz, 1986). Receptor–ligand complexes accumulate in clathrin-coated pits in the plasma membrane that pinch off to form coated vesicles. The fusion of these vesicles, after the loss of their clathrin coat, with each other or with preexisting endocytic vesicles leads to the biogenesis of an endosome. Endosomes, much like lysosomes, maintain an acidic lumen by means of proton pump ATPases (Mellman *et al.*, 1986). It is in this organelle that sorting of a number of ligands from their respective receptors takes place (e.g., ligands for asialoglycoprotein receptor and mannose receptor). One such receptor, the mannose receptor, is expressed at the cell surface of macrophages and other cells from the reticuloendothelial system (Stahl *et al.*, 1984). It recognizes mannose-containing ligands and delivers them to endosomes where receptor–ligand dissociation takes place. At this stage, receptors return to the cell surface while ligands are delivered to lysosomes. Binding of ligand to the mannose receptor is pH-dependent, with dissociation favored at low pH (Lennartz *et al.*, 1987).

A digitonin-permeabilized cell assay is described here for the study of endosome acidification and fusion. The macrophage mannose receptor is used to direct ligands along the endocytic pathway in these studies. For the acidification studies, we measure the pH-dependent dissociation of mannosylated bovine serum albumin (mannose-BSA) from the macrophage mannose receptor as a measure of endosome acidification. This permeabilized cell system could, in addition, be used for the study of other events during endocytosis such as endosome fusion and vesicle locomotion across cytoskeletal tracks.

II. Preparation of Mannose-BSA for Endosome Acidification Studies

The mannose receptor has been well characterized for its role in receptor-mediated endocytosis of high-mannose glycoproteins. This macrophage-specific receptor has been purified and reported to be a 175-kDa membrane glycoprotein (Lennartz *et al.*, 1987). Ligand binding to the mannose receptor is pH-dependent; neutral pH is required for binding, whereas low pH favors receptor–ligand dissociation. Ca^{2+} is also required for binding. Receptor–ligand complexes traverse endosomes during the internalization cycle. The low pH within this compartment's lumen permits the dissociation of ligand from receptor. As a consequence, ligand is transported to secondary lysosomes while the receptor recycles back to the cell surface. The time course for transfer of ligand to lysosomes is ~20 minutes whereas mannose receptor recycling is rapid (10–12 minutes) and takes place constitutively. Sorting of ligand from receptor is thought to occur in endosomes.

A large variety of proteins (e.g., rat preputial β-glucuronidase, horseradish peroxidase, invertase) are bound with varying affinities by the receptor. None of these naturally mannosylated proteins binds the receptor with a high enough avidity to be useful in equilibrium-binding studies. Chemical mannosylation of many proteins, on the other hand, produces high-affinity ligands for the mannose receptor (Stahl *et al.*, 1980; Wileman *et al.*, 1986). The addition of mannose residues to free amino groups on the surface of these molecules is sufficient to provide the appropriate configuration of sugars to produce a high-affinity ligand. Mannosylation of proteins that are ligands for the receptor often results in an increase of binding affinity—for example, horseradish peroxidase, bovine lactoperoxidase (Wileman *et al.*, 1986). The mannosylation of enzymes can also result in their inactivation. This can often be prevented by carrying out the coupling reaction in the presence of substrate to protect the catalytic site.

For these studies we make use of a neoglycoprotein that has high affinity for the receptor. Addition of mannose residues to BSA produces the ligand with highest affinity for the receptor known to date (K_D 2.7×10^{-10}, measured with purified receptor). Mannose-BSA is prepared as described by Lee *et al.* (1976). The mannosyl precursor, cyanomethyl-1-thioglycoside-D-mannopyranoside, can be obtained from E-Y Laboratories. The steps for coupling to protein are the following:

1. Dissolve precursor (0.4 g) in 20 ml of dry methanol to ~0.1 *M* final concentration.
2. Add 54 mg of sodium methylate to the above solution and stir at room temperature overnight.
3. Rotovap to dryness with a bath temperature <40°C.
4. Add ~150 mg of the highest grade BSA crystals to the powder, and dissolve the solids in 15 ml of sodium borate (0.25 *M*, pH 8.5). Stir at room temperature overnight.
5. Dialyze extensively against H_2O at 4°C, and store frozen until ready for use. The product is stable at 4°C as long as bacterial growth is prevented.
6. The degree of glycosylation can be determined with a standard phenol–sulfuric acid assay. To ensure that all hexoses are released, the derivatized protein (0.2 mg) is first incubated with 10 μl of 0.2 *M* mercuric acetate in 0.1 *M* acetic acid for 5 minutes at 100°C prior to the phenol–sulfuric acid assay.

7. For binding and uptake studies, mannose-BSA is radioiodinated using chloramine T (Wileman *et al.*, 1984).

The degree of mannosylation is determined by the molar ratio of mannosyl precursor to the number of free amino groups available on the protein and the length of reaction time. In general, equimolar amounts of both reagents are mixed in the reaction, but this ratio can be changed if an optimal number of mannose residues are not incorporated into protein. We have observed that the more mannose coupled to a protein, the better the protein's affinity for the mannose receptor becomes. For this reason, the reaction is often carried out for long periods of time and in the presence of molar excess of precursor.

III. Permeabilization of Cells

In order to study endosomal function within a permeabilized cell, the concentration of digitonin used should be sufficient to permeabilize the plasma membrane but should not alter the integrity of endosomal membranes. In this manner, cytosolic components can be manipulated while endosomal function remains intact. The concentration of digitonin must be optimized before performing any experiments involving functional intracellular organelles. Because digitonin in solution partitions between the aqueous and the membrane lipid phases, the concentration of detergent required depends on the area of membrane exposed to digitonin and the concentration of cholesterol within the membrane. The concentration and growth phase of the cell line chosen for these studies should be maintained constant to diminish the effect of membrane surface area on the variability of the permeabilization step.

A. Permeabilization Curve

To assess the effect of a given concentration of digitonin on the integrity of several cellular membranes, the release of various proteins can serve as markers for the permeabilization of different membrane-bound compartments within the cell. Trypan blue dye is a good marker for monitoring plasma membrane permeability to molecules of very low molecular weight. The selective release from cells of lactate dehydrogenase, an enzyme present in the cytoplasm, provides a good marker for plasma membrane permeabilization to proteins. The release of acid hydrolases is a suitable assay for the integrity of lysosomal membranes. It is important to choose an enzyme, such as β-hexosaminidase, which is not a lysosomal membrane protein. Finally, the introduction of an endocytic marker into endosomes and the subsequent quantitation of its release after cells are incubated with different concentrations of digitonin constitutes an appropriate assay for endosomal membrane integrity.

The permeabilization procedure described here was optimized for rabbit alveolar macrophages at a concentration of 10^7 cells/ml. The choice of buffer can vary considerably because the incorporation of digitonin into biological membranes is nearly independent of the buffer used. Phosphate-buffered saline (PBS) was used for these studies. Typically, a 10% solution of digitonin (obtained from Sigma) is prepared in dimethyl sulfoxide (DMSO) and then diluted to the desired final concentration in PBS. The purest digitonin available should be utilized. Sigma's digitonin is 80% pure, and this should be taken into consideration when detergent concentrations are reported. Digitonin is not very stable in aqueous solutions and, after a few hours in solution, will precipitate. Fresh aqueous solutions should always be prepared for these studies.

Cells are incubated in increasing concentrations of digitonin for at least 30 minutes at 4°C. Detergent concentrations should not exceed 0.05% to avoid the solubilizing effect of residual DMSO on the cells. Cells can be viewed under the microscope in the presence of trypan blue to quantitate the cell's permeability to the dye. Alternatively, the cells can be pelleted in a tabletop microfuge to separate the cellular pellet from the supernatant. The addition of 0.1% Triton X-100 to both supernatant and pellet solubilizes the protein contents of each fraction. Aliquots from each sample can then be analyzed for total protein or for a specific enzymatic activity. From these data, it is then possible to quantitate the percentage of total protein or enzyme released at each concentration of detergent. An example of such an optimization curve experiment is presented in Fig. 1. Each curve depicts the percentage of marker released for a given concentration of digitonin. It is important to note the striking difference in permeability between the plasma membrane and the lysosomal membrane, the latter one being much less sensitive to digitonin. The cholesterol content of lysosomal membranes is very low.

B. Assessment of Endosome Membrane Integrity after Permeabilization

There are presently no well-established molecular markers for endosomes. In fact, endosomes constitute a morphologically heterogeneous population of intracellular vesicles that may include functionally different organelle populations. Endosomes share in common the presence of

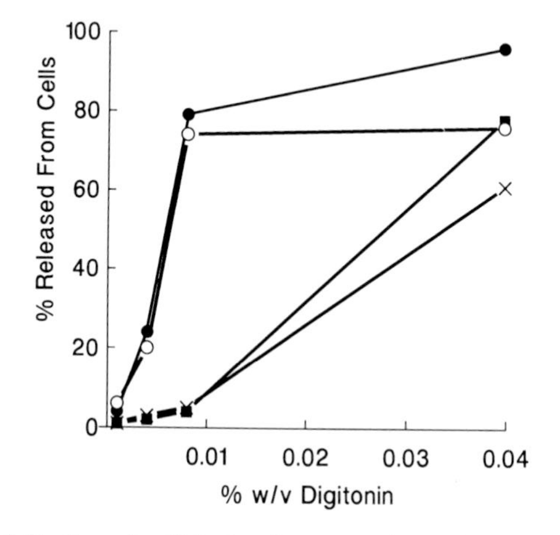

FIG. 1. Permeabilization of rabbit alveolar macrophages by digitonin. Rabbit alveolar macrophages (10^7 cells/ml) were incubated with different concentrations of digitonin in PBS at 4°C for 30 minutes. For each concentration, total protein (○), lactate dehydrogenase (●), and β-hexosaminidase (■) activity of the supernatant was measured after cells were pelleted by centrifugation (500 *g*, 5 minutes) and compared to total values for solubilized cells. Values are reported as percentage of the total released by treatment. A 5-minute uptake followed by a 5-minute chase of ^{125}I-labeled β-glucuronidase (x) by these macrophages provided a marker for the contents of endosomes. These cells were treated as already described, and percentage ligand release is reported under the same conditions. Results are the mean of at least two experiments.

internalized ligand within them at any given stage of endocytosis prior to the delivery of these ligands to other compartments (e.g., the lysosome). Consequently, the endocytic ligand constitutes the only reliable marker for endosomes. Kinetically, these compartments are often qualified as "early" or "late" endosomes (Schmid *et al.*, 1988). Morphological studies show that the structures of both early and late endosomes are also different: early endosomes have often a tubulovesicular shape, and late endosomes contain membranous inclusion bodies (multivesicular endosomes) (Harding *et al.*, 1985). It is thought that pH-dependent receptor–ligand dissociation occurs in early endosomes. Hence, this compartment is often referred to as CURL (compartment for the uncoupling of receptor from ligand) (Geuze *et al.*, 1983). The endosomal marker used for these studies must therefore be localized in this compartment to assess correctly its permeability to digitonin. Ligands that are internalized by either fluid-phase endocytosis (e.g., high molecular weight dextrans) or receptor-mediated endocytosis can in principle be good markers for endosomes. The transfer to lysosomes of ligands internalized by the latter endocytic pathway is more efficient; hence, shorter times of ligand internalization are required to load all the endosomal compartment.

The pathway followed by ligands internalized by the mannose receptor is also characterized by the temporal localization of ligand in endosomes prior to their transfer to lysosomes (Wileman *et al.*, 1985a). Ligand starts to appear in lysosomes after ~20 minutes of continuous uptake at 37°C (Wileman *et al.*, 1985b). This has been determined by subcellular fractionation studies in which the appearance of radiolabeled ligand in a lysosomal fraction is monitored. Therefore, a time course <20 minutes is recommended for the labeling of the endosomal compartment. Figure 1 shows the digitonin-permeabilization curve of endosomes when rat preputial β-glucuronidase, a standard ligand for the mannose receptor, was chosen as an endosomal marker. β-Glucuronidase was radioiodinated with chloramine T (Wileman *et al.*, 1984). Prior to the permeabilization step, rabbit alveolar macrophages were allowed to internalize ^{125}I-labeled β-glucuronidase (20 μg/ml) for 5 minutes at 37°C in HBSA (Hank's balanced salt solution buffered with 10 m*M* 2-[2-hydroxy-1,1-bis(hydroxymethyl)ethyl]-aminoethanesulfonic acid (TES) and 4-(2-hydroxyethyl)-1-piperazineethanesulfonic acid (HEPES) supplemented with 1% BSA) and washed to remove free ligand. The ligand was chased to later compartments by an additional 5-minute incubation at 37°C in HBSA. These cells were subjected to digitonin permeabilization, and the appearance of radioactivity in the supernatant provided a measure of endosomal vesicle leakage. Very little leakage of ^{125}I-labeled β-glucuronidase was observed at digitonin concentrations <0.008%. Ligands for other receptors that follow a similar intracellular route could also have been used as endosomal markers. Moreover, some ligands may only be present in a subset of endosomes, and the effects of digitonin on specific subsets can be characterized. Transferrin, for example, never leaves the early endosomes in its recycling route within the cell, making it a good marker for this compartment (Schmid *et al.*, 1988).

For rabbit alveolar macrophages, 0.008% digitonin seems to permeabilize the plasma membrane without affecting other intracellular membranes. The endosome acidification studies presented in this paper are all carried out with this concentration of digitonin.

IV. Endosome Acidification

Both endosomes and lysosomes maintain an acidic lumen by means of a proton pump ATPase (Wileman *et al.*, 1985a). The acidification of these compartments requires ATP. Studies using fluorescent ligands have shown that ligands pass into an acidic environment shortly after entering the cell (Tycko *et al.*, 1983; Galloway *et al.*, 1983). These studies require powerful imaging techniques to monitor cell fluorescence or the generation of semipurified vesicle fractions from fractionation techniques, some of which (e.g., Percoll) alter the ability to read fluorescence. We have taken advantage of the pH-dependent binding of ligands to the macrophage mannose receptor to study endosome acidification in a permeabilized-cell assay. Similar assays have been carried out with the asialoglycoprotein receptor (Wolkoff *et al.*, 1984). This assay has proved to be rapid, easy to perform, and quite sensitive to changes of endosomal pH.

A. Receptor–Ligand Dissociation Assay in Endosomes

1. Preparation of Buffers

Several buffers are required for this assay:

ATP Buffer
20 m*M* HEPES
5 m*M* NaCl
150 m*M* KCl
pH 7.4
Assay Buffer
1.0 *M* NaCl
25 m*M* HEPES
50 m*M* $CaCl_2$
0.6% BSA
0.8% Triton X-100
pH 7.0
Releasing Buffer
1.0 *M* NaCl
25 m*M* 2[*N*-morpholino]ethanesulfonic acid (MES)
10 m*M* EDTA
0.1 *M* α-methylmannoside
0.6% BSA
0.8% Triton X-100
pH 5.5
Assay Solubilization Buffer
Assay buffer with mannose-BSA (2 mg/ml)

γ-Globulin Buffer

1% γ-Globulin fraction with assay buffer that is devoid of detergent

ATP

100 m*M* ATP (disodium salt)

100 m*M* $MgCl_2$

pH 7.0

2. Ammonium Sulfate Precipitation of Receptor–Ligand Complexes

A receptor–ligand dissociation assay must distinguish free ligand from receptor-bound complexes. We have taken advantage of the fact that mannose-BSA–receptor complexes can be precipitated with much lower concentrations of ammonium sulfate than free mannose-BSA to develop a solubilization precipitation assay (Wileman *et al.*, 1985a). To show this, rabbit alveolar macrophages were incubated at 4°C for 1 hour with saturating amounts of ^{125}I-labeled mannose-BSA (5 μg/ml). After several washes to remove unbound ligand, the cells were solubilized with 0.4% Triton to release most of the cell-associated radioactivity. After removal of cell debris by centrifugation, increasing amounts of ammonium sulfate were added to the cell extract. The solubility of the extracted radioactivity was compared with that of an equivalent amount of radiolabeled ligand added at the time of precipitation of a cell extract. Figure 2 shows that 90% of receptor–ligand complexes from the cell extract precipitated at 40% saturation of ammonium sulfate while only 10% of the free ligand was precipitable. Thus, at this concentration of ammonium sulfate, receptor-bound and free mannose-BSA are easily distinguishable.

3. Characterization of pH-Dependent Dissociation of Ligand from the Macrophage Mannose Receptor

It is useful to have some information on the kinetics of receptor–ligand dissociation from the cell surface receptor. In order to determine the pH-dependent binding of mannose–BSA to the cell surface mannose receptor, rabbit alveolar macrophages (10^7/ml) are incubated in 1% paraformaldehyde in 150 m*M* NaCl, 30 m*M* $NaPO_4$ (pH 7.0) for 30 minutes at 4°C. They are then washed three times in buffered saline by centrifugation at 900 g. This fixation protocol effectively prevents cell surface receptor internalization while leaving its binding properties intact. Cells are resuspended in PBS (pH 7.4) and incubated with ^{125}I-labeled mannose-BSA (5 μg/ml) for 90 minutes at 4°C. After washing the cells to

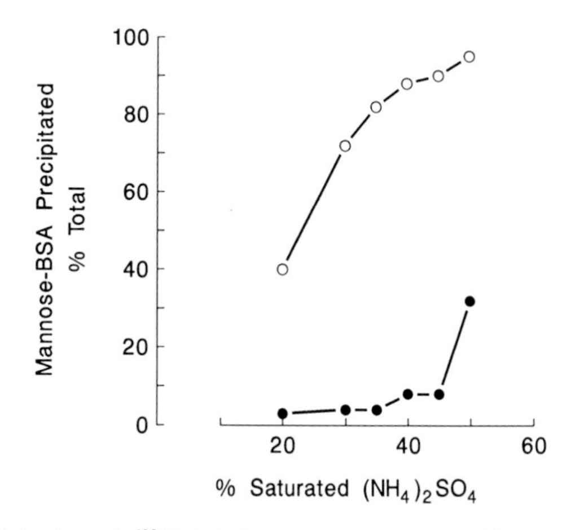

FIG. 2. Precipitation of ^{125}I-labeled mannose-BSA and ^{125}I-labeled mannose-BSA–receptor complexes by ammonium sulfate. ^{125}I-labeled mannose-BSA was bound to the surface of rabbit alveolar macrophages at 4°C. The cells were washed and then solubilized using 0.1% Triton X-100. After removal of insoluble material by centrifugation, the extracted radioactivity was precipitated by adding increasing concentrations of $(NH_4)_2SO_4$ (○). Also shown is the precipitation of ^{125}I-labeled mannose-BSA added to a macrophage Triton extract immediately before addition of $(NH_4)_2SO_4$ (●).

remove unbound ligand, they are resuspended in ATP buffer at different pH values and incubated at 37°C. The percentage of radioactive ligand that remains associated with the cells is measured by ammonium sulfate precipitation and compared to the values obtained for cells incubated at pH 7.0. Figure 3 shows a typical pH-dependent receptor–ligand dissociation curve for rabbit alveolar macrophages.

4. RECEPTOR–LIGAND DISSOCIATION ASSAY

Rabbit alveolar macrophages (1×10^7/ml) are allowed to bind ^{125}I-labeled mannose-BSA (5 μg/ml, 5×10^6 cpm/μg) at 4°C for 90 minutes in uptake buffer (HBSA). All cell surface receptors are loaded with ligand at this stage. After washing, the cells are warmed to 37°C for 5 minutes to internalize the receptor into endosomes and then immediately cooled by dilution in ice-cold uptake buffer. After several washes, cells are permeabilized with 0.008% digitonin in ATP buffer for 30 minutes at 4°C. Briefly, a standard assay is carried out as follows: 50 μl of cells are added to 150 μl of ATP buffer with ATP; addition of ATP is supplemented with equimolar amounts of $MgCl_2$. Incubations are performed at 37°C for 5 minutes. Mannose-BSA is known to be degraded in macrophage endo-

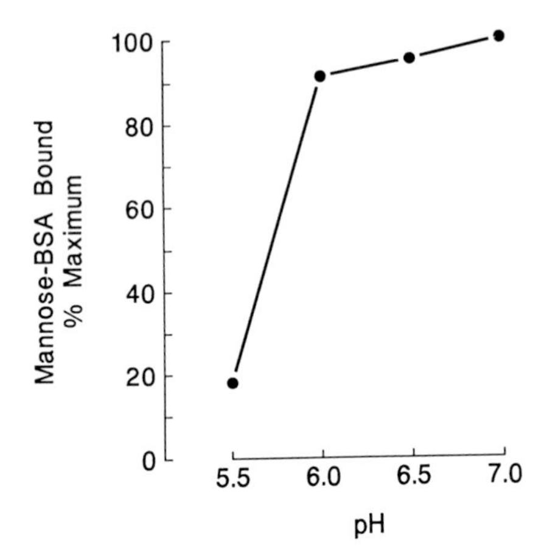

FIG. 3. pH dependency of receptor–ligand dissociation. Cells were fixed in 1% paraformaldehyde for 30 minutes at 4°C. They were then washed 3 times with 10 times their incubation volume. The cells were allowed to bind ^{125}I-labeled mannose-BSA at 4°C, washed, and warmed to 37°C in ATP buffer adjusted to the indicated pH values (pH 5.5–6.5, MES buffer; pH 7.0, HEPES buffer) for 5 minutes in the presence of 1.0 mM $CaCl_2$. Cells were then subjected to the ammonium sulfate precipitation assay to determine the extent of ligand dissociation.

somes. For this reason, long incubations should be avoided with this ligand. Very little degradation is observed after 5-minute incubations of permeabilized cells. At the end of the incubation, 200 μl of assay solubilization buffer are added, and the samples are kept on ice for 10 minutes. Samples are ammonium sulfate precipitated by adding 100 μl γ-globulin buffer and 100 μl of assay buffer followed by 400 μl of saturated ammonium sulfate. Precipitated radioactivity is collected on GF/C glass fiber filters (Whatman), and the filters are washed twice with 40% ammonium sulfate before they are counted in a γ counter. Two types of controls are always run concomitantly with each assay. Incubation of permeabilized cells containing ligand at 37°C in the absence of nucleotides provides the total value for receptor–ligand complexes before dissociation. This value is usually very close to the one observed for permeabilized cells that have been kept at 4°C. Incubation of cells at 37°C with dissociation buffer provides the 100% dissociation value; all ligand dissociates from the receptor in the presence of EDTA and α-methylmannoside at pH 5.5.

Receptor–ligand dissociation in endosomes requires ATP. When increasing concentrations of ATP are added to the reaction, a parallel increase in ligand dissociation is seen that plateaus at ~3 m*M* ATP. As Fig. 4 summarizes, ADP and AMP do not have the same effect. We have observed high levels of ATPase activity in rabbit alveolar macrophages that may result in the depletion of exogenously added ATP when long incubations at 37°C are performed (>5 minutes). ATPase activity predominates in the membrane fraction. To prevent ATP depletion, an enzymatic ATP-regenerating system can be added consisting of 1–3 m*M* ATP, 8 m*M* creatine phosphate (Sigma), and 31 units/ml of rabbit muscle creatine phosphokinase (Sigma, type I). In addition, an ATP-depletion system [5 m*M* mannose, 25 units/ml bakers' yeast hexokinase (Sigma, type F-300)] should remove any endogenous ATP in the preparation. Cytosol, though at very dilute concentrations, is normally present in the reaction mixture. Pelleting of permeabilized cells (900 g, 5 minutes) and resuspension in fresh buffer before executing the reaction may also help remove nucleotides present in the cytosol. We have not observed much deterioration of endosomal membrane integrity when these cells are pelleted after the permeabilization step.

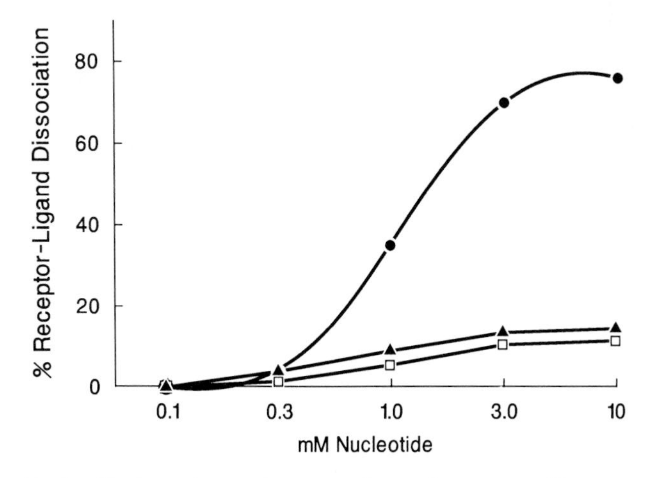

FIG. 4. Nucleotide dependence of receptor–ligand dissociation. ^{125}I-Labeled mannose-BSA was bound to the surface of rabbit alveolar macrophages at 4°C. The cells were then washed and warmed to 37°C for 5 minutes. Uptake was stopped by cooling the cells to 4°C. Cells were washed again and incubated with 0.008% w/v digitonin in ATP buffer. Digitonin-treated cells were incubated for 5 minutes at 37°C with increasing concentrations of nucleotide and an equimolar concentration of $MgCl_2$, and subjected to the solubilization and precipitation assay. The figure shows the dissociation produced by ATP (●), ADP (□), and AMP (▲). Values are reported as percentages and are the mean of at least three experiments.

B. Determination of Conditions Required for Acidification

This receptor–ligand dissociation assay can be modified for use with other pH-dependent receptor-ligand systems. Once such a receptor–ligand dissociation assay is developed, it can be used to study endosome acidification and its requirements. In particular, this assay can be used to characterize partially the endosomal proton pump ATPase and other associated proteins that may be required for acidification. Furthermore, other ionic transport mechanisms that may have a role in endosome acidification are amenable to analysis. Conditions that modify those transport processes will affect acidification. Care must be taken that the conditions tested do not affect the dissociation of ligand from receptor; otherwise, an increase or decrease in ligand dissociation may not necessarily reflect a change in acidification properties of the endosome.

1. Nucleotide Requirement

Hydrolysis of ATP is required for proton pump activity. Different nucleotides can be tested with this assay by simply substituting ATP for the nucleotide of choice. Permeabilized cell preparations are crude and may contain enzymes that can exchange phosphate groups between different nucleotides. Thus, the evaluation of the effect of different nucleotides on acidification may be difficult to assess with this assay unless cytosolic proteins are fully removed from the preparation. Pelleting the permeabilized cells as described previously may prove useful for this purpose, but all proteins may not be removed by this procedure. Nonhydrolyzable derivatives of ATP or GTP [e.g., adenyl-imidodiphosphate (PNP-AMP) and guanine 5′-(3-O-thio)triphosphate (GTPγS)] can also be tested either as competitors for the parent nucleotide or by themselves.

2. Ionic Requirements

Different ions can alter endosome acidification by affecting the proton pump directly or by modifying the properties of other ion carriers in endosomes with a modulatory role in acidification. Anions have been shown to affect proton pump ATPase activity directly when tested with purified preparations of chromaffin pump complexes (Moriyama and Nelson, 1987). Moreover, anions appear to be required to facilitate ATP-dependent acidification by moving into the vesicle as a charge-compensating ion to maintain vesicle electroneutrality (Wileman *et al.*,

1985a). When a nonpermeable anion is substituted for chloride, acidification is inhibited while a positive luminal electrochemical potential is still generated. The pump therefore appears to be electrogenic, and net positive charge is transported across the membrane. It is likely that an anion channel is present in endosomes that allows chloride ions to cross the endosomal membrane down its electrochemical gradient. The use of different anions may aid in the characterization of such an anion channel. It has been shown that early endosomes have sodium pumps that may modulate endosomal acidification (Fuchs *et al.*, 1986). Thus, the effect of inhibitors of this pump can result in changes in endosomal acidification that may be detected by the permeabilized cell assay.

The effect of both anions and cations on acidification can be tested with this assay very easily. Permeabilized-cell preparations can be treated as if they were vesicle preparations with the additional advantage that buffers can be exchanged readily by simply pelleting the cells and resuspending them in a different buffer. Sugar-based buffers should probably not be used because they may induce dissociation of ligand from sugar-specific receptors such as the mannose receptor when present at high concentrations. The standard assay contains KCl and NaCl. Buffers of different ionic composition can be prepared, keeping in mind that the final osmolarity of the solution should be maintained at ~320 mOsm. The ions tested must not affect the digitonin permeabilization step or the binding properties of the receptor. This can be tested by performing the optimization curves for digitonin permeabilization and pH-dependent dissociation of receptor–ligand complexes in the presence of these ions.

For these studies, cells can be permeabilized in the buffer of choice before carrying out the dissociation assay (Diaz *et al.*, 1989). We have tested sodium and potassium as cations and have observed no differential effect on acidification. For anions, we have tested the halide series with the exception of fluoride, which precipitates out during the solubilization step, and have observed that bromide and chloride anions are equally effective in supporting acidification, while iodide was fully inhibitory. Presumably, iodide ions are too large to pass through the anion channel. In addition, citrate, gluconate, sulfate, nitrate, and phosphate anions have been tested. Citrate and gluconate had an inhibitory effect, although the inhibition was never >60% for either ion. Sulfate, nitrate, and phosphate were stronger inhibitors of acidification, sulfate being the strongest. Nitrate ions are readily membrane permeable, yet they cause a strong inhibition of dissociation. It is likely that this decrease is caused by a direct effect on proton pump activity (Van Dyke, 1986).

3. Effect of Inhibitors

The endosomal proton pump ATPase is a member of a family of vacuolar ATPases that are distinguished from other classes by virtue of their inhibitor specificities. ATPase inhibitors have therefore been useful in the characterization of the properties of proton pumps. A number of ATPase inhibitors have been described in the literature; most of them can easily be tested for inhibition by simply adding them prior to the dissociation step. Table I lists the effect of some of these inhibitors on the dissociation assay. These agents do not interfere with the binding affinity of the mannose receptor for mannose-BSA. The specificity of these inhibitors varies. Oligomycin and azide, for example, are quite specific in their inhibition of mitochondrial ATPases. Other inhibitors, on the other hand, have a broader specificity because they modify chemical groups present in many proteins. 4,4′-Diisothiocyanostilbene-2,2-disulfonic acid (DIDS) and *N*-ethylmaleimide (NEM), for example, chemically modify amino and sulfhydryl groups of proteins, respectively. A dose–response curve is recommended for these compounds to avoid missing their inhibitory effect on the pump, because the presence of other proteins carrying the reactive groups may quench the chemical agent's activity if added at low concentrations. The permeabilized-cell assay often requires higher concentrations of these compounds because free cytoplasmic proteins are present during the assay. Of the compounds tested so far, the alkylating agent NEM and DIDS have been the only effective inhibitors of

TABLE I

Effect of Inhibitors on Receptor–Ligand Dissociation

Inhibitor	Inhibition
ATPase inhibitors	
Oligomycin (10 μg/ml)	−
Azide (0.3 m*M*)	−
Vanadate (0.3 m*M*)	−
NEM (1 m*M*)	+
DIDS (1 m*M*)	+
Proton ionophores	
Monensin (10 μM)	+
Nigericin (10 μM)	+
CCCP[a] (10 μM)	+

[a] Carbonyl cyanide *m*-chlorophenylhydrazone.

receptor–ligand dissociation. This is in agreement with other studies on endosome acidification (Mellman *et al.*, 1986).

Proton ionophores also constitute good inhibitors of vesicle acidification. If ATP-dependent receptor–ligand dissociation is driven by vesicle acidification, proton ionophores should inhibit this dissociation. They have been shown to inhibit the acidification of intracellular organelles and the transport of receptor-mediated endocytosed ligand to lysosomes (Wileman *et al.*, 1984). Proton ionophores exchange protons across membranes down their concentration gradient. This exchange is sometimes coupled to another cation; monensin exchanges Na^+ for H^+ whereas nigericin has a greater affinity for K^+. The effect of ionophores is seen at very low concentrations (1–10 μM) in that a few molecules can rapidly transport many protons across biological membranes. In the permeabilized-cell assay, all proton ionophores tested inhibited dissociation.

V. Summary and Outlook

This chapter describes a method for the study of endosome acidification using a digitonin-permeabilized cell assay. Digitonin can be used to permeabilize the plasma membrane of rabbit alveolar macrophages selectively without affecting intracellular membrane integrity. Endosomes remain functionally intact after digitonin permeabilization. This technique has proved to be an easy and rapid procedure to gain access to the cytoplasmic side of endosomes. Because other cell types are also amenable to selective digitonin permeabilization, this technique for plasma membrane permeabilization should be generally applicable.

The acidification studies described are based on the pH-dependent binding of ligands to the macrophage mannose receptor. Receptor–ligand dissociation is driven by low pH and takes place in endosomes. This dissociation becomes a measure of vesicle acidification. The differential precipitation of receptor–ligand complexes from free ligand permits the quantitation of dissociation. The results indicate that acidification requires ATP and specific anions (Cl^-). The assay also demonstrates that acidification is inhibited by proton ionophores and that NEM and DIDS, agents that chemically modify proteins, inhibited acidification. As it is designed, this permeabilized-cell assay could be used for ligands that bind to other endocytic receptors and that dissociate from these receptors in endosomes.

This technique could also be applied to the study of other endosomal functions. Studies of anion and cation fluxes across the endosomal membrane should be possible. Specifically, the transport of Ca^{2+} could be addressed by this method because the mannose receptor exhibits Ca^{2+}-dependent ligand binding. Preliminary studies indicate that such a flux exists. In addition, the putative role of the cytoskeleton in endosome and organelle motion requires the reconstitution of many cellular components, and the permeabilized-cell system should provide the necessary accessibility to the cytoplasmic contents to make these studies possible. Endosome fusion may also be reconstituted in a digitonin-permeabilized cell. Chromaffin granule exocytosis has been succesfully reconstituted by this method, indicative that the components of the plasma membrane required for its fusion to exocytic vesicles remain intact after the permeabilization step (Lelkes and Pollard, 1987; Sarafian *et al.,* 1987). Endosomes may also fuse to each other or to other intracellular vesicles (e.g., lysosomes) in permeabilized cells. Thus, there are many potential applications of this method for the study of endosome function.

Acknowledgments

We would like to thank Tom E. Wileman for his help with the initial studies on permeabilization, Rita L. Boshans for excellent technical assistance, and Janice S. Blum and Luis S. Mayorga for critical review of the manuscript. This work was supported, in part, by Health, Education and Welfare Grants CA 12858 and AI 20015.

References

Ahnert-Hilger, G., Bhakdi, S., and Gratzl, M. (1985). *J. Biol. Chem.* **260,** 12730–12734.

Brooks, J. C., and Treml, S. (1984). *Life Sci.* **34,** 669–674.

Diaz, R., Wileman, T. W., Anderson, S. J., and Stahl, P. (1989). *Biochem. J.* In press.

Dunn, L. A., and Holz, R. W. (1983). *J. Biol. Chem.* **258,** 4989–4993.

Fuchs, R., Schmid, P., Male, P., Helenius, A., and Mellman, I. (1986). *J. Cell Biol.* **103,** 439a.

Galloway, C. J., Dean, G. E., Marsh, M., Rudnick, G., and Mellman, I. (1983). *Proc. Natl. Acad. Sci. U.S.A.* **80,** 3334–3338.

Geuze, H. J., Slot, J. W., Strous, J. A. M., Lodish, H. F., and Schwartz, A. L. (1983). *Cell (Cambridge, Mass.)* **32,** 277–287.

Harding, C., Levy, M. A., and Stahl, P. (1985). *Eur. J. Cell Biol.* **36,** 230–238.

Howell, T. W., Cockcroft, S., and Gomperts, B. D. (1987). *J. Cell Biol.* **105,** 191–197.

Knight, D. E., and Baker, P. F. (1982). *J. Membr. Biol.* **68,** 107–140.

Lee, Y. C., Stowell, C. P., and Krantz, M. J. (1976). *Biochemistry* **15,** 3956–3963.

Lelkes, P. I., and Pollard, H. B. (1987). *J. Biol. Chem.* **262,** 15496–15505.

Lennartz, M. R., Wileman, T. E., and Stahl, P. (1987). *Biochem. J.* **245,** 705–711.

Mellman, I., Fuchs, R., and Helenius, A. (1986). *Annu. Rev. Biochem.* **55,** 663–700.

Moriyama, Y., and Nelson, N. (1987). *J. Biol. Chem.* **262,** 9175–9180.

Rodzial, M. M., and Haimo, L. T. (1986). *J. Cell Biol.* **103,** 2755–2764.

Sarafian, T., Aunis, D., and Bader, M. F. (1987). *J. Biol. Chem.* **262,** 16671–16676.
Schmid, S. L., Fuchs, R., Male, P., and Mellman, I. (1988). *Cell (Cambridge, Mass.)* **52,** 73–83.
Stahl, P. D., and Schwartz, A. L. (1986). *J. Clin. Invest.* **77,** 657–662.
Stahl, P. D., Schlesinger, P. H., Sigardson, E., Rodman, J. S., and Lee, Y. C. (1980). *Cell (Cambridge, Mass.)* **19,** 207–215.
Stahl, P. D., Wileman, T. E., Diment, S., and Sheperd, V. L. (1984). *Biol. Cell.* **51,** 215–218.
Steinberg, T. H., Newman, A. S., Swanson, J. A., and Silverstein, S. C. (1987). *J. Biol. Chem.* **262,** 8884–8888.
Tycko, B., Keith, C. H., and Maxfield, F. R. (1983). *J. Cell Biol.* **97,** 1762–1776.
Van Dyke, R. W. (1986). *J. Biol. Chem.* **261,** 15941–15948.
Wileman, T., Boshans, R. L., Schlesinger, P., and Stahl, P. (1984). *Biochem. J.* **220,** 665–675.
Wileman, T., Boshans, R. L., and Stahl, P. (1985a). *J. Biol. Chem.* **260,** 7387–7393.
Wileman, T., Harding, C., and Stahl, P. (1985b). *Biochem. J.* **232,** 1–14.
Wileman, T. E., Lennartz, M. R., and Stahl, P. D. (1986). *Proc. Natl. Acad. Sci. U.S.A.* **83,** 2501–2505.
Wolkoff, A. W., Klausner, R. D., Ashwell, G., and Harford, J. (1984). *J. Cell Biol.* **98,** 375–381.
Yamashiro, D. J., Fluss, S. R., and Maxfield, F. R. (1983). *J. Cell Biol.* **97,** 929–934.

Chapter 3

ATP Permeabilization of the Plasma Membrane

THOMAS H. STEINBERG

Department of Medicine
Columbia University College of Physicians and Surgeons
New York, New York 10032

SAMUEL C. SILVERSTEIN

Department of Physiology and Cellular Biophysics
Columbia University College of Physicians and Surgeons
New York, New York 10032

I. Introduction

In the last decade there has been great interest in the role of plasma membrane receptors for adenine and other purine nucleotides in cell function, and we have learned that many cells possess receptors for ATP that induce ion fluxes, phosphoinositide turnover, and other responses. An unexpected bonus resulting from these endeavors has been the discovery that in some cells extracellular ATP induces permeabilization of the plasma membrane to a variety of molecules that are normally impermeant. Because ATP permeabilization is rapid, simple to perform, and readily reversible, we and others have found it a useful method for introducing small water-soluble molecules into the cytoplasmic matrix of cells. Permeabilization appears to be mediated by ligation of a specific plasma membrane receptor for ATP^{4-}, whose physiological function is unknown at present. This review will focus on the practical aspects and the uses of ATP permeabilization as an investigative tool; examples will be drawn from our experience with mouse macrophages, that of Gomperts and colleagues with rat mast cells, and the work of Heppel and collaborators with transformed mouse fibroblasts.

II. Plasma Membrane Receptors for Adenine Nucleotides

It has been known for many years that ADP released from platelet granules mediates platelet aggregation, that ATP is a neurotransmitter in certain nonadrenergic, noncholinergic synapses, and that ATP is present in adrenergic and cholinergic synaptic vesicles where it may serve as a cotransmitter (Stone, 1981). Platelets (Meyers *et al.*, 1982), endothelial cells (Pearson and Gordon, 1979), and adrenal chromaffin cells (Douglas *et al.*, 1965) all secrete ATP; a variety of nonneural cells, including hepatocytes (Prpic *et al.*, 1982), human polymorphonuclear leukocytes (Becker and Henson, 1975), endothelial cells (McIntyre *et al.*, 1985; Pirotton *et al.*, 1987), Ehrlich ascites tumor cells (Landry and Lehninger, 1976; Dubyak, 1986), transformed mouse fibroblasts (Heppel *et al.*, 1985), mast cells (Cockcroft and Gomperts, 1979a), and macrophages (Sung *et al.*, 1985) are affected by extracellular ATP.

Extracellular ATP elicits a variety of responses from different cells including membrane depolarization (Sung *et al.*, 1985), monovalent and divalent cation fluxes (Chahwala and Cantley, 1984), phosphoinositide turnover (Prpic *et al.*, 1982; Dubyak, 1986), and release of calcium from intracellular stores (Greenberg *et al.*, 1988; Dubyak and Young, 1985); physiologic responses that ensue include histamine secretion by mast cells (Dahlquist and Diamant, 1974), release of PAF and prostacyclin by endothelial cells (McIntyre *et al.*, 1985), and vasomotor activity in canine coronary artery and other vascular beds (Su, 1985). Rozengurt and Heppel (1975) found that ATP induced membrane permeability to *p*-nitrophenyl phosphate in a variety of transformed mouse cell lines, and Rozengurt *et al.* (1977) later demonstrated efflux of nucleotides and phosphate esters when these cells were exposed to ATP. Cockcroft and Gomperts described ATP-induced membrane permeability to ^{32}P-labeled metabolites in rat mast cells (1979a) and demonstrated that ATP^{4-} was the active species of ATP (1980). The specificity and distinguishing characteristics of these various ATP-induced effects make it likely that there is more than one distinct membrane receptor for ATP.

In addition to these ATP receptors, many cells possess an ecto-ATPase activity, which may serve to limit the quantity of extracellular ATP available. Some cells, notably macrophages (Remold-O'Donnell, 1978) and neurons (Ehrlich *et al.*, 1986), also have ecto-protein kinase activities that mediate phosphorylation of membrane and added proteins in the presence of exogenous ATP. That so many cell types express surface receptors and enzymes that respond to ATP suggests an important physiologic role for exogenous ATP as a regulator of cell function. The investigator using ATP as a tool to permeabilize cells should keep in mind that ATP may affect cells in many ways in addition to permeabilizing them and should be cautious in attributing all effects observed to the permeabilizing effects of ATP.

III. ATP Permeabilization

A. Basic Considerations of ATP^{4-} Permeabilization

Mouse peritoneal macrophages (resident or elicited) and the mouse macrophagelike cell line J774 can be permeabilized to molecules as large as fura-2 (631 Da) by incubating the cells in medium containing ATP (Steinberg *et al.*, 1987b). Our standard protocol for introducing the membrane-impermeant fluorescent dye lucifer yellow into macrophages is as follows. Adherent cells or cells in suspension are incubated in Dulbecco's modified Eagle medium (DMEM) containing 5 m*M* ATP and lucifer yellow for 5 minutes at 37°C. The cells are washed at 4°C in phosphate-buffered saline (PBS), resuspended in fresh medium, and then viewed by fluorescence microscopy or used in experiments (Fig. 1). After the cells have been removed from the ATP-containing medium, the permeabilization is rapidly reversed.

We have permeabilized macrophages in a variety of media, including Dulbecco's PBS with or without divalent cations, isotonic saline buffered with HEPES, PIPES, MOPS, MES, or Tris, as well as DMEM. J774 cells incubated in isotonic choline chloride (Sung *et al.*, 1985) or potassium chloride buffered with 10 m*M* HEPES (T. H. Steinberg, unpublished results) also are permeabilized to lucifer yellow in response to ATP.

ATP^{4-} is the species of ATP that permeabilizes rat mast cells and mouse macrophages, and presumably other cells that bear appropriate receptors. ATP occurs as several species in physiologic solutions, including $MgATP^{2-}$, $CaATP^{2-}$, $HATP^{3-}$, and ATP^{4-}; the concentrations of these different species will depend on the concentrations of divalent cations and protons in the solution. Because the constants for the many equilibria involved are known, the ATP^{4-} concentration in a given situation can be determined by relying on computer programs designed to solve simultaneous linear equations. Approximations that are easier to

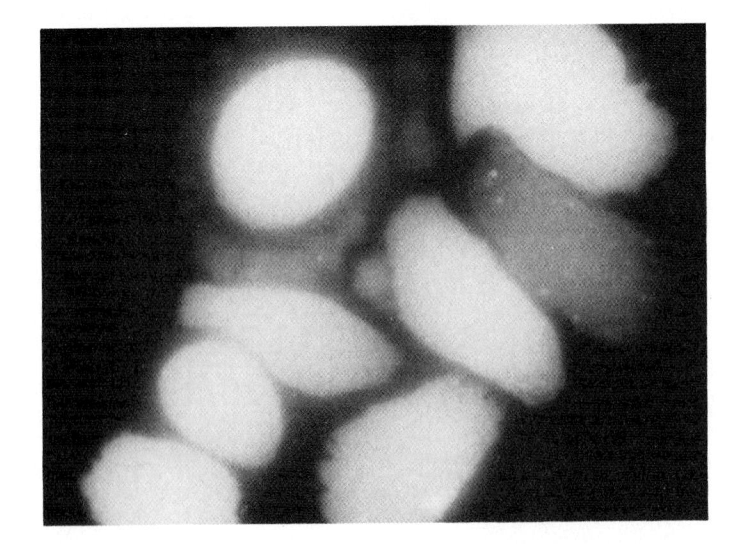

FIG. 1. The plasma membrane of J774 cells is permeabilized by ATP to the dye lucifer yellow. Adherent cells were incubated in Dulbecco's modified Eagle medium containing 0.5 mg/ml lucifer yellow and 5 m*M* ATP for 5 minutes, washed, and viewed by fluorescence microscopy. From Steinberg *et al.* (1987b), by permission.

calculate, though more limited in application, have been published by Dahlquist and Diamant (1974), by Cockcroft and Gomperts (1979b), and by Steinberg and Silverstein (1987). Such programs or equations must be employed to determine the ATP^{4-} concentration precisely. However, one can ensure that a sufficient quantity of ATP^{4-} will be present to induce permeabilization by considering the divalent cation concentrations and pH of the permeabilizing solution.

Both magnesium and calcium are chelated with high affinity by ATP, yielding the ionic species $MgATP^{2-}$ and $CaATP^{2-}$, which do not induce permeabilization. Most of the ATP present in solutions containing physiologic concentrations of divalent cations will be complexed to these cations; $MgATP^{2-}$ will be the predominant species. J774 cells can be reliably permeabilized in medium containing physiologic concentrations of calcium and magnesium by using a total ATP concentration that is several millimolar greater than the total concentration of divalent cations. We have used 5 m*M* ATP when permeabilizing mouse macrophages in medium without divalent cations or in a calcium- and magnesium-containing medium such as DMEM. In medium containing no divalent cations or in the presence of chelators such as EDTA, 200 μM ATP suffices. Figure 2 shows the concentration curves for ATP-mediated Rb^+ efflux from J774 cells in medium containing different concentrations of Mg^{2+}. The total quantity of ATP required to induce Rb^+ efflux increases as the concentration of Mg^{2+} increases, but the amount of ATP^{4-} required in each instance is the same. In rat mast cells only several micromolar ATP is necessary to induce permeabilization.

The pH of the permeabilizing solution also affects the concentration of

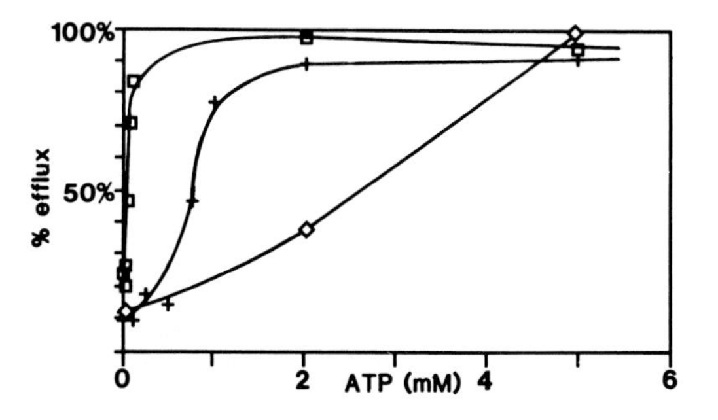

FIG. 2. Effect of extracellular Mg^{2+} concentration on ATP-induced permeabilization of J774 cells. Adherent cells preloaded with $^{86}Rb^+$ were incubated in PBS containing 1 m*M* EDTA (□, for 0 Mg^{2+}), 1 m*M* $MgSO_4$(+), or 5 m*M* $MgSO_4$ (◇) and ATP as indicated, and the percentage of Rb^+ effluxed in 10 minutes was quantitated. From Steinberg and Silverstein (1987), by permission.

ATP^{4-}. The pK for the reaction $HATP^{3-} \rightleftarrows H^+ + ATP^{4-}$ is 6.95 (Alberty, 1968); therefore, as the pH of the solution is made increasingly acidic, relatively more ATP will be protonated, and less ATP will be present in the active tetraanionic form. In accordance with these theoretical considerations, we have found that ATP permeabilization of mouse macrophages is most efficient at pH ≥8.0. Transformed fibroblasts and rat mast cells permeabilize optimally at pH 7.8. However, it is not clear whether the observed pH sensitivity of ATP permeabilization of these cells can be accounted for entirely by conversion of $HATP^{3-}$ to ATP^{4-}, or whether the ATP^{4-} receptor is itself pH-sensitive.

B. Nucleotide Specificity

Only ATP among nucleotide triphosphates is able to induce permeabilization in mouse macrophages, rat mast cells, and transformed mouse fibroblasts. Table I compares the ability of various nucleotides and nucleotide analogs to permeabilize these cell types. Cockcroft and Gomperts (1980) tested a large number of ATP analogs, and found that compounds with modification of the adenine moiety or the phosphate groups were unable to permeabilize mast cells, but that modification of

TABLE I

NUCLEOTIDES AND ANALOGS THAT MEDIATE ATP PERMEABILIZATION

	Cell type[a]		
	J774 Cells	Rat mast cells	3T6 Fibroblasts
Nucleotides			
ATP	+	+	+
GTP	−	−	−
UTP	−	−	−
CTP	−	−	−
ADP	±	−	−
AMP	−	−	−
Adenosine	−	−	−
dATP	−	+	NT[b]
AppppA	−	−	NT
ATP analogs			
ATPγS	+	+	+
AMP-PNP	+	−	+
AMP-PCP	−	−	+

[a] Permeabilization was assessed by lucifer yellow uptake in J774 cells (Steinberg *et. al.*, 1987b), and efflux of ^{32}P-labeled metabolites from rat mast cells (Cockcroft and Gomperts, 1980) and fibroblasts (Rozengurt *et al.*, 1977; Arav and Firedberg, 1985).

[b] NT, not tested.

the ribose had less of an effect. In an analysis of the responsiveness of the ATP^{4-} receptor of mast cells to ATP and ATP analogs, Tatham *et al.* (1988) found the pharmacologic profile of this receptor to be distinct from those of the described P_1 and P_2 purinergic receptors. In contrast to the results obtained in mast cells, mouse macrophages are not permeabilized by deoxyATP, but the modification of the phosphate groups has less effect: both ATP-γ-S and AMP-PNP permeabilize these cells, albeit less effectively. However, AMP-PCP is without effect. Both AMP-PNP and AMP-PCP permeabilize transformed mouse fibroblasts (Arav and Friedberg, 1985).

The ability of the nonhydrolyzable analog AMP-PNP to mediate permeabilization suggests that ATP hydrolysis is not required for permeabilization to occur. We tested this hypothesis directly by measuring Rb^+ efflux and ATP hydrolysis in J774 cells in the absence of divalent cations. Under these conditions ATP-induced permeabilization occurred, but no hydrolysis of ATP was detected. This finding confirmed that ATP permeabilization and ATP hydrolysis can be dissociated: whereas ATP^{4-} mediates permeabilization, $MgATP^{2-}$ is the substrate for ATPases.

C. Cells

In many cells extracellular ATP elicits one or more of the responses mentioned above; however, it appears that ATP-mediated permeabilization of the plasma membrane occurs in a more limited number of cell types. This permeabilization has been most fully documented in rat mast cells, transformed mouse fibroblasts, and mouse macrophages. In addition, Chinese hamster ovary (CHO) cells (Kitagawa and Akamatsu, 1981), HeLa cells, melanoma cell lines (Heppel *et al.*, 1985), and several neuroblastoma cell lines (unpublished results; F. Di Virgilio, personal communication) can be permeabilized by ATP. In many cells monovalent or divalent cation fluxes are induced by ATP, but the investigators have not reported whether permeabilization of the cells to larger molecules also occurs. Some cells do not permeabilize in response to ATP but do display other ATP-mediated effects. For example, ATP elicits a rise in intracellular calcium ion concentration ($[Ca^{2+}]_i$) in human neutrophils (Ward *et al.*, 1988), but permeabilization of the plasma membrane to the dye lucifer yellow does not occur (T. H. Steinberg, unpublished results).

D. Size of the ATP-Induced Pore

The size of the ATP-induced pore has been examined in mouse macrophages and rat mast cells. Although the ATP-induced pore in mouse macrophages and rat mast cells admits a large number of different molecules, it appears to have a fairly well-defined size limitation. Permeabilization by ATP of mouse macrophages allows a variety of molecules to enter the cells, including Na^+, K^+, Rb^+, Ca^{2+}, and the fluorescent dyes carboxyfluorescein (376 Da), ethidium bromide (394 Da), lucifer yellow (443 Da), and fura-2 (631 Da). The dyes trypan blue and Evans blue, both 869 Da, do not enter these cells even when the cells are incubated in medium containing 20 m*M* ATP (Steinberg *et al.*, 1987b). Efflux of nucleotides and sugar phosphates from rat mast cells in the presence of extracellular ATP has been reported, but these cells did not admit inulin (~5000 Da) in the presence of ATP (Bennett *et al.*, 1981).

E. Reversal of Permeabilization

The ATP-induced pore can be closed by removing ATP^{4-} from the medium or by adding a sufficient quantity of divalent cations to the medium to reduce the concentration of ATP^{4-} below the threshold required for permeabilization. When adherent J774 cells permeabilized in ATP-containing medium are washed several times and placed in fresh medium, the cells regain their normal impermeability to lucifer yellow within 5 minutes. When cells in suspension are permeabilized by 5 m*M* ATP, one can reverse permeabilization by adding excess magnesium (final concentration 10 m*M*), spinning down the cells, and resuspending them in fresh medium.

IV. Recovery of Intracellular Homeostasis following ATP Permeabilization

When cells are permeabilized by ATP, there is a rapid exchange of cytosolic and extracellular molecules that are small enough to pass through the pores. When cells are permeabilized in a medium containing isotonic sodium chloride, intracellular K^+ is lost and nearly completely replaced by Na^+ within 10 minutes (Sung *et al.*, 1985). This exchange of monovalent cations is accompanied by plasma membrane depolarization. If calcium is present in the extracellular medium, the $[Ca^{2+}]_i$ will also approach that of the extracellular medium. In cells permeabilized for <30 minutes removal of ATP from the medium allows the membrane to reexpress selective permeability to ions. Under these conditions intracellular K^+ reaccumulates and Na^+ is pumped out of the cytoplasm within 60 minutes, and the plasma membrane repolarizes. However, prolonged exposure of cells to sufficient quantities of ATP to cause permeabilization results in cell death, as will be discussed.

Morphologic changes occur in macrophages treated with ATP (Steinberg *et al.*, 1987b). Adherent cells that have spread on glass or plastic substrates round up after exposure to ATP for 5 minutes. In addition, cells exposed to ATP vacuolate after reversal of ATP permeabilization. Large phase-lucent cytoplasmic vacuoles appear 5–10 minutes after ATP is removed from the medium, are abundant for the ensuing 15 minutes, and gradually disappear.

The morphology of the lysosomal system of J774 cells is also altered by ATP permeabilization. The lysosomes of J774 cells normally appear as a network of tubular and vesicular structures that appear to be interconnected (Swanson *et al.*, 1987b), as seen by labeling this compartment with a fluorescent dye such as Texas red-conjugated ovalbumin (Steinberg *et al.*, 1987b). After ATP permeabilization, the network of tubular lysosomes is fragmented; over 30 minutes tubular structures reappear. Because microtubules form the scaffolding on which the integrity of the tubular lysosomes depends (Swanson *et al.*, 1987a), it is reasonable to assume that disruption of tubular lysosomes reflects a reversible disassembly of microtubules. However, this hypothesis has not been examined directly.

V. Assessment of Permeabilization

Probably the easiest way to assess permeabilization is to monitor the entry of fluorescent dyes into the cells. This procedure is simple, and allows one both to assess the distribution of dye by fluorescence microscopy and to quantitate dye uptake by fluorescence spectrophotometry. With the fluorescence microscope one can determine the fraction of cells in the population that have been permeabilized. The diffuse cytoplasmic distribution of dye within the cells is apparent, and one can ascertain that uptake of dye is not primarily by pinocytosis. We have found the dye lucifer yellow to be ideal for this purpose: it is water soluble, pH-insensitive, and highly fluorescent; also it can be readily seen under the fluorescence microscope using a fluorescein filter set. Carboxyfluorescein and other fluorescein derivatives can also be used, but their fluorescence bleaches more readily and they are pH-sensitive.

Another fluorescent compound that is useful for assessing membrane permeabilization is the membrane-impermeant cationic dye ethidium bromide. When this dye enters permeabilized cells, it rapidly accumulates in the nucleus by virtue of its ability to bind DNA. Because the fluorescence of ethidium bromide increases when the dye binds to DNA, it is possible to obtain "real-time" measurements of ATP permeabilization and to assess the speed with which permeabilization occurs and is reversed. Gomperts (1983) measured ethidium bromide fluorescence in mast cells suspended in medium containing 25 μM ethidium bromide (Fig. 3). The fluorescence increased within seconds when 5 μM ATP was added to the suspension, and this increase was rapidly curtailed by the addition of an excess of Mg^{2+}, which caused the conversion of ATP^{4-} to $MgATP^{2-}$ and allowed the cells' plasma membrane to regain its impermeability to the dye.

The fluorescent dye TMA-DPH can be used in a similar manner (Tatham *et al.*, 1988). This amphipathic cation is highly fluorescent when it is bound to membranes but not when it is in aqueous solution. In cells permeabilized with ATP, the TMA-DPH associates with intracellular membranes and its fluorescence increases.

Measurement of membrane potential can also yield information about the kinetics of ATP permeabilization, and is a very sensitive albeit nonspecific assay for ATP permeabilization. Sung *et al.* (1985) assessed membrane potential using the lipophilic cation tetraphenylphosphonium. Buisman *et al.* (1988) demonstrated that membrane depolarization upon

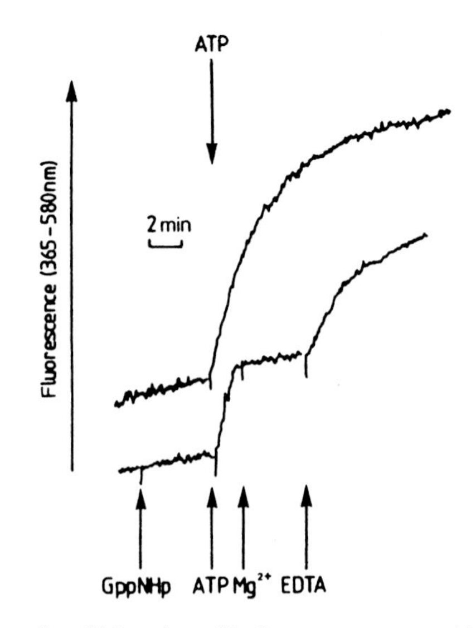

FIG. 3. The increase in ethidium bromide fluorescence caused by ATP-induced permeabilization is halted by the addition of an excess of Mg. Ethidium bromide fluorescence was measured in rat mast cells in suspension. The addition of 5 μM ATP caused a rise in fluorescence that was inhibited by the addition of 1 mM Mg^{2+}, and resumed upon addition of excess EDTA. From Gomperts (1983), by permission.

application of ATP to J774 cells occurs within 40 milliseconds using the whole-cell patch clamp technique.

Measuring the influx or efflux of radiolabeled molecules is another sensitive method to assess permeabilization. Efflux of $^{86}Rb^+$ from cells prelabeled with this ion is a very sensitive measure of permeabilization, but does not necessarily indicate permeabilization to larger molecules. Efflux of radiolabeled metabolites is a sensitive and specific assay for permeabilization (Rozengurt *et al.*, 1977). However, one cannot detect reversal of permeabilization using this assay because efflux of the total pool of radiolabel occurs rapidly.

Reversal of plasma membrane permeabilization after the concentration of ATP^{4-} has been reduced below threshold can be measured most easily by demonstrating that cells no longer admit fluorescent dyes into the cytoplasmic matrix. Assessment of membrane repolarization is of limited usefulness: if cells have been permeabilized for only a few seconds, they will repolarize rapidly when permeabilization is reversed. However, if cells have been permeabilized for a few minutes, membrane repolarization will not occur until the ionic composition of the cytosol has returned to normal, which requires 30–60 minutes (Sung *et al.*, 1985). Assessing ethidium bromide fluorescence as described previously will be a useful measurement of reversal of membrane permeabilization only if maximum nuclear accumulation of the dye has not occurred.

VI. ATP-Resistant Cells

The prolonged exposure of J774 cells to permeabilizing concentrations of ATP results in cell death, presumably due to the abnormal ionic conditions (high intracellular Na^+ and Ca^{2+}) in the cytosol, or to the loss of certain soluble intracellular molecules. More than 90% of J774 cells exposed to 10 m*M* ATP for 45 minutes will die. We exposed J774 cells to these conditions, allowed remaining cells to repopulate cultures, and repeated this cycle several times over 1–2 months. We were thereby able to select and clone several cell lines that were no longer permeabilized by ATP even when 20 m*M* ATP was used (Steinberg and Silverstein, 1987). The most extensively studied of these cell lines is ATPR B2: this variant cell line does not efflux Rb^+, accumulate lucifer yellow, or display membrane depolarization in the presence of ATP. Nevertheless, other ATP-mediated activities are present in these cells: ATP mediates a typical increase in the concentration of cytosolic Ca^{2+}, and extracellular ATP is hydrolyzed by the ecto-ATPase present on these cells. Cell lines resistant to ATP, such as ATPR B2, allow one to distinguish among the several effects of ATP^{4-} on these cells and to confirm that some of these effects occur independently of ATP-induced permeabilization; they may prove useful in identifying the receptors that mediate these effects. Similarly, Kitagawa and Akamatsu (1986) have isolated variant CHO cells that are resistant to ATP-mediated permeabilization.

VII. Uses of ATP Permeabilization

The usefulness of this method is limited primarily by the range of cells that respond to ATP in this manner and the size of the membrane pore induced by ATP. We provide several examples from our experience and the published literature for illustrative purposes.

A. Role of GTP-Binding Proteins in Signal Transduction

To test the role of guanine nucleotide-binding proteins in calcium-dependent secretion, Gomperts (1983) introduced a number of nucleotides and nucleotide analogs, including GMP-PNP, GTP, GDP, cGMP, ADP, AMP, and cAMP, into rat mast cells by ATP permeabilization. Cells were incubated for 5 minutes in medium containing $3 \mu M$ ATP, no divalent cations, $15 \mu M$ EGTA, and the solute to be introduced into the cells. Then, 2 mM Mg^{2+} was added for an additional 10 minutes, the cells were transferred to Ca^{2+}-containing medium, and histamine secretion was measured. Only GTP and the GTP analogs were able to cause Ca^{2+}-dependent histamine release in the absence of any other stimulus. Ethidium bromide staining provided a marker for cells that were successfully permeabilized by ATP: the cells that possessed fluorescent nuclei and therefore had taken up ethidium bromide and GTP were the only cells that had degranulated.

B. Role of $[Ca^{2+}]_i$ in Signal Transduction

ATP permeabilization has been used to assess the role of $[Ca^{2+}]_i$ in phagocytosis in mouse macrophages (De Virgilio *et al.*, 1988). Di Virgilio *et al.* "clamped" $[Ca^{2+}]_i$ at a very low level (5–10 nM) in mouse macrophages by introducing EGTA into the cells in the presence of ATP. The cells were then incubated with IgG-opsonized sheep erythrocytes in Ca^{2+}-free medium and phagocytosis was quantitated. Under these conditions $[Ca^{2+}]_i$ remained <10 n*M*, yet the IgG-coated erythrocytes were efficiently ingested. These studies show that a rise in $[Ca^{2+}]_i$ is not required for Fc receptor-mediated phagocytosis to occur; they also demonstrate that macrophages that have been permeabilized by ATP recover sufficiently to carry out cell functions.

C. Organic Anion Transport in Macrophages

Because ATP permeabilization is rapidly reversible, we have been able to load lucifer yellow into the cytoplasmic matrix of macrophages, remove ATP, and assess the fate of intracellular dye (Steinberg *et al.*, 1987b). We found that lucifer yellow did not remain within the cytosol; instead, it was sequestered within cytoplasmic vacuoles and secreted from the cells. The clearance of lucifer yellow from the cytoplasmic matrix was not due to incomplete resealing of the plasma membrane, because lucifer yellow in the extracellular medium was no longer able to enter the cytosol after the removal of ATP from the medium. Moreover, the cells were able to secrete the dye against a concentration gradient, suggesting that this pathway is an active transport process. Permeabilization by ATP may be used to load a number of compounds into the cytoplasmic matrix of cells to study transport processes.

VIII. Relative Merits of ATP as a Means of Membrane Permeabilization

A. Advantages

A major advantage of using ATP for permeabilization is the ease and rapidity with which large populations of cells can be rendered permeable and then resealed. More than 95% of J774 cells within a population are permeabilized by ATP (Steinberg *et al.* 1987a). Membrane depolarization occurs within 40 milliseconds upon addition of ATP, and dyes such as ethidium bromide begin to enter cells as fast as can be measured (Gomperts, 1983). In transformed mouse fibroblasts, however, there is a time lag of ~3 minutes between addition of ATP and the efflux of labeled metabolites (Heppel *et al.*, 1985).

The only limitations on the amount of a solute that can be loaded into an ATP-permeabilized cell are the size of the solute, its concentration in the extracellular medium, and the length of the time the cells withstand permeabilization before they become nonviable. In J774 cells the intracellular concentration of lucifer yellow will approach the extracellular concentration in 20–30 minutes.

Another advantage of ATP permeabilization is the rapid reversibility of its permeabilizing effect. Permeabilized cells can be resealed by the addition of Mg^{2+}, and ATP pores can be subsequently reopened by addition of EDTA. These maneuvers can be performed without washing the cells, which is clearly of benefit in experiments in which rapid changes in membrane permeability are required or in experiments utilizing cells in suspension.

The size restriction of ATP-induced pore can be advantageous because cytosolic proteins and other large soluble molecules are retained within the cell. Techniques such as electropermeabilization (Gordon and Seglen, 1982) and scrape loading (McNeil *et al.*, 1984), which create membrane holes of indeterminate size, will have a less predictable effect on the cytosolic composition.

B. Disadvantages

Using ATP for permeabilization is practicable only in selected cells and cell lines. There is not an extensive catalog of cells that can be permeabilized by ATP. As mentioned earlier, many types of cells are not susceptible to ATP permeabilization. These include human neutrophils, some lymphocytes, and 3T3 mouse fibroblasts. The size limitation of the ATP-induced pore poses an additional constraint not encountered with a number of other methods; it is not possible to introduce molecules of >800 Da into macrophages, but precise dimensional and charge specificities for molecules of approximately this size have not been thoroughly examined.

Permeabilization by ATP in sodium-based media will introduce large quantities of Na^+ into the cells. Although the normal intracellular cation composition will be restored eventually, this imbalance may produce unwanted effects on cells in the interim. In addition to sodium, other small molecules present in the extracellular medium during permeabilization will enter the cells. For experiments in which normal intracellular ionic conditions are of paramount importance, it may be necessary to allow cells a recovery period in fresh medium after ATP permeabilization, or to employ an extracellular solution that more nearly mimics the intracellular solute composition.

Another potential difficulty one may encounter when using ATP as a permeabilizing agent arises from the existence of other receptors for ATP on a number of cells, as mentioned earlier. Thus the addition of ATP to these cells will elecit a number of responses, and it may be difficult, for example, to attribute an observed cellular response to a specific second messenger introduced into the cell by ATP permeabilization. It may be possible to circumvent some of these problems by permeabilizing in the absence of divalent cations that would prevent ATP hydrolysis by the ATPase and kinase.

IX. Mechanism of ATP-Induced Permeabilization

Although the precise mechanism of ATP-mediated permeabilization is not known, the specificity of permeabilization for ATP^{4-} as opposed to other nucleotides suggests that permeabilization is initiated by ligation of a specific membrane receptor for ATP^{4-}. The selection of variant J774 clones that specifically lack the ability to be permeabilized by ATP, but express other ATP-mediated activities, also supports this hypothesis. Electrophysiologic studies suggest that the signal initiated by ligation of this receptor is not mediated by soluble intracellular messengers, because membrane depolarization occurs within 40 milliseconds of the addition of extracellular ATP, and because depolarization occurs in cells patch-clamped in the whole-cell configuration. In this configuration, the small soluble molecules present in the cytosol are dialyzed out of the cells and replaced by the solution present within the patch pipet. The molecular identity of this ATP receptor remains to be determined.

Permeabilization of transformed mouse fibroblasts by ATP has several features that distinguish is from permeabilization of macrophages and mast cells. Permeabilization of fibroblasts is sensitive to the intracellular concentration of ATP (Rozengurt and Heppel, 1979) and occurs after a 3- to 5-minute time lag during which ion fluxes occur that may induce the subsequent permeabilization (Heppel *et al.*, 1985). Heppel and colleagues (1985) have investigated the mechanism of permeabilization in these cells in detail, and have summarized their data.

X. Conclusion

Permeabilization of the plasma membrane by extracellular ATP offers a rapid and simple method for introducing small membrane-impermeant molecules into the cytoplasmic matrix of susceptible cells. Mouse macrophages, transformed mouse fibroblasts, rat mast cells, and neuroblastoma cell lines can be permeabilized by ATP, whereas human neutrophils and untransformed fibroblasts cannot. Permeabilization by ATP can be rapidly reversed by removal or chelation of ATP; however, cells may not regain normal morphology and cytosolic ionic composition for 30–60 minutes after reversal of permeabilization. Permeabilization by ATP has already been used successfully to study cell function after introduction of nucleotide analogs, calcium buffers, and fluorescent dyes into the cytoplasmic matrix, and may provide oppoutunities for similar studies in other cell types as well.

Acknowledgments

We thank P.E.R. Tatham, B.D. Gomperts, and S. K. Margolis for reviewing this manuscript.

This work was supported by Clincial Investigator Award AI 00893, a Clinical Scientist Research Fellowship from the Damon Runyon–Walter Winchell Cancer Fund, and a New York Lung Association Grants Award to T.H.S.; by USPHS grant AI 20516 to S.C.S.; and by the Cystic Fibrosis Research Development Program at Columbia University.

References

Alberty, R. A. (1968). *J. Biol. Chem.* **243,** 1337–1343.

Arav, R., and Friedberg, I. (1985). *Biochim. Biophys. Acta* **820,** 183–188.

Becker, E. L., and Henson, P. M. (1975). *Inflammation* **1,** 71–84.

Bennett, J. P., Cockcroft, S., and Gomperts, B. D. (1981). *J. Physiol. (London)* **317,** 335–345.

Buisman, H. P., Steinberg, T. H., Fischbarg J., Silverstein, S. C., Vogelzang, S. A., Ince, C., Ypey, D. L., and Leijh, P. C. J. (1988) *Proc. Natl. Acad. Sci. U.S.A.* **85,** 7988–7992.

Chahwala, S. B., and Cantley, L. C. (1984). *J. Biol. Chem.* **259,** 13717–13722.

Cockcroft, S., and Gomperts, B. D. (1979a). *Nature (London)* **279,** 541–542.

Cockcroft, S., and Gomperts, B. D. (1979b). *J. Physiol. (London)* **296,** 229–243.

Cockcroft, S., and Gomperts, B. D. (1980). *Biochem. J.* **188,** 789–798.

Dahlquist, R., and Diamant, B. (1974). *Acta Pharmacol. Toxicol.* **34,** 368–384.

Di Virgilio, F., Meyer, B. C. Greenberg, S., and Silverstein S. C. (1988). *J. Cell Biol.* **106,** 657–666

Douglas, W. W., Poisner, A. M., and Rubin, R. P. (1965). *J. Physiol. (London)* **179,** 130–137.

Dubyak, G. R. (1986). *Arch. Biochem. Biophys.* **245,** 84–95.

Dubyak, G. R., and Young, M. B. D. (1985). *J. Biol. Chem.* **260,** 10653–10661.

Ehrlich, Y. H., Davis, T. B., Bock, E., Kornecki, E., and Lenox R. H. (1986). *Nature (London)* **320,** 67–70.

Gomperts, B. D. (1983). *Nature (London)* **306,** 64–66.

Gordon, P. B., and Seglen, P. O. (1982). *Exp. Cell Res.* **142,** 1–14.

Greenberg, S., Di Virgilio, F., Steinberg, T. H., and Silverstein, S. C. (1988). *J. Biol. Chem.* **263,** 10337–10343.

Heppel, L. A., Weisman, G. A., and Friedberg, I. (1985). *J. Membr. Biol.* **86,** 189–196.

Kitagawa, T., and Akamatsu, Y. (1981). *Biochim. Biophys. Acta* **649,** 76–82.

Kitagawa, T., and Akamatsu, Y. (1986). *Biochim. Biophys. Acta* ***860,*** 185–193.
Landry, Y., and Lehninger, A. L. (1976). *Biochem. J.* ***158,*** 4247–438.
McIntyre, T. M., Zimmerman, G. A., Satoh, K., and Prescott, S. M. (1985). *J. Clin. Invest.* ***76,*** 271–280.
McNeil, P. L., Murphy, R. F., Lanni, F., and Taylor, D. L. (1984). *J. Cell Biol.* ***98,*** 1556–1564.
Meyers, K. M., Holmsen, H., and Seachord, C. L. (1982). *Am. J. physiol.* ***243,*** R454–R461.
Pearson, J. D., and Gordon, J. L. (1979). *Nature (London)* ***281,*** 384–386.
Pirotton, S., Raspe, E., Demolle, D., Erneux C., and Boeynaems, J.-M. (1987). *J. Biol Chem.* ***262,*** 17461–17466.
Prpic, V., Blackmore, P. F., and Exton, J. H. (1982). *J. Biol. Chem.* ***257,*** 11323–11331.
Remold-O'Donnell, E. (1978). *J. Exp. Med.* ***148,*** 1099–1104.
Rozengurt, E., and Heppel, L. A. (1975). *Biochem. Biophys. Res. Commun.* ***67,*** 1581–1588.
Rozengurt, E., and Heppel, L. A. (1979). *J. Biol. Chem.* ***254,*** 708–714.
Rozengurt, E., Heppel, L. A., and Friedberg, I. (1977). *J. Biol. Chem.* ***252,*** 4584–4590.
Steinberg, T. H., and Silverstein, S. C. (1987). *J. Biol. Chem.* ***262,*** 3118–3122.
Steinberg, T. H., Newman, A. S., Swanson, J. A., and Silverstein, S. C. (1987a). *J. Cell Biol.* ***105,*** 2695–2702.
Steinberg, T. H., Newman, A. S., Swanson, J. A., and Silverstein, S. C. (1987b). *J. Biol. Chem.* ***262,*** 8884–8888.
Stone, T. W. (1981). *Neuroscience* ***6,*** 523–555.
Su, C. (1985). *Annu. Rev. Physiol.* ***47,*** 665–676.
Sung, S. S., Yound J. D., Origlio, A. M., Heiple, J. M., Kaback, H. R., and Silverstein, S. C. (1985). *J. Biol. Chem.* ***260,*** 13442–13449.
Swanson, J., Bushnell, A., and Silverstein, S. C. (1987a). *Proc. Natl. Acad. Sci. U.S.A.* ***84,*** 1921–1925.
Swanson, J., Burke, E., and Silverstein, S. C. (1987b). *J. Cell Biol.* ***104,*** 1217–1222
Tatham, P. E. R., Cusack, N. J., and Gomperts, B. D. (1988). *Eur. J. Pharmacol.* ***147,*** 13–21.
Ward, P. A., Cunningham, T. W., McCulloch, K. K., Phan, S. H., Powell, J., and Johnson, K. J. (1988). *Lab. Invest.* ***58,*** 37–47.

Chapter 4

Poration by α-Toxin and Streptolysin O: An Approach to Analyze Intracellular Processes

GUDRUN AHNERT-HILGER, WOLFGANG MACH,
KARL JOSEF FÖHR, AND MANFRED GRATZL

Universität Ulm
Abteilung Anatomie und Zellbiologie
Oberer Eselsberg
D7900 Ulm, Federal Republic of Germany

I. Introduction

Permeabilized Cells—Preparations between Intact Cells and Isolated Organelles

Stimulation of a cell from the outside leads to a chain of intracellular events that finally result in a physiological response. In intact cells the single steps of such a cascade, also termed signal transduction, are difficult to resolve.

Exocytosis is a process common to various cells including neurons, endocrine and exocrine cells, mast cells, leukocytes, and lymphocytes. During the final steps of exocytosis the secretory-vesicle membrane fuses with the plasma membrane, which allows the vesicular content to leave the cell. The investigation of the intracellular processes involved in the regulation of exocytosis is hampered by the bordering plasma membrane. However, selective permeabilization of this barrier has allowed complete control of the cell interior because the extracellular and intracellular spaces thereby become continuous.

Three different techniques—high-voltage discharge (Baker and Knight, 1978; Knight and Baker, 1982), the use of detergents such as saponin and digitonin (Brooks and Treml, 1983; Dunn and Holz, 1983; Wilson and Kirshner, 1983), and the application of pore-forming toxins such as α-toxin (Ahnert-Hilger *et al.*, 1985a,b; Bader *et al.*, 1986) and streptolysin O (Ahnert-Hilger *et al.*, 1985a, 1989a,b; Howell and Gomperts, 1987; Howell *et al.*, 1987)—have been inaugurated as valuable instruments for this purpose. Here we describe the purification and handling of α-toxin and streptolysin O (SLO) for the poration of cells. This contribution also deals with the subsequent use of the permeabilized secretory cells in the analysis of intracellular Ca^{2+} sequestration and release, as well as of regulation of exocytosis.

II. Purification and Analysis of Pore-Forming Toxins

Stable transmembrane pores can be formed by specialized proteins in the plasma membrane of target cells. Examples of pore-forming proteins are the C5b–9 complex of the complement (Bhakdi and Tranum-Jensen, 1984, 1987), the cytolysin of cytotoxic T lymphocytes (Podak and Konigsberg, 1984; Henkart *et al.*, 1984; Masson and Tschopp, 1985), or various exotoxins produced by several strains of *Staphylococcus* and *Streptococcus* (for review, see Bhakdi and Tranum-Jensen, 1987; Thelestam and Blomquist, 1988). Two of the latter group, α-toxin from *Staphylococcus aureus* and SLO from β-hemolytic streptococci, have hitherto been used to permeabilize the plasma membrane selectively in order to control the composition of the cytosol.

A. α-Toxin

α-Toxin from *Staphylococcus aureus* (strain Wood 46, ATCC 10832, DSM 20491, kindly provided by S. Bhakdi, Giessen, Federal Republic of Germany) can be purified from the culture supernatant as described by Lind *et al.* (1987). After 18 hours of bacterial growth the soluble proteins of the culture medium are precipitated by the addition of 75% ammonium sulfate. The collected precipitate can be stored at −20°C for months without loss of activity. After dialysis against sodium acetate (10 mmol/liter, pH 5, containing 20 mmol/liter NaCl), the soluble material is subjected to cation exchange chromatography (Mono S column HR 5/5, Pharmacia, Freiburg, FRG). α-Toxin elutes with 170 m*M* NaCl in the column buffer. Then, it is further purified by gel filtration (Superose 12-HR, Pharmacia). Cation exchange chromatography can also be carried out using less expensive S-Sepharose. Figure 1 shows SDS–PAGE of the culture supernatant and the toxin preparations obtained after cation exchange chromatography and gel filtration. Besides the rapid procedure using the FPLC technique described by Lind *et al.* (1987), α-toxin produced in various other ways (for review, see Möllby, 1983; Füssle *et al.*, 1981) yields preparations of similar specific activity [20,000–40,000 hemolytic units (HU)/mg of protein].

The purified toxin, dialyzed against the permeabilization buffer (see Sections IV, A and B), can be lyophilized and stored at −20°C for months. Toxin solutions can also be stored at 4°C for 2–3 weeks without loss of toxicity. Commercially available α-toxin preparations can be obtained from Calbiochem (Frankfurt) and from the Institut Pasteur (Paris). They appear to be less active and contain additional high and low molecular weight protein bands. Because staphylococci produce several different toxins, it is not clear which of them contributes to the lytic activity of these preparations (Wadström, 1983; Möllby, 1983). However, these toxin preparations of relatively low titer can still be used for cell permeabilizations provided they are concentrated and further purified by two successive ammonium sulfate precipitations (55% first and 65% second), followed by dialysis against permeabilization buffer (Schrezenmeier *et al.*, 1988a,b).

B. Streptolysin O

Highly purified SLO was isolated from culture supernatants of group A β-hemolytic streptococci by ammonium sulfate and polyethylene glycol precipitation, DEAE–ion exchange chromatography, preparative isoelec-

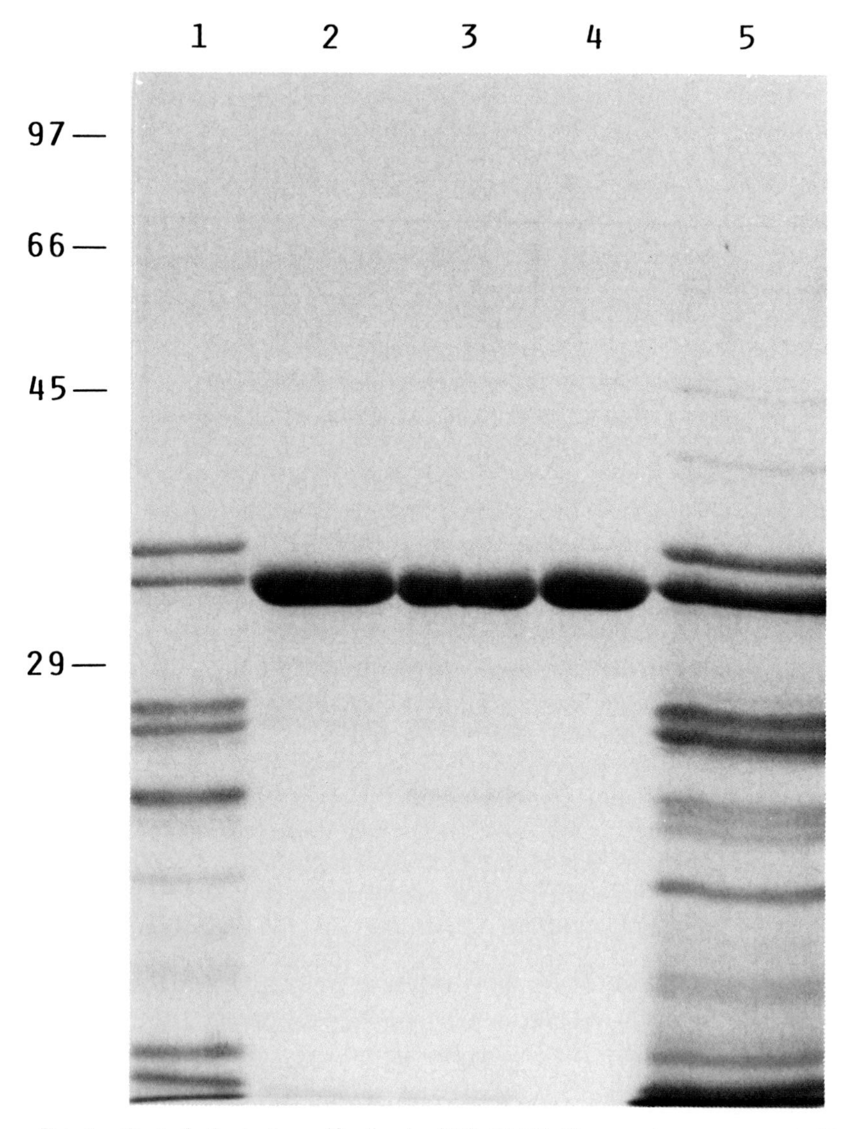

FIG. 1. Control of α-toxin purification by SDS–PAGE. The samples were separated in 12% acrylamide gel and stained with Coomassie blue. The indicated molecular mass values (kDa) correspond to the following standards run in parallel: carbonic anhydrase, 29; ovalbumin, 45; albumin, 66; and phosphorylase *b*, 97. Lanes 1 (1.5 μg) and 5 (15 μg) represent culture supernatant. Lanes 2 and 3 are peak fractions (1.5 μg) eluted by cation exchange chromatography (Mono S). Lane 4 is peak fraction (1.5 μg) after gel chromatography (Superose), which removes low molecular weight compounds. From Lind *et al.* (1987), by permission.

tric focusing, and chromatography on Sephacryl S-300 (Bhakdi *et al.*, 1984b). The excellent final product (kindly provided for the investigations in the authors' laboratory by S. Bhakdi, Giessen), using the hemolytic assay described in the following, exhibited ~250,000 HU/mg of protein. Also commercially available SLO preparations have been used for permeabilization of secretory cells. Streptolysin O from the Institute Pasteur is mainly prepared for diagnostic purposes. This product (1 "titrage") contains ~40 HU in the rabbit erythrocyte assay described in Section II,C. The low activity of this preparation can be increased by precipitation with ammonium sulfate (75%). Then it can be stored at 4°C for months and used, for example, for the permeabilization of adrenal medullary chromaffin cells in primary culture (Sontag *et al.*, 1988; Ahnert-Hilger *et al.*, 1989a,b). The toxin must be activated with 4 mmol/liter dithiothreitol (DTT) (Bhakdi *et al.*, 1984b). Another SLO preparation obtained from Sigma (St. Louis, MO), when tested as described in the following, contained ~800 HU per batch. Streptolysin O as a partially purified culture filtrate, obtained from Wellcome Diagnostics (Dartford, United Kingdom), has been used for permeabilization of rat mast cells (Howell and Gomperts, 1987; Howell *et al.*, 1987).

C. Determination of Toxin Activity

During purification and storage the toxicity of α-toxin and SLO is checked by determining the hemolytic titer. Because rabbit erythrocytes compared to that of human or bovine origin are very sensitive against both toxins, they are used routinely (see Füssel *et al.*, 1982; Möllby, 1983; Bhakdi *et al.*, 1984a,b; Bhakdi and Tranum-Jensen, 1987). For SLO, erythrocytes from other species can also be taken (Bhakdi and Tranum-Jensen, 1987).

To avoid clotting, fresh rabbit blood is immediately mixed with 4% sodium citrate. After three washes with 50 mmol/liter phosphate-buffered saline (PBS), pH 7.0, the erythrocytes are diluted 1 : 40 in the same buffer. This erythrocyte suspension (2.5%) can be used for 3–4 days when stored at 4°C. Dilutions of toxins are performed in PBS containing 0.1 bovine serum albumin (BSA) and in tests for SLO with an additional 2 mmol/liter DTT present. The latter is necessary to reduce the SH groups of SLO thereby activating the toxin (Bhakdi *et al.*, 1984b; Bhakdi and Tranum-Jensen, 1987). Then, 5μl of the appropriate toxin dilution are mixed with 50 μl of the erythrocyte suspension. Hemolysis is monitored after 40 minutes of incubation at 37°C. The samples are briefly mixed followed by centrifugation (2 minutes at 12,000 *g*). Released hemoglobin is determined spectrophotometrically at 412 nm after addition of 1 ml distilled water to

30 μl of the supernatant. Total hemolysis is determined after addition of SDS (0.2. w/v final), which gives an extinction of ~1.2. The dilution of toxin hemolyzing 50% of the erythrocytes is determined, and the reciprocal of the value obtained is taken as the number of HU per milliliter of the undiluted toxin solution (Lind *et al.*, 1987). The procedure described earlier results generally in lower values for the hemolytic activity compared to an assay in which the onset of hemolysis is determined with 2% erythrocytes (Füssle *et al.*, 1981).

III. Application for Cell Poration

α-Toxin permeabilizes cells for low molecular weight substances, whereas SLO permeabilizes cells for both high and low molecular weight substances (Table I). The properties of the SLO-permeabilized preparations, in this respect, are very similar to the cells treated with detergents such as digitonin (Table I). Besides the procedures listed in Table I, which are frequently used in the authors' laboratory, other methods including the measurement of release of amino acids from cells (Thelestam and Möllby, 1979) are also of great value.

A. Determination of Permeability for Low Molecular Weight Substances

Increased permeability of a cell can be rapidly checked by the use of membrane-impermeable dyes, which stain either various components of the cell body or the nucleus (Wilson and Kirshner, 1983; Ahnert-Hilger *et*

TABLE I

PERMEABILITY OF CELLS AFTER TREATMENT WITH α-TOXIN, STREPTOLYSIN O, AND DIGITONIN

	α-Toxin	Streptolysin O	Digitonin
Ions ($^{86}Rb^+$, ATP^{4-}, Ca^{2+}) and $^{51}CrO_4^{2-}$-labeled material	+	+	+
Dyes (azur A, trypan blue, eosin)	+	+	+
Proteins (lactate dehydrogenase, Immunoglobulins, tetanus toxin)	−	+	+

al., 1985a). Intact cells exclude the dye and are therefore not stained. The cationic phenothiazine dye azur A stains mainly the nucleus of cells provided the plasma membrane has been previously permeabilized. The cell interior of permeabilized cells can also be stained by trypan blue or eosin. Permeabilized cells and the dye solution, both in an isotonic medium containing potassium or sodium as a main cation (see media described in Sections IV, A and B), are mixed to yield a final dye concentration of ~0.2%. Because prolonged incubation inevitably leads to staining of intact cells, the percentage of stained cells must be immediately determined in a Neubauer cell-counting chamber.

Release of enzymatically loaded $^{86}Rb^+$ is a sensitive indicator for increased permeability of the plasma membrane after treatment with cytolysins. For this assay the cells ($\sim 5 \times 10^5$) are first loaded with $4–8 \times 10^{-4}$ Bq $^{86}Rb^+$ for 2 hours at 37°C in 2 ml of a physiological salt solution containing (in millimoles per liter): 150 NaCl, 1.2 Na_2HPO_4, 1 $CaCl_2$, 2.5 $MgSO_4$, 11 glucose, 10 PIPES, pH 7.2, and 0.2% BSA on culture plates of 60 mm diameter. After suspending the cells by gentle pipetting and washing, release is initiated by the addition of the permeabilizing agent to be tested. After 20 minutes the cells are centrifuged (10,000 *g* for 2 minutes) and $^{86}Rb^+$ is estimated in the supernatant as well as in the cell pellets lysed with SDS. About 6% of the $^{86}Rb^+$ provided is present within the cells at the beginning of the release experiment. This protocol was used to analyze the effects of α-toxin on PC12 and RINA2 cells (Ahnert-Hilger *et al.*, 1985a; Lind *et al.*, 1987). In a similar experimental design, $^{86}Rb^+$ release can be performed directly on the culture plates as done with adrenal medullary chromaffin cells in primary culture (Bader *et al.*, 1986). An example of $^{86}Rb^+$ release from SLO-treated PC12 cells is shown in Fig. 2.

The release of intracellular ATP from the cell is also a sensitive indicator for the permeability of the plasma membrane. The ATP released can be easily measured using the firefly assay obtained from Boehringer, Mannheim (FRG). Permeabilization with cytolysins is performed in an "intracellular" buffer [KG buffer, containing (in millimoles per liter): 150 K^+-glutamate, 0.5 EGTA, 5 NTA, 10 PIPES, pH 7.2]. The released ATP was determined in the supernatant (after centrifugation for 2 minutes at 12,000 *g*) as well as in the cell extract [performed in assay medium (in millimoles per liter): 10 Mg^{2+}-acetate, 1.5 EGTA, 50 Tris, pH 7.8] after heating to 95°C for 5 minutes (Lind *et al.*, 1987). Luciferase and its substrate luciferin were dissolved in the assay buffer as outlined in the manual of the firefly assay kit. α-Toxin (like digitonin) causes release of the ATP dose dependently from PC12 (Fig. 3) and RINA2 cells (Lind *et al.*, 1987). Similar effects have been observed with SLO (Fig. 4; the effect

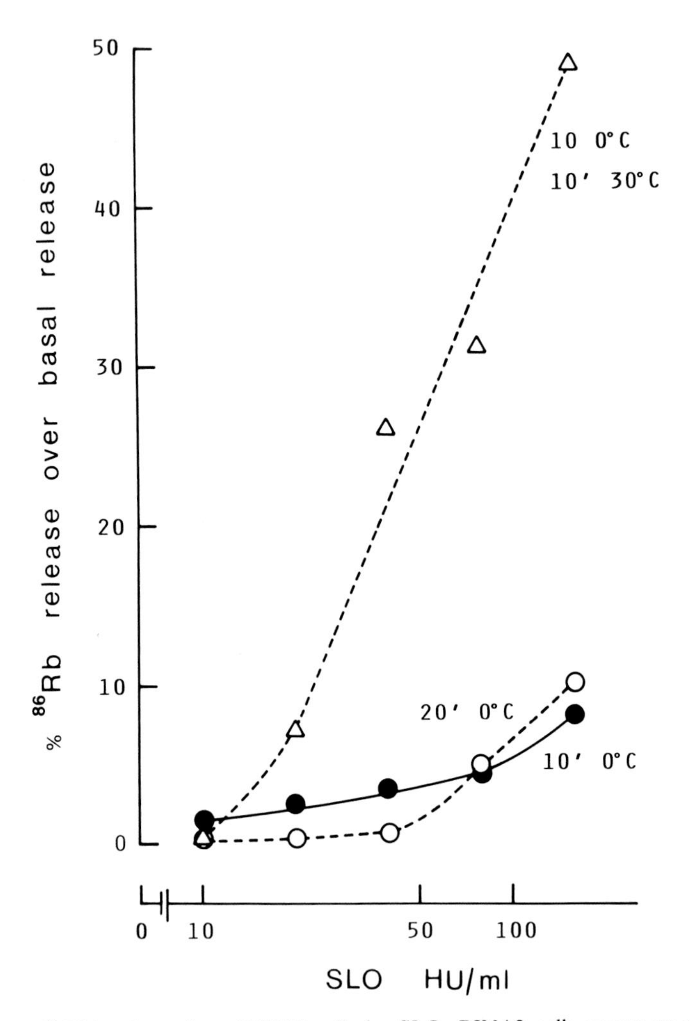

FIG. 2. $^{86}Rb^+$ release from RINA2 cells by SLO. RINA2 cells grown on polylysin (10 μg/ml)-coated plates were loaded with $^{86}Rb^+$ at 37°C for 1 hour. After three washes in the loading buffer, the cells were first incubated for 10 minutes at 0°C with different SLO concentrations and the released radioactivity determined (●). Further incubation for 10 minutes at 0°C did not increase $^{86}Rb^+$ release (○). By contrast, incubation at 37°C (△) resulted in an increased release of $^{86}Rb^+$ as a function of the SLO concentration. Basal release (2%/minute) was subtracted. The experiment clearly shows that, after binding at 0°C, pore formation with SLO can be induced by elevation of the temperature to 37°C. Similar results were obtained with PC12 cells (Ahnert-Hilger *et al.*, 1989b).

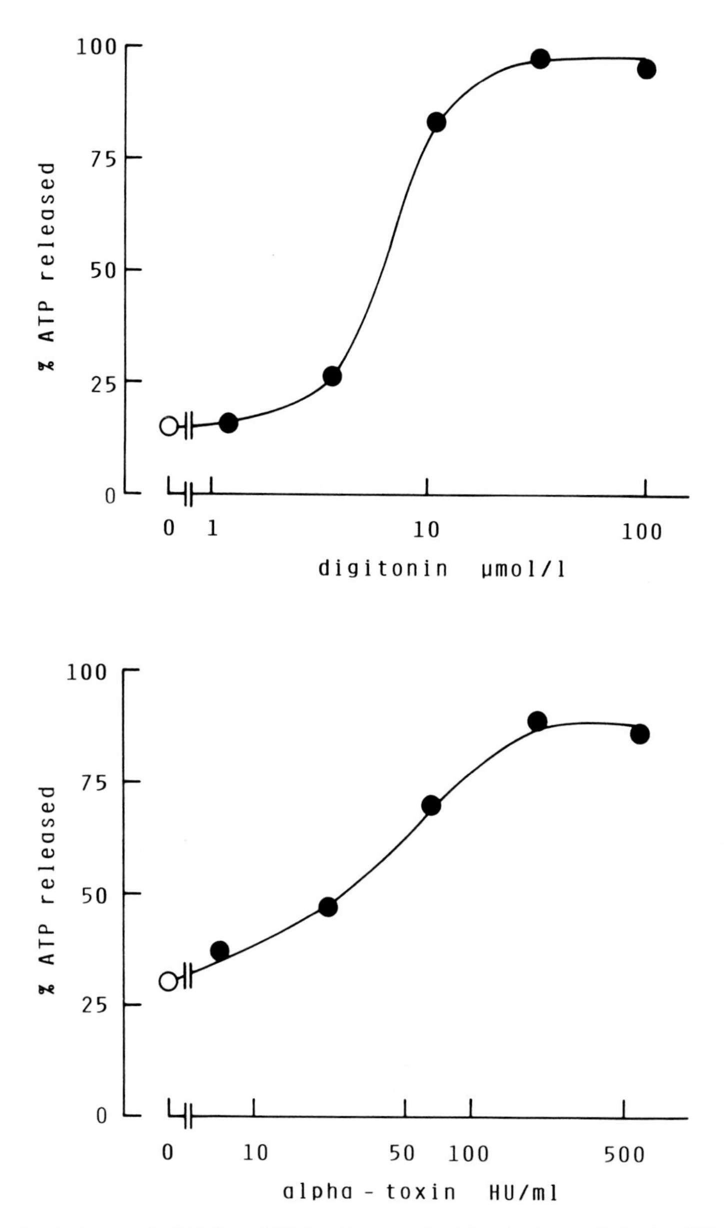

FIG. 3. Release of ATP from PC12 cells treated with α-toxin or digitonin. PC12 cells were washed and suspended in a buffer containing (in millimoles per liter): 150 NaCl, 5 KCl, 10 glucose, 10 HEPES, pH 7.4, supplemented with 0.1% BSA. After 30 minutes at 30°C the buffer was exchanged for KG buffer (see Fig. 8) containing BSA and glucose. About 2×10^6 cells were incubated at 30°C for 20 minutes (digitonin) or 30 minutes (α-toxin) with the given concentrations of the permeabilizing agent. ATP was determined in the supernatant and the lysate of the cells (see Section III, A) using the firefly assay (Lind *et al.*, 1987). Values are the mean of duplicates expressed as percentage of total (~0.5 nmol ATP per sample) present at the beginning of the experiment.

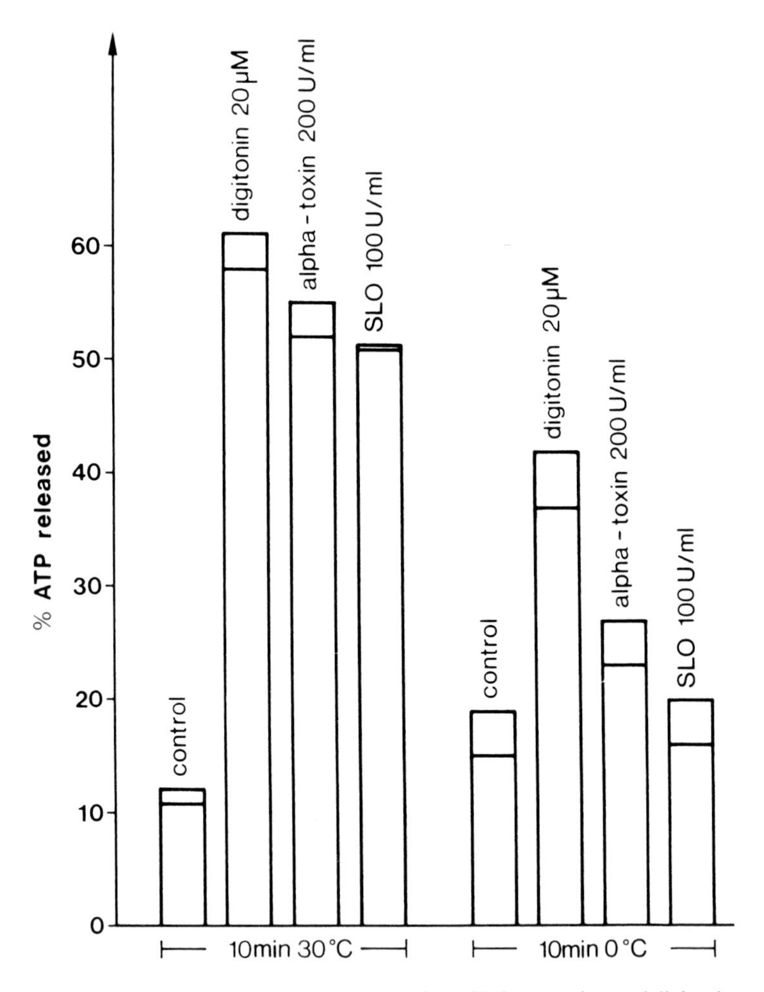

FIG. 4. Effect of temperature on ATP release from SLO-, α-toxin-, and digitonin-treated PC12 cells. PC12 cells were incubated as described in the legend to Fig. 3. After suspending them in KG buffer, the cells were incubated with the indicated compounds either at 0°C or at 30°C for 10 minutes. After centrifugation, ATP was measured in the supernatant and in the extract of the cells (see also Fig. 3).

of temperature on the toxins' attack is described in Section III, C). Thus the observations described here indicate that even the small pores of α-toxin are large enough for the free passage of nucleotides (Ahnert-Hilger and Gratzl, 1987; Lind *et al.*, 1987).

Living cells can be easily labeled with $^{51}CrO_4^{2-}$, which binds to cytoplasmic material, of which 90% have a molecular weight <4000

(Martz, 1976). Release of $^{51}CrO_4^{2-}$-labeled material indicates an increased permeability of cells such as cytotoxic T lymphocytes when treated with α-toxin or other cytolysins (Martz, 1976; Schrezenmeier *et al.*, 1988a,b). Because the ^{51}Cr-labeled material is not uniform, the exact data on the size of the pore generated by a particular cytolysin cannot be determined. Nevertheless, labeling with $^{51}CrO_4^{2-}$ is suitable to determine rapidly the effects of temperature, toxin concentration, and other parameters for optimal poration.

B. Determination of Permeability for High Molecular Weight Substances

α-Toxin, SLO, and detergents allow the passage of small molecules through the plasma membrane. The release of cytoplasmic enzymes or the introduction of antibodies directed against intracellular proteins after application of a permeabilizing agent would indicate that the pores were large enough for the free passage of proteins.

The release of cytoplasmic lactate dehydrogenase (LDH) (MW 135,000) has been demonstrated for digitonin-permeabilized adrenal medullary chromaffin cells (Dunn and Holz, 1983; Wilson and Kirshner, 1983), for SLO-permeabilized mast cells (Howell and Gomperts, 1987; Howell *et al.*, 1987), and for SLO-permeabilized PC12 cells (Ahnert-Hilger *et al.*, 1985a). This observation is consistent with data using erythrocytes, which indicate that SLO produces large transmembrane pores in the target cell (Bhakdi *et al.*, 1985; Hugo *et al.*, 1986; Bhakdi and Tranum-Jensen, 1987). On the other hand, the absence of an increased LDH release and other cytosolic proteins by α-toxin-permeabilized cells indicates the small size of the pores generated (Ahnert-Hilger *et al.*, 1985a; Bader *et al.*, 1986; Grant *et al.*, 1987; Bhakdi and Tranum-Jensen, 1987; Thelestam and Blomquist, 1988; Schrezenmeier *et al.*, 1988a). Lactate dehydrogenase was determined in the supernatant of SLO- or α-toxin-treated cells in permeabilization buffer (see Section IV, B) and in the hypotonic lysate of the cells (Ahnert-Hilger *et al.*, 1985a), using a modification (Gratzl *et al.*, 1981) of the procedure described by Kornberg (1955). Whereas α-toxin-treated cells exhibited no increased LDH release during 40 minutes of incubation (Ahnert-Hilger *et al.*, 1985a; Schrezenmeier *et al.*, 1988a), SLO induces immediate release of this cytoplasmic marker (Ahnert-Hilger *et al.*, 1985; Howell and Gomperts, 1987).

Because LDH is able to leave SLO-permeabilized cells, immunoglobulins (with roughly the same molecular weight) should be able to enter them. Indeed, SLO-treated PC12 cells accumulate antibodies directed against calmodulin as well as synaptophysin/p38 (Ahnert-Hilger *et al.*,

1989b), an integral membrane protein of small clear vesicles and secretory vesicles in neuroendocrine cells (cf. Rehm *et al.*, 1986; Navone *et al.*, 1987; Schilling and Gratzl, 1988). As shown in Fig. 5, cells permeabilized with SLO can also be stained with anticalmodulin. Digitonin also provides access for the antibody directed against calmodulin (Fig. 5), actin, or chromaffin vesicle constituents (Schäfer *et al.*, 1987). Thus the large pores generated by SLO (or digitonin) are useful to investigate the intracellular action of high molecular weight neurotoxins such as tetanus toxin or botulinum A toxin with yet-unidentified intracellular targets (Habermann and Dreyer, 1986; Ahnert-Hilger *et al.*, 1989a,b; Stecher *et al,* 1989).

C. Selective Permeabilization of the Plasma Membrane

Permeabilization with α-toxin is limited to the plasma membrane because the pores formed (diameter ~1–2 nm) are too small for the free passage of the α-toxin monomer (34 kKa) into the cell (Bhakdi *et al.*, 1981; Füssle *et al.*, 1981; Bhakdi and Tranum-Jensen, 1987). In contrast, the lesions generated by digitonin, saponin, or SLO are certainly large enough to allow access of unbound detergent molecules or of SLO monomers to the cytoplasmic space. These pores would also be highly desirable for the introduction of antibodies or other big molecules into secretory cells.

In order to limit the effects of SLO to the plasma membrane without affecting intracellular structures, an improved protocol was developed for the permeabilization of endocrine cells, which is based on previous experience obtained in the erythrocyte model (Hugo *et al.*, 1986). Incubation of cells with SLO at 0°C results only in toxin binding to the cell surface but not in pore formation. However, permeabilization occurs rapidly upon additional incubation at 30°C, as indicated by the $^{86}Rb^{+}$ release with RINA2 cells (Fig. 2) or the ATP release by PC12 cells (Fig. 4). α-Toxin can also be bound to the cells in the cold and poration of the cells triggered by warming (Fig. 4). By contrast, membrane permeabilization by digitonin is insensitive to the incubation and thus cannot be

---

FIG. 5. Access of immunoglobulins to the interior of PC12 cells permeabilized by either digitonin or SLO. PC12 cells were washed twice with KG buffer and then incubated for 10 minutes at 30°C in the same buffer without (A) or with 60 HU/ml SLO (B) or 20 μmol/liter digitonin (C). The incubation was stopped by fixation of the cells with a 4% paraformaldehyde. Immunocytochemistry was performed with a calmodulin antibody (final dilution 1 : 200) using the PAP technique. The intracellular antigen is only accessible to the antibody in the permeabilized cells (Ahnert-Hilger *et al.*, 1989b).

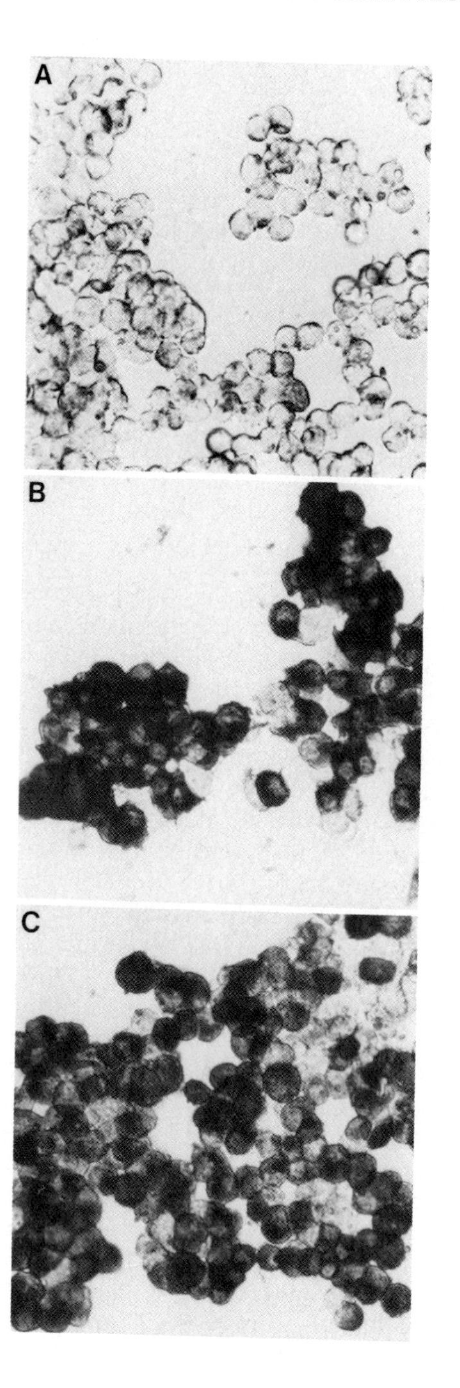
A
B
C

controlled by using different temperatures for binding and permeabilization (Ahnert-Hilger *et al.*, 1989b).

IV. Pore-Forming Toxins as Tools in the Study of Intracellular Processes

α-Toxin and SLO are powerful instruments for the poration of various cells and have been applied to the analysis of intracellular Ca^{2+} movements and the regulation of exocytosis by endocrine cells. Depending on whether low or high molecular weight compounds are to be introduced into the cells, the type of toxin required can be selected.

A. Intracellular Ca^{2+} Sequestration in Permeabilized Endocrine Cells

The intracellular free-Ca^{2+} concentration is regulated by a number of Ca^{2+}-transport systems present in the plasma membrane, mitochondria, secretory vesicles, and the endoplasmic reticulum. Permeabilized cells are suitable for investigation of intracellular systems participating in the regulation of intracellular free-Ca^{2+} concentrations. Indeed, permeabilized cells have been instrumental in the analysis of the "IP_3-sensitive pool" (Streb *et al.*, 1983; for review, see Berridge, 1987). Using a Ca^{2+}-selective electrode, dynamic changes in intracellular Ca^{2+} sequestration can be registered in a suspension of α-toxin-permeabilized cells. Highly reliable measurements of the free-Ca^{2+} concentration in the micromolar range have been made possible since the development of a suitable carrier (Simon *et al.*, 1978). A further improvement of this technique is due to a new carrier (ETH 129, kindly provided by W. Simon, ETH Zürich), which is characterized by a high specifity of Ca^{2+} over Mg^{2+} and H^+, and allows one to measure free-Ca^{2+} concentrations as low as 10^{-9} *M* (Ammann *et al.*, 1987). A Ca^{2+} electrode can be obtained from Glasmanufaktur Möller, Zürich.

In order to measure the Ca^{2+} fluxes, cells are first permeabilized with α-toxin in a medium containing (in millimoles per liter): 150 KCl, 5 NaN_3, 1 EGTA, 20 MOPS, pH 7.2. After several washings with the same buffer without EGTA, the ATP-dependent Ca^{2+} uptake into the endoplasmic reticulum can be followed with the electrode. Azide in these experiments blocks mitochondrial Ca^{2+} uptake. Increasing amounts of IP_3 can release the stored Ca^{2+} (Streb *et al.*, 1983, cited for Berridge, 1987). Figure 6 gives an example of an ATP-driven Ca^{2+} uptake followed by an IP_3-induced Ca^{2+} release by both α-toxin-permeabilized PC12 and RINA2

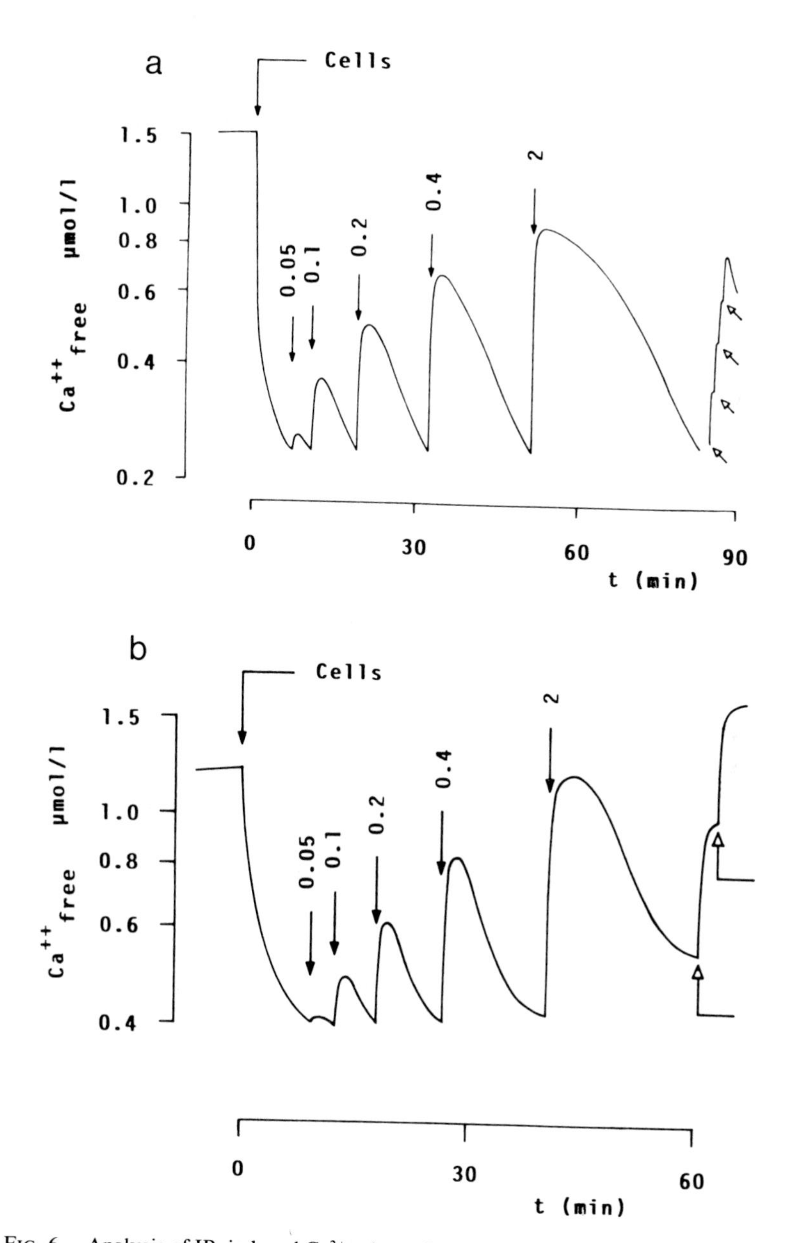

FIG. 6. Analysis of IP_3-induced Ca^{2+} release from α-toxin-permeabilized RINA2 (a) and PC12 (b) cells. Permeabilization was carried out with 300 HU α-toxin per 10^7 cells (10 minutes on ice, 30 minutes at 30°C) in a medium consisting of (in millimoles per liter) 150 KCl, 1 EGTA, 20 MOPS, pH 7.2. For measurement of Ca^{2+} fluxes, cells were incubated in the same medium containing no EGTA but 2 mmol/liter Mg^{2+}/ATP. The arrows indicate the addition of increasing amounts of IP_3 (between 0.05 and 2 nmol) corresponding to final concentrations of 0.125 μmol/liter and 5 μmol/liter, respectively. The calibration of the Ca^{2+} release was carried out by the sequential addition of 1 nmol Ca^{2+} (open arrows) at the end of the experiment.

cells. In these cell preparations Ca^{2+} flux measurements can be conducted over hours, indicating the high stability of this preparation. The rapid onset of the effects of ATP and IP_3 on the ambient free-Ca^{2+} concentration is a further indicator for the effective permeabilization of the cells by α-toxin.

In a similar experimental design, intracellular Ca^{2+} stores can be labeled with $^{45}Ca^{2+}$. These experiments were carried out with cells attached to a culture plate. Figure 7 shows the time course of ATP-dependent $^{45}Ca^{2+}$ uptake by SLO-permeabilized PC12 cells. Following loading of the ATP-dependent subcellular compartment, release experiments can be carried out under a variety of conditions.

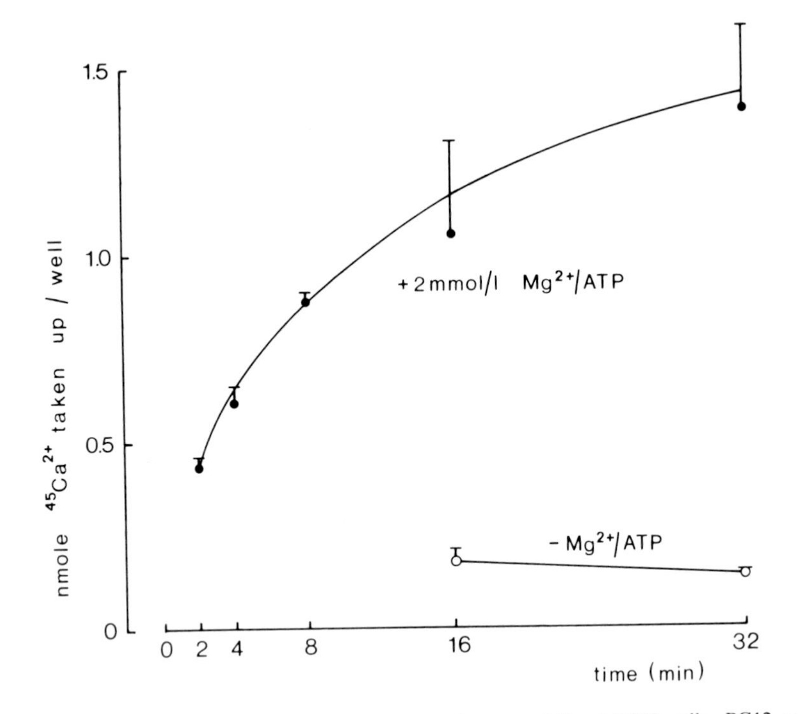

FIG. 7. ATP-dependent $^{45}Ca^{2+}$ uptake by SLO-permeabilized PC12 cells. PC12 cells were cultivated on polylysin (10 μg/ml)-coated multiwell plates (~2×10^5 per well). The cells were washed twice in a medium containing (in millimoles per liter): 150 NaCl, 1 EGTA, 10 PIPES, pH 7.2 and then in a medium containing (in millimoles per liter 150 KCl, 20 PIPES, pH 7.2. Incubation with SLO (60 HU/ml) was performed for 10 minutes at 0°C in the same buffer supplemented with 0.1% BSA and 1 mmol/liter DTT. This medium was removed and replaced by 200 μl fresh medium containing 14 nmol $^{45}Ca^{2+}$ ± 2 mmol/liter Mg^{2+}/ATP. The amount of radioactivity taken up by the cells was determined in the SDS lysate of the cells as well as in the supernatant.

B. Exocytosis by Permeabilized Secretory Cells

Toxin-permeabilized cells have been used to study the molecular requirements for exocytosis (Ahnert-Hilger *et al.*, 1985a,b, 1989a,b; Bader *et al.*, 1986; Grant *et al.*, 1987; Ahnert-Hilger and Gratzl, 1987, 1988; Howell and Gomperts, 1987; Howell *et al.*, 1987; Schrezenmeier *et al.*, 1988a,b; Sontag *et al.*, 1988). During this type of secretion the fusion of the secretory-vesicle membrane with the plasma membrane is a crucial event. Because poration of the latter with α-toxin or SLO does not affect this process, an extensive investigation of exocytosis by various secretory cells has been made possible.

In most of the studies dealing with exocytosis by permeabilized cells an "intracellular buffer" system was used, containing potassium as a main cation and glutamate as an anion. Glutamate was chosen (Baker and Knight, 1978) because of its impermeability to the chromaffin secretory-vesicle membranes (Phillips, 1977). Because the free-Ca^{2+} concentrations under resting conditions as well as during stimulation within the cells are in the micromolar range, this ion must be carefully controlled in the media used. A combination of EGTA and NTA is suitable to buffer the free-Ca^{2+} concentration between 0.1 and 100 μmol/liter. Thus a typical buffer for permeabilization contains (in millimoles per liter): 150 potassium glutamate, 0.5 EGTA, 5 NTA, 10 PIPES, pH 7.2. Added Mg^{2+} and ATP as well as the pH of the medium must be also considered, because they influence the equilibrium between Ca^{2+} and the chelators present. The free Ca^{2+} and Mg^{2+} concentrations are calculated by means of a computer program (Flodgaard and Fleron, 1974) kindly provided by T. Saermark, University of Copenhagen, using the stability constants given by Sillen and Martell (1971). Each Ca^{2+} buffer is prepared separately from stock solutions with a final check of pH and pCa by the Ca^{2+}-selective electrode (see Section IV,A). Solutions free of ATP can be stored at −20°C, whereas solutions containing ATP must be prepared freshly prior to the experiment.

Adrenal medullary chromaffin cells or pheochromocytoma cells from rat (PC12; Greene and Tischler, 1982) take up labeled catecholamines, store them in vesicles, and release them upon stimulation. Thus the assay for exocytosis in permeabilized chromaffin cells includes the loading of the cells with $[^3H]$dopamine (PC12) or $[^3H]$norepinephrine (adrenal medullary chromaffin cells), washing, and permeabilization. Stimulation is carried out with micromolar concentrations of Ca^{2+}, which then triggers the release of the stored labeled catecholamines. Table II summarizes the procedure for either PC12 cells or adrenal medullary chromaffin cells

TABLE II

ASSAY FOR EXOCYTOSIS

Release of catecholamines by permeabilized PC12 cells (chromaffin cells in primary culture)

1. Loading of cells with tritium labeled dopamine (norepinephrine) for 1–2 hours
2. Washing with Ca^{2+} free balanced salt solutions several times
3. Suspension of the cells in permeabilization buffer = KG-buffer containing (in millimoles per liter): 150 K^+-glutamate, 10 PIPES, 5 NTA, 0.5 EGTA, pH 7.2, and 0.1% BSA (+2 mmol/l Mg/ATP and 1 mmol/l free Mg^{2+}).
4. Treatment with pore-forming toxins: α-toxin: 30 min at 30°C or 37°C SLO: 5 min at 0°C (1 or 2 min at 30°C or 37°C)
5. Centrifugation at low speed and removal of supernatant (if done on plates, removal of supernatant)
6. Stimulation with micromolar amounts of free Ca^{2+} (+ATP) for 10 min at 30°C or 37°C
7. Centrifugation and counting of released catecholamines in the supernatant
8. Solubilization of cells with SDS and counting of the remaining catecholamines

using α-toxin or SLO. α-Toxin-permeabilized cells respond to micromolar concentrations of Ca^{2+} for >1 hour (Ahnert-Hilger and Gratzl, 1987), whereas SLO-treated cells respond for ~40 minutes (Ahnert-Hilger *et al.*, 1989a,b). Later they become insensitive to stimulation with Ca^{2+} (Sarafian *et al.*, 1987). In a comparable experimental design, exocytosis can be measured by α-toxin-permeabilized cytotoxic T lymphocytes. Here the release of a vesicular serine esterase (Pasternack *et al.*, 1986; Henkart *et al.*, 1984) is taken as a measure for exocytosis. This release is triggered with increasing Ca^{2+} concentrations, provided ATP and GTPγS are present (Schrezenmeier *et al.*, 1988a,b). Furthermore, mast cells previously permeabilized with small amounts of SLO release histamine upon addition of Ca^{2+} and a nucleotide (Howell and Gomperts, 1987; Howell *et al.*, 1987).

Besides exocytosis, the observed release of secretory product by permeabilized cells may be due to leakiness of secretory vesicles or even loss of intact secretory vesicles from the cells. Detergents like saponin or digitonin may destroy the secretory-vesicle membrane or support the escape of intact secretory vesicles through the large membrane lesions generated (Brooks and Carmichael, 1983; Bader *et al.*, 1986). The parallel release of low and high molecular weight secretory products is a convincing indication for exocytosis, provided that under the same conditions large cytoplasmic constituents do not leak out. Such a situation has been found with α-toxin-permeabilized adrenal medullary chromaffin cells, which release labeled catecholamines and the vesicular protein chromogranin A but not cytoplasmic LDH in the presence of micromolar concentrations of Ca^{2+} (Bader *et al.*, 1986). Similarly, exocytosis is indicated by the parallel release of catecholamines and dopamine β-hydroxylase from electrically permeabilized adrenal medullary chromaffin cells (Knight and Baker, 1982) and the release of vesicular serine esterase by α-toxin-permeabilized cytotoxic T lymphocytes (Schrezenmeier *et al.*, 1988a,b), but not of cytoplasmic proteins. In all these preparations the pores in the plasma membrane are too small to allow a direct escape of the vesicular proteins.

Another approach takes the metabolism of catecholamines in the cytoplasm of α-toxin-permeabilized PC12 cells as an indicator of whether or not the secretory products leave the cell by exocytosis. Since cytoplasmic enzymes such as LDH remain entrapped in these cells after permeabilization, the enzymes involved in the metabolism of catecholamines are also retained. Thus, the discharge of vesicular dopamine into the cytoplasm (e.g., by nigericin) results in its enzymatic oxidation to 3,4-dihydroxyphenylacetic acid (DOPAC) and 3,4-dihydroxyphenylethanol (DOPET). In contrast, if catecyholamines are released by exocytosis, the cytoplasm is avoided and no metabolism can occur. The pattern of catecholamines and their metabolites released by α-toxin-permeabilized PC12 cells is in accordance with these predictions. Ca^{2+} only results in a release of dopamine and norepinephrine, whereas the release of the metabolies resembles that seen with either unstimulated or intact control cells (Ahnert-Hilger *et al.*, 1987; Fig. 8). Therefore, it can be concluded that permeabilized PC12 cells release catecholamines by exocytosis with micromolar amounts of free Ca^{2+}.

The three types of secretory cells permeabilized by pore-forming toxins differ in their molecular requirements for exocytosis. Whereas in PC12 cells permeabilized with α-toxin or SLO, Ca^{2+} alone is sufficient to release the stored catecholamines (Figs. 9, 10; Ahnert-Hilger and Gratzl, 1987; Ahnert-Hilger *et al.*, 1985a, 1989a,b), adrenal medullary chromaffin cells require additional Mg^{2+}/ATP (Bader *et al.*, 1986; Grant *et al.*, 1987). The situation is even more complicated in the permeabilized cytotoxic T lymphocyte, where Ca^{2+} and ATP alone result only in a small release of the serine esterase. Only when a G protein is activated by GTPγS, can a full exocytotic response be obtained (Schrezenmeier *et al.*, 1988a,b). In SLO-permeabilized mast cells, Ca^{2+} plus a nucleotide must be present (Howell *et al.*, 1987; Howell and Gomperts, 1987).

In permeabilized PC12 cells the Ca^{2+}-stimulated dopamine release is not affected by the composition of the medium. Potassium can be exchanged for sodium as well as glutamate for chloride (Ahnert-Hilger *et al.*, 1985a, 1987). This holds also for cytotoxic T lymphocytes (Schrezen-

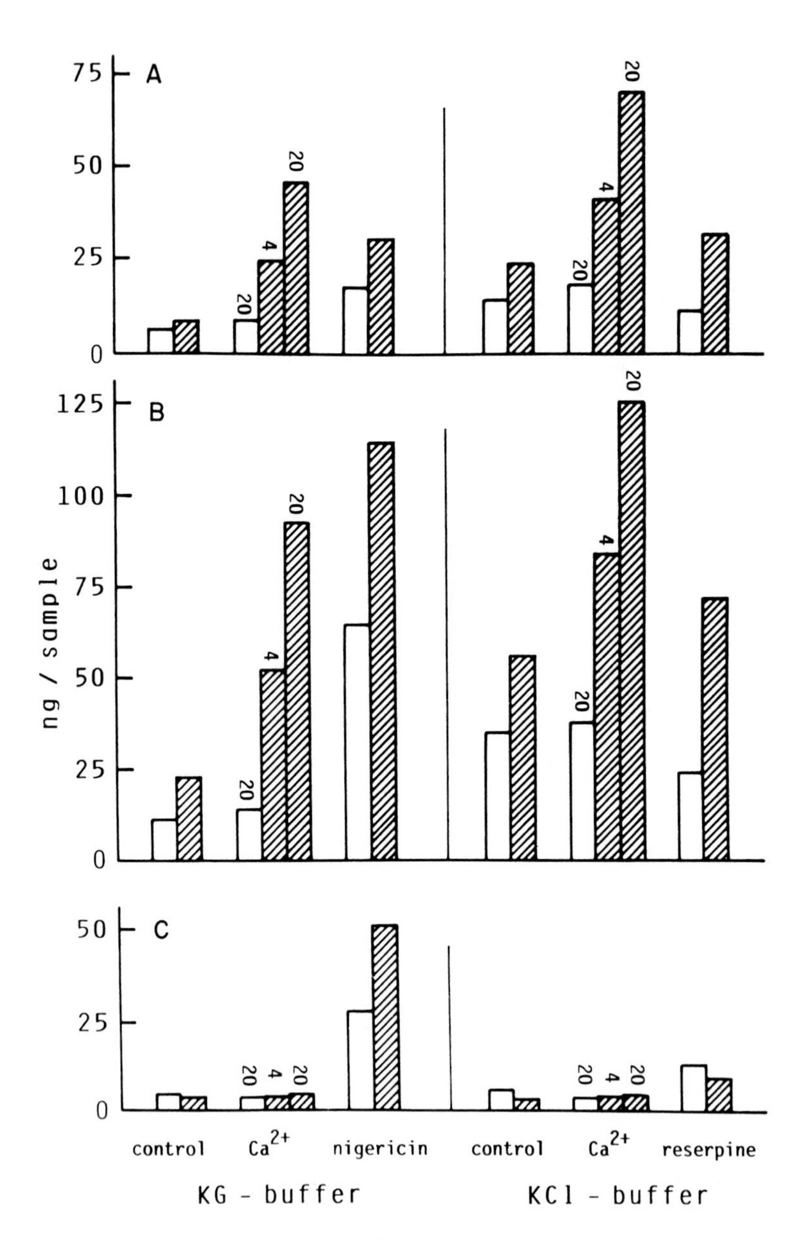

FIG. 8. Effects of Ca^{2+}, nigericin, and reserpine on the release of (A) norepinephrine, (B) dopamine, and (C) its metabolite DOPAC by permeabilized and intact PC12 cells. PC12 cells were first permeabilized with α-toxin in KG buffer or KCl buffer (K^+ glutamate was exchanged for KCl), and then incubated with buffer (controls), with 4 or 20 μmol/liter free Ca^{2+}, with nigericin (1 μmol/liter), or with reserpine (0.2 μmol/liter) for 10 minutes at 30°C. Catecholamines were determined in the supernatant and the lysate of cells by the HPLC technique. The hatched bars represent permeabilized cells, the open ones intact cells. The samples contained 560±32 μg of protein. From Ahnert-Hilger *et al.* (1987), by permission.

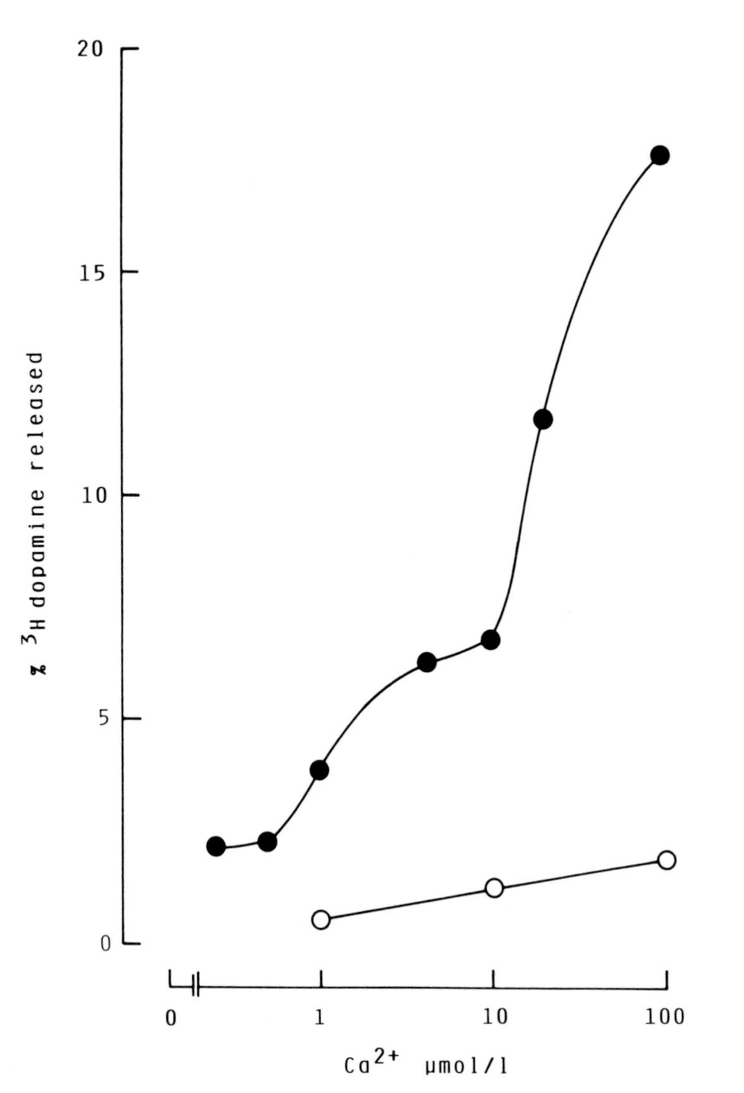

FIG. 9. Ca^{2+} dependency of [^{3} H]dopamine release by α-toxin permeabilized PC12 cells. Cells were loaded with [^{3} H]dopamine, washed, and treated with α-toxin (●) or KG buffer alone (○), as described in Fig. 8. The permeabilization medium was exchanged for a fresh one containing the amount of the free-Ca^{2+} concentration given in the abscissa. Each point represents the mean of two samples. The release in the absence of Ca^{2+} (7.5% for α-toxin-treated cells and 8% for intact cells) was subtracted. Micromolar amounts of Ca^{2+} release dopamine only from α-toxin-permeabilized cells. From Ahnert-Hilger and Gratzl (1987), by permission.

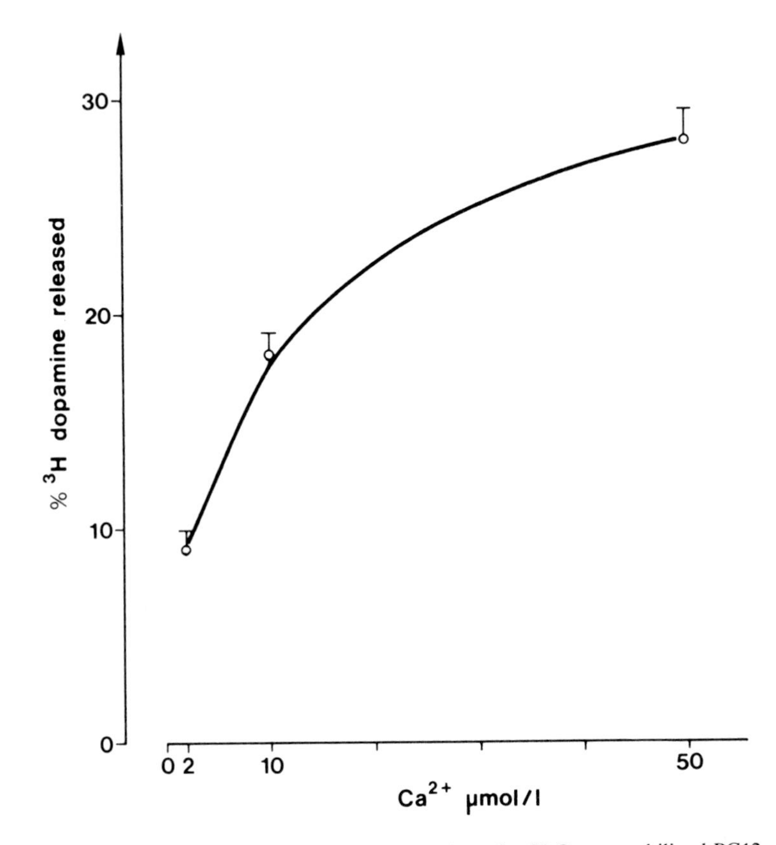

FIG. 10. Ca^{2+} dependency of [^{3}H]dopamine release by SLO-permeabilized PC12 cells. Preloaded PC12 cells were treated and incubated with SLO as given in Fig. 8, prior to stimulation with the given free-Ca^{2+} concentrations (abscissa). Values represent the Ca^{2+}-stimulated release of three samples (±SD).

meier *et al.*, 1988a,b). By contrast, in electrically permeabilized adrenal medullary chromaffin cells exocytosis has been reported to be inhibited by chloride (Knight and Baker, 1982). Even when all ions are replaced by sucrose or the pH is varied between 6.6 and 7.2, the Ca^{2+}-stimulated release remains unchanged (Ahnert-Hilger *et al.*, 1985a; Ahnert-Hilger and Gratzl, 1987).

Permeabilized PC12 cells are a very useful preparation to study the modulation of exocytosis by intracellular regulator systems such as protein kinase C and G proteins, because exocytosis by these cells can be triggered by Ca^{2+} alone. In these cells Mg^{2+} in the millimolar range increases the Ca^{2+}-induced exocytosis (Ahnert-Hilger and Gratzl, 1987), whereas in the adrenal medullary chromaffin cells Mg^{2+} alone has no effect but must be complexed with ATP to sustain exocytosis (Bader *et al.*, 1986; Knight and Baker, 1982). Very high amounts of Mg^{2+} have even been reported to be inhibitory in these cells (Knight and Baker, 1982). Activation of protein kinase C by the diacylglycerol analog 1-oleyl-2-acetylglycerol (OAG) or the phorbol ester 1-*O*-tetradecanoylphorbol-13-acetate (TPA) ameliorates Ca^{2+}-induced exocytosis by both permeabilized PC12 cells (Peppers and Holz, 1986; Ahnert-Hilger *et al.*, 1987) and adrenal medullary chromaffin cells (Knight and Baker, 1983). In PC12 cells this effect was shown to be dependent on the presence of Mg^{2+}/ATP (Ahnert-Hilger and Gratzl, 1987). In adrenal medullary chromaffin cells the modulatory role of protein kinase C is difficult to analyze because exocytosis by these cells also depends on ATP.

G proteins are specialized membrane proteins involved in the transduction of various signals (Gilman, 1987). In permeabilized PC12 cells, activation of G proteins by GTPγS results in an incomplete inhibition of exocytosis in the presence of Mg^{2+}. The amounts of GTPγS (between 2 and 100 μM) did not interfere with the free-Ca^{2+} concentration in the buffer system as measured by the Ca^{2+}-sensitive electrode. When cells were pretreated during the loading period (see Fig. 8) with pertussis toxin (11 μg/ml, obtained from List Biological Laboratories, Campbell, CA), the inhibitory effect of GTPγS could be overcome Fig. 11). In a similar experimental design, cells were pretreated with cholera toxin (100 μg/ml; obtained from Sigma, Munich) which did not alter the GTPγS-induced inhibition of exocytosis. Thus, it can be concluded that exocytosis by PC12 cells can be modulated by a pertussis toxin-sensitive G protein (Ahnert-Hilger *et al.*, 1987). Similar results have been reported for freshly isolated bovine adrenal medullary chromaffin cells (Knight and Baker, 1985), whereas a stimulatory effect of GTPγS has been obtained in adrenal medullary chromaffin cells from chicken (Knight and Baker, 1985), from bovine tissue (Bittner *et al.*, 1986), and in mast cells (Howell and Gomperts, 1987; Howell *et al.*, 1987; Neher, 1988). Besides the fact that different cell preparations were used, these contradictory results may reflect the modulation of exocytosis by different G proteins. Figure 12 summarizes our current data concerning the regulation and modulation of exocytosis in PC12 cells.

Experimental data from our laboratory show that the large pores generated by SLO are also well suited to study the intracellular effects of tetanus toxin, a neurotoxin of 150 kDa from clostridium tetani. This toxin when applied extracellulary inhibits exocytosis from neurons but not from endocrine cells (cf. Habermann and Dreyer, 1986; Knight, 1986). Upon intracellular injection, however, tetanus toxin becomes capable of inhibit-

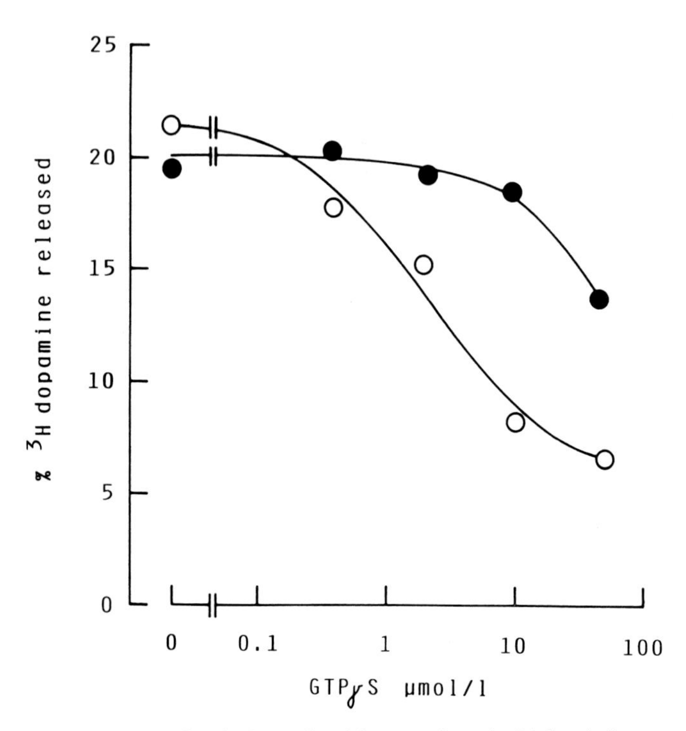

FIG. 11. Pretreatment (●) of PC12 cells with pertussis toxin (11.5 µg/ml) overcomes the inhibitory effects of GTPγS on exocytosis. PC12 cells were treated with or without (○) pertussis toxin for 4 hours during the loading period. Then the cells were washed and permeabilized with α-toxin described in Fig. 8. Free Mg^{2+} (1 mmol/liter) and the indicated amounts of GTPγS were also present. The medium was exchanged for a fresh one containing the same constituents and 10 µmol/liter free Ca^{2+}. From Ahnert-Hilger *et al.* (1987), by permission.

ing exocytosis from adrenal medullary chromaffin cells (Penner *et al.*, 1986), indicating that it attacks a step during exocytosis which is common to neurons and endocrine cells. In contrast to the study using single injected cells, SLO-permeabilized cells allow the application of defined dosis of toxin to a great number of cells and further biochemical analysis of the mechanism of tetanus toxin action. Tetanus toxin consists of a heavy and a light chain covalently linked by a disulfide bond (Habermann and Dreyer, 1986). Chain separation by reduction of this bond initiates the inhibitory action on exocytosis of the toxin in SLO-permeabilized chromaffin and PC12 cells (Ahnert-Hilger *et al.*, 1989a,b). Also cleavage of the disulfide bond linking the heavy and light chain of botulinum A toxin is necessary for its inhibitory action on exocytosis (Stecher *et al.*, 1989). Even the light chain of tetanus toxin alone is fully active whereas the heavy chain has no effect on exocytosis (Ahnert-Hilger *et al.*, 1989b).

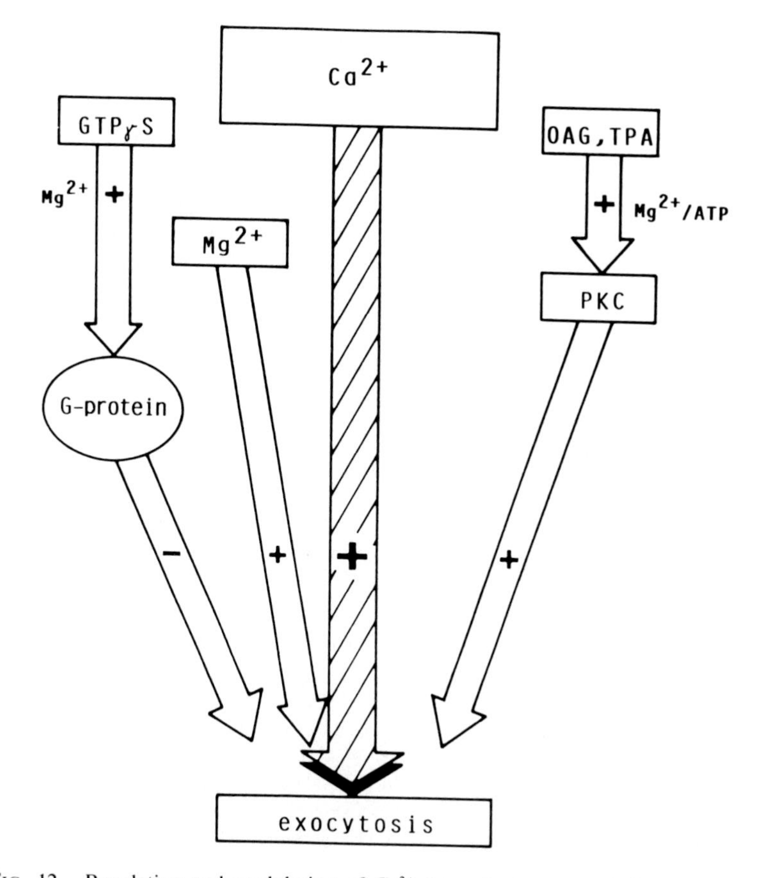

FIG. 12. Regulation and modulation of Ca^{2+}-dependent exocytosis by permeabilized PC12 cells. In permeabilized PC12 cells exocytosis can be stimulated by Ca^{2+}. Ca^{2+}-stimulated exocytosis can be augmented by Mg^{2+}, and activators of the protein kinase C provided Mg^{2+}/ATP is present. Ca^{2+}-induced exocytosis is inhibited by GTPγS in the presence of Mg^{2+} because of activation of a pertussis-sensitive G protein.

Thus tetanus toxin has to be reduced before the light chain can exert its biological effect on exocytosis.

V. Concluding Remarks

In contrast to other permeabilizing procedures, the pores inserted into the plasma membrane with the aid of bacterial toxins are stabilized by a proteinaceous ringlike structure. Depending on the aim of an experiment, the cells can be made permeable either for large or for small molecules by selection of a suitable bacterial pore-forming protein. This allows permanent access to the cells' interior in order to investigate intracellular processes as diverse as fusion of secretory vesicles with the inner surface of the cell membrane, contraction, metabolism of hormones, fluxes of ions, or glucose metabolism. Besides the different secretory cells already referred to in this article, poration by α-toxin or SLO was carried out with hepatocytes (McEwen and Arion, 1985), rat basophilic leukemia cells (Hohman, 1988), fibroblasts (Thelestam and Möllby, 1979), and smooth muscle cells (Cassidy *et al.*, 1978). This indicates that the novel approach of permeabilization of cells by channel-forming toxins has already become a widely used tool for the investigation of a variety of intracellular processes in situ.

Acknowledgments

The work of the authors was supported by the Deutsche Forschungsgemeinschaft (Gr 681) and the Forschungsschwerpunkt 24 of the state of Baden-Württemberg. We are grateful to Mr. J. Fassberg for his helpful advice and Mrs. B. Mader for typing the manuscript.

References

Ahnert-Hilger, G., and Gratzl, M. (1987). *J. Neurochem.* ***49***, 764–770.
Ahnert-Hilger, G., and Gratzl, M. (1988). *Trends Pharmacol.* ***9***, 195–197.
Ahnert-Hilger, G., Bhakdi, S., and Gratzl, M. (1985a). *J. Biol. Chem.* ***260***, 12730–12734.
Ahnert-Hilger, G., Bhakdi, S., and Gratzl, M. (1985b). *Neurosci. Lett.* ***58***, 107–110.
Ahnert-Hilger, G., Bräutigam, M., and Gratzl, M. (1987). *Biochemistry* ***26***, 7842–7848.
Ahnert-Hilger, G., Weller, U., Dauzenroth, M. E., Habermann, E., and Gratzl, M. (1989a). *FEBS Lett.* ***242***, 245–248.
Ahnert-Hilger, G., Bader, M. F., Bhakdi, S., and Gratzl, M., (1989b). *J. Neurochem.* **52,** 1751–1758.
Ammann, D., Bührer, T., Schefer, U., Müller, M., and Simon, W. (1987). *Pflüegers Arch.* ***409***, 223–228.
Bader, M. -F., Thiersé, D., Aunis, D., Ahnert-Hilger, G., and Gratzl, M. (1986). *J. Biol. Chem.* ***261***, 5777–5783.
Baker, P., and Knight, D. (1978). *Nature (London)* ***276***, 620–622.
Berridge, M. (1987). *Annu. Rev. Biochem.* ***56***, 159–193.
Bhakdi, S., and Tranum-Jensen, J. (1984). *Philos. Trans. R. Soc. London, Ser. B* ***306***, 311–324.
Bhakdi, S., and Tranum-Jensen, J. (1987). *Rev. Physiol. Biochem. Pharmacol.* ***107***, 147–223.
Bhakdi, S., Füssle, R., and Tranum-Jensen, J. (1981). *Proc. Natl. Acad. Sci. U.S.A.* ***78***, 5475–5479.
Bhakdi, S., Muhly, M., and Füssle, R. (1984a). *Infect. Immun.* ***46***, 318–323.
Bhakdi, S., Roth, M., Sziegoleit, A., and Tranum-Jensen, J. (1984b). *Infect. Immun.* ***46***, 394–400.
Bhakdi, S., Tranum-Jensen, J., and Sziegoleit, A. (1985). *Infect. Immun.* ***47***, 52–60.

Bittner, M., Holz, R., and Neubig, R. (1986). *J. Biol. Chem.* ***261,*** 10182–10188.
Brooks, J., and Carmichael, S. (1983). *Mikroskopie* ***40,*** 347–356.
Brooks, J., and Treml, S. (1983). *J. Neurochem.* ***40,*** 468–474.
Cassidy, P., Hoar, P. E., and Kerrick, W. G. L. (1978). *Biophys. J.* ***21,*** 44a.
Dunn, L., and Holz, R. (1983). *J. Biol. Chem.* ***258,*** 4983–4993.
Flodgaard, H., and Fleron, P. (1974). *J. Biol Chem.* ***249,*** 3465–3474.
Füssle, R., Bhakdi, S., Sziegoleit, A., Tranum-Jensen, J., Kranz, T., and Wellensiek, H. -J. (1981). *J. Cell Biol.* ***91,*** 83–94.
Gilman, A. (1987). *Annu. Rev. Biochem.* ***56,*** 615–649.
Grant, N. J., Aunis, D., and Bader, M. F. (1987). *Neuroscience* ***23,*** 1143–1155.
Gratzl, M., Krieger-Brauer, H., and Ekerdt, R. (1981). *Biochim. Biophys. Acta* ***649,*** 355–366.
Greene, L., and Tischler, A. (1982). *Adv. Cell. Neurobiol.* ***3,*** 373–407.
Habermann, E., and Dreyer, F. (1986). *Curr. Top. Microbiol. Immunol.* ***129,*** 93–179.
Henkart, P. A., Millard, P. J., Reynolds, C. W., and Henkart, M. P. (1984). *J. Exp. Med.* ***160,*** 75–93.
Hohman, R. J. (1988). *Proc. Natl. Acad. Sci. U.S.A.* ***85,*** 1624–1628.
Howell, T. W., and Gomperts, B. (1987) *Biochim. Biophys. Acta* ***927,*** 177–183.
Howell, T. W., Cockcroft, S., and Gomperts, B. D. (1987). *J. Cell Biol.* ***105,*** 191–197.
Hugo, F., Reichweiss, J., Arvand, M., Krämer, S., and Bhakdi, S. (1986). *Infect. Immun.* ***54,*** 641–645.
Knight, D. (1986). *FEBS Lett.* ***207,*** 222–226.
Knight, D., and Baker, P. (1982). *J. Membr. Biol.* ***68,*** 107–140.
Knight, D., and Baker, P. (1983). *FEBS Lett.* ***160,*** 98–100.
Knight, D., and Baker, P. (1985). *FEBS Lett.* ***189,*** 345–349.
Kornberg, A. (1955). *In* "Methods in Enzymology" (S. P. Colowick and N. O. Kaplan, eds.) Vol. 1, pp. 441–443. Academic Press, New York.
Lind, I., Ahnert-Hilger, A., Fuchs, G., and Gratzl, M. (1987). *Anal. Biochem.* ***164,*** 84–89.
Martz, E. (1976). *Cell. Immunol.* ***26,*** 313–321.
Masson, D., and Tschopp, J. (1985). *J. Biol. Chem.* ***260,*** 9069–9072.
McEwen, B. F., and Arion, W. J. (1985). *J. Cell Biol.* ***100,*** 1922–1929.
Möllby, R. (1983). *In* "Staphylococci and Staphylococcal Infections" (C. S. F. Easmon and C. Adlam, eds.), Vol. 2, pp. 619–669. Academic Press, New York.
Navone, F., Jahn, R., Di Gioia, G., Stukenbrok, H. Greengard, P. and De Camilli, P. (1987). *J. Cell Biol.* ***103,*** 2511–2527.
Neher, E. (1988). *J. Physiol. (London)* ***395,*** 193–215.
Pasternack, M. S., Verret, C. R., Lice, M. A., and Eisen, H. N. (1986). *Nature (London)* ***322,*** 740–743.
Penner, R., Neher, E., and Dreyer, F. (1986). *Nature* ***324,*** 26–28.
Peppers, S., and Holz, R. (1986). *J. Biol. Chem.* ***261,*** 14665–14670.
Phillips, J. H. (1977). *Biochem. J.* ***168,*** 289–297.
Podack, E. R., and Konigsberg, P. J. (1984). *J. Exp. Med.* ***160,*** 695–710.
Rehm, H., Wiedenmann, B., and Betz, H. (1986). *EMBO J.* ***5,*** 535–541.
Sarafian, T., Aunis, D., and Bader, M. F. (1987). *J. Biol. Chem.* ***262,*** 16671–16676.
Schäfer, T., Karli, U., Gratwohl, E., Schweizer, F., and Burger, M. (1987). *J. Neurochem.* ***49,*** 1697–1797.
Schilling, K., and Gratzl, M. (1988). *FEBS Lett.* ***233,*** 22–24.
Schrezenmeier, H., Ahnert-Hilger, G., and Fleischer B. (1988a). *J. Exp. Med.* ***168,*** 817–822.
Schrezenmeier, H., Ahnert-Hilger, G., and Fleischer, B. (1988b). *J. Immunol.* ***141,*** 3785–3790.

Sillen, L., and Martell, A. (1971). *Supplement **1**, Spec. Publ.—Chem. Soc.*, No. 25.
Simon. W., Ammann, D., Oehme, M., and Morf, W. (1978). *Ann. N. Y. Acad. Sci.* ***307,*** 52–70.
Sontag, J. -M., Aunis, D., and Bader, M. -F. (1988). *Eur. J. Cell. Biol.* ***46,*** 316–326.
Stecher, B., Gratzl, M., and Ahnert-Hilger, G. (1989). *FEBS Lett.* (in press).
Streb, H., Irvine, R. F., Schulz, J., and Berridge, M. J. (1983). *Nature (London)* ***306,*** 67–69.
Thelestam, M., and Blomquist, L. (1988). *Toxicon* ***26,*** 51–65.
Thelestam, M., and Möllby, R. (1979). *Biochim. Biophys. Acta* ***557,*** 156–169.
Wadström, T. (1983). *In* "Staphylococci and Staphylococcal Infections" (C. S. Easmon and C. Adlam, eds.), Vol. 2, pp. 671–704. Academic Press, New York.
Wilson, S., and Kirshner, N. (1983). *J. Biol. Chem.* ***258,*** 4994–5000.

Chapter 5

Preparation of Semiintact Chinese Hamster Ovary Cells for Reconstitution of Endoplasmic Reticulum-to-Golgi Transport in a Cell-Free System

C. J. M. BECKERS, D. S. KELLER, AND W. E. BALCH

Department of Molecular Biophysics and Biochemistry
Yale University
New Haven, Connecticut 06405

I. Introduction

The enzymology underlying trafficking of protein between organelles of the secretory pathway [endoplasmic reticulum (ER), Golgi complex, and the plasma membrane] remains to be elucidated. A number of lines of evidence indicate that interorganelle communication is dissociative: transport between subcellular compartments occurs by the budding and fusion of carrier vesicles. At least four vesicle budding–fusion cycles are required for delivery of protein to the cell surface: (1) export from the ER to the cis (first)-Golgi compartment, (2–3) inter-Golgi transport between the cis, medial, and trans-Golgi compartments, and (4) transport from the trans-Golgi compartment to the cell surface. Three important questions need to be addressed concerning this process:

1. What are the biochemical mechanisms for the (selective) budding of a carrier vesicle and its content from each compartment?
2. What are the mechanisms that ensure specificity of delivery of content to the target membrane (fusion)?
3. What are the sorting signals found on each type of protein that regulate transfer at this stage of the pathway?

Without rigorous control of these processes, the cell would quickly randomize its intracellular content.

In order to understand the biochemical mechanisms responsible for the budding and fusion of carrier vesicles at each stage of the exocytic pathway, it will be necessary to reconstitute each step using cell-free systems. Translocation of protein into the ER (Walter *et al.*, 1984), and transport between subcompartments of the Golgi apparatus (Balch *et al.*, 1984a,b; Melancon *et al.*, 1987; Wattenberg and Rothman, 1986; Orci *et al.*, 1986) have been successfully reconstituted *in vitro* using purified organelles. In both cases, the ER and the Golgi have been extensively fragmented or vesiculated during their preparation from whole cells. It is apparent that at least for reconstitution of translocation or inter-Golgi transport, intact organelles are not essential. In contrast, it is currently not possible to reconstitute the next step in secretion from microsomes: transport of protein from the ER to the cis compartment of the Golgi (Balch *et al.*, 1987; Beckers *et al.*, 1987). Because the ER is a single extensive reticular-cisternal network occupying the entire cytoplasm of the living cell, a more intact form may required for recovery of its transport function *in vitro*.

To approach the general problem of preparation of intact subcellular organelles amenable to study *in vitro*, a new approach has been developed. Our laboratory (Beckers *et al.*, 1987) and others (Simons and Virta, 1987; Brands and Feltkamp, 1988) have pioneered techniques in which cells grown in tissue culture can be selectively sheared to remove only a fragment of the plasma membrane. These semiintact (Beckers *et al.*,

1987), perforated (Simons and Virta, 1987), or wet-cleaved (Brands and Feltkamp, 1988) cells lose their soluble content but clearly retain organelles of the secretory pathway (ER, Golgi), nuclei and mitochondria within a cytoplasmic cytoskeletal matrix. The interior of the semiintact cell is accessible to a wide range of biochemical reagents and macromolecules.

In order to measure transport between the intact ER and the Golgi compartments in these semiintact cells, we have developed an assay based on the G protein of vesicular stomatitis virus (VSV). G protein is glycoprotein that transits the secretory pathway during cell infection in an identical fashion to host plasma membrane proteins. A temperature-sensitive virus strain, strain tsO45, expresses a mutant form of G protein. When tsO45 G protein is synthesized at the restrictive temperature (40°C), it will not exit the ER. However, upon shift of cells to the permissive temperature (32°C), G protein is rapidly transported to the cell surface. Incubation of semiintact cells containing tsO45 G protein in the ER *in vitro* with the soluble components lost during preparation and ATP will reconstitute the transport along the exocytic pathway (Beckers *et al.*, 1987). The procedures involved in the preparation of semiintact Chinese hamster ovary (CHO) cells, and the assay for measurement of transport of protein between the ER and the Golgi are presented in this article.

II. Growth of Cells

A CHO cell line clone 15B was maintained in monolayer (10-cm dishes) in α-MEM (GIBCO, Grand Island, NY) supplemented with 10% fetal calf serum (FCS). Clone 15B is missing the enzyme *N*-acetylglucosamine transferase I (Tr I), which is essential for maturation of the $man_5GlcNAc_2$ oligosaccharide form of G found in the cis-Golgi compartment to the complex structure containing *N*-acetylglucosamine (GlcNAc), galactose, and sialic acid. In 15B cells G protein accumulates on the cell surface in the $man_5GlcNAc_2$ form. For preparation of semiintact cells it is essential that healthy cells are 90–95% confluent (2000 cells/mm^2), but still maintain a well-spread morphology. Tightly packed cells (>3000 cells/mm^2) will result in a considerably reduced recovery of semiintact cells during their preparation.

Clone 15B cells are seeded and grown to a density of 1.5×10^7 cells per 10-cm dish. Cells are released by trypsinization and reseeded at a cell density of 3×10^6 per 10-cm dish (1 : 5 split). Semiintact cells are prepared after an additional 48 hours of incubation using monolayers that have just reached confluency (1.5×10^7 cells per 10-cm dish).

III. Preparation of Semiintact Cells from Adherent Cells

Two different procedures can be used to prepare semiintact cells from cells grown on tissue culture dishes: a swelling procedure that sensitizes the plasma membrane of intact cells to shear, or nitrocellulose stripping, a procedure in which a portion of the plasma membrane is removed from each cell through selective attachment of the surface of each cell to nitrocellulose. In the latter technique, cells are first overlaid with a sheet of nitrocellulose. Under the appropriate conditions the nitrocellulose sheet will adhere tightly to the surface of the cells. Disruption occurs when the nitrocellulose sheet is subsequently peeled away from the plate. The nitrocellulose technique offers the advantage that cells are not exposed to low ionic strength buffers. Both procedures are rapid and highly reproducible when applied to the appropriate density of adherent, confluent cell populations. Typically these procedures will convert >95% of the total cell population to semiintact cells as measured by loss of refractility using phase-contrast light microscopy and trypan blue permeability.

A. Method 1—Swelling Technique

To prepare semiintact cells using the swelling technique, confluent CHO cells are washed three times with a 5-ml volume of an ice-cold H–KCl *swelling buffer* containing 10 m*M* HEPES–KOH and 15 m*M* KCl (pH 7.2), and incubated in a 5-ml volume for 10 minutes on ice. We generally work in a cold room to ensure that the cells are never exposed to temperatures >4°C. After 10 minutes the swelling buffer is aspirated and replaced with 3 ml of a H–KOAc *breaking buffer* containing 50 m*M* HEPES–KOH and 90 m*M* potassium acetate (KOAc), pH 7.2. This buffer may be supplemented as necessary with up to 5 m*M* $MgCl_2$ without any change in the yield of semiintact cells. Cells are immediately scraped from the plate using a pliable rubber policeman (Macalaster-Bicknell Inc., cat. no. 36300-0014, New Haven, CT). It is important to scrape the cells from the plate using rapid, vigorous strokes with the rubber policeman in order to assure a high semiintact cell index. Released cells are pelleted at $800 \times g$ for 5 minutes, and washed once in the H–KOAc breaking buffer. The cells from each 10-cm plate are resuspended in 0.4 ml of the breaking buffer. Just prior to use for assay of transport (see later), cells are very gently dounced using *slow, even strokes* in a 1-ml dounce (Wheaton Scientific Inc., cat. no. 357538, Millville, NJ) and the tight "A" pestle. These conditions promote efficient exchange of the buffer for residual soluble cytoplasmic protein in the semiintact cells, rendering transport to be consistently cytosol-dependent. The number of strokes in this step will vary from a few (5) to many (40) for a given dounce–pestle combination because there is considerable variability during their manufacture. Vigorous douncing will lead to complete cell disruption and will result in loss of transport activity between the ER and the cis-Golgi compartment. After douncing, semiintact cells containing G protein in the ER are pelleted for 3 minutes at $800 \times g$ (or 10 seconds in a microfuge), the supernatant is discarded, and cells resuspended in 0.4 ml of ice-cold H–KOAc buffer. Cells can be maintained on ice for up to 4 hours with full recovery of activity. One plate using this procedure yields sufficient material for 80 assays (see later).

B. Method 2—Nitrocellulose Technique

In the following protocol, it is essential that all procedures are carried out in a cold room. To prepare semiintact cells using the nitrocellulose technique, confluent tissue culture cells are washed three times with H–KOAc breaking buffer. A nitrocellulose filter (Bio-rad Laboratories, cat. no. 162-0115, Richmond, CA) equal in diameter to the inner dimensions of the 10-cm tissue culture dish is preequilibrated with the H–KOAc breaking buffer to saturate the entire filter. The filter is wetted by inserting the newly cut edge of the filter into a dish containing the breaking buffer to initiate wicking of the buffer into the filter, followed by slowly lowering the filter into the dish. Just before use, the filter is blotted between two pieces of Whatman 3MM filter paper under a heavy weight (5–10 kg) for 1 minute to remove excess buffer. At this time an edge (1–2 mm in width) of the filter is bent 90° to facilitate handling of the nitrocellulose with fine-tipped forceps (we use stainless-steel electron-microscopy forceps). For attachment of the nitrocellulose filter to cells, the tissue culture dish is drained for 30–45 seconds in a vertical position in a cold room to remove excess buffer. The nitrocellulose filter, which has been blotted dry, is inserted at one edge of the plate and lowered slowly and evenly onto the cells without trapping air. After 10 minutes, the filter is removed by lifting one edge with forceps while the opposite edge is held in place using the tip of a second pair of forceps. A smooth, rapid motion will effectively shear the upper surface of the cells as the nitrocellulose filter is removed. The filter is subsequently transferred to the tissue culture dish lid (with the side of the filter containing the attached cells facing up), and the dish and the filter are immediately overlayed with 3 ml of the H–KOAc breaking buffer. Either the dish or the filter can be scraped with a rubber policeman to release the attached cells as described formerly. Under the correct conditions of confluency, the majority of the cells will remain attached to the surface of the plate. However, in some cases where a majority of the cells are transferred to the nitrocellulose filter, semiintact cells can be readily prepared by scraping the filter with the rubber policeman. The suspended semiintact cells are washed as described before, resuspended in 0.4 ml of the H–KOAc breaking buffer and gently dounced prior to assay. Alternative procedures for use of nitrocellulose filters have been described by Brands and Feltkamp (1988) and Simons and Virta (1987).

C. Quantitation of Semiintact Cell Recovery

The fraction of cells converted to the semiintact form can be easily quantitated by phase-contrast light microscopy, or trypan blue permeability. Under phase contrast (400× magnification), semiintact cells appear to be morphologically normal, but have lost the refractility normally observed for intact cells. Semiintact cells appear as a phase-dark image. Alternatively, cells may be tested for permeability to trypan blue by mixing 10 μl of the washed, suspended semiintact cells with 1 μl of a 1% stock solution of trypan blue, a membrane-impermeant chromatin-binding dye. Semiintact cells (but not intact cells) will bind the dye in the nucleus yielding a dark blue-staining cell morphology. Other approaches involve assay of soluble cytoplasmic proteins such as lactate dehydrogenase to assess extent of soluble cytoplasmic protein loss, or use of antibody probes to cytoplasmic determinants to assess accessibility as described previously for G protein (Beckers *et al.*, 1987; Simons and Virta, 1987; Hughson *et al.*, 1988).

IV. Preparation of Semiintact Cells from Suspension Cells

Suspension cells can be converted to semiintact cells using the swelling conditions described previously, followed by a gentle-dounce protocol. A suspension containing 2×10^7 cells per 0.4 ml is incubated for 10 minutes on ice in the H–KCl swelling buffer. Subsequently, the cells are transferred to a 1-ml dounce as described before, the buffer adjusted to 50 m*M*

HEPES (pH 7.2) and 90 m*M* KOAc by the addition of 40 μl from a 250 m*M* HEPES (pH 7.2), 0.9 *M* KOAc concentrated stock solution. Cells are gently dounced using a tight pestle until the semiintact cell index reaches 90%. Further homogenization will result in loss of activity. Cells are subsequently pelleted at 800 $\times$ *g* for 3 minutes, washed once in the H–KOAc breaking buffer, and resuspended in 0.4 ml of H–KOAc. Cells are stored on ice until use.

V. Incubation Conditions to Achieve Transport between the ER and Golgi *in Vitro*

A. Infection of Cells with Vesicular Stomatitis Virus (VSV)

Confluent 15B CHO cells are infected with 10–20 plaque-forming units of strain tsO45 VSV per cell in 1 ml of an infection medium containing α-MEM and 25 m*M* HEPES (pH 7.4) for 45 minutes at the permissive temperature (32°C) with gentle rocking to ensure even virus distribution. Each dish is then supplemented with an additional 4 ml of α-MEM containing 10% FCS and incubated at 32°C for 4.5 hours.

B. Labeling of Cells

In the following procedure it is important always to maintain the cells at 40°C during the labeling procedure to ensure localization of tsO45 G protein to the ER. Tissue culture dishes containing infected cells are transferred to a water bath (40°C) containing a perforated metal plate with the water height adjusted to wet the bottom of the dish, but not allow the dish to float. The cells on the plate are washed three times with 2.5-ml portions of a labeling medium (Joklik's minimal essential medium, JMEM) lacking methionine and containing nonessential amino acids (GIBCO), and 20 m*M* HEPES (pH 7.4) equilibrated to 40°C (the restrictive temperature, a condition that inhibits transport of G protein from the ER to the Golgi). For each plate, cells are preincubated for 5 minutes at 40°C in the methionine-lacking labeling medium. The preincubation medium is aspirated and 1 ml of a labeling medium equilibrated to 40°C is added. This medium is immediately supplemented with 50 μCi of [^{35}S]methionine. Cells are rocked briefly at 30-second intervals to ensure a uniform distribution of the labeling medium over the cells for a total time period of 10 minutes. Unlabeled methionine is added to a final concentration of 2.5 m*M*, and the incubation is continued for an additional 5 minutes at 40°C prior to removal of the labeling medium by aspiration, transfer of each dish to an ice-cold, wetted metal surface, and immediate addition of 5 ml of the ice-cold H–KOAc breaking buffer to chill cells rapidly and prevent transport from the ER. Semiintact cells can be prepared from these infected, labeled cells as described before.

C. Incubation Conditions

Five microliters of infected, labeled semiintact cells are incubated in a total volume of 40 μl in a 1.5-ml microcentrifuge tube containing the following supplements (final concentrations): 25 m*M* HEPES–KOH (pH 7.2), 125 m*M* KOAc, 2.5 m*M* magnesium acetate, 1 m*M* ATP, 5 m*M* creatine phosphate (CP), and 0.2 IU of rabbit muscle creatine phosphokinase (CPK). The ATP-regenerating system (ATP–CP–CPK) was made fresh daily by mixing 20 μl CPK (2000 IU/ml, stored at −80°C), 100 μl of 200 m*M* CP, and 100 μl of 40 m*M* ATP (Na form, neutralized with NaOH). The ATP-regenerating system was added as a concentrate to achieve the indicated final concentrations. We have noted that some brands of 1.5-ml microcentrifuge tubes contain residual chemicals that inhibit transport. We use 1.5-ml centrifuge tubes purchased from Sarstedt (cat. no. 72.690, Newton, NC).

Transport requires the addition of a cytosol fraction prepared from CHO cell homogenates as described previously (Balch *et al.*, 1987; Beckers *et al.*, 1987). For preparation of cytosol, twenty 15-cm diameter tissue culture dishes of confluent 15B CHO cells are harvested by gently scraping each plate with a pliable rubber policeman, and are washed once in a buffer containing 10 m*M* triethanolamine and 150 m*M* KCl (pH 7.4), and once in H–KOAc breakage buffer. The pellet is resuspended at a 1 : 4 ratio of cell pellet to buffer and homogenized using a stainless-steel ball-bearing homogenizer (Balch and Rothman, 1985). A cytosol fraction is prepared by centrifugation of the homogenate in a SW 50.1 rotor for 90 minutes at 49,000 rpm, and the supernatant removed without disturbing the pellet. Where necessary, cytosol can be rapidly desalted by passage through a G25 gel filtration column equilibrated with 25 m*M* HEPES (pH 7.2) and 125 m*M* KOAc. Cytosol can be rapidly frozen in 100-μl portions by immersion in liquid nitrogen, and stored at −80°C without loss of activity for at least 2 months. Generally, transport is optimal in the presence of 25–50 μg of cytosol per 40 μl reaction volume as described previously. Transport is initiated by transfer of cells to 32°C. Transport is terminated by transfer of cells to ice.

D. Postincubation with Endo D

Transport of G protein from the ER to the cis-Golgi compartment in 15B CHO cells results in the processing the $man_9GlcNAc_2$ oligosaccharide of G protein acquired in the ER to the $man_5GlcNAc_2$ structure by α-1,2-mannosidase I, a cis-Golgi compartment enzyme. The $man_5GlcNAc_2$ oligosaccharide species, in contrast to all other carbohydrate-processing intermediates, is uniquely sensitive to the enzyme endoglycosidase D (endo D), which cleaves the oligosaccharide chain from G protein (Beckers *et al.*, 1987). Appearance of the endo D-sensitive form of G protein can be determined by postincubation of semiintact cells in the presence of detergent and endo D as described next.

Upon completion of transport and transfer to ice, each incubation is supplemented by the addition of 20 μl of a carrier CHO cell suspension prepared by resuspension of a confluent plate of cells in 2 ml of the H–KOAc breaking buffer. Subsequently, each tube containing cells is pelleted for 15 seconds in the microfuge, and the supernatant aspirated. The pellet is resuspended in 50 μl of a buffer containing 50 m*M* $NaPO_4$ (pH 6.5), 2.5 m*M* EDTA (to inhibit soluble α-1,2-mannosidase activity), and 0.2% Triton X-100 (to solubilize membranes). After mixing, 2 μl (0.25 mU) of an endo D stock solution [prepared by resuspending 0.1 IU of endo D (Boehringer Mannheim, cat. no. 752 991, Indianapolis, IN) in 1 ml of 10 m*M* Tris-HC1 (pH 7.4) and 200 m*M* NaCl] is added to each incubation mixture. Endo D is stored in 100-μl portions at −80°C until use. Each sample is incubated overnight at 37°C. Endo D processing is terminated by the addition of 10 μl of a 5× concentrate of a gel sample buffer (Laemmli, 1970) containing 12.5 ml of a 1 *M* Tris buffer (pH 6.8), 20 ml of glycerol, 0.5 g dithiothreitol (DDT), 4 g of SDS, and 4 mg of bromophenol blue, in a final volume of 40 ml of H_2O, and boiled for 3 minutes.

Samples are analyzed on 7.5% polyacrylamide gels (SDS–PAGE) (McNeil *et al.*, 1984), treated for autoradiographic enhancement by incubation for 30 minutes in 100 ml of a 30% methanol solution containing 0.125 *M* salicylic acid (Na form; pH 7.0), and autoradiographed for 16 hours (Kodak XAR-5 film). The fraction of G protein processed to the endo D-sensitive form, which migrates significantly faster during gel electrophoresis, can be determined by densitometry of the exposed autoradiogram (Beckers *et al.*, 1987).

VI. Discussion

We have described two approaches to prepare semiintact cells from cells grown on tissue culture dishes that allow direct access to subcellular organelles retained by the cell. Both techniques are rapid and highly reproducible when applied to a healthy population of confluent CHO cells. The semiintact cells used for assay of ER to Golgi transport described herein, which remain perforated throughout use, are unrelated to the cells prepared by the method referred to as "scrape-loading" (Hughson *et al.*, 1988). In the latter case, the subpopulation of cells scraped from the plate that reseal their plasma membrane lesion are used for further experimentation.

For cells grown on tissue culture dishes, it is important that they are not overconfluent in order to obtain a high yield of semiintact cells using either technique. In all cases where the technology is extended to different cell lines, it will be important to optimize growth conditions to achieve appropriate levels of confluency, and to optimize conditions for swelling, or attachment of cells to the nitrocellulose filter. The ionic strength of the swelling buffer should be considered to be a critical determinant in the preparation of a population of semiintact cells that can be perforated efficiently during scraping without release of intact cells or extensive cell disruption. Factors affecting the nitrocellulose protocol include residual buffer remaining on cells prior to addition of filter, the amount of buffer retained by the filter after blotting, and the binding time. The nitrocellulose technique offers the advantage that it is unnecessary to perturb cells osmotically prior to perforation. Simons and Virta (1987) and Brands and Feltkamp (1988) have described the use of nitrocellulose to prepare perforated cells from MDCK cells and BHK cells, respectively. It is anticipated that these techniques are applicable to a wide range of adherent cell lines.

Semiintact cells can also be made from cells grown in suspension (or released from tissue culture dishes) by swelling in a low osmotic strength buffer followed by a more *gentle* douncing protocol than that used to prepare cell homogenates. This differs from standard cell homogenization protocols only in that the end point for terminating douncing is optimized by quantitating the level of plasma membrane perforation rather than complete cell disruption.

For both confluent and suspension cells, optimal transport is generally observed by maintaining freshly prepared semiintact cells for up to 4 hours on ice. Conditions for establishing frozen stocks of semiintact cells that retain 100% of their activity before storage have not been successfully developed.

In semiintact cell preparations, 50% of the total G protein is transported from the ER to the Golgi compartment *in vitro*. In order to reconstitute reproducibly cytosol-dependent transport, we have found it is important to dounce the cells gently following release from the plate. We presume that the gentle douncing is necessary for an efficient exchange of the buffer with the soluble cytosolic pool. Excessive douncing will result in cell disruption and loss of activity. Some preparations of semiintact cells are inactive even in the presence of cytosol and ATP, but they transport G protein when supplemented with additional acceptor Golgi compartments (Beckers *et al.*, 1987). These results suggest that at least in some cases Golgi compartments can be released from the cell prior to release or vesiculation of the ER. When establishing conditions for measurement of transport *in vitro* between the ER and the Golgi using either different cell lines or preparative procedures different from those described herein, it may be necessary to supplement the assay with cytosol, ATP, and an acceptor Golgi-containing membrane fraction (Balch *et al.*, 1984a). If our understanding of the variables that control the functional integrity of the ER is correct, we should be able to remove these organelles from semiintact cells in an intact form.

Transport of protein from the ER to the cis-Golgi compartment can be readily detected in the mutant 15B CHO cells, which are defective in the processing of G-protein oligosaccharides past the $man_5GlcNAc_2$ form. Semiintact cells prepared from a wild-type cell line allow for the detection of additional inter-Golgi transport events. When semiintact cells are prepared from the parent, wild-type cell line of 15B, transport to the trans-Golgi compartment can be reconstituted *in vitro* with high efficiency by incubation in the presence of UDP-*N*-acetylglucosamine, UDP-galactose, and CMP-sialic acid, sugar–nucleotide precursors essential for formation of the complex oligosaccharide structure found on G protein in the trans-Golgi compartment (C. J. M. Beckers and W. E. Balch, unpublished). These results suggest that all of the early stages of delivery of protein through the trans-Golgi compartment can be studied in semiintact cells. In addition, transport from the trans-Golgi to the cell surface has been partially reconstituted in MDCK cells (Bennett *et al.*, 1988). It is apparent that semiintact cells may provide a generally applicable model system for study of all steps of the exocytic pathway.

In a more general sense, use of this approach may be applicable to a broad range of problems in cell biology. Semiintact cells have provided a major technical advance for immunoelectron microscopy (Hughson *et al.*, 1988). Cells can be incubated with antibodies directed to cytoplasmic determinants prior to fixation for electron microscopy. This "pre-embedment" technique (Hughson *et al.*, 1988) circumvents the problems associated with denaturation of antigenic epitopes by the fixative and detergent conditions that are necessary to penetrate the plasma membrane of intact cells. Semiintact cells can be subsequently postfixed in glutaraldehyde and embedded using plastic. This approach renders unnecessary the use of immunocryoelectron-microscopic techniques and the associated los of morphological detail. Since both the interior of the cell containing intact organelles, and the exterior of the cell are jointly accessible to a wide range or reagents and macromolecules, semiintact cells may provide a useful model system for study of other important problems in cell biology, including endocytosis, signal transduction, organization of the cytoskeletal matrix, and gene activation.

References

Balch, W. E., and Rothman, J. E. (1985). *Arch. Biochem. Biophys.* **240,** 413–425.

Balch, W. E., Dunphy, W. G., Braell, W. A., and Rothman, J. E. (1984a). *Cell* (*Cambridge, Mass.*) **39,** 405–416.

Balch, W. E., Glick, B. S., and Rothman, J. E. (1984b). *Cell* (*Cambridge, Mass.*) **39,** 525–536.

Balch, W. E., Wagner, K. R., and Keller, D. S. (1987). *J. Cell Biol.* **104,** 749.

Beckers, C. J. M., Keller, D. S., and Balch, W. E. (1987). *Cell* (*Cambridge, Mass.*) **50,** 523.

Bennett, M., Wandinger-Ness, A., and Simons, K. (1988). *EMBO J.* (in press).

Brands, R., and Feltkamp, C. A. (1988). *Exp. Cell. Res.* **176,** 309.

Hughson, E., Wandinger-Ness, A., Gausepohl, H., Griffiths, G., and Simons, K. (1988). *EMBO J.* **6** (in press).

Laemmli, U. K. (1970). *Nature* (*London*) **227,** 680–685.

McNeil, P. L., Murphy, F. F., Lanni, F., and Taylor, D. L. (1984). *J. Cell Biol.* **98,** 1556–1564.

Melancon, P., Glick, B. S., Malhotra, V., Weidman, P. J., Serafini, T., Gleason, M. L., Orci, L., and Rothman, J. E. (1987). *Cell* (*Cambridge, Mass.*) **51,** 1071.

Orci, L., Glick, B. S., and Rothman, J. E. (1986). *Cell* (*Cambridge, Mass.*) **46,** 71.

Simons, K., and Virta, H. (1987). *EMBO J.* **6,** 2241.

Walter, P., Gilmore, R., and Blobel, G. (1984). *Cell* (*Cambridge, Mass.*) **38,** 5–8.

Wattenberg, B. W., and Rothman, J. E. (1986). *J. Biol. Chem.* **261,** 2208–2221.

Chapter 6

Perforated Cells for Studying Intracellular Membrane Transport

MARK K. BENNETT, ANGELA WANDINGER-NESS, IVAN DE CURTIS, CLAUDE ANTONY, AND KAI SIMONS

European Molecular Biology Laboratory
D-6900, Heidelberg, Federal Republic of Germany

JÜRGEN KARTENBECK

Deutsches Krebsforschungszentrum
Heidelberg, Federal Republic of Germany

I. Introduction

Intracellular membrane transport consists of the movement of membrane proteins and lipids from one compartment to another, in most cases

METHODS IN CELL BIOLOGY, VOL. 31

by means of vesicular carriers. In the case of the biosynthetic pathway (for review, see Pfeffer and Rothman, 1987), membrane components are synthesized in the endoplasmic reticulum (ER) and are subsequently transported through the sequential compartments of the Golgi complex. From the last Golgi compartment, the trans-Golgi network (TGN), the membrane components are delivered to their final destinations in lysosomes, secretory granules, or the plasma membrane. A similar series of membrane compartments and transport steps may be involved in the endocytic pathway (Helenius *et al.*, 1983; Farquhar, 1985). A molecular description of these intracellular transport steps will require the identification of (1) the signals present on proteins required for retention in or targeting to a particular compartment, and (2) the cellular machinery responsible for the interpretation of such signals, including the receptors involved and the mechanisms for the formation, targeting, and fusion of specific transport vesicles. Studies of these problems have resulted in the identification of the mannose-6-phosphate signal and its receptor, responsible for lysosomal targeting (for reviews, see von Figura and Hasilik, 1986; Kornfeld, 1986), and the KDEL sequence responsible for the retention of proteins in the ER (Munro and Pelham, 1987). In addition, the role of clathrin in receptor-mediated endocytosis has been clearly established (Goldstein *et al.*, 1985).

In recent years, cell-free systems have been established to study the molecular requirements of specific membrane transport steps. Transport within the Golgi complex (Balch *et al.*, 1984; Rothman, 1987), from the ER to the Golgi (Balch *et al.*, 1987), from the Golgi to the plasma membrane (Woodman and Edwardson, 1986), and within the endocytic pathway (Davey *et al.*, 1985; Gruenberg and Howell, 1986; Braell, 1987) have all been reconstituted *in vitro*. The intra-Golgi transport system has been most thoroughly developed and has resulted in the identification of an *N*-ethylmaleimide-sensitive factor (Block *et al.*, 1988) and a GTP-binding protein (Melancon *et al.*, 1987), which are involved in vesicular transport from one Golgi cisterna to another. One drawback of the cell-free systems is the disruption of cellular organization that occurs during homogenization and the potential damage to the organelles during their isolation. These factors may be very important for the efficient reconstitution of certain membrane transport steps.

Two systems have subsequently been developed to permeabilize the plasma membrane of cells, making the cytoplasmic compartment available to manipulation, while leaving the overall morphology and intracellular organization intact (Simons and Virta, 1987; Beckers *et al.*, 1987). These perforated or semiintact cells have been generated by either (1) binding a nitrocellulose filter to the surface of the cells and subsequently removing the filter and, presumably, fragments of the plasma membrane, or (2) swelling the cells in a hypotonic buffer and then scraping them from their substrate. These systems have been successfully used to reconstitute transport from the ER to the Golgi complex (Simons and Virta, 1987; Beckers *et al.*, 1987) and from the Golgi complex to the cell surface (Simons and Virta, 1987; de Curtis and Simons, 1988), and have resulted in the identification and characterization of putative exocytic transport vesicles (Bennett *et al.*, 1988). The nitrocellulose procedure has the added advantage that cell–cell and cell–substrate interactions are maintained following perforation, which may be important for the application of the perforated-cell approach to other experimental problems.

In this chapter we will first describe the methods used for the generation of perforated cells using the nitrocellulose filter procedure, with the aim of providing a general approach that can be adapted to different cell types. We will then present our results on the characterization of the perforated cells and their use in the reconstitution of membrane transport.

II. Generation and Characterization of Perforated Cells

The basic procedure for the generation of perforated cells is quite simple. A monolayer of cells grown on a solid substrate is overlaid with a nitrocellulose filter. The filter is allowed to bind under controlled conditions for a specified period of time and is then peeled away from the cells. The cells remain attached to the substrate but have holes introduced into their plasma membrane. These holes allow manipulation of the cytoplasmic composition in cells whose overall morphology and organellar organization remain intact.

A. Perforation Procedures

A number of considerations are important in establishing an efficient and reproducible system for the perforation of cells using the nitrocellulose filter procedure. Perhaps the most important consideration is the strength of the cell attachment to the substrate on which they are grown. This will determine the amount of binding between the cells and the nitrocellulose filter that can be allowed to occur. If the cells are weakly adherent, the entire cell layer will be removed by the nitrocellulose filter instead of small fragments of the plasma membrane. For example, BHK cells were perforated more reproducibly when grown on polylysine-coated coverslips as a result of increased adherence to the substrate.

A second consideration is the method of binding the nitrocellulose filter to the cell layer. In this case, the amount of drying is critical in determining the amount of binding: with insufficient drying, the cells will not be perforated; with excessive drying, the cells will become detached. It is therefore important to establish conditions that allow for reproducible and even drying. A number of variables including the time, temperature, humidity, and air circulation should be controlled to obtain the most reproducible results. A third possibility for the optimization of perforation conditions is to modify the cell surface or nitrocellulose filter to increase the interaction between the two. Sambuy and Rodriguez-Boulan (1988) have coated MDCK cells with alternating layers of cationic colloidal silica and polyacrylic acid and then bound the cells to polylysine-coated nitrocellulose or glass. Such a procedure resulted in the efficient isolation of the apical plasma membrane from these cells. Intracellular membrane transport was not studied in this system.

The remainder of this section lists three different protocols for preparing perforated cells that have subsequently been used to reconstitute membrane transport (Section III). These examples illustrate how the perforation procedure has been modified for different cell types grown on different substrates and should provide a starting point for the application of this procedure to other cell types.

1. MDCK Cells Grown on Glass Coverslips

MDCK cells (strain I or II) were seeded on untreated glass coverslips (18 × 18 mm) in 35-mm-diameter culture dishes and grown for 2 days. At this time the cell layer had just reached confluence. Excess moisture was removed from the cell layer by touching the edge of the coverslip to Whatman 3MM filter paper. The coverslp was placed on a smooth plastic surface with the cells facing up. A nitrocellulose acetate filter (0.45 μm pore size, Millipore) was moistened in PBS containing 0.9 m*M* $CaCl_2$ and 0.5 m*M* $MgCl_2$ [PBS(+)], blotted against filter paper, and then carefully placed on top of the coverslip. A hair dryer (50 cm above the coverslip) blowing unheated air (22°–24°C) was turned on for 4 minutes. The filter was carefully peeled away from the cell layer. This procedure resulted in the permeabilization of 95% of the cells to antibodies. The cells retained the capacity for transport from the ER to Golgi (see Section III,B) and from the TGN to the plasma membrane.

2. MDCK Cells Grown on Polycarbonate Filters

MDCK cells were seeded on premounted polycarbonate filters (0.4 μm pore size, 24 mm diameter, Transwell 3412, Costar) and grown for 2 days. The following manipulations were performed in a cold room (4°C) with cold buffers. The filters were rinsed twice with KOAc buffer [25 m*M* HEPES–KOH (pH 7.4), 115 m*M* potassium acetate, 2.5 m*M* $MgCl_2$] and cut from the holder with a scalpel. Excess buffer was removed by blotting the edge of a Whatman No. 1 filter paper and the filter culture placed in a 50-mm culture dish in a 20°C water bath. A nitrocellulose acetate filter was soaked in KOAc buffer, blotted for 1 minute between two pieces of Whatman filter paper, and carefully placed directly on top of the filter culture. The nitrocellulose filter was covered with a Whatman filter and gently smoothed with a bent Pasteur pipet to remove any excess moisture and to maximize the contact between the cells and the nitrocellulose filter. The Whatman filter was removed and the cells allowed to bind to the nitrocellulose filter for 90 seconds. The nitrocellulose filter was then wetted by the addition of 200 μl KOAc buffer. The excess buffer was aspirated, and the nitrocellulose filter was separated from the filter culture. This procedure resulted in the permeabilization of at least 95% of the cells to both small molecules and antibodies. These cells lost most of their cytosolic proteins and were inefficient in both ER to Golgi and TGN to plasma membrane transport. However, the cells released putative exocytic transport vesicles into the incubation medium, from which the vesicles could be collected and characterized (see Section III,D).

3. BHK Cells Grown on Glass Coverslips

BHK cells wre cultured for 2 days to ~90% confluence in 35-mm-diameter dishes on poly-L-lysine-coated glass coverslips (18 × 18 mm). The coverslip was washed twice with ice-cold washing buffer [50 m*M* HEPES–KOH (pH 7.4), 75 m*M* KCl], excess buffer removed by blotting the edge on a filter paper, and placed on a smooth plastic surface with the cells facing up. A nitrocellulose filter that had been previously soaked in washing buffer was briefly blotted between two layers of filter paper to remove excess buffer and then carefully placed on top of the coverslip. After 30 seconds, the filter was carefully removed from the coverslip. This procedure resulted in the permeabilization of 90% of the cells to small molecules and ~50% of the cells to antibodies. The cells were capable of membrane transport from the ER to the Golgi complex and from the TGN to the plasma membrane (see Section III,C).

B. Assessment of Permeabilization

For screening a variety of perforation conditions, a simple method was developed to determine the number of cells perforated. Following perforation, the cells were incubated in PBS(+) containing the nuclear stain Hoechst 33258 (2.5 μg/ml) for 5 minutes at room temperature. The cells were then either fixed and mounted on a slide and observed by fluorescence microscopy, or observed directly in PBS(+) with the fluorescence microscope using a water immersion objective. The percentage of cells permeabilized was determined by comparing the nuclear-staining pattern with the number of cells visible by phase-contrast microscopy. Figure 1A shows the Hoechst-staining pattern of BHK cells perforated as just described. In this case 95% of the cells were stained. Figure 1B presents the control in which intact cells were stained using the same procedure. A similar fast and easy screening assay for permeabilization

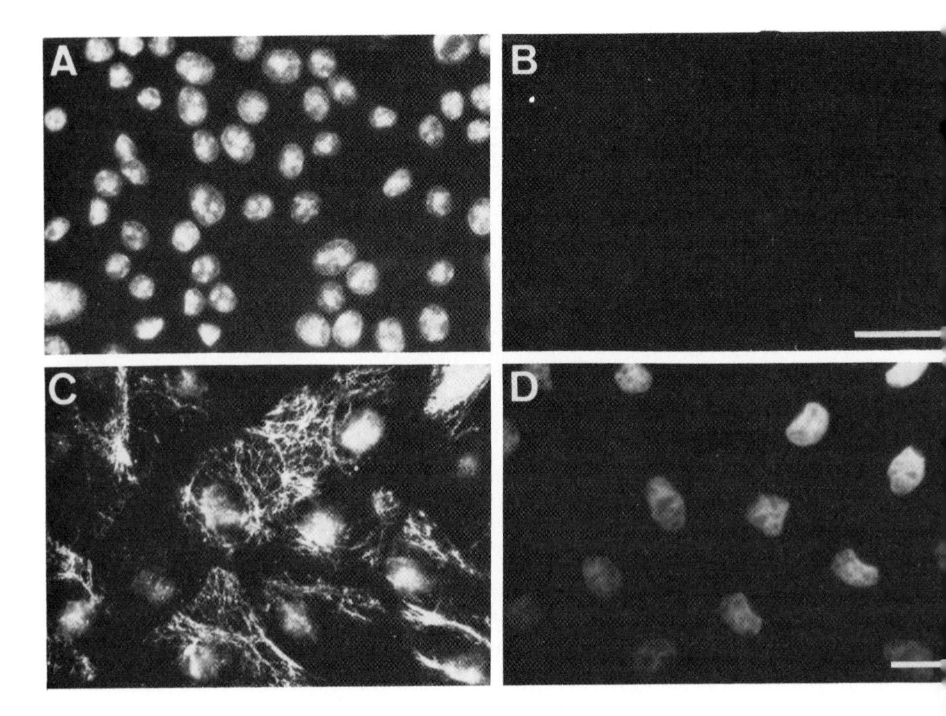

FIG. 1. Assessment of permeabilization. (A) Perforated BHK cells stained for 5 minutes with Hoechst 33258 as described in the text. (B) Intact BHK cells stained with Hoechst 33258. (C) Perforated MDCK cells stained with anticytokeratin antibody as described in the text. (D) Same field as in C stained with Hoechst dye to visualize nuclei. Bars = 10 μm.

was developed using the filamentous actin stain rhodamine–phalloidin, which, in the absence of detergent, stained perforated but not intact cells. Both the nuclear and actin stains provided simple procedures for the assessment of permeabilization to small molecules. Immunofluorescence staining was performed to determine the extent of permeabilization to large protein molecules. Antibodies were chosen that recognize only intracellular epitopes exposed to the cytoplasmic compartment. For example, MDCK cells grown on glass coverslips were perforated and stained with a rabbit antiserum directed against cytokeratins. The antigen–antibody complexes were visualized by indirect immunofluorescence using rhodamine-labeled anti-rabbit IgG (Fig. 1C). The total cell number was determined from the same microscope field by staining the cell nuclei with Hoechst 33258 (Fig. 1D). In this example, 100% of the cells were perforated.

An indirect method to assess the extent of perforation is to monitor the distribution of a soluble cytoplasmic marker such as lactate dehydrogenase (M_r 140,000). In the case of MDCK cells grown on polycarbonate filters, 90% of the soluble cytosolic marker was recovered in the incubation medium following perforation and incubation at 37°C for 1 hour. This suggests that at least 90% of the cells were perforated.

C. Morphology of the Perforated Cells

One of the advantages of the perforated-cell approach is that the damage is limited to the plasma membrane while the intracellular organization remains intact. Using MDCK cells grown on glass, it was shown that the organization of the Golgi complex, centrosomes, and endosomes was not altered by the perforation procedure as detected at the level of fluorescence microscopy (Simons and Virta, 1987). In order to get a description of the effects of cell perforation on cellular morphology at the ultrastructural level, the cells were examined by scanning and transmission electron microscopy. Scanning electron microscopy was performed to observe the effects of the perforation procedure on the plasma membrane, whereas transmission electron microscopy was used to study the organization of intracellular organelles.

1. Scanning Electron Microscopy

Scanning electron microscopy was performed on BHK cells grown on glass and MDCK cells grown on polycarbonate filters. The cells were fixed as an intact monolayer or immediately following perforation. The samples were then processed for scanning electron microscopy using standard procedures (Schroeter *et al.*, 1984). Scanning electron micrographs of filter-grown MDCK cells before (Fig. 2A) and after (Fig. 2B) perforation revealed dramatic differences. Prior to perforation, the apical surface of the monolayer was completely covered with a dense array of microvilli. The density of the microvilli made it difficult to detect the borders between the cells. At higher magnification (inset to Fig. 2A) the border between two cells is detected by the intertwining of microvilli. Following perforation, the microvilli were almost completely absent and the surface of the monolayer was relatively uneven. Cells were also seen that displayed a membranous "bubble" on the apical surface. These bubbles, shown in the inset in Fig. 2B, may represent a sheet of microvillar membrane that was pulled away from the cell (see later). The situation for BHK cells grown on glass coverslips was quite different. Prior to perforation (Fig. 2C), the cells displayed a typical fibroblastic morphology with their greatest thickness at the position of the nucleus and gradual flattening toward the cell periphery. In addition, there were few microvilli. Following perforation (Fig. 2D), the BHK cells looked similar with the exception that some disruption of the plasma membrane was detected in the vicinity of the nucleus. The membrane in the periphery of the cell appeared to be intact.

2. Transmission Electron Microscopy

Transmission electron microscopy was performed on MDCK cells cultured on glass or polycarbonate filters. Following perforation the cells were fixed immediately and processed for transmission electron microscopy as described previously (Simons and Virta, 1987). Transmission electron microscopy of MDCK cells grown on glass (Fig. 3A) revealed that the overall cellular organization was preserved following perforation. An intact Golgi complex and ER were clearly seen. Discontinuities were seen in the apical membrane and the cytoplasm was only lightly stained, as compared to intact cells, presumably because of loss of soluble cytoplasmic components. The results for MDCK cells grown on polycarbonate filters were quite similar (Fig. 3B). The intracellular organization was preserved in spite of the dramatic changes in the plasma membrane detected by scanning electron microscopy. In addition, the intercellular junctions were preserved along the lateral aspects of the cells. However, the electrical resistance across the monolayer decreased to background values following perforation; this was caused by occasional ruptures in the monolayer that gave rise to current leaks. In 13% of the cells observed, a "bubble" was seen along the apical surface (Fig. 3C) which may correspond to the bubbles seen by scanning electron microscopy.

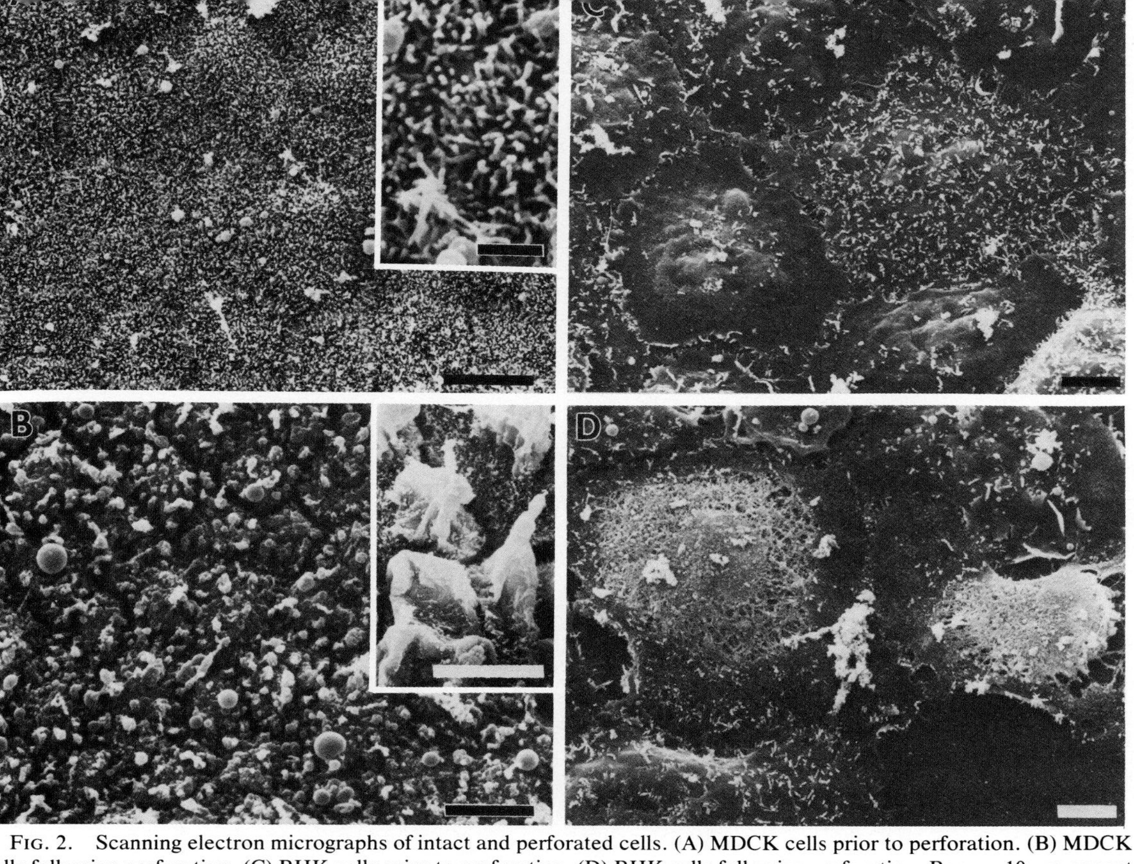

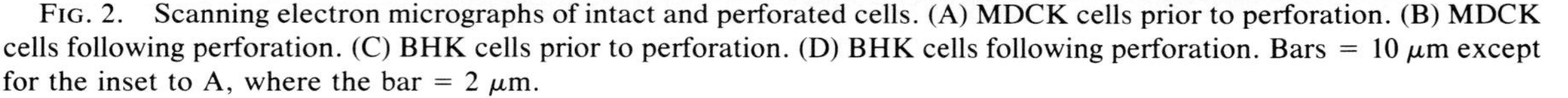

FIG. 2. Scanning electron micrographs of intact and perforated cells. (A) MDCK cells prior to perforation. (B) MDCK cells following perforation. (C) BHK cells prior to perforation. (D) BHK cells following perforation. Bars = 10 μm except for the inset to A, where the bar = 2 μm.

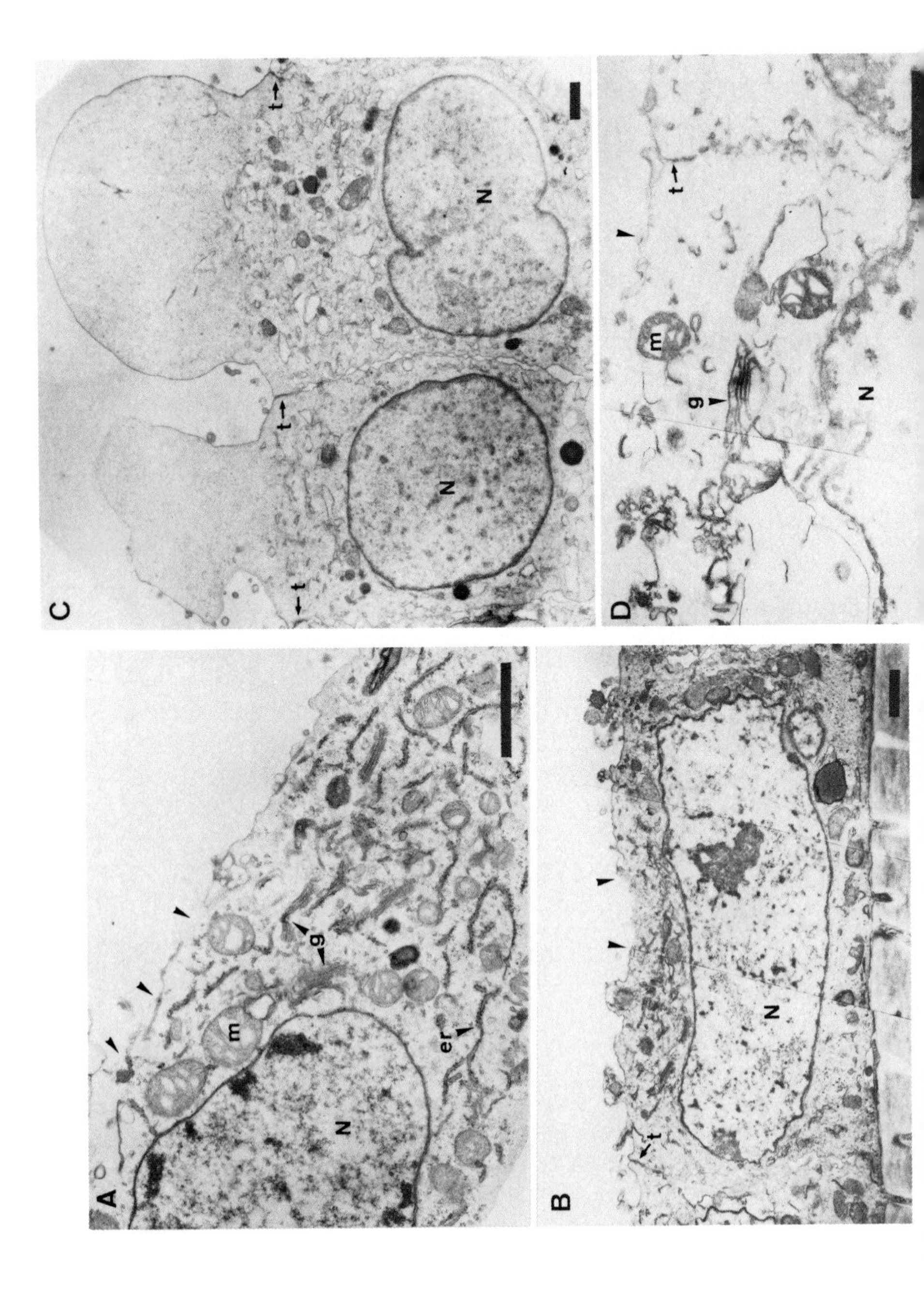
C
t
t
t
N
N
D
t
m
g
N
A
N
m
g
g
er
B
t
N

Most of the cells with bubbles (84%) were not perforated, as judged by cytoplasmic density and plasma membrane discontinuities. In addition, the cytoplasm within the bubble was devoid of membranous organelles, suggesting that a barrier was maintained at the former apical boundary of the cell. In the remaining 16% of cells with a bubble, the bubble appeared to be cracking and the content leaking out. This suggests that the bubble may be a fragile structure that represents an intermediate in the perforation of some of the cells. The morphology of the cells following incubation under conditions that promote membrane transport was also examined (Fig. 3D). Apical bubbles were no longer seen, and all of the cells were judged to be perforated. The cells were essentially devoid of cytoplasmic staining as a result of leakage of soluble components during the incubation (compare with Fig. 3B). However, the organellar organization was maintained during the incubation, as indicated by the presence of intact nuclei, Golgi complex, and intercellular junctions.

These morphological results indicate that the perforation procedure results in damage to the plasma membrane and the loss of soluble cytoplasmic components, while the overall organization of the intracellular organelles remains fairly well preserved, even after long incubations (60 minutes at 37°C). The discontinuities seen in the plasma membrane indicate that in most cases perforation results from the removal of fragments of the plasma membrane by the nitrocellulose filter. Indeed, Simons and Virta (1987) detected a plasma membrane marker attached to the nitrocellulose filter. The bubbles seen by scanning and transmission electron microscopy on MDCK cells grown on polycarbonate filters may represent an intermediate in the perforation of some cells that results from the flattening of the microvillar membrane into a sheet. The absence of bubbles on BHK cells or MDCK cells grown on glass coverslips is probably due to differences in the morphology of the cells prior to perforation or in the conditions used for perforation.

FIG. 3. Transmission electron micrographs of perforated MDCK cells. (A) MDCK cell grown on glass fixed immediately following perforation. (B) MDCK cell grown on polycarbonate filter fixed immediately following perforation. (C) Perforated filter-grown MDCK cells displaying apical "bubbles." (D) Perforated filter-grown cell following 1 hour at 37°C in GGA buffer (Section III,D). Arrows indicate discontinuities in the plasma membrane. N, Nucleus; g, Golgi complex; m, mitochondria; er, endoplasmic reticulum; t, tight junction. Bars = 1 μm.

III. Reconstitution of Intracellular Membrane Transport

The perforated cells described earlier were used to reconstitute transport from the ER to the Golgi complex and from the TGN to the cell surface. In this section, we will discuss first the transport markers that have been used to monitor membrane transport and then present some results on the reconstitution of membrane transport, taking one example from each of the three perforation systems described previously.

A. Membrane Transport Markers

Several viral membrane glycoproteins as well as a fluorescent lipid analog were used as markers to follow membrane transport. Vesicular stomatitis virus G protein (VSV G) as well as the Semliki forest virus (SFV) spike protein p62 were used to monitor transport from the ER to the Golgi complex. Cells were infected with the appropriate virus and pulse-labeled for a short time to ensure that newly synthesized protein remained in the ER. The perforated cells were then incubated under a variety of conditions and transport of the labeled protein to the Golgi complex detected by the development of endoglycosidase H (endo H) resistance and by the acquisition of terminal glycosylation.

Transport from the TGN to the plasma membrane was assayed using a number of markers. In each case, a 20°C block in membrane transport was used to accumulate the transport markers in the TGN (Matlin and Simons, 1983). Three different viral spike glycoproteins [fowl plague virus (FPV) hemagglutinin (HA), SFV p62, and VSV G] were used as markers. Two of the viral spike glycoproteins, SFV p62 and FPV HA, undergo a proteolytic-processing event late in the transport pathway (Klenk *et al.*, 1974; Green *et al.*, 1981). Hemagglutinin is cleaved into HA1 and HA2, whereas p62 is cleaved into E2 and E3. The cleavages occur at dibasic sites and are catalyzed by an enzyme with a specificity similar to that of enzymes responsible for prohormone maturation (Docherty and Steiner, 1982). The cleavages occur to a low extent during the 20°C block, but proceed rapidly upon warming the cells to 37°C. However, the cleavages appear to occur prior to delivery of the proteins to the plasma membrane. Therefore, this proteolytic-processing step was used as a marker for a post-Golgi transport event, perhaps transport vesicle formation. Appearance of the spike glycoproteins on the cell surface was monitored by surface immunoprecipitation with antibodies directed against their exoplasmic spike domains. The typical procedure for a TGN-to-plasma membrane transport experiment was as follows. The cells were infected with virus, pulse-labeled, and incubated at 20°C to accumulate the labeled membrane protein in the TGN. The cells were then perforated and incubated at 37°C under different conditions. The amount of viral protein cleavage and surface appearance was then determined.

The fluorescent lipid analog C6-NBD-ceramide was also used to monitor transport from the TGN to the plasma membrane. This analog rapidly partitions into all the cell membranes and is metabolized into C6-NBD-sphingomyelin and C6-NBD-glucosylceramide in the Golgi complex (Lipsky and Pagano, 1983; van Meer *et al.*, 1987). This conversion traps the fluorescent marker on the luminal side of the Golgi membrane, where it is accumulated at 20°C. Following perforation and incubation under the appropriate conditions, transport of the fluorescent marker to the cell surface was detected by fluorescence microscopy.

B. Transport from the ER to the Golgi Complex

Transport of viral glycoproteins from the ER to the Golgi complex in perforated cells was reconstituted in MDCK cells (Simons and Virta, 1987) and BHK cells grown on glass. The data presented here are from MDCK.

MCDK II cells were grown on glass coverslips for 2 days, infected with VSV for 3.5 hours, and perforated as described previously (Section II,A). Following perforation, the cell layer was immediately overlaid with 80 μCi [^{35}S]methionine in 250 μl medium containing one-tenth he normal concentration of methionine. After 10 minutes at 37°C the coverslip was washed with transport buffer A [78 mM KCl, 50 mM HEPES–KOH (pH 7.0), 4 mM $MgCl_2$, 10 mM EGTA, 8.37 mM $CaCl_2$, 1 μM dithiothreitol (DTT)] at 0°C. The cells were then placed in 500 μl of transport buffer A containing 1 mM methionine supplemented with either an ATP-regenerating system (0.5 mM ATP, 4 mM creatine phosphate, 20 μg/ml creatine phosphokinase) or an ATP-depleting system (50 U/ml apyrase) and incubated at 37°C for 40 minutes. Following the incubation, the ^{35}S-labeled G protein was immunoprecipitated by the addition of an antibody directed against the cytoplasmic domain of VSV G (Kreis, 1986) to the coverslip in the absence of detergent for 30 minutes at 0°C. This ensured that only the VSV G protein in the perforated cells was analyzed. The unbound antibody was removed by washing the monolayer three times with PBS containing 0.2% gelatin. The immunoprecipitation was completed by solubilizing the cells in 1% Triton X-100 and adding protein A–Sepharose 4B as an immunoadsorbant. The immunoprecipitates were then tested for sensitivity to digestion with endo H. One-half of each sample, 100 μl suspended in 0.2 *M* sodium citrate (pH 5.5), was treated with 5 μl endo H (1 U/ml) at 37°C for 12 hours. The other half of the sample was incubated at 37°C for 12 hours in the absence of endo H. The protein was released from the Sepharose beads with SDS, concentrated by trichloroacetic acid precipitation, and resolved on a 10% polyacrylamide gel as previously described (Pfeffer *et al.*, 1985). The results are presented in Fig. 4. Immediately following the pulse-labeling, the VSV G protein was completely sensitive to endo H (lanes 1 and 2), as is characteristic of newly synthesized VSV G, which is present in the ER. When the perforated cells were incubated at 37°C in the absence of ATP (lanes 3 and 4), the VSV G protein remained sensitive to endo H digestion. However, when the perforated cells were incubated in the presence of ATP, the VSV G protein acquired complete resistance to endo H and underwent a shift in mobility characteristic of terminal glycosylation (lanes 5 and 6). These properties were the same as those of VSV G from intact cells that were pulse-labeled and incubated at 37°C (lane 7). When intact control cells were subjected to immunoprecipitation with the antibody directed against the cytoplasmic domain of G in the absence of detergent, no G protein was precipitated (lane 8). This

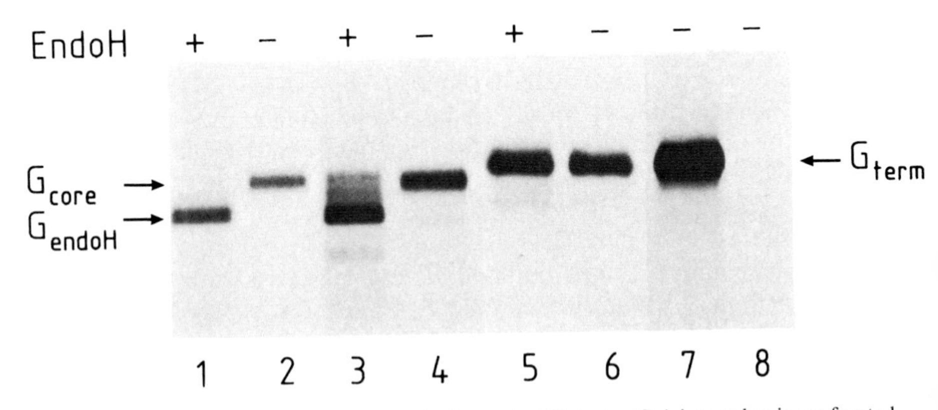

FIG. 4. Transport of VSV G protein from the ER to the Golgi complex in perforated MDCK cells. Cells infected with VSV were perforated, pulse-labeled with [^{35}S]methionine for 10 minutes at 37°C (lanes 1 and 2), and chased for 40 minutes at 37°C in transport buffer A in the absence of ATP (lanes 3 and 4) or in the presence of ATP (lanes 5 and 6). The G protein was immunoprecipitated with antibody against the cytoplasmic domain of G, added to the cell layers before solubilization (lanes 1–6 and 8). In lane 7 the cells were solubilized before adding the antibody. The samples in lanes 1, 3, and 5 were treated with endo H. The positions of the core-glycosylated (G_{core}), endo H-digested (G_{endoH}), and terminally glycosylated (G_{term}) forms of VSV G are indicated. For experimental details, see text.

demonstrated that the antibody had access to the cytoplasmic domain of VSV G in the perforated cells but not in intact cells.

These results demonstrate that the VSV G protein is efficiently transported from the ER to at least the medial Golgi compartment (where conversion to endo H resistance occurs) in the perforated MDCK cells in an ATP-dependent manner. Similar results were obtained using BHK cells infected with SFV (I. deCurits, unpublished observation). In addition, Beckers *et al.* (1987) have reported the reconstitution of ER-to-Golgi transport in semintact CHO cells, which is dependent on both cytosol and ATP. In contrast, transport from the ER to the Golgi in MDCK cells grown on polycarbonate filters was inefficient following perforation. Using the same transport conditions described earlier for MDCK cells grown on glass, only a small portion of the VSV G acquired endo H resistance and none reached a state of terminal glycosylation (M. Bennett, unpublished observation). This difference will be considered further in the discussion (Section IV).

C. Transport from the TGN to the Plasma Membrane

Transport of a viral glycoprotein from the TGN to the plasma membrane was reconstituted in perforated BHK cells (de Curtis and Simons, 1988). BHK cells were grown on polylysine-coated coverslips for 2 days, infected with SFV for 3.5 hours, and then pulse-labeled with 50 μCi [^{35}S]methionine for 5 minutes at 37°C. The cells were washed three times and then incubated at 19.5°C for 90 minutes in medium containing 2.5 m*M* unlabeled methionine. This incubation allowed the accumulation of pulse-labeled p62 glycoprotein in the TGN. The cells were then perforated as described previously (Section II,A) and incubated in 0.4 ml KCl transport buffer [25 m*M* HEPES–KOH (pH 7.4), 115 m*M* KCl, 2.5 m*M* $MgCl_2$, 2.5 m*M* methionine, 12 m*M* glucose, 10 μM free Ca^{2+}] for 60 minutes at 37°C. Following the incubation, the cells were either subjected to surface immunoprecipitation (to detect the viral spike protein that reached the cell surface), or solubilized and then subjected to immunoprecipitation (to measure the total viral spike protein and amount of p62 cleavage). Surface immunoprecipitation was performed by adding antibody directed against the exoplasmic domains of p62 and E2 to the transport buffer at the end of the incubation and incubating on a rocking platform for 30 minutes on ice. Unbound antibody was removed by washing the coverslips once with PBS(+) containing 0.2% BSA and twice with PBS(+) alone. The cells were then solubilized with 0.3 ml lysis buffer (containing 1% Nonidet P-40) and the antigen–antibody complexes precipitated with rabbit anti-mouse IgG and protein A–Sepharose. For the analysis of total viral protein, the cells were first solubilized in lysis buffer. The antibody was then added and the immunoprecipitation continued as described. The immunprecipitates were analyzed on 10% polyacrylamide gels. The results are presented in Fig. 5. When either intact or perforated cells were analyzed prior to incubation at 37°C, the majority of the viral spike protein was in the p62 (uncleaved) form (lanes 1 and 3), and was not on the cell surface (lane 6). This demonstrated that both cleavage and surface appearance were efficiently arrested during the 19.5°C block of transport. Upon warming the cells to 37°C for 60 minutes cleavage of p62 into E2 occurred (lanes 2 and 4). In the perforated cells the cleavage efficiency was 60–75% of that found in intact cells. The appearance of the E2 cleavage product on the cell surface was also reconstituted in the perforated cells (lanes 5 and 7). The inclusion of glucose and low concentrations of Ca^{2+} in the transport buffer was essential for the reconstitution of both p62 cleavage and E2 surface appearance in perforated but not in intact cells. In contrast, p62 cleavage, but not E2 surface appearance was inhibited by nonhydrolyzable analogs of ATP and GTP in the perforated BHK cells (de Curtis and Simons, 1988). This result demonstrates that the cleavage and delivery steps can be dissociated and is strong evidence that cleavage precedes surface appearance.

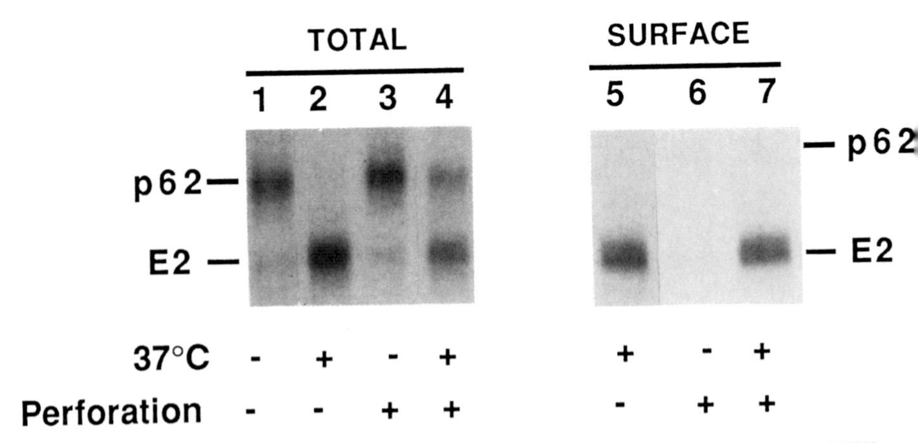

FIG. 5. Transport of SFV glycoprotein from the TGN to the plasma membrane. BHK cells were infected with SFV, pulse-labeled, and incubated at 19.5°C for 90 minutes. The cells were either perforated or left intact, and the viral protein analyzed by immunoprecipitation either before or after incubation at 37°C for 60 minutes in KCl transport buffer, as indicated. Lanes 1–4 represent the total viral glycoprotein pattern, and lanes 5–7 represent the viral glycoprotein present on the cell surface. For experimental details, see text.

Transport from the TGN to the plasma membrane was also reconstituted in perforated MDCK cells grown on glass coverslips (Simons and Virta, 1987). In this case transport was monitored by the appearance of the C6-NBD-lipid marker on the basolateral plasma membrane by fluorescence microscopy. The transport was found to be dependent on ATP and inhibited by apyrase. Transport from the TGN to the plasma membrane of perforated MDCK cells grown on filters will be discussed in the next section.

D. Release of Putative Exocytic Transport Vesicles

In order to study membrane transport from the TGN to the distinct apical and basolateral plasma membrane domains of MDCK cells, perforated MDCK filter cultures were used. In these studies, FPV HA was used as an apical membrane marker, VSV G as a basolateral membrane marker, and the metabolites of C6-NBD-ceramide as a marker for both domains.

The cleavage of HA was used as a late transport event to establish conditions for the reconstitution of transport. The cells were infected with FPV for 3.5 hours, pulse-labeled for 4 minutes at 37°C with 50 μCi [^{35}S]methionine, and incubated at 20°C for 90 minutes in medium containing 0.9 m*M* unlabeled methionine to accumulate the labeled viral glycoprotein in the TGN. The cells were perforated and incubated at 37°C for 60 minutes under different conditions and the amount of HA cleavage monitored. Using GGA transport buffer [25 m*M* HEPES–KOH (pH 7.4), 38 m*M* potassium gluconate, 38 m*M* potassium glutamate, 38 m*M* potassium aspartate, 2.5 m*M* $MgCl_2$, 2 m*M* EGTA], an ATP-dependent cleavage of HA that was 20–25% the efficiency of intact cells was obtained. We next measured the appearance of HA on the cell surface by surface immunoprecipitation. None could be detected. This indicated that a step in the transport from the TGN to the cell surface, as detected by the cleavage of HA, was reconstituted, but that the final step in the process, delivery of the marker to the plasma membrane, did not occur. This raised the possibility that transport vesicles were formed during the incubation, but that their fusion with the plasma membrane was prevented. In order to test whether these putative transport vesicles were released from the perforated cells, the incubation medium fraction was separated from the filter culture and each analyzed separately for the amount of HA2 cleavage product. The results are presented in Table I. In the absence of ATP, 18% of the HA2 was released from the perforated cells, whereas in the presence of ATP 40% was released, a 2.1-fold stimulation by ATP.

TABLE I

TRANSPORT MARKERS RELEASED FROM PERFORATED MDCK CELLS

Transport Marker	Release from Perforated MDCK II Cells (%)		Stimulation by ATP (-fold)
	−ATP	+ATP	
FPV HA2	18.0 ± 9.0	37.9 ± 6.0	2.1
VSV G	8.1 ± 3.3	21.7 ± 4.1	2.7
C6-NBD-Sphingomyelin	8.0 ± 2.3	22.3 ± 1.3	2.8

One possible explanation for the release of HA2 was that the extensive damage to the apical membrane during the perforation procedure prevented fusion of transport vesicles with the apical membrane and thus led to their release. Therefore, the distribution of the basolateral membrane marker VSV G and the lipid marker C6-NBD-sphingomyelin (which is delivered in approximately equal amounts to the two domains) was monitored. Pulse-labeled VSV G or C6-NBD-sphingomyelin were accumulated in the TGN at 20°C as described earlier for HA. The cells were perforated and incubated at 37°C for 60 minutes in GGA buffer in the presence or absence of ATP. As with HA2, there was an ATP-stimulated release of these two transport markers (Table I). The percentage of these markers released was less than that of HA2 both in the presence and absence of ATP, but the fold stimulation was similar. Control experiments in which the distribution of the Golgi marker enzymes galactosyltransferase and sialyltransferase were monitored indicated that general vesiculation of the Golgi complex was not occurring. This is supported by the transmission electron microscopy results in which intact organelles including Golgi complex were observed following perforation and incubation (Fig. 3D).

In order to characterize further these putative transport vesicles, we investigated their orientation and sedimentation properties. Transport vesicles derived from the TGN would be expected to contain the viral spike protein in an orientation such that the spike domain would be internal and the cytoplasmic domain would be exposed on the vesicle surface. This prediction was tested by determining the accessibility of trypsin to the different domains of the VSV G released from the perforated cells. The results are presented in Fig. 6. Following digestion with trypsin (Try) in the absence of detergent, the VSV G exhibited a

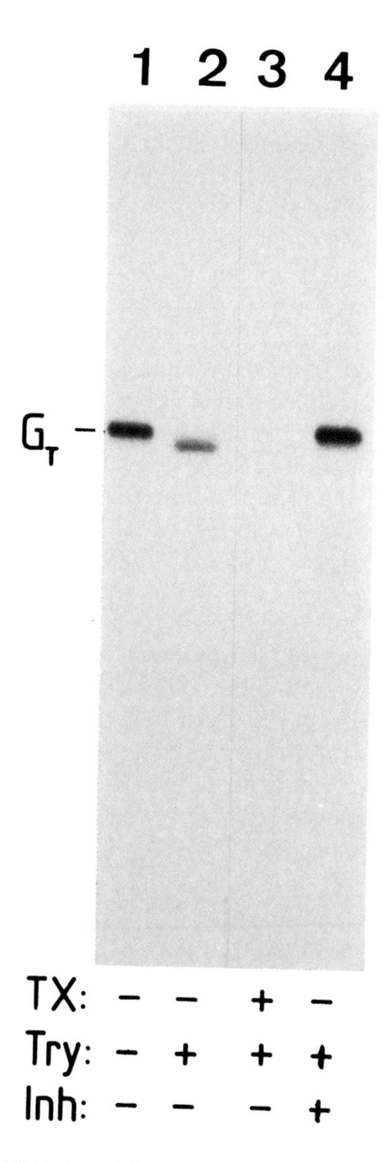

FIG. 6. Topology of VSV G in vesicles released from perforated MDCK cells. MDCK filter cultures were infected with VSV, pulse-labeled, and incubated at 20°C for 90 minutes. The cells were perforated, incubated at 37°C for 60 minutes, and the incubation medium collected and analyzed. The medium sample was either left untreated (lane 1), digested with trypsin (0.25 mg/ml; lane 2), digested with trypsin in the presence of 1% Triton X-100 (lane 3), or treated with trypsin in the presence of trypsin inhibitor (1.25 mg/ml; lane 4). Following the treatments, the VSV G protein was immunoprecipitated and analyzed on a 10% polyacrylamide gel.

slight increase in mobility on the gel (lane 2) as compared to untreated VSV G (lane 1). This shift is consistent with the removal of the cytoplasmic domain of the VSV G (Katz *et al.*, 1977). When trypsin inhibitor (Inh) was included in the incubation, the shift did not occur (lane 4), and when the digestion was done in the presence of detergent (1% Triton X-100, TX), the VSV G was completely degraded (lane 3). Quantitation of this result indicated that 80% of the VSV G released from the perforated cells was contained in sealed membranous vesicles with the orientation expected of transport vesicles derived from the TGN. A similar conclusion concerning the orientation of vesicles containing FPV HA2 was reached from studies on the accessibility of antibodies directed against the exoplasmic spike domain of HA.

Finally, the behavior of the putative transport vesicles was monitored by equilibrium sedimentation. In this experiment, the C6-NBD-ceramide was used to monitor the sedimentation. However, identical results were obtained when the sedimentation of the viral spike proteins was analyzed. The NBD-labeled vesicles were first separated from soluble components by centrifugation through a 0.3 *M* sucrose cushion for 60 minutes at 100,000 *g*. The pellet was resuspended and layered on top of a linear 0.3–1.5 *M* sucrose gradient. The gradient was subjected to centrifugation at 100,000 *g* for 14 hours to allow the vesicles to reach their equilibrium density. The gradient was fractionated and the fractions assayed for the presence of NBD fluorescence. As shown in Fig. 7, the putative transport vesicles migrated as a well-defined peak with a density of 1.09–1.11 g/ml. These results demonstrate that perforated MDCK cells release vesicles with the properties expected of exocytic transport vesicles.

IV. Discussion

We have presented our methods for the generation of perforated cells and the applications of these cells in the reconstitution of membrane transport. Our experience has been with MDCK cells grown on glass or polycarbonate filters and BHK cells grown on glass. In addition, Beckers *et al.* (1987) have used the nitrocellulose filter procedure to perforate CHO cells grown on plastic. The perforation procedures were modified for each of these systems, and our experience suggests that the procedures could be adapted for the perforation of other cell types. The different systems provided different results concerning the requirements for membrane transport. MDCK and BHK cells grown on glass provided

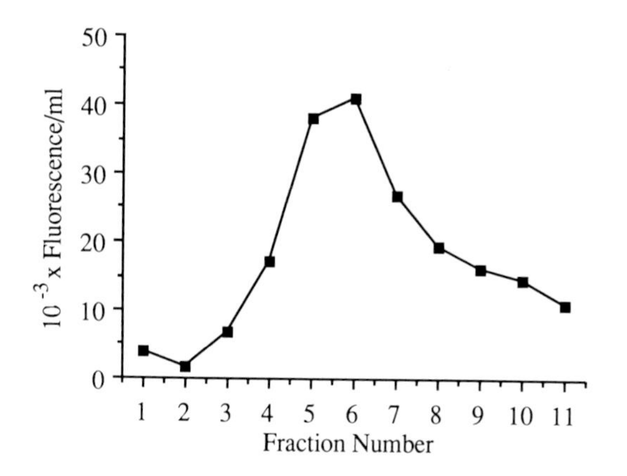

FIG. 7. Sedimentation of NBD-labeled vesicles. An MDCK filter culture was labeled with C6-NBD-ceramide at 20°C, perforated, and incubated at 37°C for 60 minutes in GGA transport buffer. The particulate material from the incubation medium was collected by centrifugation through a 0.3 *M* sucrose cushion (in 10 m*M* HEPES, pH 7.4, 2 m*M* EGTA, 1 m*M* DTT) at 100,000 *g* for 60 minutes. The particulate material was resuspended in 200 μl GGA transport buffer and resolved on linear 0.3–1.5 *M* sucrose gradient by centrifugation at 100,000 *g* for 14 hours. The gradient fractions were analyzed for NBD fluorescence.

very similar results. In both cases, transport from the ER to the Golgi and from the TGN to the plasma membrane proceeded efficiently. The main difference between the two systems was that in BHK cells transport was dependent on glucose whereas in MDCK cells transport was dependent on ATP. This difference may reflect a difference in the metabolism in the two cells. In BHK cells, glucose may serve as a more efficient energy source than ATP. In both cases, transport was reconstituted in the absence of added cytosolic protein. This is in contrast with the cell-free assays for membrane transport in which addition of cytosolic protein is required (Balch *et al.*, 1984, 1987; Davey *et al.*, 1985; Woodman and Edwardson, 1986). The perforated cells grown on glass substrates may retain enough cytosolic protein to continue membrane transport efficiently. No attempt was made in these experiments to remove the cytosol by washing the cells prior to incubation to establish a cytosol dependence. However, this has been done by Beckers *et al.* (1987). They showed that cytosol is necessary for membrane transport from the ER to the Golgi in their semintact CHO cell system.

Membrane transport in perforated MDCK filter cultures, in contrast to MDCK on glass, was inefficient, both from the ER to the Golgi and from the TGN to the plasma membrane. The main difference between the two systems was the substrate on which the cells were grown. The cells grown on glass coverslips were perforated shortly after reaching confluence and therefore, because of variability in the thickness of different regions of the cell, presented an irregular apical surface for the binding of the nitrocellulose filter. In contrast, the filter-grown cells form a monolayer of cuboidal cells of more uniform thickness with a high density of apical microvilli. These morphological differences, along with the stronger adherence of the cells to the filter substrate, allowed perforation conditions to be used that resulted in more extensive damage to the apical membrane. Scanning electron-microscopic analysis of the filter-grown cells before and after perforation showed dramatic morphological changes including the loss of microvilli over the entire apical surface (Fig. 2A and B). This may cause the filter-grown cells to lose their cytosolic proteins more rapidly or to a greater extent than the cells grown on coverslips. Indeed, we found that the filter-grown cells lost the majority of their cytosolic protein following perforation and estimate that the cytosol was diluted over 200-fold with incubation buffer. It is possible that a cytosolic protein required for membrane fusion was diluted below a critical concentration in the perforated filter culture so that all fusion events required for intracellular membrane transport were inefficient or blocked. It may be possible to reconstitute vesicle fusion in the filter-grown cells by the addition of a cytosolic fraction, such as has been done in the cell-free transport assays. The lack of fusion in the filter-grown MDCK cells has in fact been an advantage, because it has allowed the identification and characterization of transport vesicles derived from the TGN.

The main advantage of the perforated-cell approach is that the intracellular organization is preserved in cells whose cytoplasmic compartment is available to manipulation. These cells have proved useful in the reconstitution of intracellular membrane transport. We have concentrated on the biosynthetic-transport pathway, but a similar approach could be applied to the endocytic pathway. However, the perforated-cell system is not limited to applications in membrane transport. The perforated-cell system also has applications in immunocytochemistry. Because cytoplasmically disposed epitopes are accessible to antibodies in the perforated cells, the antigens need not be fixed prior to labeling. Fixation can be performed following antigen–antibody interaction. This method has been successfully used to localize viral glycoproteins at the ultrastructural level (Hughson *et al.*, 1988). Perforated cells could also be used to study the interactions of cytoplasmic components with a cytosolic face of cellular membranes. For example, the interaction of receptors, adhesion and junctional molecules with cytoskeletal elements could be studied. Thus, perforated cells provide an intermediate between intact cells and cell homogenates or purified cellular components with many potential cell biological applications.

ACKNOWLEDGMENTS

The authors wish to express their thanks to Hilkka Virta for expert technical assistance in the provision of virus stocks and antibodies. M. K. B. and A. W. N. were supported by U.S. Public Health Service National Research Service Awards GM 11726-02 and GM 11209-01, respectively, from the National Institutes of General Medical Sciences.

REFERENCES

Balch, W. E., Dunphy, W. G., Braell, W. A., and Rothman, J. E. (1984). *Cell (Cambridge, Mass.)* **39,** 405–416.

Balch, W. E., Wagner, K. R., and Keller, D. S. (1987). *J. Cell Biol.* **104,** 749–760.

Beckers, C. J. M., Keller, D. S., and Balch, W. E. (1987). *Cell (Cambridge, Mass.)* **50,** 523–534.

Bennett, M. K., Wandinger-Ness, A., and Simons, K. (1988). *EMBO J.* **7,** 4075–4085.

Block, M. R., Glick, B. S., Wilcox, C. A., Wieland, F. T., and Rothman, J. E. (1988). *Proc. Natl. Acad. Sci. U.S.A.* **85,** 7852–7856.

Braell, W. A. (1987). *Proc. Natl. Acad. Sci. U.S.A.* **84,** 1137–1141.

Davey, J., Hurtley, S. M., and Warren, G. (1985). *Cell (Cambridge, Mass.)* **43,** 643–652.

de Curtis, I., and Simons, K. (1988). *Proc. Natl. Acad. Sci. U.S.A.* **85,** 8052–8056.

Docherty, K., and Steiner, D. F. (1982). *Annu. Rev. Physiol.* **44,** 625–638.

Farquhar, M. G. (1985). *Annu. Rev. Cell Biol.* **1,** 447–452.

Goldstein, J. L., Brown, M. S., Anderson, R. G. W., Russell, D. W., and Schneider, W. J. (1985). *Annu. Rev. Cell Biol.* **1,** 1–39.

Green, J., Griffiths, G., Louvard, D., Quinn, P., and Warren, G. (1981). *J. Mol. Biol.* **152,** 663–698.

Gruenberg, J. E., and Howell, K. E. (1986). *EMBO J.* **5,** 3091–3001.

Helenius, A., Mellman, I., Wall, D., and Hubbard, A. (1983). *Trends Biochem. Sci.* **10,** 447–452.

Hughson, E., Wandinger-Ness, A., Gausepohl, H., Griffiths, G., and Simons, K. (1988). *In* "Molecular Biology and Infectious Diseases" (M. Schwartz, ed.). Elsevier/North-Holland, Amsterdam and New York, pp. 75–89.

Katz, F. N., Rothman, J. E., Knipe, D. M., and Lodish, H. F. (1977). *J. Supramol. Struct.* **7,** 353–370.

Klenk, H.-D., Wöllert, W., Rott, R., and Scholtissek, C. (1974). *Virology* **57,** 28–41.

Kornfeld, S. (1986). *J. Clin. Invest.* **77,** 1–6.

Kreis, T. E. (1986). *EMBO J.* **5,** 931–941.

Lipsky, N. G., and Pagano, R. E. (1983). *Proc. Natl. Acad. Sci. U.S.A.* **80,** 2608–2612.

Martlin, K. S., and Simons, K. (1983). *Cell (Cambridge, Mass.)* **34,** 233–243.

Melancon, P., Glick, B. S., Malhotra, V., Weidman, P. J., Serafini, T., Gleason, M. L., Orci, L., and Rothman, J. E. (1987). *Cell (Cambridge, Mass.)* **51,** 1053–1062.

Munro, S., and Pelham, H. R. B. (1987). *Cell (Cambridge, Mass.)* **48,** 899–907.

Pfeffer, S. R., and Rothman, J. E. (1987). *Annu. Rev. Biochem.* **56,** 829–852.

Pfeffer, S. R., Fuller, S. D., and Simons, K. (1985). *J. Cell Biol.* **101,** 470–476.

Rothman, J. E. (1987). *J. Biol. Chem.* **262,** 12502–12510.

Sambuy, Y., and Rodriguez-Boulan, E. (1988). *Proc. Natl. Acad. Sci. U.S.A.* **85,** 1527–1533.

Schroeter, D., Spiess, E., Paweletz, N., and Benke, R. (1984). *J. Electron Microsc. Tech.* **1,** 219–225.

Simons, K., and Virta, H. (1987). *EMBO J.* **6,** 2241–2247.

van Meer, G., Stelzer, E. H. K., Wijnaendts-van Resandt, R. W., and Simons, K. (1987). *J. Cell Biol.* **105,** 1623–1635.

von Figura, K., and Hasilik, A. (1986). *Annu. Rev. Biochem.* **55,** 167–193.

Woodman, P. G., and Edwardson, J. M. (1986). *J. Cell Biol.* **103,** 1829–1835.

Chapter 7

Reconstitution of Protein Transport Using Broken Yeast Spheroplasts

DAVID BAKER AND RANDY SCHEKMAN

Department of Biochemistry
University of California
Berkeley, California 94720

I. Introduction

To investigate the molecular mechanisms underlying intercompartmental protein transport, *in vitro* reconstitution is essential. Reconstitution of transport in extracts from *Saccharomyces cerevisiae* is particularly useful because individual transport factors can be characterized through *in vitro* analysis of mutants defective in protein transport. Here we describe a method for preparing transport-competent yeast extracts and the use of these extracts in reconstituting transport from the endoplasmic reticulum (ER) to the Golgi apparatus. Our results using these methods have been discussed in detail elsewhere (Baker *et al.*, 1988).

II. Preparation of Transport-Competent Membranes

A. General Comments

"Semiintact" mammalian cells, which have lost most of their cytoplasm but retain intact organelles, support a variety of intercompartmental protein transport reactions (see Beckers *et al.*, Chap. 5, this volume; Simons and Virta, 1987; Beckers *et al.*, 1987). Yeast spheroplast ghosts can be prepared by osmotic or freeze–thaw lysis. Dilution of spheroplasts into buffers of low osmotic strength permeabilizes the plasma membrane with little rupture of internal organelles (W. Balch, personal communication). Ghosts can also be prepared by controlled freeze-thawing. The principal advantage of this method is convenience: spheroplasts can be prepared in large quantities and then frozen in aliquots that can be thawed when needed. The high protein concentration inside an intact spheroplast apparently serves as a good cryoprotectant: membranes prepared by freeze–thaw lysis of spheroplasts can support efficient protein transport reactions. In contrast, freezing of semiintact cells reduces transport efficiency, and they are generally prepared fresh for every experiment (W. Balch, personal communication).

B. Freeze–Thaw Lysis

A brief review of the stresses placed on cells during freeze-thawing is useful before a detailed description of the lysis procedure. The extent of lysis is determined largely by the rates of freezing and thawing (Grout and Morris, 1987). The principal source of damage during freeze-thawing is intracellular ice formation. When a cell suspension is cooled, the surrounding medium (having a larger volume and hence an increased probability of ice crystal nucleation) begins to freeze first. This creates an osmotic stress on the cells, and water flows out across the plasma membrane. When the temperature is lowered sufficiently, any remaining intracellular water freezes. If the rate of freezing is sufficiently slow, most of the intracellular water is lost by dehydration and little intracellular ice forms. During rapid freezing, however, low temperatures are reached before significant dehydration has occurred and internal water freezes. Slow freezing leads to less intracellular ice formation than rapid freezing and hence is less stressful. Conversely, rapid thawing, which allows less time for potentially damaging ice recrystallization, is less stressful than slow thawing. By adjusting the rates of freezing and thawing the desired degree of lysis can be achieved.

C. Method

1. Preparation of Actively Metabolizing Spheroplasts

Yeast cultures are grown in YPD medium [1% Bacto yeast extract (Difco), 2% Bacto peptone, and 5% glucose] to 2–4 OD_{600} U/ml (1 OD_{600} U is ~10^7 cells). Wild-type strains are grown at 30°C, temperature-sensitive (ts) strains at 24°C. Cells are harvested by centrifugation (1000 *g*, 5 minutes, 24°C) and resuspended at 50 OD_{600} U/ml in 10 m*M* dithiothreitol (DTT), 100 m*M* Tris-HCl, pH 9.4. After 5 minutes at 24°C, cells are harvested by centrifugation and resuspended at 50 OD_{600} U/ml in 0.7 *M* sorbitol, 0.75 × YP (1% Bacto yeast extract, 2% Bacto peptone), 0.5% glucose, 10 m*M* Tris-HCl, pH 7.5 (25 ml 2.8 *M* sorbitol, 75 ml YP, 1 ml 50% glucose, and 1 ml 1 *M* Tris-HCl, pH 7.5, per 100 ml of medium) and cells walls are digested with 20 U lyticase per OD_{600} U cells (Scott and Schekman, 1980). Zymolyase (Miles) can probably be used in place of the lyticase. The cell suspension is incubated at 30°C (24°C for ts strains) until the OD_{600} of a 1 : 100 dilution in H_2O drops to <10% of the initial value. The incubation to form spheroplasts should not be longer than 20 minutes; we have had poor results with longer incubations. The rate of synthesis of at least one protein, invertase, drops dramatically under spheroplasting conditions (Manfred Schleyer, unpublished observations). To allow protein synthesis to resume, spheroplasts are harvested by centrifugation, resuspended at 6 OD_{600} U/ml in 0.7 *M* sorbitol, 0.75 × YP, 1% glucose, and incubated with gentle shaking for 20 minutes at 30°C (24°C for ts strains).

2. Freezing Conditions

Two variables are important: the freezing rate and the buffer composition. In the absence of sophisticated equipment to control the cooling rate

(Grout and Morris, 1987), we have used simple but reproducible methods. Rapid, intermediate, and slow rates of freezing are achieved by direct immersion in liquid N_2, suspension over liquid N_2 vapor, and placement in an insulated box in a −85°C freezer, respectively. The lysis buffer contains 400 m*M* sorbitol, 150 m*M* KOAc, 2 m*M* MgOAc, 0.5 m*M* EGTA, and 20 m*M* HEPES, pH 6.8. The Mg^{2+} is required to stabilize nuclei, the pH and salt concentrations approximate physiological conditions, and the EGTA chelates any Ca^{2+} released during lysis. Higher sorbitol concentrations reduce the lysis efficiency and lower concentrations lead to osmotic lysis.

Biosynthetically active spheroplasts are harvested by centrifugation (1000 *g*, 5 minutes 4°C), washed, and resuspended at 300 OD_{600} U/ml in lysis buffer at 4°C. Aliquots of the spheroplast suspension are transferred to 1.5-ml Eppendorf microcentrifuge tubes and frozen in lqiuid N_2 vapor (see later). Smaller aliquots freeze more rapidly than large aliquots, and the degree of lysis can be controlled through the aliquot size. Aliquots of 200 μl are used to prepare membranes for the ER–Golgi transport assay; smaller aliquots (50–100 μl) give more efficient lysis but lower transport efficiency.

The following procedure is used to freeze spheroplasts in liquid N_2 vapor: A Styrofoam ice bucket is filled with liquid N_2 to a height of ~6 cm. A cardboard freezer box divider is suspended at a level even with the top of the bucket with wire attached to each corner, ~10 cm above the liquid N_2. Eppendorf tubes containing 50- to 200-μl portions of spheroplast suspension are placed in the cardboard rack, and the ice bucket is covered and then sealed with tape. After 40 minutes the tubes are transferred to a −85°C freezer. Membranes have been stored for 2 months at −85°C without loss of activity. Tubes are thawed by immersion in a 25°C water bath with shaking and then placed on ice.

D. Characterization of Freeze–Thaw Lysates

1. Phase-Contrast Microscopy

The extent of lysis can be easily monitored by phase-contrast microscoy. Intact spheroplasts have a characteristic bright halo resulting from the sharp difference in refractive index between spheroplast and surrounding buffer (Fig. 1A). Freeze-thawing produces ghosts which can be differentiated from intact spheroplasts by the absence of the halo, presumably because of loss of cytoplasmic contents (Fig. 1C,E). Ghosts with similar appearance are produced when spheroplasts are diluted into low-osmotic support buffers.

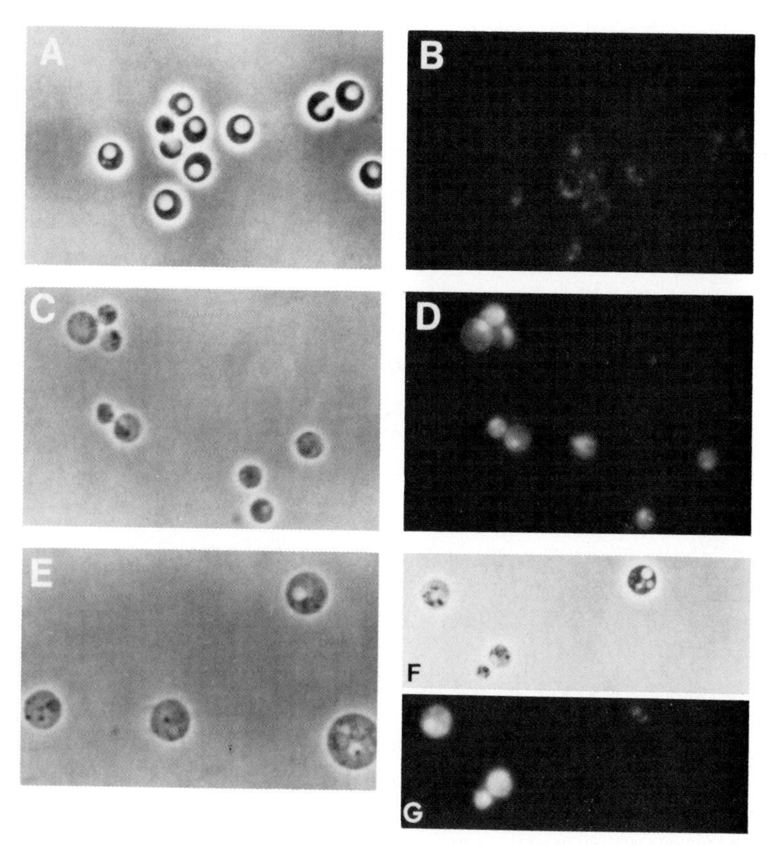

FIG. 1. Morphology of broken spheroplasts as revealed by light microscopy. Spheroplass before (A,B) and after (C–G) freezing in liquid N_2 vapor were suspended in lysis batter containing 20 μg/ml DAPI and examined using phase-contrast (A,C,E,F) or fluorescence (B,D,G) microscopy. The exposure time was 6 seconds in B and 3 seconds in (D) and (G). (A,B,C,D,F,G) ×400 (E) ×630.

2. DAPI STAINING

The integrity of the nuclei and the permeability of the plasma membrane can be quickly assessed using the DNA-binding fluorescent dye 4,6-diamidino-2-phenylindole (DAPI) (Sigma). Addition of 20 μg/ml DAPI to unfixed, intact spheroplasts gives rise to the faint staining pattern seen in Fig. 1B. Freeze–thaw ghosts stained under identical conditions produce a much stronger signal (Fig. 1D; the exposure time in Fig. 1B is twice that of Fig. 1D). A field containing ghosts and an intact spheroplast is shown in Fig. 1F and G for direct comparison of DAPI-staining intensity. The increase in staining intensity is most likely due to permeabilization of the plasma membrane and not the nuclear envelope: invertase accumulated within the nuclear envelope/ER in a *sec18* mutant is not released during freeze-thawing, and the nuclear envelope appears intact by electron microscopy (see later).

3. Marker Enzyme Fractionation

The degree of permeabilization of the plasma membrane and the ER can be assessed by monitoring the release of the cytoplasmic and secreted forms of invertase. In a *sec18* mutant at the nonpermissive temperature, core-glycosylated invertase accumulates in the ER (Esmon *et al.*, 1981). *sec18* Spheroplasts were prepared as described previously and shifted to 0.7 *M* sorbitol, 0.75 × YP, 0.1% glucose for 20 minutes at 24°C to derepress invertase synthesis and then to 37°C to accumulate core-glycosylated invertase in the ER. The spheroplasts were then suspended in lysis buffer and frozen at different rates. After thawing, lysates were centrifuged for 10 seconds at 12,000 *g* in a Fisher microcentrifuge and the pellet and supernatant fractions analyzed by native gel electrophoresis followed by an invertase activity stain. Freezing over liquid N_2 vapor caused release of cytosolic but little core-glycosylated invertase from the sedimentable to the supernatant fractions (Fig. 2, lanes 2P, 2S). Both forms of invertase were sedimentable in samples containing spheroplasts before freezing (lanes 1P, 1S) or spheroplasts frozen slowly in an insulated freezer box (lanes 3P, 3S). Approximately 40% of the core-glycosylated invertase was released into the supernatant following rapid freezing in liquid N_2 (data not shown). Thus slow freezing rates leave spheroplasts intact, intermediate freezing rates permeabilize the plasma membrane but not the ER, and rapid freezing rates permeabilize both the plasma membrane and the ER. Spheroplasts remain viable when frozen sufficiently slowly: invertase accumulated in the ER was quantitatively secreted when spheroplasts frozen in an insulated box were thawed and incubated in YPD plus 0.7 *M* sorbitol (data not shown). The extent of lysis of spheroplasts and vacuoles during preparation of wild-type membranes for the ER–Golgi assay was monitored by immunological detection of the cytosolic enzyme phosphoglycerate kinase (PGK) (serum provided by J. Thorner, Dept. of Biochemistry, UCB) and the vacuolar protease carboxypeptidase Y (CPY) (Stevens *et al.*, 1982). Immunoblotting was used to show that ~50% of the PGK remained soluble after 10 seconds of centrifugation in a microcentrifuge while >90% of the CPY sedimented. Vacuoles were sensitive to rapid freezing: >60% of the CPY remained soluble when spheroplasts were lysed by direct immersion in liquid N_2 (data not shown).

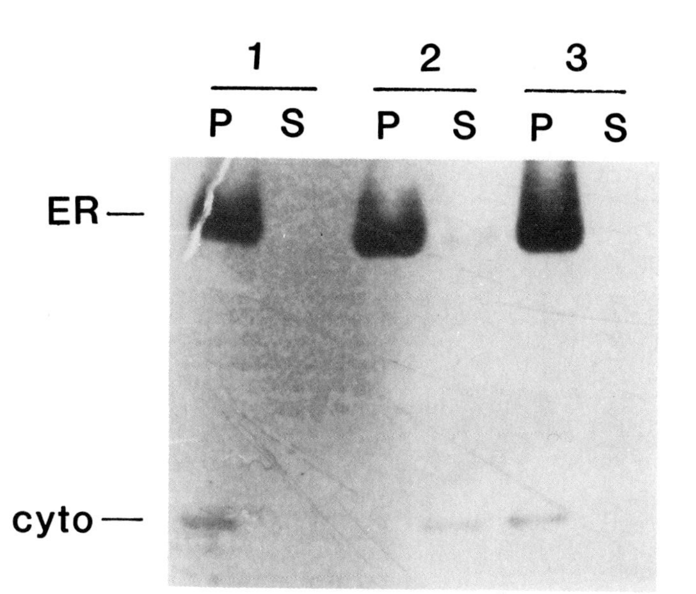

FIG. 2. Invertase fractionation. Pellet and supernatant fractions were prepared from *sec18* spheroplasts before freezing (1), after freezing in liquid N_2 vapor (2), or after freezing in an insulated box (3), as described in the text. The fractions were adjusted to Triton X-100 5 m*M* 2-[*N*-morpholino]ethane sulfonic acid (MES) (pH 6.5), and heated for 15 minutes at 50°C to dissociate invertase oligomers. Glycerol and bromphenol blue were then added to final concentrations of 10% and 0.1%, and portions were subjected to native gel electrophoresis (5 OD/U equivalents of extract per lane) through a 7% polyacrylamide gel buffered with 0.1 *M* Tris-HCl, pH 7.5. The gel was run in 120 m*M* Tris-borate (pH 7.5) for 3 hours at 10 mA. Invertase activity was detected following electrophoresis by incubating the gel first in 0.1 *M* sucrose, 0.1 *M* NaOAc (pH 5.1) for 30 minutes at 30°C, and then in 0.1% 2,3,5-triphenyltetrazolium chloride, 0.5 *M* NaOH for 4 minutes at 90°C. P, Pellet fraction; S, supernatant fraction; ER, core-glycosylated invertase; cyto, cytoplasmic invertase.

4. ELECTRON MICROSCOPY

Examples of the morphology of spheroplasts broken by freezing in 200-μl aliquots over N_2 vapor are presented in Fig. 3B–D. Much of the cytoplasm is released (compare the density of ribosomes in Fig. 3A to that in 3B, C, and D), but major organelles such as nuclei and vacuoles are largely intact. Ribosome-studded membranes, likely to be ER, are observed (see Fig. 3C above the nucleus). An example of breaks in the plasma membrane through which most of the cytosol may have escaped is indicated in Fig. 3B, upper right.

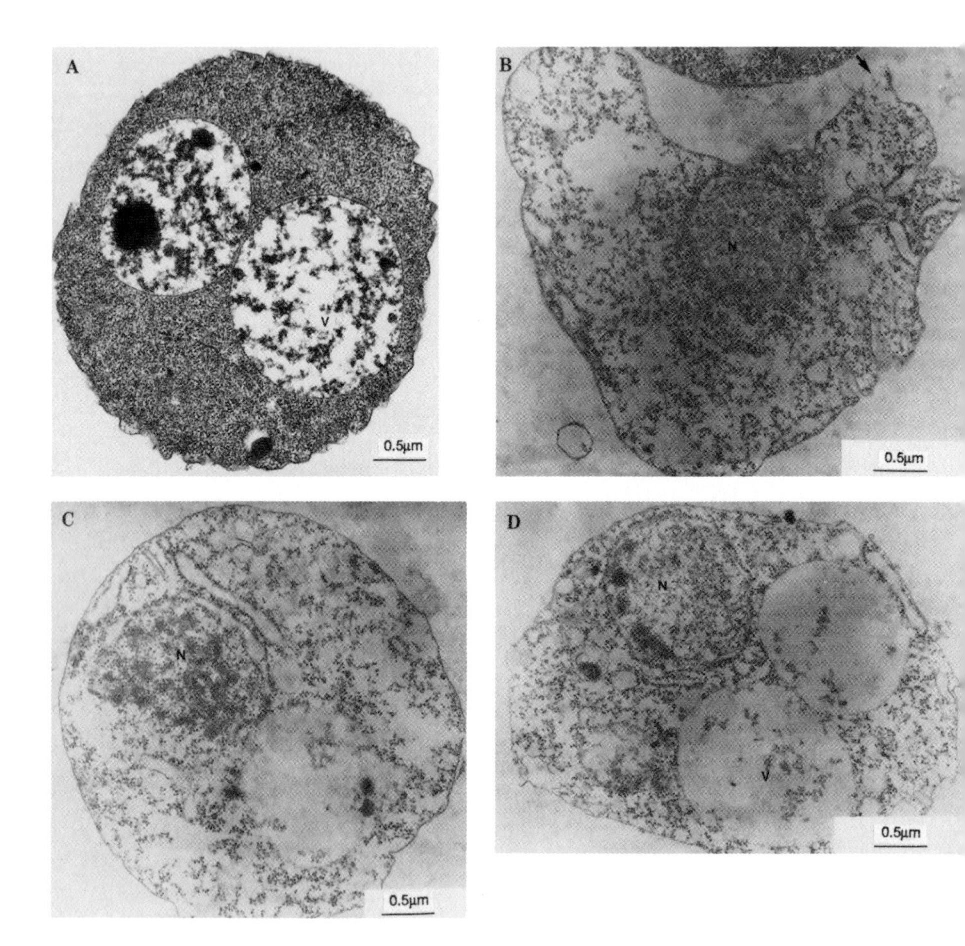

FIG. 3. Morphology of broken spheroplasts as revealed by electron microscopy. Spheroplasts were processed for electron microscopy before (A) and after (B–D) freeze-thawing. The arrow in (B) marks a break in the plasma membrane. N, Nucleus; V, vacuole. (Baker *et al.*, 1988). Reproduced with permission from CELL Press.

III. Reconstitution of Protein Transport

A. Design of ER–Golgi Transport Assay

Transport of the α-factor precursor from the ER to the Golgi apparatus has been reconstituted using freeze–thaw lysates. Transport is measured through the coupled addition of outer-chain carbohydrate to [^{35}S]methionine-labeled α-factor precursor translocated into the ER of broken spheroplasts. The *in vitro* reaction contains membranes prepared as described before, [^{35}S]methionine-labeled α-factor precursor, a yeast S100 fraction, and an ATP-regenerating system. Reaction products are detected by immunoprecipitation with an antibody specific for $\alpha 1 \rightarrow$ 6-linked mannose, which recognizes only outer-chain epitopes (Ballou, 1970; Esmon *et al.*, 1981).

B. Preparation of Translation-Competent Yeast Lysate

The translation lysate was prepared according to a standard procedure (Moldave and Gasior, 1983), except that cells were broken by agitation with glass beads (Deshaies, 1988). Protease-deficient (*pep4*) strains are grown at 30°C in 12 liters of YPD to 1.5–5.0 OD_{600} U/ml. Cells are harvested with a continuous-flow rotor (Sharples Corp., Philadelphia) and washed twice by resuspending in 250 ml distilled water [all solutions used from here on are treated with 0.1% diethylpyrocarbonate (Sigma) and autoclaved to quench RNase activity] and sedimenting for 5 minutes at 5000 rpm in a GSA rotor (Sorvall). Washed cell pellets (26–45 g wet weight) are resuspended in a minimal volume (~25 ml) of buffer A (100 m*M* potassium acetate, 2 m*M* magnesium acetate, 2 m*M* DTT, 20 m*M* HEPES, pH 7.4) supplemented with 0.5 m*M* phenylmethylsulfonyl fluoride (PMSF), and disrupted at 4°C by agitation in a 100-ml bead beater (Biospec Products, Bartlesville, OK) as described by the manufacturer. The crude lysate is centrifuged at 6500 rpm for 6 minutes in an HB-4 rotor (Sorvall) at 4°C, and the resulting supernatant fraction is centrifuged for 30 minutes at 100,000 *g* in a Ti45 rotor (Beckman) at 4°C. The clear, yellow-colored supernatant fraction (S100) is collected and subjected to chromatography at 4°C on a 100-ml Sephadex G-25 column (12 × 3.8 cm) equilibrated with buffer A plus 25% glycerol (BA-G). The column is eluted with BA-G, and 2-ml fractions are collected. Fractions with an A_{260} of >30 are pooled, diluted to a final A_{260} of 60 with BA-G, and frozen as 1-ml aliquots in liquid nitrogen. Typically, 40–50 ml of translation-competent extract are obtained from a single preparation. To degrade endogenous mRNA, thawed lysates are adjusted to 0.8 m*M* $CaCl_2$, and 500 units/ml staphylococcal nuclease S7 (Boehringer Mannheim Biochemicals) is added. Nuclease digestion is terminated after 25 minutes at 20°C, by addition of EGTA to a final concentration of 2 m*M*.

C. Preparation of α-Factor mRNA

Plasmid pDJ100, which contains the gene encoding prepro-α-factor (MFα1), was digested with *Xba*I and transcribed *in vitro* by SP6 polymerase as described (Hansen *et al.*, 1986). A large-scale transcription reaction contained, in a final volume of 800 μl: 80 μg pDJ100 converted to linear DNA with *Xba*I, 6 m*M* $MgCl_2$; 2 m*M* sperimidine, 0.5 m*M* ATP, 0.5 m*M* CTP, 0.5 m*M* UTP, 0.1 m*M* GTP, 0.5 m*M* GpppG (New England Biolabs), 10 m*M* DTT, 1000 units/ml RNAsin (Amersham), and 500 U/ml of SP6 RNA polymerase. The reaction was incubated for 60 minutes at 40°C, and the nucleic acid was isolated by ethanol precipitation.

The reaction yielded ~1 mg of mRNA, which was either dissolved in H_2O for immediate use or stored as a precipitate in EtOH at −20°C.

D. Preparation of [^{35}S]Methionine-Labeled Prepro-α-factor

[^{35}S]Methionine-labeled prepro-α-factor sufficient for >80 experiments can be prepared in a large-scale translation reaction. The reaction contains, in a final volume of 5.4 ml: 30 m*M* HEPES–KOH (pH 7.4), 160 m*M* KOAc, 2.8 m*M* $Mg(OAc)_2$, 25 m*M* creatine phosphate (CP) (Sigma), 200 μg/ml creatine phosphokinase (CPK) (Boehringer Mannheim), 0.5 m*M* ATP, 100 μ*M* GTP, 3 m*M* DTT, 40 μ*M* each amino acid (except methionine, 0.4 μ*M*), [^{35}S]methionine (2.7 mCi) (Amersham), 500 U/ml RNAsin (Amersham), 1.8 ml micrococcal nuclease-treated yeast S100 (60 OD_{200} U/ml), and 0.5 mg pDJ100 mRNA. The translation mixture is incubated for 40 minutes at 20°C and desalted by gel filtration on a 30-ml Sephadex G-25 column equilibrated in reaction buffer [20 m*M* HEPES (pH 6.8), 150 m*M* KOAc, 250 m*M* sorbitol, 5 m*M* MgOAc]. The peak of trichloroacetic acid (TCA)-precipitable cpm coincides with the first peak of total cpm and is well separated from the second, larger peak of total cpm (unincorporated [^{35}S]methionine). Fractions in the first peak are pooled and aliquots frozen in liquid N_2 and stored at −85°C. The pool contains 1–2.5 × 10^5 cpm/μl, of which 40–60% are TCA-precipitable. No significant loss of activity has been detected over 1 month of storage.

E. Preparation of Cytosol

The *in vitro* transport reaction requires supplementation with a yeast cytosol fraction containing ~20 mg/ml protein in reaction buffer. Protein concentration is determined using the Bradford protein assay with BSA as a standard (Bradford, 1976). The method of preparation may not be critical; we have used glass bead breakage of whole cells. For small-scale prepartions, 4000 OD_{600} U of cells are harvested by centrifugation (5 minutes, 3000 *g*) and washed twice in reaction buffer at 4°C. Cells are resuspended in 2 ml of reaction buffer containing 1 m*M* DTT and 0.5 m*M* PMSF in a 30-ml corex tube. Glass beads (4 g; 0.5 mm; Biospec) are added and the cells are lysed by ten 30-second periods of agitation on a VWR Vortexer 2 at full speed, spaced by 30-second intervals on ice. The extent of lysis can be assessed quickly by phase-contrast microscopy: broken cells lose the refractile appearance and halo of intact cells. To obtain a cytosol fraction of 20 mg/ml protein, >50% of the cells should be broken. The homogenate is clarified by centrifugation at 3000 *g* for 5 minutes and the supernatant fraction further centrifuged at 100,000 *g* for 30 minutes. The resultant S100 fraction is frozen in aliquots in liquid N_2. The preparation can be scaled up 10-fold with proportional increases in volumes, except that cells are broken in a 100-ml bead beater chamber. For study of the roles of small molecules in transport, the cytosol fraction may be desalted by gel filtration on a Sephadex G-25 (coarse) column equilibrated with reaction buffer.

F. *In Vitro* Transport Reaction

For each experiment an aliquot of membranes is thawed by immersion in a 25°C water bath and the required amount of membranes (20 μl per reaction) washed three times with 1 ml each reaction buffer by brief (~10 seconds) centrifugation in a Fisher microcentrifuge and resuspension in 1 ml of reaction buffer at 4°C.

Coupled translocation–transport reactions contain 5 μl of prepro-α-factor translation product (150,000 TCA-precipitable cpm), 60 μg additional cytosol, 50 μM GDP-mannose (Sigma), 1 mM ATP, 40 mM CP, 200 μg/ml CPK, and 20 μl (original volume) membranes washed as before and resuspended in reaction buffer to bring the reaction volume to 25 μl. The GDP-mannose, ATP, CP, and CPK are added from a 10× stock prepared in reaction buffer and stored in small aliquots at −85°C. After 45 minutes at 20°C, reactions are terminated by addition of 40 ml of Laemmli sample buffer (Laemmli, 1970) and heated for 5 minutes at 95°C.

The results of such an incubation are shown in Fig. 4. Roughly half of the prepro-α-factor was converted to the 30-kDa ER form (lane 2). About 25% of the pro-α-factor was converted to forms of heterogeneous mobility that migrated more slowly than the ER form (lane 2), and which were not present when SDS was added at the beginning of the incubation (lane 1). The species in a broad region beginning slightly above the 30-kDa ER form are precipitable with an antibody specific for $\alpha 1 \rightarrow 6$-linked mannose (lane 4).

To avoid the necessity of running SDS gels after each experiment and to obtain more quantitative data, the amount of α-factor precursor receiving carbohydrate modification can be determined by precipitation with anti $\alpha 1 \rightarrow 6$-Man serum or concanavalin A (Con A)–Sepharose 4B

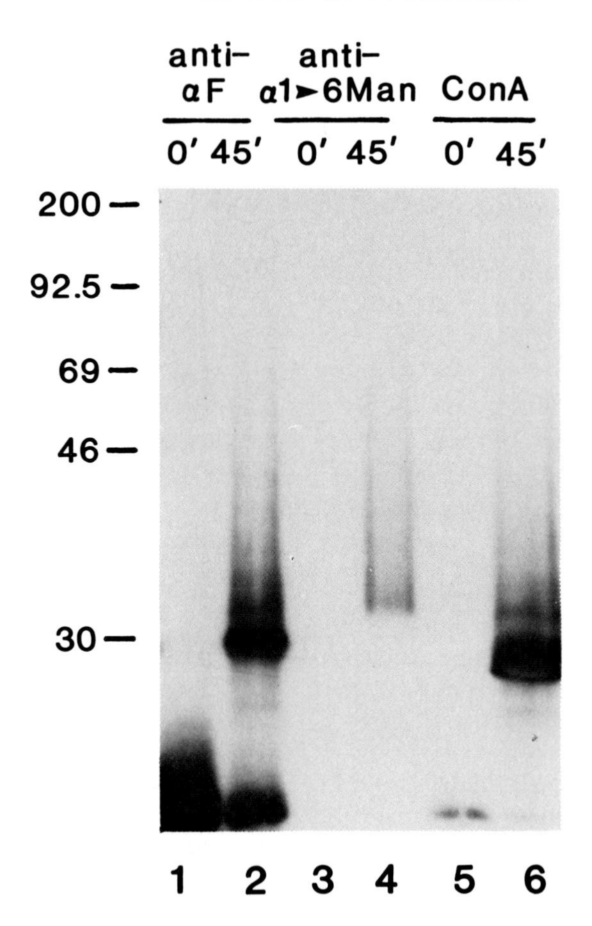

FIG. 4. Prepro-α-factor incubated with gently lysed yeast is converted to the 30-kDa ER form and to a more slowly migrating form that is immunoprecipitable with anti-α1 → 6-Man serum. Gently lysed yeast and [^{35}S]methionine-labeled prepro-α-factor were incubated together in the presence of GDP-mannose and an ATP-regenerating system. Laemmli sample buffer was added at the time indicated and the quenched reactions heated for 5 minutes at 95°C. Portions (20 μl) of each reaction were precipitated with anti-α-factor serum, anti-α1 → 6-Man serum, or Con A. The immunoprecipitates were electrophoresed on an 11.25% polyacrylamide gel. The gel was incubated with Amplify, dried, and exposed to X-ray film for 48 hours. (Baker *et al.*, 1988). Reproduced with permission from CELL Press.

(Sigma), followed by scintillation counting (for procedure, see later). Con A precipitation measures the total amount of prepro-α-factor translocated during an incubation, because Con A reacts with all glycosylated species (Fig. 4, lane 6). The ratio of anti-α1 → 6-Man precipitable pro-α-factor to Con A-precipitable pro-α-factor, the "transport efficiency," is typically ~0.25 with some variation between different preparations and experiments. The anti-$\alpha 1 \rightarrow$ 6-Man serum gives a slight underestimate of the amount of α-factor receiving outer-chain modification, as it does not precipitate the portion of heterogeneous material just above the ER form.

Two-stage reactions can be used to study transport independent of translocation. The procedure takes advantage of the fact that transport, but not translocation, is blocked at 10°C. The first stage of two-stage reactions contains 8 μl prepro-α-factor translation product, 50 μM GDP-mannose, 1 mM ATP, 40 mM CP, 200 μg/ml CPK, and 20 μl (original volume) membranes washed as before and resuspended in reaction buffer to bring the reaction volume to 25 μl. The reaction mix is incubated for 15 minutes at 10°C and the membranes washed two times by centrifugation and resuspension in reaction buffer at 4°C. A complete second-stage incubation contains, in a final volume of 25 μl: washed pro-α-factor-containing membranes, 80 μg cytosol, 1 mM ATP, 50 μM GDP-mannose, 40 mM CP, and 200 μg/ml CPK. Typically, 8–10 reactions are combined in one tube during the first stage, and the washed pro-α-factor-containing membranes are aliquoted to individual tubes with various additions at the beginning of the second-stage incubation. After 45 minutes at 20°C the reactions are terminated by addition of Laemmli sample buffer and heated for 5 minutes at 95°C.

G. Immunoprecipitation

Optimal immunoprecipitation with low background values is obtained when protein A–Sepharose, rather than Staph A cells are used to collect antibody complexes by centrifugation. A 20- to 30-μl portion of a terminated reaction is mixed with 1 ml of immunoprecipitation (IP) buffer (150 mM NaCl, 1% Triton X-100, 0.1% SDS, 15 mM Tris-HCl, pH 7.5) and 35 μl of a 20% v/v suspension of protein A–Sepharose CL-4B (Pharmacia), 8–10 μl of anti-$\alpha 1 \rightarrow$ 6-Man serum (Ballou, 1970; Esmon *et al.*, 1981), or 8 μl of anti-α-factor serum (Rothblatt and Meyer, 1986) in an Eppendorf tube and mechanically rotated for 2 hours at room temperature or overnight at 4°C. The immunoprecipitates are collected by centrifugation (a 1-second pulse in a Fisher microcentrifuge) and washed twice with 1 ml of IP buffer, twice with 1 ml of 2 M urea, 200 mM NaCl, 1% Triton X-100, 100 mM Tris-HCl (pH 7.5), once with 1 ml of 500 mM NaCl, 1% Triton X-100, 20 mM Tris-HCl (pH 7.5), and once with 1 ml of 50 mM NaCl, 10 mM Tris-HCl (pH 7.5).

The washed immunoprecipitates are analyzed either by electrophoresis and autoradiography or directly by scintillation counting. Immunoprecipitates to be analyzed by electrophoresis are heated at 95°C in Laemmli sample buffer and electrophoresed on an 11.25% SDS–polycarylamide gel (Laemmli, 1970). Gels are treated with Amplify after fixation, dried, and exposed to X-ray film. Immunoprecipitates to be analyzed by scintillation counting are heated for 5 minutes at 95°C in 150 μl of 2% SDS, then transferred to 8-ml scintillation vials. After adding 5 ml of Aquasol (New England Nuclear), the samples are counted in a scintillation counter.

For Con A precipitations, 10–15 μl of a terminated reaction are added to 1 ml of 500 m*M* NaCl, 1% Triton X-100, 20 m*M* Tris-HCl (pH 7.5), and 30 μl of a 20% v/v suspension of Con A–Sepharose (Sigma), and samples are rotated for 2 hours at room temperature or overnight at 4°C. The washes and subsequent analysis are as described for antibody precipitations.

IV. Future Prospects

Use of an *in vitro*-synthesized transport substrate translocated into the ER of spheroplasts broken by freeze-thawing provides a rapid and sensitive assay for measuring intercompartmental transport *in vitro*. Protein transport reactions other than ER–Golgi transport may be reconstituted using a similar approach. Mitochondrial precurosrs added to freeze–thaw lysates are imported efficiently into mitochondria (S. Emr, personal communication). Intra-Golgi transport may be detected through proteolytic maturation of pro-α-factor (Julius *et al.*, 1984) or acquisition of later outer-chain epitopes. Transport to the vacuole may be studied by using the precursor of the vacuolar protease carboxypeptidase Y as the *in vitro*-synthesized transport substrate. As the efficiency of *in vitro* transport reactions may be <100%, reconstitution of late stages of the secretory pathway may not be feasible using *in vitro*-translated substrates. Instead, substrates may be accumulated *in vivo* using kinetic, temperature, or mutant blocks prior to freeze–thaw lysis. Intact spheroplasts may produce a background in this type of assay, and either the freezing protocol must give quantitative lysis or unbroken spheroplasts surviving the freeze-thawing must be lysed osmotically. Use of a membrane fraction that is frozen from the beginning eliminates the necessity of preparing fresh membranes for every experiment and improves the pace of research.

Acknowledgments

We thank Ray Deshaies for the translation lysate, Carolyn Comell for expert assistance with the electron microscopy, Robin Wright for assistance with the fluorescence microscopy, Jon Rothblatt for comments on the manuscript, and Peggy McCutchan Smith for help in preparing the manuscript.

REFERENCES

Baker, D., Hicke, L., Rexach, M., Schleyer, M., and Schekman, R. (1988). *Cell (Cambridge, Mass.)* **54,** 335–344.
Ballou, C. (1970). *J. Biol. Chem.* **245,** 1197–1203.
Beckers, C. J. M., Keller, D. S., and Balch, W. E. (1987). *Cell (Cambridge, Mass.)* **50,** 523–534.
Bradford, M. M. (1976). *Anal. Biochem.* **72,** 248–254.
Deshaies, R. J. (1988). Ph.D. Thesis, University of California, Berkeley.
Esmon, B., Novick, P., and Schekman, R. (1981). *Cell (Cambridge, Mass.)* **25,** 451–460.
Grout, B. W. W., and Morris, G. J. (1987). "The Effects of Low Temperatures on Biological Systems." Edward Arnold, London.
Hansen, W., Garcia, P. D., and Walter, P. (1986). *Cell (Cambridge, Mass.)* **45,** 397–406.
Julius, D., Schekman, R., and Thorner, J. (1984). *Cell (Cambridge, Mass.)* **36,** 309–318.
Laemmli, U. K. (1970). *Nature (London)* **227,** 680–685.
Moldave, K., and Gasior, E. (1983). *In* "Methods in Enzymology" (R. Wu, L. Grossman, and K. Moldave, eds.), Vol. 101, pp. 644–650. Academic Press, New York.Rothblatt, J., and Meyer, D. I. (1986). *EMBO J.* **5,** 1031–1036.
Scott, J., and Schekman, R. (1980). *J. Bacteriol.* **142,** 414–423.
Simons, K., and Virta, H. (1987). *EMBO J.* **6,** 2241–2247.
Stevens, T., Esmon, B., and Schekman, R. (1982). *Cell (Cambridge, Mass.)* **30,** 439–448.

Chapter 8

Reconstitution of Transport from the ER to the Golgi Complex in Yeast Using Microsomes and Permeabilized Yeast Cells

HANNELE RUOHOLA,[1] ALISA KASTAN KABCENELL, AND SUSAN FERRO-NOVICK

Department of Cell Biology
Yale University School of Medicine
New Haven, Connecticut 06510

I. Introduction

A detailed dissection of the events involved in the transit of proteins from the endoplasmic reticulum (ER) to the Golgi complex in yeast

[1] Present address: Department of Biochemistry, University of Helsinki, Finland.

requires the development of an assay that faithfully reproduces these steps *in vitro*. To undertake this task we have developed a new assay that utilizes translocation into microsomes or the ER of permeabilized yeast cells (PYC) to reconstitute transport to the Golgi complex. We have chosen the yeast *Saccharomyces cerevisiae* to develop this system because it is a simple eukaryote with a secretory pathway that is analogous to that present in mammalian cells yet is accessible to biochemical and genetic manipulation.

Classical genetic studies have resulted in the isolation of a collection of mutants that block the exit of proteins from the lumen of the ER to the Golgi apparatus. These studies have shown that mutations in 11 genes (SEC or BET) can affect transit at this stage of the pathway (Novick *et al.,* 1980; Newman and Ferro-Novick, 1987). The *bet* and *sec* mutants are temperature-sensitive for growth and protein transport. At 37°C, core-glycosylated precurosrs to secreted, vacuolar, and membrane proteins accumulate in an extensive network of ER. Molecular genetic approaches have permitted the identification of several SEC and BET gene products. The isolation of mutants that disrupt transit at this stage of the pathway and the availability of antibodies to the Sec and Bet proteins will provide powerful tools for analyzing the transit of proteins from the ER to the Golgi complex.

To develop an *in vitro* assay that can reconstitute transport from the ER to the Golgi complex in yeast, we have used the secreted pheromone α-factor as a marker protein in our assay. This protein is initially synthesized as a 19-kD preproprotein. In the lumen of the ER the signal sequence of prepro-α-factor is cleaved and three N-linked oligosaccharide units are added, yielding a proprotein of 26 kD. These oligosaccharide chains are then extended in the Golgi complex, resulting in a highly branched structure consisting of 50–150 mannose residues. Addition of this outer-chain carbohydrate results in the high molecular weight or Golgi form of α-factor, which migrates as a heterogeneous smear on SDS–polyacrylamide gels. Proteolytic processing of this proprotein is thought to begin in the Golgi complex. The mature form of α-factor, a 13-amino acid peptide, is ultimately secreted into the medium (Julius *et al.,* 1984).

The transit of prepro-α-factor to the Golgi apparatus *in vitro* is assayed in two stages (Ruohola *et al.,* 1988). In the first stage of the reaction, ~50–70% of the preproprotein synthesized in a yeast translation lysate is translocated into a donor compartment retained within PYC. In the second stage of the reaction, the 26-kD ER form of α-factor is transported to the Golgi complex. Transit to the Golgi apparatus is dependent on the presence of ATP and a 3000 *g* supernatant (S3 fraction) of a yeast lysate

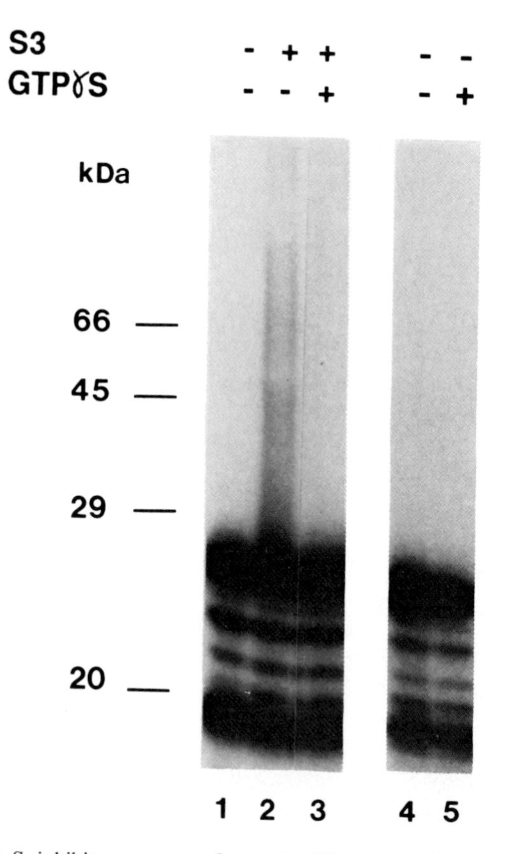

FIG. 1. GTPγS inhibits transport from the ER to the Golgi complex. Wild-type permeabilized cells, containing the ER form of α-factor, were incubated in the absence (lane 1) or presence (lane 2) of an S3 fraction. In the presence of 1×10^{-4} *M* GTPγS (lane 3), transport was inhibited. Prepro-α-factor was translocated *in vitro* in the absence (lane 4) and presence (lane 5) of 1×10^{-4} *M* GTPγS. Reproduced from the *Journal of Cell Biology*, 1988, **107**, 1465–1476, by copyright permission of the Rockefeller University Press.

(Fig. 1, compare lanes 1 and 2). This supernatant contains soluble and membrane-bound factors that include the acceptor compartment.

The assay we developed has certain advantages. First, the function of the donor, acceptor, and soluble fractions can be assayed separately. The donor compartment is provided by PYC, whereas the soluble factors and acceptor Golgi compartment is supplied by the S3 fraction. Although we have shown that permeabilized cells serve as an efficient donor in this reaction, they are not essential because yeast microsomes can also perform the same function. Second, transport from the ER to the Golgi complex is very efficient; generally 48–66% of the ER form of α-factor is converted to the Golgi form.

The assay we developed consists of three sequential steps:

1. *In vitro* translation of prepro-α-factor
2. *In vitro* translocation of prepro-α-factor into the ER lumen of PYC or yeast microsomes
3. *In vitro* transport of pro-α-factor from the ER to the Golgi complex

We have described each step in more detail in Sections II–IV.

II. *In Vitro* Translation of Prepro-α-factor in a Yeast Lysate

A. *In Vitro* Transcriptions of the Gene Encoding Prepro-α-factor

In vitro transcription of prepro-α-factor and the preparation of yeast translation lysates were performed as described before (Hansen *et al.*, 1986), with slight modifications. To transcribe the gene encoding prepro-α-factor, plasmid pDJ100 (obtained from D. Julius and described in Hansen *et al.*, 1986) was linearized with *Xba*I (Boehringer Mannheim) and transcribed with SP6 RNA polymerase as described before (Hansen *et al.*, 1986). The RNA obtained from transcribing 30 μg of DNA was resuspended in 100 μl of RNase-free water and used to translate prepro-α-factor *in vitro*.

B. Preparation of Translation Lysates

For the preparation of yeast translation lysates, cells were grown overnight in YP medium (2% Bacto peptone, 1% yeast extract) containing 2% glucose to an A_{599} = 2.0. Approximately 4000 A_{599} units of cells were then converted to spheroplasts during a 30-minute incubation at 37°C in 20 ml of spheroplasting medium I [1.4 *M* sorbitol, 0.4% β-mercaptoethanol, 30 mg of Zymolyase 100T (ICN Immunobiologicals) and one-half volume of 0.1 *M* potassium phosphate, pH 7.4]. The spheroplasts formed during this incubation were pelleted, resuspended in 200 ml of regeneration medium [YM-5 (Hartwell, 1967) with 1.4 *M* sorbitol], and incubated at 37°C for 1 hour. Regenerated spheroplasts were washed with 20 ml of 1.4 *M* sorbitol and then lysed in 8 ml of lysis buffer I [100 m*M* potassium acetate, 2 m*M* magnesium acetate, 2 m*M* dithiothreitol (DTT), 0.5 mM phenylmethylsulfonylfluoride (PMSF), 2× protease inhibitor cocktail (Pic, described by Waters and Blobel, 1986) and 1/50 volume of 1 *M* HEPES–KOH, pH 7.4] by douncing (40 strokes) the sample in a motor-driven Potter homogenizer. The lysate was centrifuged at 27,000 *g* for 15 minutes at 4°C in a Ti50 rotor, and the supernatant was recentrifuged at 100,000 *g* for 30 minutes. The final supernatant was desalted on a Sephadex G-25 column and fractions with an A_{260} >0.3 were pooled and treated with staphylococcal nuclease (150 U/ml) for 20 minutes at 20°C in the presence of 1 m*M* $CaCl_2$ and 1× Pic. The reaction was terminated by the addition of EGTA (4 m*M* final concentration). Samples were flash-frozen in liquid nitrogen and stored in −80°C.

C. *In Vitro* Translation of Prepro-α-factor

Prepro-α-factor was translated in a reaction volume of 25 μl. The final concentration of components was as follows: 10 μl of translation lysate, 20 m*M* HEPES–KOH (pH 7.4), 150 m*M* potassium acetate, 3 m*M* magnesium acetate, 2.8 m*M* DTT, 1 m*M* ATP, 0.1 m*M* GTP, 20 m*M* creatine phosphate (CP), 0.2 mg/ml creatine phosphate kinase (CPK), 30 μ*M* of each amino acid except methionine, 20 μCi of [^{35}S]methionine (Amersham, 1000 Ci/mmol), 0.2 mg/ml yeast tRNA (Boehringer Mannheim), 10 U RNAsin (Promega, 40,000 U/ml), 8% glycerol, 75 m*M* sucrose, 1× Pic, and 3 μl of RNA (1 : 10 dilution of the preparation described before). The reaction was allowed to proceed for 60 minutes at 20°C.

III. *In Vitro* Translocation into the ER Lumen of PYC or Microsomes

A. Preparation of PYC

Permeabilized yeast cells (PYC) were prepared by removing the yeast cell wall enzymatically and by osmotically shocking spheroplasts. Cells were grown overnight at 24°C to early log phase and then harvested by centrifugation in a clinical tabletop centrifuge at room temperature. The pellet was resuspended in YP medium containing 0.1% glucose (A_{599} = 1.5), and the resuspended cells were incubated at 24°C for 30 minutes. Following this incubation the cells (150 A_{599} units) were pelleted again and resuspended in 50 ml of spheroplasting medium II [YP medium, 0.1% glucose, 50 m*M* potassium phosphate, 1.4 *M* sorbitol, 50 m*M* β-mercaptoethanol (pH 7.5), and 1 mg Zymolyase 100T]. Spheroplasts formed during a 30-minute incubation at 37°C (or 90 minutes at 24°C for mutant-permissive experiments) were resuspended in 100 ml of regeneration medium (YP medium, 0.1% glucose, and 1 *M* sorbitol) and incubated for an additional 90 minutes at 37°C (or 90 minutes at 24°C for mutant-permissive experiments). If the incubations just described were performed in medium containing 2% glucose, the regenerated spheroplasts did not lyse well. We observed that it was important to maintain the same ratio of cells to volume during these incubations in order to prevent glucose starvation of cells. Regenerated spheroplasts formed during this time were pelleted and resuspended in 10 ml of permeabilization buffer (0.1 *M* potassium acetate, 0.2 *M* sorbitol, 2 m*M* magensium chloride, and 1 : 50 volume of 1 *M* HEPES–KOH, pH 7.2) by pipeting the cells up and down five times. The permeabilized cells were pelleted at 3000 *g* for 5 minutes, and the pellet was resuspended in 50 μl of translocation buffer (250 m*M* sucrose, 2 m*M* DTT, 1 m*M* EGTA, and 1/50 volume of 1 *M* HEPES–KOH, pH 7.4) supplemented with 1× Pic. The protein concentration of the cells (19 mg/ml) was measured by the method of Bradford (1976) using ovalbumin as a standard.

B. Morphological Analysis of PYC

To assess the degree of permeabilization morphologically, thin sections of PYC were compared to spheroplasts. Permeabilized yeast cells were incubated in fixation buffer (1% paraformaldehyde, 0.8% glutaraldehyde, 0.1 *M* potassium acetate, 0.2 *M* sorbitol, 2 m*M* magnesium chloride, and 1 : 10 volume of 1.5 *M* cacodylate buffer, pH 7.2) for 3.5 hours, while spheroplasts were incubated in fixation buffer containing 1 *M* sorbitol. The samples were then washed three times with 0.15 *M* cacodylate buffer (pH 7.2) and postfixed for 1 hour with 1% osmium tetroxide, 0.1 *M* cacodylate buffer (pH 7.2) at 4°C. Samples were washed three times with 0.15 *M* sodium chloride and stained for 2 hours at 4°C with uranyl acetate in 0.5% sodium chloride. After washing the samples with 0.15 *M* sodium chloride, they were resuspended in 2% Bacto-agar and formed into blocks. The blocks were dehydrated with ethanol and embedded in Spurr medium (Polysciences, Inc.). Thin sections were stained with uranyl acetate and lead citrate and viewed in a Phillips 301 electron microscope at 80 kV.

Electron-microscopic analysis of thin sections revealed that the protocol we employed depleted the cells of cytoplasmic contents; however, nuclear and vacuolar contents appered to be retained (compare Fig. 2C with Fig. 2A and B). A biochemical assessment of these cell preparations was consistent with morphological analysis and indicated that most of the hexokinase B, a cytoplasmic enzyme, was released from the cells; however, the majority of the vacuolar constituent, carboxypeptidase Y, was retained.

C. Preparation of Yeast Microsomes

Microsomes used as the donor compartment in the transport assay were prepared as described earlier (Hansen *et al.*, 1986), with modifications. Cells were grown and spheroplasted (using 20 mg of Zymolyase 100T) as described earlier for the preparations of translation lysates. Spheroplasts generated from approximately 4000 A_{599} units were centrifuged at 5000 *g* for 10 minutes through a 5-ml sorbitol cushion (1.7 *M* sorbitol and 1/50 volume of 1 *M* potassium phosphate, pH 7.4) in an SS-34 rotor. The pellet was resuspended in 5 ml of lysis buffer II (500 m*M* sucrose, 1 m*M* EDTA, 2 m*M* DTT, 3 m*M* magnesium acetate, 0.5 m*M* PMSF, 2× Pic, and 1 : 50 volume of 1 *M* HEPES–KOH, pH 7.4) by douncing the sample 10 times in a Wheaton dounce. The lysate was centrifuged for 10 minutes at 15,900 *g* in an SS-34 rotor and the supernatant was saved (supernatant I). The pellet was resuspended again in 5 ml of lysis buffer II, dounced 10 times more and centrifuged for 10 minutes at 15,900 *g*. The supernatant (supernatant II) was combined with supernatant I and recentrifuged for 10 minutes at 15,900 *g* for 10 minutes. The final supernatant (supernatant III) was centrifuged onto a sucrose cushion (2.0 *M* sucrose, 2 m*M* DTT, 2× Pic, and 1 : 50 volume of 1 *M* HEPES–KOH, pH 7.4) using a Ti50 rotor at 100,000 *g* for 1 hour. The microsomes at the interface were collected, treated with staphylococcal nuclease as described before (Hansen *et al.*, 1986), and washed by passage through a Sephacryl S-500 column (Sigma) as described earlier (Walter and Blobel, 1983).

D. Translocation into PYC and Microsomes

Prepro-α-factor was translocated into the ER lumen of PYC (60 μg of protein) or microsomes (16 μg of protein) in the presence of a yeast translation lysate supplemented with 20 m*M* CP, during a 20-minute incubation at 20°C. At the end of the incubation, the translocation reaction was terminated by chilling samples to 4°C. The amount of prepro-α-factor translocated was estimated by densitometric scanning of light exposures of autoradiograms. The percentage of prepro-α-factor translocated was defined as the amount of 19-kDa primary translation product converted to the 26-kDa ER form of α-factor. Routinely 50–70% of the prepro-α-factor was translocated.

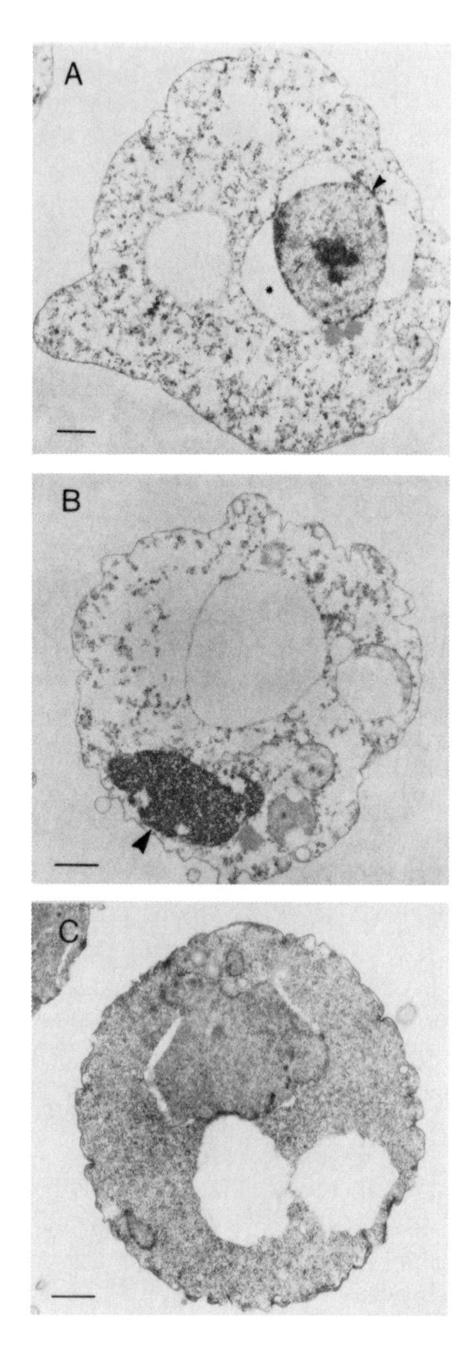
A
B
C

IV. *In Vitro* Transport of Pro-α-Factor from the ER to the Golgi Complex

A. Preparation of S3, HSS, and HSP Fractions

In vitro transport from the ER to the Golgi complex in yeast requires the presence of an S3 fraction. This S3 fraction can be subfractionated into a high-speed superantant (HSS) and high-speed pellet (HSP) fraction. Neither the HSS or HSP alone can support transport; however, transport is fully restored when these fractions are combined. To prepare the S3 fraction cells were grown, spheroplasted, and regenerated as described for the preparation of permeabilized cells. Regenerated spheroplasts prepared from 150 OD_{599} U of cells were lysed with 412 μl of 20 m*M* HEPES–KOH, pH 7.2, and centrifuged at 3000 *g* for 5 minutes. The supernatant, referred to as the S3 fraction, was supplemented with 1× Pic. The HSS and HSP pellet fractions were prepared from an S3 fraction centrifuged at 100,000 *g* for 1 hour, and the pellet was resuspended in an equal volume of 20 m*M* HEPES–KOH, pH 7.2. The protein concentration of each fraction was measured using the method of Bradford, S3 fraction (14 mg/ml), HSS (8 mg/ml), and HSP (6 mg/ml).

B. *In Vitro* Transport

Permeabilized cells or microsomes, containing radiolabeled pro-α-factor in the lumen of the ER, were washed (25 μl per reaction) with transport buffer (115 m*M* potassium acetate, 2.5 m*M* magnesium chloride, 0.2 *M* sorbitol, 1× Pic, and 1 : 50 volume of 1 *M* HEPES–KOH, pH 7.2) and then resuspended in 25 μl of the same buffer. Prior to the assay, concentrated HEPES buffer, potassium acetate, magnesium chloride, and sorbitol were added to the S3, HSS, and HSP fractions. The final concentration of buffer in each fraction was 115 m*M* potassium acetate, 2.5 m*M* magnesium chloride, 0.2 *M* sorbitol, 1× Pic, and 1 : 28 volume of 1 *M* HEPES–KOH, pH 7.2. We have observed that transport was more efficient if potassium acetate was used as the salt in the transport buffer instead of potassium chloride.

FIG. 2. Morphology of permeabilized yeast cells and spheroplasts. (A), (B) Permeabilized cells. (C) Spheroplasts for comparison. The asterisk marks the lumen of the nuclear envelope; the small arrow denotes a nucleus, and the large arrow points to a vacuole. Bar = 0.5 μm. Reproduced from the *Journal of Cell Biology*, 1988, **107,** 1465–1476, by copyright permission of the Rockefeller University Press.

The transport reaction (~82–136 μl total volume) contained 25 μl of permeabilized cells (60 μg of protein), 50 μl of an S3 fraction (0.7–1.0 mg of protein), 0.92 m*M* GDP-mannose (Sigma), and an ATP-regenerating system (1 m*M* ATP, 0.1 m*M* GTP, 20 m*M* CP, 0.2 mg/ml CPK). Samples were incubated at 20°C for 90 minutes and the reaction was terminated by chilling the samples to 4°C. After the reaction, samples were boiled in 1% SDS, diluted 50- to 100-fold with 1 ml of dilution buffer [0.2 *M* sodium chloride, 2% Triton X-100, 12.5 m*M* potassium phosphate (pH 7.6)], plus aprotinin (100 units/ml) and centrifuged for 15 minutes in an Eppendorf microfuge at 4°C. The supernatant was removed and incubated with 2 μl of anti-α-factor antibody for 12–18 hours. Samples were washed as described before (Newmann and Ferro-Novick, 1987), and solubilized immunoprecipitates were electrophoresed in a 12.5% SDS–polyacrylamide slab gel and fluorographed (Chamberlain, 1979). The efficiency of transport was estimated from the decrease in the ER form of α-factor and concomitant increase in the high molecular weight form. This was determined by densitometric scanning of lightly exposed autoradiograms or by excising and solubilizing protein bands from gels and liquid scintillation counting (Ruohola *et al.*, 1988). The increase in counts in the Golgi form of α-factor was largely accounted for by the decrease in counts in the ER form of α-factor.

V. Summary

We have developed a highly efficient *in vitro*-transport assay that couples translocation across the ER membrane and transport to the Golgi complex using the secreted pheromone α-factor as a marker protein. Radiolabeled prepro-α-factor of high specific radioactivity is obtained by *in vitro*-translating this protein in a yeast lysate. Prepro-α-factor synthesized *in vitro* is then translocated directly into microsomes or the ER of permeabilized yeast cells. Conversion of the 26-kDa ER form of pro-α-factor to the high molecular weight Golgi form is dependent on the presence of ATP and soluble and membrane-bound factors. Differential centrifugation and fractionation on a sucrose gradient have shown that the ER and Golgi forms of α-factor are enriched in separate compartments after the transport reaction. These and other findings (see Ruohola *et al.*, 1988, for a more complete discussion) indicate that conversion to the high molecular weight form of α-factor is the result of authentic intercompartmental transport.

Permeabilized mammalian cells have been used to reconstitute transport from the ER to the Golgi complex. In these systems (Becker *et al.*, 1987; Simons and Virta, 1987), a viral membrane glycoprotein protein (vesicular stomatitis virus G protein) is used as the marker protein. This protein is radiolabeled with [^{35}S]methionine during virus infection, either before or after the cells are permeabilized. Radiolabeled G protein, residing in the ER, is then transported to the Golgi complex in the presence of an ATP-regenerating system. In the mammalian system the donor and acceptor compartments are retained within the permeabilized cells (Simons and Virta, 1987); however, on occasion the addition of an exogenous acceptor compartment is required (Beckers *et al.*, 1987). The assay we developed (Ruohola *et al.*, 1988) differs from the mammalian assay (Beckers *et al.*, 1987) in that we introduce radiolabeled marker protein into the ER *in vitro* during translocation rather than during virus infection. In addition, in our assay the acceptor Golgi compartment is always provided exogenously to the permeabilized cells. Therefore, if acceptor membranes are present in the PYC, they are not utilized. Because the permeabilized cells and the S3 fraction are prepared differently, the conditions used to prepare the cells may lead to inactivation or loss of the acceptor compartment.

The *in vitro* assay will enable us to purify components involved in transporting proteins from the lumen of the ER to the Golgi complex. Antibody prepared to purified components can be used to clone the genes that code for these proteins. To study the *in vivo* role of biochemically identified components, the isolated genes can be mutagenized *in vitro* and reintroduced into the yeast genome. The *in vitro* assay we developed has other advantages. Various drugs and analogs that cannot penetrate whole cells can now be tested in the assay to determine their effect on intracellular protein transport. For example, we have recently shown that the nonhydrolyzable GTP analog GTPγS inhibits transport from the ER to the Golgi complex in yeast (Ruohola *et al.*, 1988). This finding implicates the involvement of a GTP-binding protein at this stage of the pathway. Translocation, the first stage of our two-stage reaction, is not affected by GTPγS (Fig. 1, compare lanes 4 and 5).

Because the function of donor, acceptor, and soluble fractions can be assayed separately in our *in vitro* assay, we have been able to determine the defective fraction in a yeast secretory mutant that blocks transport from the ER to the Golgi complex *in vivo* (Ruohola *et al.*, 1988). The analysis we did with this mutant can be extended to the other *sec* and *bet* mutants that block transport from the ER to the Golgi complex in yeast. This will enable us to determine the defective fraction in each of these mutants and should serve as a first step in dissecting their molecular lesion.

Acknowledgments

We thank Peter Novick for suggestions regarding the permeabilization of yeast cells. This work was supported by a grant from the National Institutes of Health (GM 35421) to S.F-N.H.R. is a pre-doctoral fellow of the Nordic Yeast Research Program and A.K.K. is a post-doctoral fellow of the Jane Coffin Childs Memorial Fund.

References

Beckers, C. J. M., Keller, D. S., and Balch, W. (1987). *Cell (Cambridge, Mass.)* **50,** 523–534.

Bradford, M. M. (1976). *Anal. Biochem.* **73,** 248–254.

Chamberlain, J. P. (1979). *Anal. Biochem.* **98,** 132–135.

Hansen, W., Garcia, P. D., and Walter, P. (1986). *Cell (Cambridge, Mass.)* **45,** 397–406.

Hartwell, L. (1967). *J. Bacteriol.* **93,** 1662–1670.

Julius, D., Brake, A., Blair, L., Kunisawa, R., and Thorner, J. (1984). *Cell (Cambridge, Mass.)* **37,** 1075–1089.

Newman, A. P., and Ferro-Novick, S. (1987). *J. Cell Biol.* **105,** 1587–1594.

Novick, P., Field, C., and Schekman, R. (1980). *Cell (Cambridge, Mass.)* **21,** 205–215.

Ruohola, H., Kabcenell, A. K., and Ferro-Novick, S. (1988). *J. Cell Biol.* **107,** 1465–1476.

Simons, K., and Virta, H. (1987). *EMBO J.* **6,** 2241–2247.

Walter, P., and Blobel, G. (1983). *In* "Methods in Enzymology" (S. Fleischer and B. Fleischer, eds.), Vol. 96, pp. 84–93. Academic Press, New York.

Waters, M. G., and Blobel, G. (1986). *J. Cell Biol.* **102,** 1543–1550.

Chapter 9

Delivery of Macromolecules into Cells Expressing a Viral Membrane Fusion Protein

HARMA ELLENS[1]

Department of Pharmacology
University of California
San Francisco, California 94143

STEPHEN DOXSEY AND JEFFREY S. GLENN

Department of Biochemistry and Biophysics
University of California
San Francisco, California 94143

JUDITH M. WHITE

Departments of Pharmacology and Biochemistry and Biophysics
University of California
San Francisco, California 94143

---

[1] Present address: Smith, Kline and French Laboratories, Drug Delivery Group L720, King of Prussia, Pennsylvania 19406.

I. Introduction

One approach to study intracellular events is to deliver macromolecules (e.g., antibodies, nucleic acids) into the cytoplasm of living cells and to assess their effects on specified functions. Therefore, considerable effort has gone into devising simple, efficient, and nontoxic strategies for introducing macromolecules into cells. Several methods are currently available, ranging from microneedle injection to the use of permeabilized cells (Celis, 1984; McNeil *et al.*, 1984; Neumann *et al.*, 1982; Potter *et al.*, 1984; see also Sowers, 1987, and chapters in this volume). Ultimately the choice of a particular method rests on the nature and amount of the molecule to be delivered, the target cell type, the number and percentage of recipient cells, and the length of time the cells must remain viable after delivery. Here we will review a macromolecular delivery technique that employs cells that express a potent membrane fusion protein, the influenza virus hemagglutinin (HA).[2] The basic strategy is outlined in Fig. 1. Briefly, the protein or nucleic acid of interest is loaded into either red blood cells (RBC) or liposomes. After binding of the delivery vesicles to the HA-expressing cells, fusion is induced by briefly treating the cultures at pH 4.8. The cells are then returned to normal growth medium. As a result of the fusion step, the contents of the RBC or liposomes are delivered into the cell cytoplasm. In this article we will discuss the methods for generating HA-expressing cell lines, for loading RBC and liposomes, and for fusing the delivery vesicles to target cells. We will conclude with an evaluation of the advantages and disadvantages of this method.

[2] Abbreviations: CAT, chloramphenicol acetyltransferase; CHO, Chinese hamster ovary; DHFR, dihydrofolate reductase; FACS, fluorescence-activated cell sorter; HA, hemagglutinin; HA_0, the fusion-incompetent hemagglutinin precursor; HRP, horseradish peroxidase; MES, 2(*N*-morpholino)ethanesulfonic acid; PC, phosphatidylcholine; PE, phosphatidylethanolamine; RBC, red blood cell; SV40, simian virus 40.

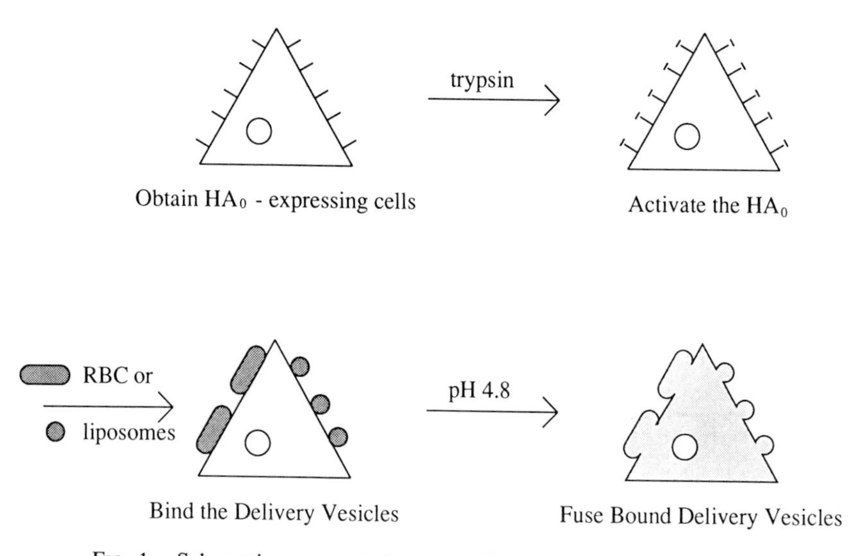

FIG. 1. Schematic representation of the HA-mediated delivery method.

A. Delivery Strategies Based on Viral Membrane Fusion Proteins

Viral membrane fusion proteins have been used for macromolecular delivery for some time (for reviews, see Celis, 1984; Schlegel and Rechsteiner, 1986). Soon after it was realized that viruses have specific "fusion proteins" that mediate delivery of the viral genome into the host cell, schemes were developed to exploit these proteins. Viral membrane fusion proteins have been used as delivery agents in three primary configurations: (i) the intact virus has been used as the fusogen, (ii) the viral spike glycoproteins have been reconstituted in artificial lipid vesicles, and (iii) target cells have been established that express the viral fusion protein on their surface. Fusion-dependent delivery requires the encapsulation of the macromolecule in a membrane-bound vesicle. For this purpose both RBC and artificial lipid vesicles have been used. In the first delivery strategy (Furusawa *et al.*, 1974; Schlegel and Rechsteiner, 1975, 1986; Loyter *et al.*, 1975), macromolecules are introduced into RBC by hypotonic lysis, and the loaded RBC are then fused to the target cells using intact Sendai virus as the "fusogen." This method is effective with cells in suspension, but the efficiency with monolayer cells is lower (Schlegel and Rechsteiner, 1986). In the second type of delivery strategy, the fusion protein is reconstituted into the membrane of the delivery vesicle. This approach works better with monolayer cells. Reconstitution of viral envelope glycoproteins into artificial lipid vesicles (often referred to as virosomes) has the distinct advantage of creating delivery vesicles that can fuse with many different cell types, because most eukaryotic cells carry receptors for the envelope glycoproteins of viruses such as Sendai and influenza. One drawback of this approach is that large quantities of virus must be grown in order to purify sufficient quantities of the fusion protein. In addition, because the resultant virosomes are usually small, ~100–250 nm (Harmsen *et al.*, 1985; Stegmann *et al.*, 1987), the amount of encapsulated material that can be delivered is limited. Gould-Fogerite and Mannino (1985) have developed a special technique for reconstitution of Sendai and influenza envelope glycoproteins that yields large vesicles with a high encapsulation efficiency. The amount of material that can be delivered to cells with these larger vesicles is currently being investigated (Gould-Fogerite *et al.*, 1988). The remainder of this chapter will focus on the third delivery scenario. Here the viral fusion protein is constitutively expressed in the plasma membrane of the recipient cells. Although, in principle, any viral fusion protein could be used for this purpose, the influenza HA is the only one that has been employed in this fashion.

B. Use of the Influenza Virus HA

The influenza virus HA is the best-characterized viral fusion protein (for recent reviews, see Wiley and Skehel, 1987; Doms *et al.*, 1989). The HA is a trimeric molecule in which each monomer is composed of an HA_1 and an HA_2 subunit. The HA is synthesized as an HA_0 precursor and requires cleavage into the HA_1 and HA_2 subunits in order to be active as a fusion protein. Upon exposure to mildly acidic pH, HA undergoes an irreversible conformational change that exposes the apolar and highly conserved N-terminal segment of HA_2, the "fusion peptide" (Skehel *et al.*, 1982; Doms *et al.*, 1985; White and Wilson, 1987; Wharton *et al.*, 1988). It is thought that this conformational change triggers the fusion event and that the newly exposed fusion peptide interacts with the target bilayer (White *et al.*, 1986; Gething *et al.*, 1986; Doms *et al.*, 1989). In addition to its fusion function, the HA is responsible for binding virions to target cells. The receptor-binding site is located near the top of the globular head domain in the HA_1 subunit; this binding site interacts with specific cell surface sialyloligosaccharide structures that can be attached to either protein or lipid molecules (Wilson *et al.*, 1981; Paulson *et al.*, 1979; Weis *et al.*, 1988).

Because the HA possesses both receptor-binding activity and fusion activity, it is eminently suited for cytoplasmic delivery purposes. Moreover, HA-mediated delivery can be controlled at two levels: first, HA_0 must be cleaved from its precursor to its active form; the HA must then be treated at low pH to manifest fusion activity (Doxsey *et al.*, 1985, 1987).

II. Development of Cell Lines that Express the Influenza HA

The influenza HA can be expressed in eukaryotic cells either by viral infection or by generating cell lines that stably express the HA gene. A wide variety of cell types can be infected with influenza virus, and CV-1 cells can be infected with recombinant simian virus 40 (SV40)-HA vectors (Gething and Sambrook, 1981; White *et al.*, 1982a; Doxsey *et al.*, 1985). Infected cells have been used in delivery experiments to study several important cell biological questions (Doxsey *et al.*, 1987; van Meer and Simons, 1983, 1986; van Meer *et al.*, 1985, 1986). However, viral infections are ultimately lytic, and there is only a limited time window in which experiments can be performed: 12–16 hours for CV-1 cells infected with SV40-HA vectors (Doxsey *et al.*, 1985, 1987).

A major advance has been the development of cell lines that permanently express high levels of the HA. The obvious advantages of these cells are that they require no advance preparation (i.e., infection) and that they remain viable following delivery. Several HA-expressing murine cell lines were generated by transfecting cells with a vector containing the transforming fragment of bovine papilloma virus and the HA gene from the Japan strain of influenza virus (Sambrook *et al.*, 1985). The cells were cotransfected with a plasmid containing the neomycin resistance gene and selected for growth in the presence of the antibiotic G418. Cells expressing high levels of HA were isolated with the aid of a fluorescence-activated cell sorter (FACS) based on their ability to bind fluoresceinated RBC (3T3HA-b, Doxsey *et al.*, 1985; gp4, J. Sambrook and M.-J. Gething, unpublished). Further selection for cells that remained adherent throughout the delivery protocol and that bound about twice as many RBC per cell as their parents generated cell lines superior for delivery (3T3HA-b2, S. J. Doxsey, unpublished; gp4F, Ellens *et al.*, 1989). Because these latter lines arose from single-cell clones, they uniformly express high levels of the HA. Recent work has shown that cells suitable for delivery can be directly cloned as described previously but without prior FACS sorting (J. White, unpublished).

Don Wiley and colleagues have developed a Chinese hamster ovary (CHO) cell line that expresses high levels of the HA (A/Aichi/1968). These cells have been used for both RBC and liposome-mediated delivery (D. Wiley, personal communication; Glenn *et al.*, 1989). This cell line, referred to here as CHO-HA, was made by cotransfecting a CHO cell line deficient in the enzyme dihydrofolate reductase (DHFR) with the genes encoding HA and DHFR. These genes were subsequently amplified by treating the cells with methotrexate (D. Wiley, personal communication). It should now be possible to isolate chromosomal DNA from the CHO-HA cells and to transfect the amplified HA genes into a variety of other cell types (Gros *et al.*, 1986). The resultant cells should be targets for both RBC and liposome-mediated delivery.

III. Red Blood Cell-Mediated Delivery

Red blood cells have been used extensively as macromolecular carriers for a number of delivery strategies (Rechsteiner and Schlegel, 1986). Red blood cells are useful delivery vesicles because they are easy to obtain in a highly pure form, they possess a large internal volume (90 μm^3), and they are easy to load with macromolecules (Schlegel and Rechsteiner, 1986). Red blood cells are especially suited for our approach because they possess numerous sialic acid residues, which mediate binding to the surface of HA-expressing cells (Gething and Sambrook, 1981).

A. Loading RBC

Several methods for loading RBC have been developed (see Schlegel and Rechsteiner, 1986). We use the preswell technique developed by Rechsteiner (1982) because it requires small volumes of the test molecule, takes a short time to complete, and is reproducible. Although this method has been described previously, it is reviewed here with attention to critical technical details (see Section VI for a detailed protocol).

Freshly drawn human blood is optimal for RBC loading, although unwashed day-old blood can also be used. Red blood cells obtained from blood stored for several weeks (e.g., from blood banks) tend to leak after loading and are less stable in general. The blood is first washed to remove serum proteins and leukocytes, which appear as a white band on top of the red cell pellet. The RBC are then swollen in a hypotonic solution and the supernatant is carefully aspirated. It is important to remove as much of the supernatant as possible to ensure optimal RBC lysis and loading. The RBC are lysed by adding the test molecule (in 10 m*M* Tris) and incubating the RBC on ice for 2 minutes. During this incubation, a fraction of the test molecule is trapped and ~60–70% of the hemoglobin is released. The RBC are then resealed by returning them to isotonic conditions. Although it has been reported that loaded RBC can be stored (Schlegel and Rechsteiner, 1986), we find that delivery is best if freshly loaded RBC are used.

The efficiency of RBC loading increases with decreasing molecular size (Schlegel and Rechsteiner, 1986; Doxsey *et al.,* 1985, and Doxsey, unpublished). We have found that virtually 100% of the RBC receive horseradish peroxidase (HRP) and ~80% receive IgG. Differences in the amount of protein loaded into individual RBC are sometimes observed. Quantitative analysis of RBC loading has demonstrated that HRP (MW 40,000) and molecules of lower molecular weight are entrapped with high efficiency (e.g., 2×10^7 HRP molecules per RBC). Molecules of IgG (MW 150,000) load less efficiently (e.g., 2×10^6 IgG molecules per RBC). These values were obtained with starting protein concentrations of 20–30 mg/ml. The concentration of HRP inside the RBC approaches the maximum possible. [The maximum loading value is half of the initial HRP concentration, because the HRP solution is diluted 1 : 1 with RBC during the loading step (see Section VI for details).] The internal IgG concentration is about half of the theoretically expected value. Any IgG that is not trapped in the RBC can be recovered by gel filtration HPLC and reused in subsequent delivery experiments (Doxsey *et al.,* 1985).

B. Red Blood Cell-Mediated Delivery

Cytoplasmic delivery of macromolecules contained within the RBC is achieved by fusing the RBC to the plasma membrane of the target cell (see Section VI for a detailed protocol). To obtain efficient delivery, the target cells must express high levels ($\geq 10^6$ HA trimers per cell) of the mature, fusion-competent HA molecule on their surface. In all of the infected and permanently transformed cell lines in use, HA is expressed in the fusion-inactive form (HA_0). Activation of the precursor HA_0 can easily be accomplished by mild trypsinization of the cells. If the HA_0 is not cleaved, the target cells will bind RBC, but acid-dependent fusion of the two membranes will not occur (Doxsey *et al.,* 1985). Once the HA on the target cells has been activated, loaded RBC are added and allowed to bind. The number of RBC that bind to each target cell depends on the amount of surface area available for binding and the amount of cell surface HA (S. J. Doxsey, unpublished).

When maximal binding is attained, the cells are briefly exposed to low pH to trigger fusion between the RBC and target cell membranes. Following fusion, up to 95% of 3T3HA-b2 cells (Fig. 2) receive small proteins (e.g., HRP and RNase). The efficiency of IgG delivery is generally lower; ~80% of 3T3HA-b2 and SV40-HA-infected CV-1 cells receive IgG (Doxsey *et al.,* 1987; S. J. Doxsey, unpublished). The lower efficiency of IgG delivery probably reflects the lower efficiency of IgG loading into RBC. There is some variability in the amount of protein delivered into individual target cells (Doxsey *et al.,* 1985, 1987).

Generally 60–70% of the bound RBC do not fuse. Therefore, in order to quantitate the number of macromolecules delivered to each target cell it is necessary to remove the unfused RBC. Treating the cells with high concentrations of neuraminidase releases unfused RBC. With 3T3HA-b cells $\sim 6 \times 10^7$ HRP molecules (the contents of three RBC) are delivered per cell. With SV40-HA-infected CV-1 cells $\sim 1.8 \times 10^8$ molecules of HRP and $\sim 1.4 \times 10^7$ molecules of IgG are delivered per cell (Doxsey *et al.,* 1985). Delivery of HRP to the most recently developed HA-expressing cell lines [gp4F, 3T3HA-b2 (Fig. 2) and CHO-HA] appears to be more efficient than with the original HA-expressing cells (J. White and S. J. Doxsey, unpublished).

The viability of cells after RBC-mediated fusion has been tested by a variety of methods (Doxsey *et al.,* 1985, 1987). Growth and division of target cells after fusion with RBC remains unchanged. In general, protein synthesis, endocytosis, and membrane traffic are not impaired, although some inhibition of these events (~15%) is occasionally seen.

C. Applications of RBC-Mediated Delivery

In the first cell biological application of the technique, antibodies to clathrin, a molecule associated with the cytoplasmic surfaces of coated vesicles and other intracellular membranes, were delivered into the cytoplasm of CV-1 cells infected with an SV40-HA vector. The results demonstrated that antibodies to clathrin heavy chains inhibited endocytic processes but had no detectable effect on the constitutive transport of a newly synthesized protein in the exocytic pathway. This study provided evidence for a direct role for clathrin in the early stages of endocytosis (Doxsey *et al.,* 1987). In another study, oligonucleotides complementary to the mRNA encoding the endogenous HA (antisense RNA) were loaded into RBC and introduced into the cytoplasm of SV40-HA-infected CV-1 cells. New synthesis of HA was inhibited, presumably by the formation of

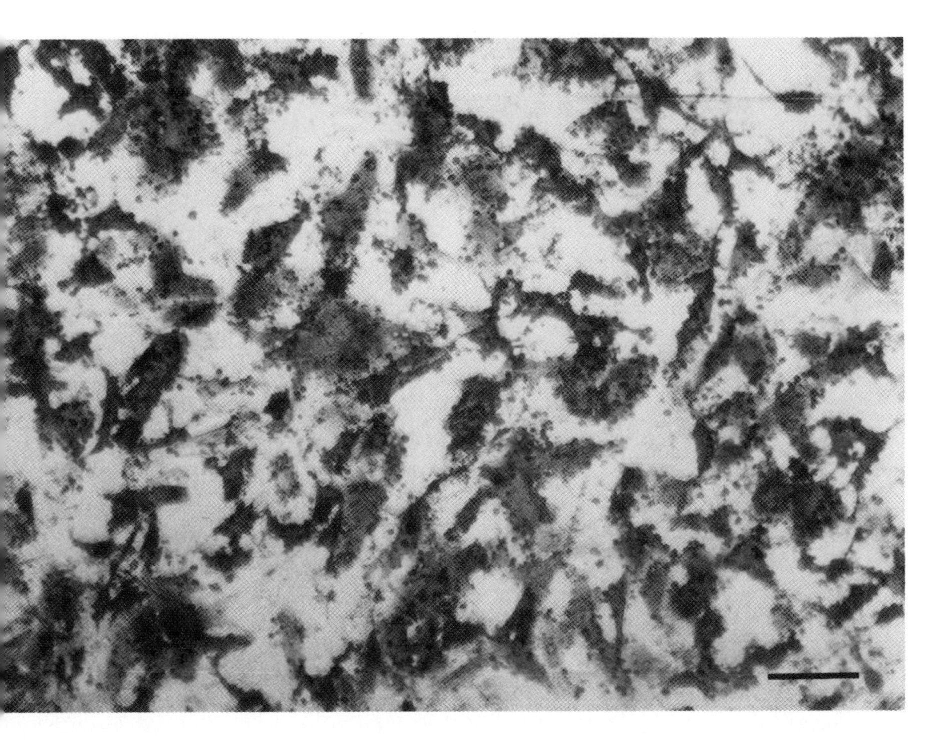

FIG. 2. Delivery of horseradish peroxidase (HRP) into 3T3HA-b2 cells. Red blood cells loaded with HRP were fused to the target cells, and the monolayer was processed for HRP cytochemistry as described (Doxsey *et al.*, 1985). Reaction product was found within the cytoplasm of 95% of the cells. Bar = 20 μm.

duplexes between the HA RNA and the antisense RNA (S. Froshauer, personal communication). The availability of several stable target cell lines and the potential to develop others should now broaden the range of intracellular phenomena that can be investigated using this approach.

IV. Liposome-Mediated Delivery

Although RBC are very well suited for the delivery of proteins and some low molecular weight nucleic acids, they are less suited for the delivery of high molecular weight nucleic acids (Schlegel and Rechsteiner, 1986). Therefore, we have developed an HA-mediated delivery protocol for high molecular weight nucleic acids using liposomes as the delivery vesicle. The method was developed particularly for RNA, because the existing method using DEAE–dextran is not highly efficient (Glenn *et al.,* 1989). Liposomes are also useful for introducing defined exogenous lipid molecules into the plasma membrane of HA-expressing cells (van Meer and Simons, 1983, 1986; van Meer *et al.,* 1985, 1986).

A. Liposomes as Delivery Vesicles

The major advantages of using liposomes as delivery vesicles are that they (i) have well-defined compositions, (ii) are simple to prepare, (iii) are virtually nontoxic, (iv) protect nucleic acids from degradation, and (v) can be used to encapsulate nucleic acids independent of size (up to MW $\sim 10^6$; Mannino *et al.,* 1979). In addition, parameters such as liposome size and lipid composition can be varied to some extent to suit particular applications. If, for example, the object is to introduce lipid molecules into the plasma membrane, small liposomes can be made (25–200 nm) to create a large surface area to volume ratio. If the object is to deliver entrapped molecules, one can prepare large unilamellar liposomes (300–600 nm) to ensure a large encapsulated volume (Szoka and Papahadjopoulos, 1978; MacDonald and MacDonald, 1975; Hope *et al.,* 1985). The lipid composition can be varied to modify the charge, fluidity, and stability of the bilayer. Glycolipids (Jonah *et al.,* 1978; Mauk *et al.,* 1980) or membrane glycoproteins (e.g., MacDonald and MacDonald, 1975) can be incorporated in the lipid bilayer to provide the liposomes with a receptor for a viral fusion protein such as the influenza HA. Antibodies can be covalently coupled to the liposome surface (Heath *et al.,* 1980a,b, 1981; Huang *et al.,* 1980; Martin *et al.,* 1981; Leserman *et al.,* 1980), thereby conferring specific target cell selectivity and improving the overall magnitude of liposome binding.

B. Choice of Liposome Composition

The effect of the target membrane composition on the low pH-induced HA-mediated fusion between influenza virus and liposomes has been investigated extensively (for a review, see Bentz and Ellens, 1988). Overall it has been found that this fusion is largely independent of the lipid composition of the target membrane. Phosphatidylethanolamine (PE) appears to have a slight enhancing effect (Maeda *et al.,* 1981; White *et al.,* 1982b). We therefore include PE in addition to phosphatidylcholine (PC) and cholesterol in the delivery vesicles.

As mentioned previously, influenza virus agglutinates human RBC by binding to cell surface sialic acid residues (e.g., Paulson *et al.,* 1979; Rogers *et al.,* 1983; Weis *et al.,* 1988). Glycophorin, the major glycoprotein on the red cell surface, is primarily responsible for the virus binding. Influenza virus can also use the sialyloligosaccharide chain of gangliosides as a specific receptor. Suzuki *et al.* (1985) showed that adding the ganglioside GM_3-NeuAc to asialoerythrocytes completely restored binding of influenza A/Aichi/2/68. Therefore, to provide liposomes with an HA receptor, one can employ either glycophorin or appropriate gangliosides. We have chosen to use glycophorin, because one can prepare relatively large glycophorin liposomes (see below).

C. Liposome Preparation

There are several recent review articles that describe various ways to prepare liposomes (Gregoriadis, 1984). We will emphasize a few important laboratory practices for handling lipids and liposomes, and then briefly discuss several ways of making glycophorin liposomes.

Liposomes can be prepared from phospholipids isolated from natural sources or from commercially available phospholipids. It is advisable to use highly purified lipid preparations, because contaminants such as lysophospholipids, fatty acids, and metal ions can be toxic to cells and can interfere with liposome stability. The lipids should be checked occasionally by thin-layer chromatography for the presence of impurities (see Kates, 1972). Lipids from natural sources contain polyunsaturated fatty acids that are particularly susceptible to peroxidation. Peroxidation products can form during storage of lipids or during preparation and storage of liposomes (Konings, 1984); they may be toxic to cells (Pietronigro *et al.*, 1977) and/or they may interfere with lipsome stability. To minimize peroxidation, lipids should be stored at low temperatures (−70°C) in an inert atmosphere (e.g., under argon). The degree of peroxidation of the lipid can be chekced by measuring the ratio of absorbance at 233 nm and 215 nm (Klein, 1970). Lipids are stored in chloroform or chloroform–methanol. Preparation of liposomes generally starts with evaporation of the organic solvent using a rotary evaporator connected to a water aspirator or in a stream of nitrogen or argon. Removal of the organic solvent results in the formation of a thin lipid film on the sides of a glass tube. Care should be taken to remove the organic solvent completely. This can best be accomplished by exposing the lipid film for several hours to a high vacuum, such as that generated in a freeze-drying apparatus.

Glycophorin can be purified in large quantities (35–50 mg from 450 ml of human blood; Marchesi and Andrews, 1971) or obtained commercially. Various methods have been employed to reconstitute glycophorin into lipid bilayers (Tosteson *et al.*, 1973; Redwood *et al.*, 1975; Juliano and

Stamp, 1976; van Zoelen *et al.*, 1978a,b), including detergent dialysis (Grant and McConnell, 1974; Ong *et al.*, 1981; Mimms *et al.*, 1981) and rehydration of lipid–protein films (MacDonald and MacDonald, 1975). Glycophorin liposomes prepared by detergent dialysis are usually small, ranging from ~30 nm for liposomes prepared with cholate (Ong *et al.*, 1981) to 240 nm for liposomes prepared with octylglucoside (Mimms *et al.*, 1981). Evidence for a transmembrane "right-side out" orientation of glycophorin in liposomes prepared from cholate-mixed micelles was obtained from neuraminidase and trypsin digestion experiments. Experiments using ^{1}H-NMR provided direct evidence for an interaction of glycophorin with the hydrophobic portion of the lipid bilayer (Ong *et al.*, 1981).

The liposomes used in our delivery experiments are prepared essentially according to MacDonald and MacDonald (1975), because this method yields liposomes of a larger diameter (500–600 nm) than the detergent dialysis procedure. We typically start out with 2 μmol PC, 2 μmol PE, and 1 μmol cholesterol dissolved in 2.5 ml chloroform and 1.25 ml methanol. To this we add 100 μl of a 5 mg/ml glycophorin suspension. The organic solvent is evaporated in a rotary evaporator, and the resulting lipid–protein film is subsequently dried under high vacuum for at least 2 hours. The lipid–protein film is then rehydrated with ~0.7 ml of an aqueous solution containing the macromolecule to be encapsulated. This procedure can be scaled down when reagents such as specific RNA molecules are limiting (Glenn *et al.*, 1989). The liposome suspension that forms is a mixture of glycophorin liposomes, liposomes devoid of glycophorin, and aggregates of unincorporated glycophorin. The glycophorin-containing liposomes can be isolated based on their density. The liposome suspension is layered onto 5% sucrose and centrifuged for 90 minutes at 125,000 *g*. Under these conditions the glycophorin liposomes pellet. Liposomes with little or no glycophorin concentrate at the 5% sucrose interface while glycophorin aggregates remain distributed throughout the sucrose solution. The pellet containing the glycophorin liposomes is washed several times in phosphate-buffered saline prior to use. The glycophorin liposomes can also be purified more quickly by pelleting them through 1 *M* sodium chloride (MacDonald and MacDonald, 1975).

The absence of free, unincorporated glycophorin in the liposome pellet can be demonstrated by Sepharose CL-2B column chromatography of liposomes labeled with either the lipid label [^{14}C]cholesteryl oleate or with ^{125}I-labeled glycophorin. Both the [^{14}C]cholesteryl oleate and the ^{125}I-labeled glycophorin elute in the void volume; unincorporated ^{125}I-labeled glycophorin is retarded on this column (MacDonald and Macdonald,

1975; Ong *et al.*, 1981). The fraction of lipid and glycophorin present in the final liposome pellet is somewhat variable from preparation to preparation. Typically we find ~50% of the lipid and ~25% of the glycophorin in the final pellet. Thus, the glycophorin–phospholipid mole ratio in the liposomes is on the order of 1 : 500. MacDonald and MacDonald (1975) have shown that this ratio can be varied by changing the initial ratio in the starting material. There has been some debate about the orientation of glycophorin in liposomes prepared by hydration of a lipid–protein film (MacDonald and MacDonald, 1975; van Zoelen *et al.*, 1978c). In our hands, the glycophorin in these liposomes is oriented fairly symmetrically.

Electron micrographs of the liposomes prepared by the MacDonald procedure show that their size varies from 0.2 to 2 μm and that they are unilamellar (MacDonald and MacDonald, 1975). This is remarkable, because liposomes prepared by direct hydration of a lipid film are usually multilamellar. Apparently the presence of glycophorin in the lipid film changes the mechanics of liposome formation upon hydration. We have measured the size of the glycophorin liposomes by dynamic light scattering. Their *Z* average diameter ranges from ~500 to 600 nm.

The glycophorin liposomes prepared by the procedure of MacDonald and MacDonald (1975) are permeable to molecules with MW ≤900 (van der Steen *et al.*, 1983; van Hoogevest *et al.*, 1983, 1984). The reason for this permeability is unknown. With respect to this, we were able to encapsulate only very small amounts of uridine triphosphate and the fluorescent marker aminonaphthalene trisulfonic acid. De Kroon *et al.* (1985) encountered a similar phenomenon upon encapsulation of Tb $(\text{citrate})_3^{6-}$, dipicolinate (DPA), and the Tb $(\text{DPA})_3^{3-}$ complex in glycophorin-PS liposomes. However, the glycophorin liposomes retain molecules of higher molecular weight. They are therefore good vehicles for delivery of macromolecules.

D. Liposome-Mediated Delivery

The target cell line that we have used for most of our studies is the HA-expressing 3T3 cell line, gp4F, described in Section II. Before liposome binding and fusion, the cells are exposed to a brief trypsin and neuraminidase treatment (10 μg/ml and 1 mg/ml, respectively, in DME H21 without serum; 4 minutes at room temperature) to cleave HA_0 into HA_1 and HA_2 and to clear the surface, to some extent, from sialic acid residues. Recent work suggests that the amount of neuraminidase in this treatment can be reduced to 0.1–0.2 mg/ml (J. White, unpublished). After 4 minutes, the trypsin is quenched with soybean trypsin inhibitor

(20 μg/ml in DME H21, 10% serum). The cells are then allowed to recover from this treatment for 1–1.5 hours in 37°C CO_2 incubator. During this time there is minimal endocytosis of the HA (Lazarovits and Roth, 1988). Liposomes are allowed to bind to the target cells for 30 minutes at room temperature in bicarbonate-free RPMI 1640 medium buffered to pH 7.4 with 10 m*M* HEPES and supplemented with 0.2% BSA and 35 m*M* NaCl to achieve an osmolarity of ~290 mOsm. Within the 30-minute binding period, the cells are centrifuged twice for 5 minutes at 500 *g*; the plates are rotated 180° between the two centrifugations. This centrifugation protocol increases liposome binding by a factor of three. Fusion is induced by a 90 second incubation at pH 4.8 and 37°C.

To demonstrate cytoplasmic delivery of the liposome contents the A chain of ricin was encapsulated into liposomes at a concentration of 0.23 mg/ml, which extrapolates to ~400 ricin A chain molecules per liposome. Ricin is a toxin that inhibits protein synthesis by catalytically inactivating the 60 S ribosomal subunit (Sperti *et al.*, 1973; Benson *et al.*, 1975). The A chain of ricin by itself has no means of gaining access to the cytoplasm (Olsnes and Sandvig, 1985). Therefore, any inhibition of protein synthesis activity must be the result of a fusion event between the liposome and plasma membranes. Figure 3 shows the inhibition of protein synthesis that occurred when the delivery protocol was carried out with varying concentrations of ricin A chain-containing liposomes (0–20 nmol phospholipid per milliliter). Protein synthesis activity was measured 2 hours after the fusion step.

Ricin is a very potent toxin. It is believed that as little as one molecule of ricin A is sufficient to inhibit protein synthesis activity completely (Eiklid *et al.*, 1980). In the experiment shown in Fig. 3, at concentrations ≥20 nmol phospholipid per milliliter the inhibition of protein synthesis was ~90%. (In other experiments, the maximum inhibition varied from 70 to 90%.) At lower liposome concentrations the inhibition was less (e.g., 50% at 1 nmol phospholipid per milliliter). If the fusion of one liposome results in complete inhibition of protein synthesis, then the data in Fig. 3, in combination with the data on liposome binding (at 2 nmol phospholipid per milliliter there are about 100 liposomes bound per cell), indicate that only a fraction of the bound liposomes fuse. Detailed quantitation of the number of liposomes fused per cell will be presented elsewhere (Ellens *et al.*, 1989).

E. Applications of Liposome-Mediated Delivery

We have shown that we can deliver a toxin encapsulated in liposomes to 70–90% of the gp4F cells in a culture. However, if the object is to

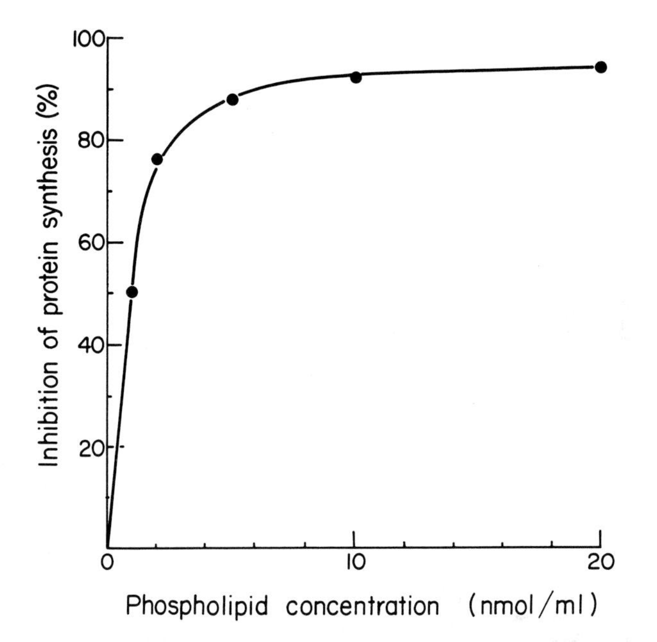

FIG. 3. Delivery of ricin A chain to HA-expressing 3T3 cells using the liposome-mediated delivery technique. Solutions containing various concentrations of ricin A chain-loaded liposomes were added to gp4F cells. The liposomes were bound and fused to the cells as described in the text. Two hours after delivery, the amount of [^{35}S]methionine incorporated into protein was determined. The values represent the percentage of protein synthesis observed in cultures treated in the same manner but maintained at pH 7.4 to prevent fusion.

deliver large amounts of protein, RBC are clearly superior delivery vesicles. On the other hand, the liposome delivery method seems to be well suited for delivering RNA. Sample data depicting the expression of the enzyme chloramphenicol acetyltransferase (CAT) following liposome-mediated delivery of mRNA-encoding CAT is shown in Fig. 4. For this experiment, sp6 transcripts of CAT mRNA were loaded into liposomes, the liposomes were fused to gp4F cells and, at various times, the cells were assayed for CAT activity. Using this system we express $\sim 10^4$ CAT molecules per cell at the time of maximal expression (6 hours postdelivery). We have recently increased the expression levels about two orders of magnitude by using an RNA amplification vector and the CHO-HA cell line described in Section II. In our hands this method is significantly more efficient than DEAE–dextran-mediated RNA delivery (Glenn *et al.*, 1989). This method should therefore be useful for expressing foreign mRNAs, for delivering antisense RNA, and for studying

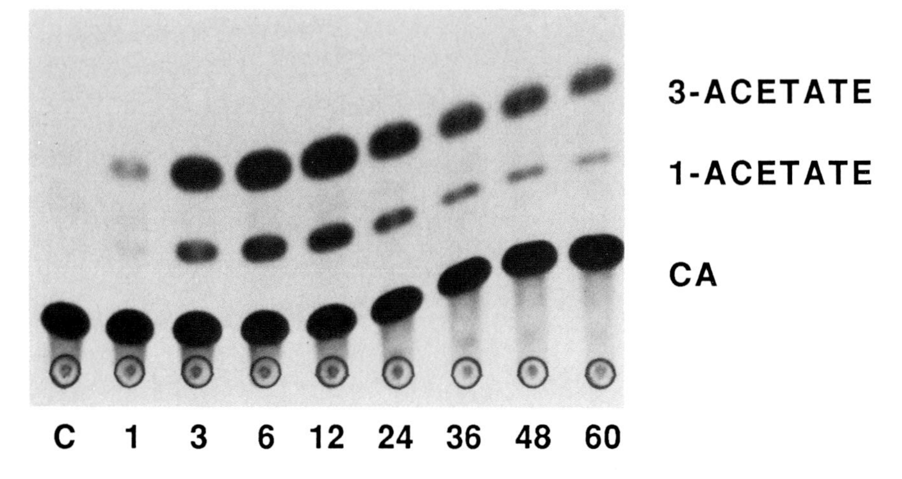

FIG. 4. Delivery of mRNA-encoding chloramphenicol acetyltransferase (CAT) to HA-expressing 3T3 cells using the liposome-mediated delivery technique. Sp6 transcripts of mRNA-encoding CAT were loaded into liposomes. The liposomes were fused to gp4F cells essentially as described in the text. At the indicated times (hours), cells were harvested and analyzed for CAT activity. C, No enzyme control; CA, chloramphenicol; 1-acetate, chloramphenicol acetylated in the 1-position; 3-acetate, chloramphenicol acetylated in the 3-position.

such questions as mRNA stability and the mechanisms of replication of small RNA genomes.

V. Advantages and Disadvantages of the Method

The power of the delivery technique described here is that it permits quantitative biochemical analysis of the fate and effect of a variety of delivered molecules within the physiological environment of the living cell. The RBC delivery technique has been employed to study the role of clathrin in the movement of intracellular membranes (Doxsey *et al.*, 1987), and the liposome delivery technique is presently being used to study the replication of a small-RNA virus (J. Glenn and J. White, unpublished). There are several advantages to the technique: (1) Nearly all of the cells in a population receive macromolecules. (ii) Large numbers of molecules can be delivered per cell. (iii) The same general approach and recipient cells can be used for the delivery of proteins, nucleic acids, and other substances. (iv) The loading of RBC and liposomes is straightfoward; the unencapsulated macromolecules can be reclaimed and re-used. (v) The method can be scaled to the experimental needs from small cluster wells to large plates. (vi) With practice, the procedures are highly reproducible. (vii) Most importantly, the method is not toxic to the cells. After delivery the cells remain viable and can be further passaged.

A limitation of the technique is the need to obtain cells that express high levels of the HA. However, several HA-expressing cell lines suitable for delivery are currently available. Moreover, by using appropriate expression vectors (Gething and Sambrook, 1981; Sambrook *et al.*, 1985; Kaufman and Sharp, 1982) or by transfecting cells with chromosomal DNA containing the HA gene in an amplified form (Gros *et al.*, 1986; D. Wiley, personal communication), it should now be possible to generate a wide variety of target cell types. Such cells should expand the realm of questions that can be addressed using the HA-mediated delivery technique.

VI. Protocols

A. Red Blood Cell Loading (Rechsteiner, 1982)

1. Wash Blood

1. Obtain 10 ml of fresh human blood in a 15-ml conical tube and add anticoagulant (e.g., heparin).
2. Wash three times in PBS.
   a. Spin 2000 *g* for 5 minutes at 4°C.
   b. Aspirate the 'buffy coat' of white blood cells (remove ~1 ml packed RBC per wash).
   c. Resuspend in 10 ml PBS.
   d. Repeat steps a–c (1–2 ml of packed RBC should remain after washing).

2. Swell RBC

1. Accurately prepare a 50% suspension of RBC.
2. Add 300 μl of the suspension to 11.4 ml of swelling buffer [6 ml PBS, 5.4 ml water (53% PBS, 208 mOsm); we find that >3% deviation results in suboptimal loading].
3. Spin 1000 *g* for 10 minutes at 4°C (pellet should be 1.5 times the volume of unswollen RBC; supernatant should be clear to light pink).

3. Lyse and Load RBC

1. Aspirate supernatant carefully and completely (use a yellow pipetman tip or drawn-glass pipet).
2. Add 150 μl of the solution containing the test macromolecule (in 10 m*M* Tris, pH 7.6; other pH values should be acceptable).
3. Vortex and incubate on ice for 2 minutes.
4. Add 22 μl of 10× Hanks buffer (see Section VI,A,4).
5. Incubate 45–60 minutes at 37°C.
6. Wash two times in PBS as before and use immediately (supernatant of first wash should be red, $OD_{540} \approx 1.8$).
7. The loading procedure can be scaled down by a factor of two.

4. Buffers

Hanks V
- NaCl, 105.50 g
- KCl, 6.27 g
- Na_2HPO_4, 12.86 g
- KH_2PO_4, 1.75 g
- Bring to 1 liter with H_2O

Hanks VI
- $MgSO_4 \cdot 7H_2O$, 1.22 g
- $MgCl \cdot 6H_2O$, 2.16 g
- $CaCl_2 \cdot 2H_2O$, 3.50 g
- Bring to 1 liter with H_2O

Final 10× Hanks
- 1.0 ml Hanks V
- 0.3 ml 20 mg/ml $NaHCO_3$
- 0.1 ml 20 mg/ml glucose
- Just before use add 1 ml Hanks VI

B. Red Blood Cell-Mediated Delivery

1. Activate Cell Surface HA_0 (to Produce Fusion-Active HA)

1. Use cells at 50–75% confluency (to avoid cell–cell fusion).
2. Wash cells in PBS.
3. Add trypsin (10 μg/ml), incubate 10 minutes at room temperature (RT).
4. Add soybean trypsin inhibitor from a 100× stock to achieve a final

concentration that is 5× the molar concentration of trypsin. Incubate 2–3 minutes at RT.

5. Wash cells in PBS.

2. Bind Loaded RBC

1. Prepare a ~1% suspension of loaded RBC (bring pellet up in 10 ml).
2. Add RBC to target cells (use 1 ml per 35-mm dish).
3. Agitate plates every 3–5 minutes for 15 minutes.
4. View cells in microscope to ensure complete binding of RBC.

3. Fuse Loaded RBC

1. Aspirate RBC suspension.
2. Add fusion medium prewarmed to 37°C (PBS, 10 m*M* HEPES, 10 m*M* MES, pH to 4.8 at 37°C; use ≥2 ml per 35-mm plate.
3. Incubate in 37°C water bath for 2 minutes.
4. Aspirate fusion medium.
5. Add complete medium.
6. Return to incubator for 45–60 minutes.
7. Wash two or three times in PBS before use.

Acknowledgments

J. White and S. Doxsey would like to express their gratitude to Dr. Ari Helenius in whose laboratory the red blood cell delivery technique was developed. We thank Joe Sambrook and Mary-Jane Gething for the parent gp4 cell line and Don Wiley for the CHO-HA cell line. J. White acknowledges support from the National Science Foundation (PCM 8318570) and the Pew Memorial Trust. H. Ellens was supported by the American Cancer Society; S. Doxsey is supported by a Damon Runyon–Walter Winchell postdoctoral fellowship; J. Glenn is supported by the Medical Scientist Training Program (UCSF). We thank Beth Rupprecht for manuscript preparation.

References

Benson, S., Olsnes, S., Pihl, A. Skorve, J., and Abraham, A. K. (1975). *Eur. J. Biochem.* **59,** 573–580.

Bentz, J., and Ellens, H. (1988). *Colloids Surf.* **30,** 65–112.

Celis, J. E. (1984). *Biochem. J.* **223,** 281–291.

de Kroon, A. I. P. M., van Hoogevest, P., Geurts van Kessel, W. S. M., and de Kruyff, B. (1985). *Biochemistry* **24,** 6382–6389.

Doms, R. W., Helenius, A., and White, J. (1985). *J. Biol. Chem.* **260,** 2973–2981.

Doms, R. W., White, J., Boulay, F., and Helenius, A. (1989). *In* "Cellular Membrane Fusion" (J. Wilschut and D. Hoekstra, eds.). Dekker, Amsterdam (in press).

Doxsey, S. J., Sambrook, J., Helenius, A., and White, J. (1985). *J. Cell Biol.* **101,** 19–27.
Doxsey, S. J., Brodsky, F., Blank, G., and Helenius, A. (1987). *Cell* (*Cambridge, Mass.*) **50,** 453–463.
Eiklid, K., Olsnes, S., and Phil, A. (1980). *Exp. Cell Res.* **126,** 321–326.
Ellens, H., Mason, D., and White, J. (1989). In preparation.
Furusawa, M., Nishimura, T, Yamaizumi, M., and Okada, Y. (1974). *Nature* (*London*) **249,** 449–150.
Gething, M.-J., and Sambrook, J. (1981). *Nature* (*London*) **293,** 620–625.
Gething, M.-J., Doms, R. W., York, D., and White, J. M. (1986). *J. Cell Biol.* **102,** 11–23.
Glenn, J., Ellens, H., and White, J. (1989). In preparation.
Gould-Fogerite, S., and Mannino, R. J. (1985). *Anal. Biochem.* **148,** 15–25.
Gould-Fogerite, S., Mazurkiewicz, J. E., Bhisitkul, D., and Mannino, R. J. (1988). *In* "Advances in Membrane Biochemistry and Bioenergetics" (C. H. Kim, H. Tedeschy, J. J. Diwan, and J. C. Salerno, eds.), pp. 569–586. Plenum, New York.
Grant, C. W. M., and McConnell, H. M. (1974). *Proc. Natl. Acad. Sci. U.S.A.* **71,** 4653–4657.
Gregoriadis, G., ed. (1984). "Lipsome Technology," Vols. 1-3. CRC Press, Boca Raton, Florida.
Gros, P., Fallows, D. A., Croop, J. M., and Housman, D. E. (1986). *Mol. Cell. Biol.* **6,** 3785–3790.
Harmsen, M. C., Wilschut, J., Scherphof, G., Hulstaert, C., and Hoekstra, D. (1985). *Eur. J. Biochem.* **149,** 591–599.
Heath, T. D., Fraley, R. T., and Papahadjopoulos, D. (1980a). *Science* **210,** 539–541.
Heath, T. D., Robertson, D., Birbeck, M. S. C., and Davies, A. J. S. (1980b). *Biochim. Biophys. Acta* **599,** 42–62.
Heath, T. D., Macher, B. A., and Papahadjopoulos, D. (1981). *Biochim. Biophys. Acta* **640,** 66–81.
Hope, M. J., Bally, M. B., Webb, G., and Cullis, P. R. (1985). *Biochim. Biophys. Acta* **812,** 55–65.
Huang, A., Huang, L., and Kennel, S. J. (1980). *J. Biol. Chem.* **255,** 8015–8018.
Jonah, M., Cerny, E. A., and Rahman, Y. E. (1978). *Biochim. Biophys. Acta* **541,** 321–323.
Juliano, R. L., and Stamp, D. (1976). *Nature* (*London*) **261,** 235–238.
Kates, M. (1972). *In* "Laboratory Techniques in Biochemistry and Molecular Biology" (T. S. Work and E. Work, eds.). North-Holland Publ., New York.
Kaufman, R. J., and Sharp, P. A. (1982). *J. Mol. Biol.* **159,** 601–621.
Klein, R. A. (1970). *Biochim. Biophys. Acta* **210,** 486–489.
Konings, A. W. T. (1984). *In* "Liposome Technology" (G. Gregoriadis, ed.), Vol. 1, pp. 139–161. CRC Press, Boca Raton, Florida.
Lazarovits, J., and Roth, M. (1988). *Cell* (*Cambridge, Mass.*) **53,** 743–752.
Leserman, L. D., Barbet, J., Kourilsky, F., and Weinstein, J. N. (1980). *Nature* (*London*) **288,** 602–604.
Loyter, A., Zakai, N., and Kulka, R. G. (1975). *J. Cell Biol.* **66,** 292–304.
MacDonald, R. I., and MacDonald, R. C. (1975). *J. Biol. Chem.* **250,** 9206–9214.
Maeda, T., Kawasaki, K., and Ohnishi, S. I. (1981). *Proc. Natl. Acad. Sci. U.S.A.* **78,** 4133–4137.
Mannino, R. J., Allebach, E. S., and Strohl, W. A. (1979). *FEBS Lett.* **101,** 229–232.
Marchesi, V. T., and Andrews, E. P. (1971). *Science* **174,** 1247–1248.
Martin, F., Hubbel, W. L., and Papahadjopoulos, D. (1981). *Biochemistry* **20,** 4429–4438.
Mauk, M., Gamble, R., and Baldeschwieler, J. (1980). *Science* **207,** 309–311.

McNeil, P. L., Murphy, R. F., Lanni, F. L., and Taylor, D. L. (1984). *J. Cell Biol.* **98,** 1556–1564.

Mimms, L. T., Zampighi, G., Nozaki, Y., Tanford, C., and Reynolds, J. A. (1981). *Biochemistry* **20,** 833–840.

Neumann, E., Schaefer-Ridder, M., Wang, Y., and Hofscheider, P. H. (1982). *EMBO J.* **7,** 841–845.

Olsnes, S., and Sandvig, K. (1985). *In* "Endocytosis" (I. Pastan and M. C. Willingham, eds.), pp. 195–234. Plenum, New York.

Ong, R. L., Marchesi, V. T., and Prestegard, J. H. (1981). *Biochemistry* **20,** 4283–4292.

Paulson, J. C., Sadler, J. E., and Hill, R. L. (1979). *J. Biol. Chem.* **254,** 2120–2124.

Pietronigro, D. D., Jones, W. B. G., Kalty, K., and Demopoulos, H. B. (1977). *Nature (London)* **267,** 78–79.

Potter, H., Weir, L., and Leder, P. (1984). *Proc. Natl. Acad. Sci. U.S.A.* **75,** 145–149.

Rechsteiner, M. C. (1982). *In* "Techniques in Somatic Cell Genetics" (J. Shay, ed.), pp. 385–398. Plenum, New York.

Rechsteiner, M. C., and Schlegel, R. A. (1986). *In* "Microinjection and Organelle Transplantation Techniques: Methods and Applications" (J. E. Celis, A. Graessmann, and A. Loyter, eds.), pp. 89–116. Academic Press, Orlando, Florida.

Redwood, W. R., Jansons, V. K., and Patel, B. C. (1975). *Biochim. Biophys. Acta* **406,** 347–361.

Rogers, G. N., Paulson, J. C., Daniels, R. S., Skehel, J. J., Wilson, I. A., and Wiley, D. C. (1983). *Nature (London)* **304,** 76–78.

Sambrook, J., Rodgers, L., White, J., and Gething, M.-J. (1985). *EMBO J.* **4,** 91–103.

Schlegel, R. A., and Rechsteiner, M. C. (1975). *Cell (Cambridge, Mass.)* **3,** 371–379.

Schlegel, R. A., and Rechsteiner, M. C. (1986). *In* "Microinjection and Organelle Transplantation Techniques: Methods and Applications" (J. E. Celis, A. Graessmann, and A. Loyter, eds.), pp. 67–87. Academic Press, Orlando, Florida.

Skehel, J. J., Bayley, P. M., Brown, E. B., Martin, S. R., Waterfield, M. D., White, J. M., Wilson, I. A., and Wiley, D. C. (1982). *Proc. Natl. Acad. Sci. U.S.A.* **79,** 968–972.

Sowers, A. E., ed. (1987). "Cell Fusion." Plenum, New York.

Sperti, S., Montanaro, L., Mattioli, A., and Stirpe, F. (1973). *Biochem. J.* **136,** 813–815.

Stegmann, T., Morselt, H., Booy, F., van Breeman, J., Scherphof, G., and Wilschut, J. (1987). *EMBO J.* **6,** 2651–2659.

Suzuki, Y., Matsunaga, M., and Matsumoto, M. (1985). *J. Biol. Chem.* **260,** 1362–1365.

Szoka, F., and Papahadjopoulos, D. (1978). *Proc. Natl. Acad. Sci. U.S.A.* **75,** 4194–4198.

Tosteson, M. T., Lau, F., and Tosteson, D. C. (1973). *Nature (London), New Biol.* **243,** 112–114.

van der Steen, A. T. M., Taraschi, T. F., Voorhaut, W. F., and de Kruyff, B. (1983). *Biochim. Biophys. Acta* **733,** 51–64.

van Hoogevest, P., van Duyn, G., Batenburg, A. M., de Kruyff, B., and de Gier, J. (1983). *Biochim. Biophys. Acta* **734,** 1–17.

van Hoogevest, P., du Maine, A. P. M., de Kruyff, B., and de Gier, J. (1984). *Biochim. Biophys. Acta* **771,** 119–126.

van Meer, G., and Simons, K. (1983). *J. Cell Biol.* **97,** 1365–1374.

van Meer, G., and Simons, K. (1986). *EMBO J.* **5,** 1455–1464.

van Meer, G., Davoust, J., and Simons, K. (1985). *Biochemistry* **24,** 3593–3602.

van Meer, G., Gumbiner, B., and Simons, K. (1986). *Nature (London)* **322,** 639–641.

van Zoelen, E. J. J., de Kruyff, B., and van Deenen, L. L. M. (1978a). *Biochim. Biophys. Acta* **508,** 97–108.

van Zoelen, E. J. J., van Dyck, P. W. M., de Kruyff, B., Verkley, A. J., and van Deenen, L. L. M. (1978b). *Biochim. Biophys. Acta* **514,** 9–24.

van Zoelen, E. J. J., Verkley, A. J., Zwaal, R. F. A., and van Deenen, L. M. (1978c). *Eur. J. Biochem.* **86,** 539–546.

Weis, W., Brown, J. H., Cusack, S., Paulson, J. C., Skehel, J. J., and Wiley, D. C. (1988). *Nature (London)* **333,** 426–431.

Wharton, S., Ruigrok, R., Martin, S., Skehel, J., Bayley, P., Weis, W., and Wiley, D. C. (1988). *J. Biol. Chem.* **263,** 4474–4480.

White, J., and Wilson, I. (1987). *J. Cell Biol.* **105,** 2887–2896.

White, J., Helenius, A., and Gething, M. J. (1982a). *Nature (London)* **300,** 658–659.

White, J., Kartenbeck, J., and Helenius, A. (1982b). *EMBO J.* **1,** 217–222.

White, J., Doms, R., Gething, M.-J., Kielian, M., and Helenius, A. (1986). *In* "Virus Attachment and Entry into Cells" (R. L. Crowell and K. Lonberg-Holm, eds.), pp. 54–59. Am. Soc. Microbiol., Washington, D.C.

Wiley, D. C., and Skehel, J. J. (1987). *Annu. Rev. Biochem.* **56,** 365–394.

Wilson, I. A., Skehel, J. J., and Wiley, D. C. (1981). *Nature (London)* **289,** 366–373.

Part II. Cell-Free Reconstitution of Transport: Proteins That Interact with the Endodomain of Membranes

The mixing of defined subcellular fractions and suitable soluble components has resulted in surprisingly extensive reconstitution of many transport events. The most outstanding examples are vectorial transport to the cisternal space of the RER and import of mitochondrial proteins. A further major landmark was the elaborate cell-free analysis of Golgi-to-Golgi transport (Balch *et al.*, 1983). With these conspicuous successes as precedent, investigators seeking maximal control over the environment of other organelles have now accomplished cell-free simulation of exocytosis (Crabb and Jackson, 1985; Woodman and Edwardson, 1986) and, in selected cases, fusion to lysosomes (Caltstiel and Branton, 1983). Reconstitution of early steps along the coated-pit endocytic path have been successfully documented, both as in the included articles [see also Howell *et al.*, Chap. 15, this volume (Davey *et al.*, 1985; Braell, 1987)] and by the independent approach described by Anderson (Moose *et al.*, 1987). Despite the drama and sense of conquest inherent in such studies, in all these cases there remains some uncertainty about the accuracy of simulation. This persistent doubt is one of the factors responsible for the development of permeabilized-cell models, as discussed in Part I of this volume.

Because critical determinants for governing membrane–membrane interaction and fusion must lie on the endodomain of the membranes in question, there is a growing need to characterize this surface of isolated organelles. Moreover, this membrane surface presumably interacts with cytoskeletal components, at least some of which must have already been identified in other contexts. A number of antibodies have been prepared that recognize such superficial determinants, some of which are only reversibly associated with membranes (Chicheportiche and Tartakoff, 1988). In addition to the two studies of superficial components discussed in Chapters 12 and 13 by Pollard and Pearse (which should be read in conjunction with Sandvig *et al.*, Chapter 16 in Vol. 32), a major effort has gone into the characterization of synapsin I, a superficial protein of synaptic vesicles that appears to be crucial for exocytosis (DeCamilli and Greengard, 1986), and into the characterization of a superficial protein of yeast secretory vesicles that binds GTP and shows homology to the *ras* oncoprotein (Goud *et al.*, 1988).

An issue of persistent concern in such reconstitution experiments is the relation between membrane–membrane interaction and organelle translocation. Despite many indications of vesicle–cytoskeletal interactions (Vallee, 1986), the literature on vesicle translocation has not yet made it clear to what extent identified motors (kinesin, dynein, etc.) play a general role in the events of transport along the secretory and endocytic pathways. Nevertheless, studies such as those of Haimo and Rozdzial (Chap. 1, this volume) document important microtubule–pigment granules interactions.

References

Altstiel, L., and Branton, D. (1983). Fusion of coated vesicles with lysosomes: Measurement with a fluorescence assay. *Cell (Cambridge, Mass.)* **32,** 921–929.

Balch, W., Fries, E., Dunphy, W., Urbani, L., and Rothman, J. (1983). Transport-coupled oligosaccharide processing in a cell-free system. *In* "Methods in Enzymology" (S. Fleischer and B. Fleischer, eds.), Vol. 96, pp. 37–46. Academic Press, New York.

Braell, W. (1987). Fusion between endocytic vesicles in a cell-free system. *Proc. Natl. Acad. Sci. U.S.A.* **84,** 1137–1141.

Chicheportiche, Y., and Tartakoff, A. (1988). The use of antibodies for analysis of the secretory and endocytic paths of eukaryotic cells. *Subcell. Biochem.* **12,** 243–276.

Crabb, J., and Jackson, R. (1985). In vitro reconstitution of exocytosis from plasma membrane and isolated secretory vesicles. *J. Cell Biol.* **101,** 2263–3373.

Davey, J., Hurtley, S., and Warren, G. (1985). Reconstitution of an endocytic fusion event in a cell-free system. *Cell (Cambridge, Mass.)* **43,** 643–652.

DeCamilli, P., and Greengard, P. (1986). Synapsin I: A synaptic vesicle-associated neuronal phophoprotein. *Biochem. Pharmacol.* **35,** 4349–4357.

Goud, B., Salminen, A., Walwouth, N., and Novick, P. (1988). A GTP-binding protein required for secretion rapidly associates with secretory vesicles and the plasma membrane in yeast. *Cell (Cambridge, Mass.)* **53,** 753–768.

Moose, M., Mahaffey, D., Brodsky, F., and Anderson, R. (1987). Assembly of clathrin-coated pits onto purified plasma membranes. *Science* **236,** 558–563.

Vallee, R., ed. (1986). "Methods in Enzymology," Vol. 134. Academic Press, Orlando, Florida.

Woodman, P., and Edwardson, J. M. (1986). A cell-free assay for the insertion of a viral glycoprotein into the plasma membrane. *J. Cell Biol.* **103,** 1829–1835.

Chapter 10

Reconstitution of Intracellular Vesicle Fusion in a Cell-Free System after Receptor-Mediated Endocytosis

LUIS S. MAYORGA, RUBEN DIAZ, AND PHILIP D. STAHL

Department of Cell Biology and Physiology
Washington University School of Medicine
St. Louis, Missouri 63110

I. Introduction

A. Fusion Events during Receptor-Mediated Endocytosis

The endocytic pathway is responsible for the internalization and delivery of ligands to intracellular compartments of most eukaryotic cells. The pathway involves a number of membrane-bound structures that together allow for the sorting of ligands to different intracellular destinations and for the recycling of receptors to the cell surface. It is not clear, however, how the endocytosed material is transported from one compartment to another. In the exocytic pathway the traffic of material among structures like the endoplasmic reticulum and the Golgi apparatus, which are relatively stable organelles, seems to be mediated by transport vesicles that bud from one compartment and fuse with another (Lodish *et al.*, 1987; Orci *et al.*, 1986). On the other hand, after endocytosis, fluid and extracellular molecules are found in vesicular structures of variable size, with morphology and composition that change with time. This variability in size and shape may reflect fusion of the subcellular membranes that make up the endocytic compartments as materials move along the pathway (Wileman *et al.*, 1985).

Among the fusion events that can be postulated during the receptor-mediated endocytosis are (i) the pinching off of membrane from the cell surface during the formation of clathrin-coated vesicles from coated pits (i.e., domains in the plasma membrane where receptor–ligand complexes accumulate), (ii) the fusion of early endosomes with each other and perhaps with preformed structures (e.g., Golgi-derived or endosome-derived vesicles) to form larger vesicles, (iii) the pinching off of receptor-enriched vesicles from endosomes after dissociation of receptor–ligand complexes, (iv) the fusion of these recycling vesicles with plasma membrane, and (v) the fusion of vesicles containing ligand in their lumen with lysosomes. Fusion has also been postulated between elements of the endocytic and exocytic pathway (e.g., fusion of endosomes with trans-Golgi reticulum; Fishman and Fine, 1987; Duncan and Kornfeld, 1988).

The mechanisms and regulation of these fusion events are poorly understood at present. In order to address these problems, several *in vitro* assays have been developed to reconstitute fusion of endosomes in cell-free systems. Davey *et al.* (1985) have studied fusion between endocytic vesicles containing Semliki Forest virus whose coat glycoproteins were labeled with tritiated sialic acid and endocytic vesicles containing Fowl plague virus, which expresses neuraminidase activity on its surface. Liberation of sialic acid served as a measurement of fusion between these vesicles. Gruenberg and Howell (1986) have used vesicular stomatitis virus (VSV) G protein, localized in purified endosomes, as a probe of fusion with vesicles containing lactoperoxidase incorporated by fluid-phase endocytosis. Fusion was assessed by measuring ^{125}I incorporation into VSV G protein. Braell (1987) has designed an assay based on two molecules, avidin-β-galactosidase and biotinylated IgG, which are incorporated into endosomes by fluid-phase endocytosis and that form an enzymatically active complex when they are present in the same compartment. Finally, radioactive transferrin and antitransferrin antibody have been used by Woodman and Warren (1988) as probes to assess fusion between endosomes.

The fusion assay that we describe here is based on two proteins: mannosylated monoclonal IgG (Man-IgG) specific for dinitrophenol (DNP) groups, and DNP-derivatized β-glucuronidase. These two molecules are efficiently internalized by cell lines that express the macrophage mannose receptor and form a stable immune complex when they are localized in the same compartment. This complex can be quantified using a fluorescent substrate for β-glucuronidase. The assay is very sensitive, reproducible, and relatively inexpensive.

B. General Principles to Assess Fusion of Vesicles

Vesicle fusion can be defined as that event which leads to the formation of a single membrane-bound compartment from two or more vesicles without significant loss of their intravesicular contents. Vesicle membranes and contents mix after fusion reaction. Most assays take advantage of this mixing to assess vesicular fusion. Among the methods developed to determine the occurrence of vesicle fusion, fluorescence techniques have proved useful. In particular, fluorescence unquenching or resonance energy transfer between two fluorescent probes have been used to study liposome–liposome and virus–liposome fusion (Wilschut *et al.*, 1985; Nir *et al.*, 1986). Until now, however, the methods developed to study fusion between organelles rely on (1) enzymatic differences between compartments and (2) complex formation.

If one set of vesicles contains an enzymatic activity and the other a suitable substrate, the formation of the product of the enzymatic reaction depends on the fusion of these vesicles. This approach has been extensively used to study the vesicle-mediated transport of newly synthesized proteins between the endoplasmic reticulum and the Golgi (Beckers *et al.*, 1987), and between different Golgi cisternae (Balch *et al.*, 1984). It has also been used to study lysosome–lysosome fusion in intact cells (Ferris *et al.*, 1987), and endosome–endosome fusion in cell-free systems (Davey *et al.*, 1985; Gruenberg and Howell, 1986). Because these assays are based on an enzymatic reaction, they have the advantage of being very sensitive. However, the enzymatic activity may be affected differently by the various conditions imposed on the assay. Care must be taken, therefore, to control for such changes in enzymatic activity because they would not reflect changes in fusogenic activity.

When two molecules with high affinity for each other are loaded in two different sets of vesicles, the formation of a complex between the molecules depends on fusion events between these vesicles. This approach has been used to study endosome–endosome fusion (Braell, 1987; Woodman and Warren, 1988) and is the one we use in our method (Diaz *et al.*, 1988). As long as the binding reaction between both probes is not affected by the luminal environment of the vesicles, the complex formation provides an accurate estimate of fusion. Newly formed complex can be detected in at least two ways. The complex may be enzymatically active or radioactive.

Whenever vesicle fusion is tested by some interaction between the contents of the vesicles, a potential source of artifact is that the same product of reaction can be formed in the extravesicular compartment if the probes leak from the vesicles. In some instances the dilution effect after leakage is enough to prevent the reaction, but it is always advisable to quench the activity of the probes in the extravesicular compartment.

II. Generation of Probes

A. Mannosylated Monoclonal Anti-DNP IgG (Man-IgG)

Dinitrophenol (DNP) has strong antigenic properties when it is bound to a macromolecule. It has been extensively used to raise polyclonal and monoclonal antibodies; some of them are commercially available. We purify mouse monoclonal anti-DNP IgG (HDP-1) from the culture media of a hybridoma cell line, which was a generous gift from Dr. Julian Fleischman (Washington University, St. Louis, MO) using the method of Otsuka *et al.* (1984). The method relies on the affinity of the antibody for NBD (7-nitrobenz-2-oxa-1,3-diazole), a fluorescent analog of DNP. In brief, the spent medium is precipitated in 50% saturated ammonium sulfate. The pellet is resuspended in PBS (150 m*M* NaCl, 10 m*M* phosphate buffer, pH 7.2) and dialyzed against PBS. Approximately

300 mg of protein are applied to a 5-ml NBD–alkylamine agarose column. The column is washed with 20 ml of 0.5 *M* NaCl in 10 m*M* phosphate buffer (pH 7.2), and with 20 ml of PBS, and the antibody is eluted with 10 ml of PBS containing 1.5 m*M* NBD–ε-aminocaproic acid. The antibody is then extensively dialyzed against PBS.

Immobilized diaminodipropylamine agarose beads (Pierce Chemical Company) are used to prepare NBD–alkylamine agarose. Five milliliters of agarose are washed with 50 ml of 50 m*M* sodium borate (pH 7.3)–acetonitrile (1 : 1) in a scintered-glass funnel. The beads should not dry during the wash. The excess buffer is removed and 5 ml of the same buffer added. Then, while stirring, 0.5 ml of 0.5 *M* NBD-Cl (4-chloro-7-nitrobenz-2-oxa-1,3-diazole) in acetonitrile is added. The mixture is incubated for 10 minutes at room temperature with stirring. At this point the color of the resin should be golden yellow. The beads are washed with the borate–acetonitrile buffer until no more yellow color appears in the eluate. Finally, they are washed with PBS.

To elute the column, NBD–ε-aminocaproic acid is used as a competitive analog; because of its low-affinity binding to the antibody, it can be easily eliminated from the antibody preparation by dialysis. To synthesize this compound, a 20 m*M* solution of NBD-Cl is prepared in acetonitrile. A 50-ml aliquot of this solution is mixed with 50 ml of 100 m*M* ε-aminocaproic acid in 50 m*M* sodium borate buffer (pH 8). After 2 hours at room temperature, the acetonitrile is evaporated using a rotary evaporator. Then, 10 ml of concentrated ammonium hydroxide are added and the volume adjusted to 100 ml. The solution is transferred to a 1-liter separatory funnel and washed by adding 125 ml of chloroform and 250 ml methanol, mixing, and then adding 125 ml of chloroform and 125 ml of water. The bottom layer is discarded, and the top layer is washed five times with 150 ml of chloroform. The upper phase is then acidified with concentrated HCl until the pH drops to 1 and then extracted three times with 125 ml of chloroform. The organic fractions are then pooled and evaporated in a rotary evaporator. The NBD–ε-aminocaproic acid is redissolved in 25 ml of chloroform–method (1 : 1) and stored at −20°C. The required amount of NBD–ε-aminocaproic acid to elute the antibody from the column is evaporated in a glass flask and redissolved in PBS. The concentration of the solution can be calculated using a molar-extinction coefficient of 24,000 at 475 nm.

Mannosylated anti-DNP IgG is prepared as described by Lee *et al.* (1976). The mannosyl precursor, cyanomethyl-1-thioglycoside-D-mannopyranoside, can be obtained from E·Y Laboratories. The protocol for coupling to protein is the following: dissolve 0.2 g of the precursor in 10 ml of dry methanol to a final concentration of ~0.1 *M*. Add 27 mg of sodium methylate to this solution and stir at room temperature overnight. The mxiture is then dried in a rotovap, keeping the temperature below 40°C. To the solid residue add 4 ml of 0.25 *M* sodium borate buffer (pH 8.5) containing 4 mg of anti-DNP IgG and 0.1 m*M* NBD–ε-aminocaproic acid to protect the antigen-binding site of the antibody. Stir at room temperature overnight and dialyze extensively in PBS at 4°C. Mannosylated anti-DNP IgG is stable at 4°C for weeks and can be stored frozen at −20°C. Repetitive freezing and thawing can cause antibody aggregation.

B. Dinitrophenol-Derivatized β-Glucuronidase (DNP-β-Glucuronidase)

β-Glucuronidase can be purified in large quantities from female rat preputial glands using the method of Keller and Touster (1975). All the procedures are performed at 4°C. Approximately 15 g of rat preputial glands (with the fat carefully removed) are homogenized in 120 ml of 0.1 *M* Tris-acetate buffer (pH 7.8) using a Polytron homogenizer. The homogenate is centrifuged at high speed (150,000 *g*, 30 minutes). The enzyme is precipitated by very slowly adding 44 g of solid ammonium sulfate to 100 ml of the supernatant with constant stirring. The enzyme is collected by centrifugation at 30,000 *g* for 20 minutes. The pellet is dissolved in 75 ml of 0.02 *M* Tris-acetate buffer (pH 7.8) and fractionated by ethanol precipitation. A first fraction is obtained by mixing, drop by drop while stirring, 0.4 ml of chilled (4°C) ethanol per milliliter of the dissolved pellet and discarded by centrifugation at 30,000 *g* for 20 minutes. The enzyme is then precipitated from the supernatant by slowly adding 0.8 ml of cold ethanol per milliliter of supernatant. The pellet is extracted with 10 ml of 0.02 *M* Tris-acetate buffer (pH 7.8), using a glass–glass homogenizer and centrifuged as before. The extraction is repeated once more and the extracts combined. Normally this procedure is sufficient for obtaining a relatively pure preparation. To assess purity, the intermediate and final products can be tested for β-glucuronidase activity using the enzymatic assay described later. Sodium dodecyl sulfate (SDS)–gel electrophoresis of the purified enzyme should show a single band of MW 60,000. The enzyme is a homotetramer of MW 240,000. Further purification can be achieved by ion exchange chromatography on Whatman DE52 resin. The enzyme is dialyzed against 5 m*M* Tris-acetate buffer (pH 7.4) containing 30 m*M* NaCl and loaded onto the column equilibrated with the same buffer. The column is then washed with 70 m*M* NaCl in the same buffer and the enzyme eluted with Tris-acetate buffer containing 150 m*M* NaCl. The enzyme is very stable at −20°C in 50% (v/v) glycerol buffered to neutral pH.

β-Glucuronidase can be derivatized with DNP without losing its enzymatic activity. To obtain an almost fully active preparation that is ~80% immunoprecipitable with the anti-DNP antibody, the enzyme, 2 mg/ml in 0.05 *M* pyrophosphate-citrate buffer (pH 8), and dinitrophenyl fluoride (4% v/v) are mixed at room temperature for 60 minutes with strong shaking. The unreacted oil can be pelleted by centrifugation in a microcentrifuge. The enzyme in the supernatant is dialyzed against PBS and stored at −20°C in 50% (v/v) glycerol. Its enzymatic activity and immunoprecipitation properties should be tested before use in a fusion assay. Bovine serum albumin (BSA) can be DNP-derivatized using the same procedure.

Dinitrophenol-β-glucuronidase is bound with high affinity by the antibody, and this affinity is relatively independent of pH. At pH 5 the amount of enzyme that is immunoprecipitated is only 20% lower than that obtained at pH 7. This is important because the probes are supposed to interact in endocytic compartments which are acidic. Another important feature of these probes is that they are not degraded in early steps after endocytosis; no TCA-soluble radioactivity appears in the incubation media after 5 minutes of uptake of either radiolabeled probe followed by a 30-minute chase.

To visualize the probes in the electron microscope they can be coupled to colloidal gold using standard techniques (Roth, 1983). We have used two sizes of gold coated with the probes to follow their intracellular route and to assess morphologically *in vitro* fusion between endosomes (Fig. 1). The probes can also be radiolabeled using chloramine T without losing their biological activities. A 100-μg portion of protein is mixed with 1 mCi of ^{125}I in a minimum volume of 0.1 *M* phosphate buffer, pH 7.5. The reaction is started by the addition of 30 μl of chloramine T (0.3 mg/ml) in phosphate buffer. After 10 minutes at 4°C, the reaction is stopped with 300 μl of β-mercaptoethanol (1 μl diluted in 10 ml of phosphate buffer) and the protein is dialyzed successively in 4 liters of 1 *M* NaCl, 10 m*M* KI, 5 m*M* Tris-Cl (pH 7.5), and in 4 liters of 150 m*M* NaCl, 5 m*M* Tris-Cl (pH 7.5), or alternatively, passed over a G-25 gel filtration column to remove unreacted iodine.

III. Vesicle Preparation

The two probes, obtained as described earlier, are recognized by the mannose receptor of macrophages. In principle, almost any cell type expressing this receptor could be used. Mannose receptor-mediated endocytosis is observed in several cells of the reticuloendothelial system,

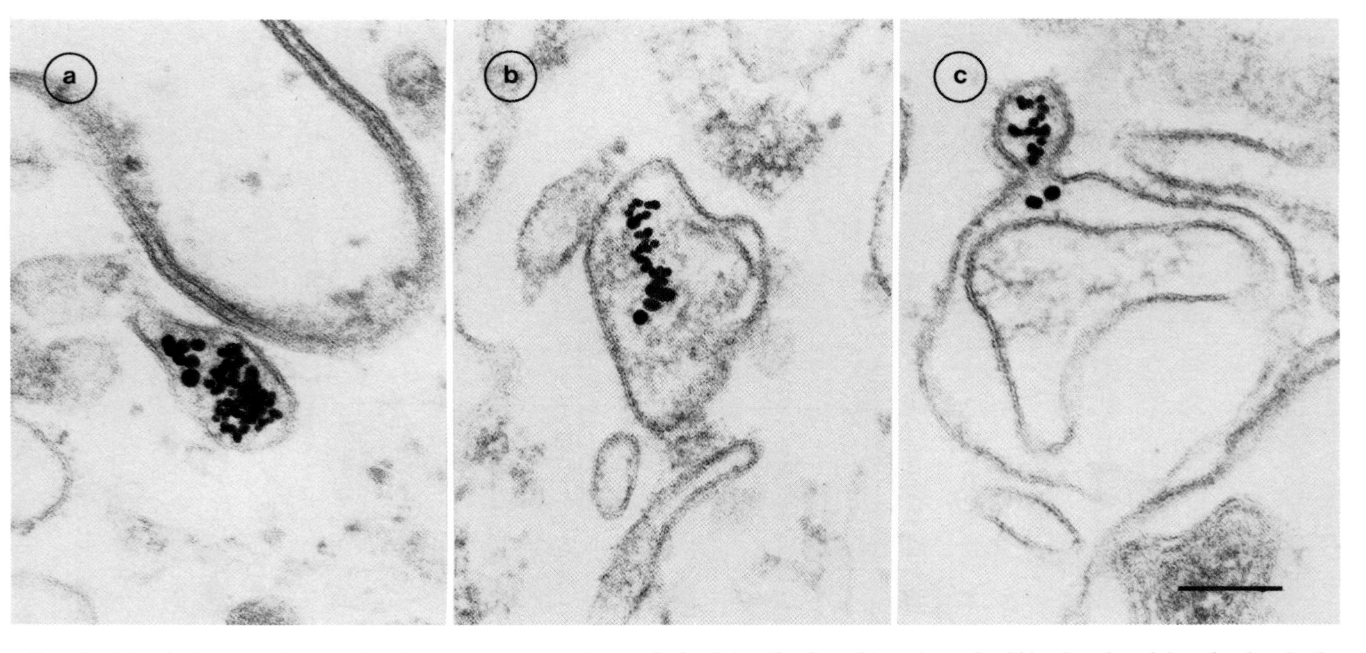

FIG. 1. Morphological evidence of endosome–endosome fusion. (a, b) Colocalization of two sizes of gold in closed vesicles after incubation at 37°C. (c) Two vesicles containing different sizes of gold in an early stage after fusion: 20-nm colloidal gold coated with man-IgG and 10 nm gold coated with DNP-β-glucuronidase. Uptake and fusion conditions were similar to those described in the text. Vesicles were fixed and pelleted. The pellet was embedded in plastic, sectioned, and analyzed by transmission electron microscopy. Bar = 100 nm.

including Kupffer cells, bone marrow, and alveolar macrophages (Lennartz *et al.*, 1987). The cell that we have chosen for our work has been the murine macrophagelike line J774-E clone. This clone has been characterized in our laboratory and it expresses the mannose receptor (Diment *et al.*, 1987). J774-E clone is grown in monolayer in minimum essential medium containing Earle salts and supplemented with 10% fetal calf serum.

A. Uptake of Ligands

Standard preparations of vesicles are obtained from cells that have been allowed to internalize the probes for 5 minutes at 37°C. The uptake is performed in Hanks balanced salt solution buffered with 10 m*M* HEPES and 10 m*M* 2-[2-hydroxy-1,1-bis(hydroxymethyl)ethyl]-aminoethanesulfonic acid (TES) and supplemented with 10 mg/ml BSA (uptake medium). Cells are washed twice with this medium, pelleted (800 *g*, 5 mintues), and resuspended in warm medium (2×10^7 cells/ml) containing 10 μg/ml of Man-IgG or 20 μg/ml DNP-β-glucuronidase. After 5 minutes at 37°C the cells are diluted with cold uptake medium and washed at 4°C, first with the same medium, then with 5 m*M* EDTA in PBS, and finally with homogenization buffer (250 m*M* sucrose, 0.5 m*M* EGTA, 20 m*M* HEPES, pH 7.0). After these washes >90% of the cell-associated ligand is intracellular. The Man-IgG is offered to the cells in a lower concentration because its affinity for the mannose receptor is higher than that of DNP-β-glucuronidase.

B. Homogenization and Fractionation

Standard methods of homogenization (e.g., glass–glass homogenizer or nitrogen cavitation) are not very efficient for macrophages. We have developed an inexpensive method for generating homogenates with good vesicle preservation from these cells. The apparatus consists of two 3-ml plastic syringes with 27-gauge needles connected by plastic tubing (Micro-line; i.d. 0.25 mm, o.d. 0.76 mm). A small metal plate with a hole in the center is used to hold the syringe so that the liquid can be forced from one syringe to the other (see Fig. 2). All the following steps are carried out at 4°C. Cells are resuspended in homogenization buffer (5×10^7 cells/ml) and loaded in one syringe. The suspension is then forced to pass to the other syringe as fast as possible. From 8 to 12 passes should be sufficient to homogenize the cells. The number of passes should be optimized for each operator in order to obtain maximal cell disruption with minimal release of intravesicular components (Fig. 2). In our

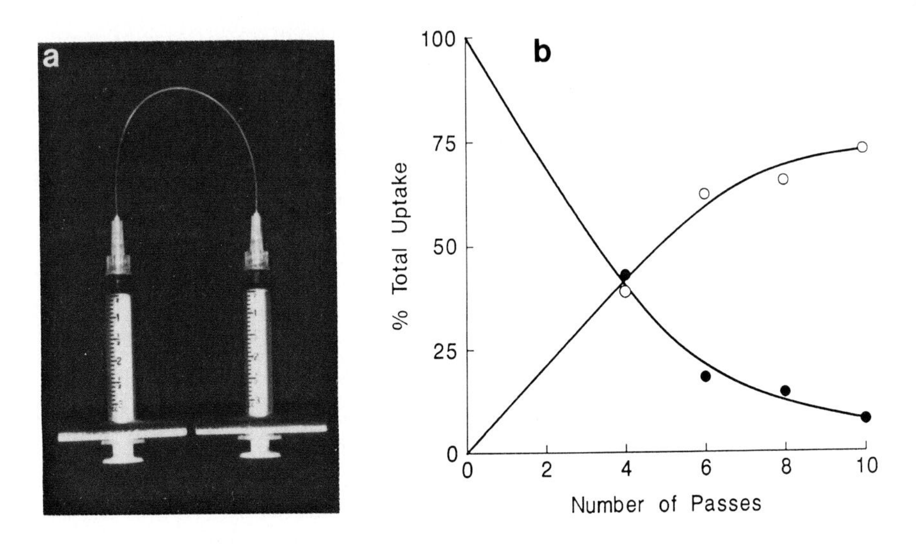

FIG. 2. (a) Photograph of the homogenizer device and (b) a test for its efficiency. Cells were allowed to internalize ^{125}I-labeled β-glucuronidase for 10 minutes at 37°C, washed, and forced to pass from one syringe to the other several times. Aliquots were collected at different steps of homogenization and centrifuged first at 800 *g* for 5 minutes, and then at 40,000 *g* for 15 minutes. Radioactivity in the first pellet corresponds to ligand associated with intact cells (●), while radioactivity in the second pellet corresponds to ligand present inside free endosomes (○). The final supernatant contains ligand that has leaked from damaged vesicles.

conditions <10% of the cells exclude trypan blue and <30% of the internalized ligands is in the extravesicular compartment after homogenization. The homogenate is then centrifuged at low speed (800 *g*, 5 mintues) to eliminate unbroken cells and nuclei. This preparation can be kept at 4°C for several hours or stored frozen in liquid nitrogen. A good homogenate for fusion assays should have a protein concentration not lower than 4 mg/ml.

Endosomes containing the probes can be further purified from this crude preparation. Cytosolic factors can be eliminated by pelleting the vesicles at 15,000 *g* for 30 minutes. Most of the fusogenic vesicles are sedimented at this velocity, and they remain fusogenic after resuspension in appropriated medium. To purify endosomes on Percoll gradients, vesicles from 5–10 × 10^7 cells are pelleted at 40,000 *g* for 15 minutes. The pellet is resuspended in 1 ml of homogenization buffer, homogenized by douncing to eliminate aggregates, and mixed with 30 ml of Percoll (density 1.05 g/ml or 1.04 g/ml) prepared in 0.25 *M* sucrose and containing 1 mg/ml BSA. Gradients are formed by centrifugation at 40,000 *g* for 45 minutes in a fixed-angle rotor. At density 1.05 g/ml the endosomes are recovered in light fractions together with plasma membrane markers, and separated from lyosomal markers (Fig. 3a). At 1.04 g/ml density, endosomes equilibrated in the bottom fractions while plasma membrane vesicles are recovered in light fractions (Fig. 3b). In order to remove the Percoll and concentrate the vesicles, endosomal fractions have to be centrifuged at 100,000 *g* for 2 hours and the pellet resuspended in homogenization buffer. Unfortunately, after all these manipulations fusion among these vesicles is less efficient.

C. Effect of Chase

Dinitrophenol-β-glucuronidase is easily chased to later endocytic compartments using the following protocol: cells are allowed to internalize the

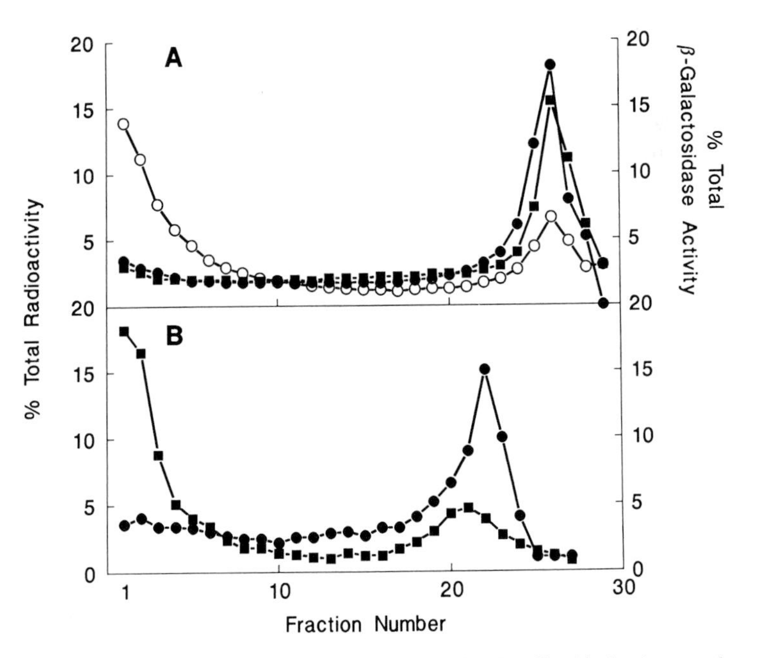

FIG. 3. Subcellular fractionation of radioactive probes localized in 5-minute endosomes on Percoll gradients. (A) Distribution of endosomes in the dense Percoll (1.05 g/ml starting density). β-Galactosidase activity (○) was used as a lysosomal marker. ■, DNP-β-glucuronidase; ●, Man-IgG. (B) Distribution in the light Percoll (1.04 g/ml starting density). Iodinated mouse IgG, aggregated with rabbit anti-mouse IgG (●) to make it a ligand for the macrophage Fc receptor, was bound to the cells at 4°C for 2 hours and used as a marker for plasma membrane-derived vesicles. ■, DNP-β-glucuronidase.

ligand for 5 minutes at 37°C; they are then washed once with 5 m*M* EDTA in PBS to eliminate the surface-associated ligand, and once with uptake medium. The cells are resuspended in warm uptake medium (2×10^7 cells/ml) and incubated at 37°C for the required chase time. The chase is stopped with cold uptake medium and the cells are washed with homogenization buffer and homogenized as described. After a 30-minute chase most of the enzyme sediments in heavy fractions in a Percoll gradient (starting density 1.05 g/ml), together with the enzymatic activity of lyosomal enzymes and separated from endosomal markers (Fig. 4). Very early during the chase, DNP-β-glucuronidase-containing vesicles lose their ability to fuse with 5-minute endosomes. Mannosylated anti-DNP IgG is not transported to lysosomes as efficiently; 60% of the antibody reappears in the media without degradation after a 30-minute chase. The basis for this observation is under investigation.

The low fusion capability of late endocytic compartments has been pointed out in other *in vitro*-reconstitution systems of endosome–endosome fusion (Braell, 1987; Gruenberg and Howell, 1987). It seems that vesicles that pinch off from the plasma membrane are highly fusogenic, and that this property is transient. Fusion at this stage is also specific; that is, early endosomes will not fuse with compartments farther along the endocytic pathway.

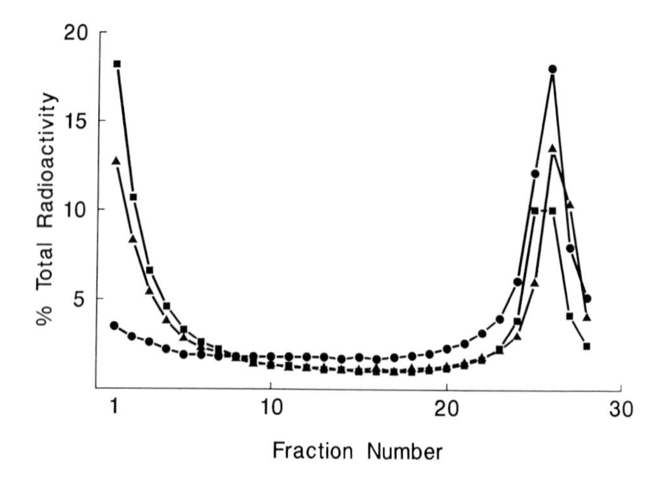

FIG. 4. Subcellular distribution of ^{125}I-labeled DNP-β-glucuronidase internalized for 5 minutes and chased for different periods of time on Percoll gradients (1.05 g/ml starting density): ●, 0-minute chase; ▲, 10-minute chase; ■, 30-minute chase.

IV. General Requirements for Endosome–Endosome Fusion

In the standard assay for endosome–endosome fusion reconstitution, vesicles containing Man-IgG are mixed with vesicles containing DNP-β-glucuronidase in a final volume of 20 μl at 37°C for 30 minutes in a buffer containing 0.25 *M* sucrose, 0.5 m*M* EGTA, 20 m*M* HEPES (pH 7.0), 1.5 m*M* $MgCl_2$, 50 m*M* KCl, 1 m*M* dithiothreitol (DTT), and 50 μg/ml DNP-BSA. The medium also contains an ATP-regenerating system consisting of 1 m*M* ATP, 8 m*M* creatine phosphate, and 31 units/ml creatine phosphokinase. Fusion reactions can be stopped by cooling at 4°C. To precipitate the immune complex formed, vesicles are solubilized by addition of 150 μl of solubilization buffer [1% Triton X-100, 0.2% methylbenzethonium chloride, 1 m*M* EDTA, 0.1% BSA, 150 m*M* NaCl, 10 m*M* Tris-HCl (pH 7.4)] containing 50 μg/ml of DNP-BSA and 2 μl of *Staphylococcus* A (Staph A, 10% suspension) coated with rabbit anti-mouse IgG. Staph A (IgGsorb, The Enzyme Center) is coated by incubation of 2 μl of a 10% bacteria suspension with 1 μl of rabbit anti-mouse IgG (rabbit IgG fraction, 4 mg/ml, Organon Teknika Corporation) for 30 minutes at room temperature, followed by three washes with solubilization buffer. The samples are incubated at 4°C for 30 minutes, diluted with 1 ml of solubilization buffer, and pelleted at 1500 *g* for 5 minutes. The Staph-A bound immunoprecipitates are then washed twice with 1 ml of solubilization buffer. To quantify the immune complex formed, pellets are resuspended in 100 μl of solubilization buffer and an equal volume of β-glucuronidase substrate is added (4 m*M* 4-methylumbelliferyl-β-D-glucuronide in 0.1 *M* acetate buffer, pH 4.5). Samples are then incubated at 37°C for 1–2 hours and the reaction stopped with 1 ml of glycine buffer (133 m*M* glycine, 67 m*M* NaCl, 83 m*M* Na_2CO_3 adjusted to pH 9.6 with NaOH). The fluorescence of umbelliferone is measured in a spectrofluorometer at 366 nm excitation, 450 nm emission, or in a fluorometer with appropriate filters.

Several control tubes can be added to the assay. Tubes containing detergent during the fusion reaction measure the extravesicular formation of immune complex. The presence of DNP-BSA in the media keeps the activity of these tubes fairly low. Tubes with detergent but without DNP-BSA give a measurement of the total immune complex in the system. Frequently we express fusion efficiency as a percentage of this amount, because it represents the activity that should result from the complete mixing of the compartments containing the probes. It overestimates this amount because some of the ligands are not in sealed vesicles.

The proportion of a given probe in closed vesicles can be estimated by measuring the proportion of immune complex that is obtained when vesicles containing the ligand are mixed with a solution of the complementary probe in the presence and absence of detergent.

The rate of fusion increases with the total amount of vesicles present in the system. However, at very high concentrations some components of the reaction (e.g., ATP) can be depleted. For postnuclear homogenates, good fusion is obtained by mixing 30–50 μg of total proteins from each preparation in a 20-μl final volume.

A. Time and Temperature

Under the conditions described earlier, the amount of immune complex increases with time, reaching a plateau after 30–45 minutes of incubation at 37°C (Fig. 5). At 4°C the vesicles are rather stable and no fusion can be observed after prolonged incubations. The enzymatic activity of immunoprecipitates obtained from samples incubated at this temperature is minimal and represents mostly background activity of the immunoprecipitation procedure. Very low fusogenic activity is observed at temperatures <20°C. Fusion increases linearly with subsequent increases in temperature, and a maximum is reached at 37°C.

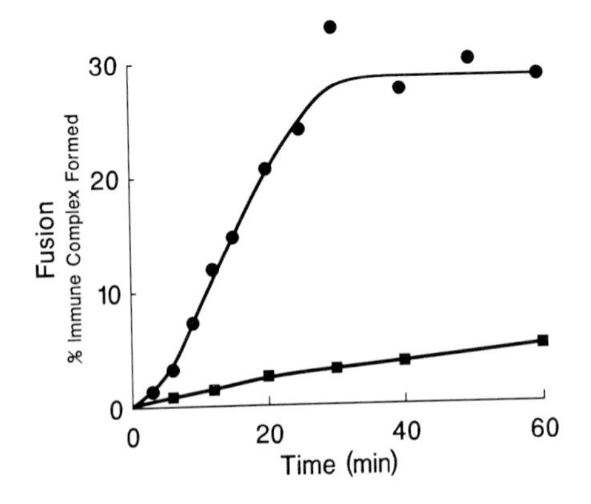

FIG. 5. Time course of endosome–endosome fusion *in vitro*. Increase of immune complex formed upon incubation at 37°C in the presence of an ATP-regenerating system (●) or an ATP-depleting system (■). The activity in the immunoprecipitates was compared with the activity obtained when detergent was included in the fusion reaction, in the absence of DNP-BSA.

B. Salts and pH

No fusion is observed if sucrose is substituted for KCl in the buffer. The salt requirement is not specific for KCl because NaCl or potassium gluconate can substitute for KCl without lost of activity. Buffers with higher salt concentrations, like the one used by Braell (1987) in his assay, also support fusion in our system. However, cell disruption is harder to attain in these buffers, and some aggregation of vesicles may occur when the homogenates are frozen and thawed.

In the presence of millimolar concentratiosn of Ca^{2+}, fusion is observed even in the presence of an ATP-depleting system (5 m*M* mannose, 25 units/ml of hexokinase). Similarly, fusion in the absence of ATP is observed when EGTA is excluded from the homogenization buffer. This suggests that a fusion mechanism that does not require energy, and that is mediated in some way by Ca^{2+}, may exist in our system. Magnesium does not promote fusion at concentrations <5 m*M*. ATP-dependent fusion is only slightly affected by changes in pH between 6.5 and 7.4. However, ATP-independent fusion increases below pH 6.5. Ca^{2+} is more efficient in promoting fusion at low pH, consistent with the possibility that some fusogenic protein, like those associated with secretory granules (Creutz, 1981), may have a role in the calcium-mediated fusion.

C. Energy

When EGTA is included in the homogenization buffer and in the fusion reaction, ATP addition is an absolute requirement for fusion; almost no activities are obtained in the presence of an ATP-depleting system or even when the regenerating system is not included in the reaction. It appears that ATP is consumed during the reaction, because adenyl-imidodiphosphate, a nonhydrolyzable ATP analog, does not support fusion. Fusion is observed in the presence of GTP, but the use of crude preparations does not permit determination of the exact nucleotide requirement for fusion.

V. Cytosol and Membrane-Associated Factors

Fusion of endosomes not only requires the appropriate temperature, ionic strength, and energy supply, but also cytosolic and membrane-associated factors, presumably proteins.

A. Cytosolic Factors

Vesicles can be separated from cytosol by centrifugation (15,000 *g*, 30 minutes) and resuspended in fusion buffer. No fusion is observed with this vesicular preparation, unless cytosol is added to the resuspension buffer. The factor(s) required seems to be a protein, because cytosol is inactivated by trypsin digestion (0.4 mg/ml for 2 hours at 4°C followed by 0.8 mg/ml soybean trypsin inhibitor), boiling for 5 minutes, and ultrafiltration (Centricon microconcentrator, 10-kDa cutoff), whereas dialysis does not affect its activity.

Cytosol is inactivated by the alkylating agent *N*-ethylmaleimide (1 m*M* NEM for 30 minutes at 4°C followed by 2 m*M* DTT to quench the excess of NEM). This suggests that an NEM-sensitive factor has an important role in endosome–endosome fusion. This factor seems to be mostly soluble, because NEM treatment of vesicles only produces a partial inhibition when normal cytosol is present in the assay.

Fibroblast (L929) and rabbit alveolar macrophage cytosol may substitute for J774-E cytosol in the fusion assay. Brain and liver extracts are also active. Cytosol can be obtained by high-speed centrifugation of postnuclear homogenates. These preparations can be stored at −80°C after freezing in liquid nitrogen.

B. Membrane-Associated Factors

Only early endosomes are active in our *in vitro* assay. Other vesicular compartments (e.g., late endosomes and lysosomes) do not fuse with early endosomes under the conditions described earlier. This specificity of fusion is probably provided by factors present in the membrane of the fusogenic vesicles. Membrane factors can be assayed independently of cytosolic factors by using a vesicular fraction obtained by centrifugation. Incubation of vesicles with trypsin (20 μg/ml for 30 minutes at 4°C, followed by 40 μg/ml of soybean trypsin inhibitor) completely blocks fusion. *N*-Ethylmaleimide treatment of the vesicles (1 m*M* for 30 minutes at 4°C, followed by 2 m*M* DTT to block the unreacted NEM) only decreases the fusion activity by 20%, indicating that the cytosol added to the assay can supply the NEM-sensitive factor required for the fusion reaction.

VI. Conclusions

We present an assay that permits the *in vitro* reconstitution of fusion events early in the endocytic cycle. For this purpose we have developed two probes, which have proved suitable for these studies because both are efficiently internalized by receptor-mediated endocytosis and, when present in the same compartment, form a complex that can be easily quantified by enzymatic analysis.

Several assays have been developed in the last few years to assess fusion between endosomes in cell-free systems. All of them have produced similar results even though they use (i) different probes, (ii) different ways to assess fusion, (iii) different cell lines, and (iv) different ways to enter into the endocytic pathway. This suggests that the fusion process that is being reconstituted in the test tube is a relevant step during endocytosis and not an artifact of a particular *in vitro* assay. The basic requirements of this fusion step have been defined, and they show some common features with the transport of proteins from the endoplasmic reticulum to the Golgi apparatus, and between different Golgi cisternae. Probably the most interesting similarities are the requirements of energy, cytosol, and an NEM-sensitive factor. Some differences are also evident; while Golgi transport is blocked by NEM treatment of the vesicles, endosomes still fuse after this treatment (Balch and Rothman, 1985; Glick and Rothman, 1987).

The probes that we have developed may potentially be used to study other fusion steps along the intracellular route followed by endocytosed ligands. Reconstitution of these events in cell-free systems will help clarify the role of fusion in the endocytic pathway. *In vitro* assays may also be applied to the study of the molecular mechanisms involved in the specific recognition between fusogenic vesicles and membrane fusion.

Acknowledgments

We would like to thank Elizabeth M. Peters and Lia E. B. Mayorga for expert technical assistance and Rita L. Boshans for critical review of the manuscript. This work was supported, in part, by Health, Education and Welfare Grants CA 12858 and AI 20015. L. M. is a Fellow of the National Research Council of Argentina.

References

Balch, W. E., and Rothman, J. E. (1985). *Arch. Biochem. Biophys.* **240,** 413–425.

Balch, W. E., Dunphy, W. G., Braell, W. A., and Rothman, J. E. (1984). *Cell (Cambridge, Mass.)* **39,** 405–416.

Beckers, C. J. M., Keller, D. S., and Balch, W. E. (1987). *Cell (Cambridge, Mass.)* **50,** 523–534.

Braell, W. M. (1987). *Proc. Natl. Acad. Sci. U.S.A.* **84,** 1137–1141.

Creutz, C. E. (1981). *J. Cell Biol.* **91,** 247–256.

Davey, J., Hurtley, S. M., and Warren, G. (1985). *Cell (Cambridge, Mass.)* **43,** 643–652.

Diaz, R., Mayorga, L., and Stahl, P. (1988). *J. Biol. Chem.* **263,** 6093–6100.

Diment, S., Leech, M. S., and Stahl, P. D. (1987). *J. Leuk. Biol.* **42,** 485–490.
Duncan, J. R., and Kornfeld, S. (1988). *J. Cell Biol.* **106,** 617–628.
Ferris, A. L., Brown, J. C., Park, R. D., and Storrie, B. (1987). *J. Cell Biol.* **105,** 2703–2712.
Fishman, J. B., and Fine, R. E. (1987). *Cell (Cambridge, Mass.)* **48,** 157–164.
Glick, B. S., and Rothman, J. E. (1987). *Nature (London)* **326,** 309–312.
Gruenberg, J. E., and Howell, K. E. (1986). *EMBO J.* **5,** 3091–3101.
Gruenberg, J. E., and Howell, K. E. (1987). *Proc. Natl. Acad. Sci. U.S.A.* **84,** 5758–5762.
Keller, R. K., and Touster, O. (1975). *J. Biol. Chem.* **250,** 4765–4769.
Lee, Y. C., Stowell, C. P., and Krantz, M. J. (1976). *Biochemistry* **15,** 3956–3963.
Lennartz, M. R., Wileman, T. E., and Stahl, P. D. (1987). *Biochem. J.* **245,** 705–711.
Lodish, H. F., Kong, N., Hirani, S., and Rasmussen, J. (1987). *J. Cell Biol.* **104,** 221–230.
Nir, S., Klappe, K., and Hoekstra, D. (1986). *Biochemistry* **25,** 8261–8266.
Orci, L., Glick, B. S., and Rothman, J. E. (1986). *Cell* **46,** 171–184.
Otsuka, F. L., Welch, M. J., McElvany, K. D., Nicolotti, R. A., and Fleischman, J. B. (1984). *J. Nucl. Med.* **25,** 1343–1349.
Roth, J. (1983). *In* "Techniques in Immunocytochemistry" (G. R. Bullock and P. Petrusz, eds.), Vol. 2, pp. 217–284. Academic Press, London.
Wileman, T., Harding, C., and Stahl, P. (1985). *Biochem. J.* **232,** 1–14.
Wilschut, J., Nir, S., Scholma, J., and Hoekstra, D. (1985). *Biochemistry* **24,** 4630–4636.
Woodman, P. G., and Warren, G. (1988). *Eur. J. Biochem.* **173,** 101–108.

Chapter 11

Fusion of Endocytic Vesicles in a Cell-Free System

PHILIP G. WOODMAN AND GRAHAM WARREN

The Imperial Cancer Research Fund
Lincoln's Inn Fields
London WC2A 3PX, England

I. Introduction

The pathway of receptor-mediated endocytosis in mammalian cells has been established by morphological and kinetic studies of the internalization of a variety of ligands (Brown *et al.*, 1983). Ligands bind to specific receptors on the cell surface. These receptors are located in, or migrate upon ligand binding to, clathrin-coated areas of membrane, which then invaginate to form coated vesicles. After uncoating, these vesicles are targeted to the endosome and membrane fusion occurs. The nature of the interaction that promotes recognition between organelles and the mechanism of fusion are not known. Information should be obtained by applying biochemical techniques to the problem, as has been achieved for other intracellular transport events (Balch *et al.*, 1984). With this goal in mind, we have reconstituted the fusion of endocytic vesicles involved in the receptor-mediated endocytosis of transferrin (Woodman and Warren, 1988), the mechanism by which cells accumulate ferric iron.

The transferrin cycle has been particularly well characterized. In A431 cells, a line rich in transferrin receptors, transferrin bound at 4°C is internalized rapidly upon warming (Bleil and Bretscher, 1982; Hopkins and Trowbridge, 1983). Such an efficient and synchronized burst of endocytosis makes this an attractive system to work with. The principle of the fusion assay is described in Fig. 1. Endocytic vesicles containing ^{125}I-labeled transferrin (^{125}I-transferrin), prepared from donor cells to which ^{125}I-transferrin has been bound and internalized, are mixed with vesicles carrying endocytosed antitransferrin antibody, prepared from acceptor cells. Vesicle fusion permits the formation of a radiolabeled immune complex, which can be recovered from the incubation by the addition of fixed *Staphylococcus aureus* (Staph A) cells in a detergent-containing buffer. The assay is rapid and simple. It is also sensitive, owing to the high specific activity of ^{125}I-transferrin. Furthermore, because the assay measures the formation of a stoichiometric complex, the assay should enable the efficiency of vesicle fusion to be determined accurately.

II. Methods

A. Materials

Creatine phosphate (CP) and creatine phosphokinase (CPK) are obtained from Boehringer Mannheim. All other reagents, unless specified, are obtained from Sigma.

B. Cells

A431 cells were maintained in Dulbecco's modified Eagle medium (DMEM) supplemented with 10% (v/v) fetal calf serum and 100 U/ml of both penicillin and streptomycin. All media and supplements can be obtained from Flow Laboratories (Richmansworth, England).

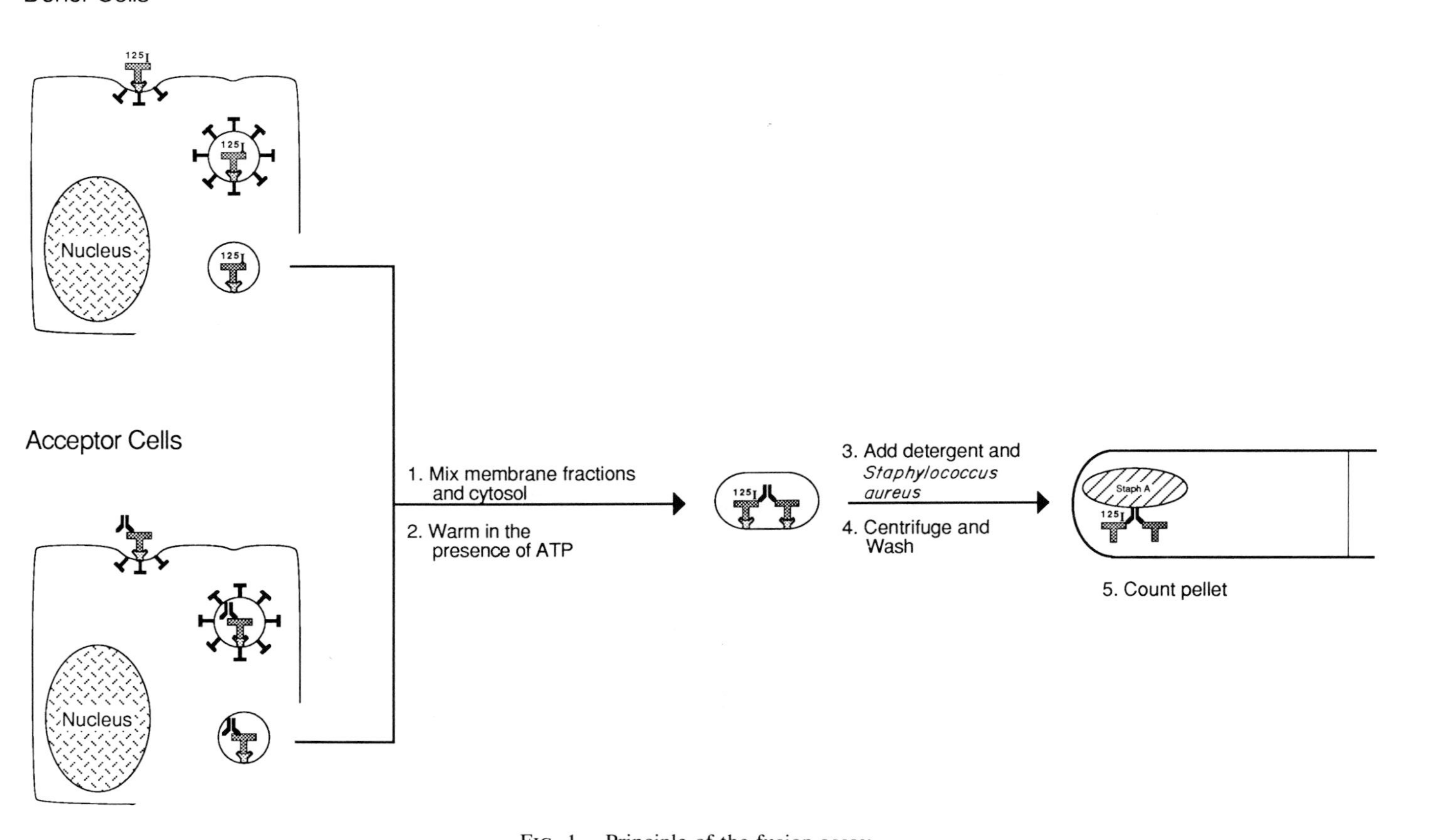

FIG. 1. Principle of the fusion assay.

C. Antibodies

The success of the assay relies on the production of sera with high titers of antibodies against native human transferrin.

1. Rabbit Antitransferrin

Inject native human transferrin, emulsified in an equal volume of Freund's complete adjuvant, into a single popliteal lymph node (25 μg; 0.25 ml) and intradermally at six sites (75 μg; 0.75 ml total) (Goudie *et al.*, 1966) into a Dutch Half-lop rabbit. The exposed lymph node can be identified by injection of 1 ml of a solution of Evans blue dye (0.5% w/v, in PBS) into the footpad of the animal 1 hour before operating. Boost after 3 weeks with antigen, in Freund's incomplete adjuvant, injected ip (50 μg; 0.5 ml) and sc (two sites: 50 μg; 0.5 ml total), and again after a further week (sc, two sites: 100 μg; 1.0 ml total). First bleeds can be taken 1 week after the second boost, and thereafter at weekly intervals. All sera are treated at 56°C for 30 minutes and stored at 4°C. Titers of sera may be assessed by immunoprecipitation (see Section II,H for details) of ^{125}I-transferrin onto Staph A cells; typically, 1 μl of serum can precipitate 0.2 μg of transferrin.

1. Sheep Antitransferrin

Alternatively, we have used a sheep antitransferrin antiserum obtained from the Scottish Antibody Production Unit (SAPU; Carluke, Scotland). Results (not shown) indicate that this antibody recognizes more epitopes on human transferrin than the rabbit antibody does, and therefore more active acceptor preparations can be produced (see Section II,F). The serum is a pool of selected bleeds from sheep immunized with native human transferrin. Serum (1 μl) can immunoprecipitate 1 μg of ^{125}I-transferrin onto Staph A cells that have been pretreated with rabbit antisheep antibody (see Section II,H for details of the immunoprecipitation protocol).

D. Radiolabeling

We have found that the Iodogen method can be used to radioiodinate transferrin to a high specific activity (Fraker and Speck, 1978) with minimal loss of specific binding activity. Evaporate 20 μl of a solution of Iodogen (1,3,4,6-tetrachloro-3α,6α-diphenylglycoluril; Pierce Chemical Co., Rockford, IL) (0.5 mg/ml in chloroform) in a glass tube under a stream of nitrogen. To this, add 50 μg human transferrin in 20 μl 0.1 *M* sodium phosphate buffer, pH 7.2, followed by 2 mCi (20 μl) $Na^{125}I$ (16 mCi/μg; Amersham). Incubate the reactants for 15 minutes on ice before stopping the reaction by diluting with 200 μl of the phosphate buffer and removing from the Iodogen tube. The protein is separated from free iodine on a 10-ml Sephadex G-50 column run in phosphate buffer. Collect the void volume and dialyze overnight against 200 volumes of phosphate buffer. The specific activity is ~3×10^7 dpm/μg. Store at 4°C for up to 2 weeks.

E. Donor Preparations

Seed A431 cells at ~2×10^7 cells on a 24 × 24 cm tissue culture dish, supplied by GIBCO (Paisley, Scotland). Grow to near confluence (~10^8 cells). Wash four times with ice-cold Dulbecco's phosphate-buffered saline, pH 7.4 (PBS), and incubate, with rocking, for 2 hours at 4°C with 1 μg/ml ^{125}I-transferrin (specific activity ~3×10^7 dpm/μg) in 15 ml binding medium [DMEM containing 20 m*M* HEPES, pH 7.4, and 0.2% w/v bovine serum albumin (BSA)]. Discard the medium and wash the cells three more times with ice-cold PBS before applying 50 ml fresh, prewarmed binding medium. Incubate the cells for 10 minutes at 31°C in a water bath, then wash four times in ice-cold buffer (100 m*M* KCl, 85 m*M* sucrose, 20 m*M* HEPES, pH 7.4; KSH). After draining the dishes for 2 minutes at 4°C to remove excess buffer, scrape the cells off with a rubber policeman to yield a suspension in ~1.5 ml and homogenize on ice by passing the suspension six times through a 8.02-mm bore in a stainless-steel block containing an 8.004-mm-diameter ball (Balch *et al.*, 1984). Cell breakage can be assessed by trypan blue staining. It is important to obtain 80–90% cell breakage with minimum damage to nuclei. Centrifuge the homogenate for 10 minutes at 500 g_{av} at 4°C to obtain a postnuclear supernatant. Apply this directly to a discontinuous sucrose gradient of 2.5 ml 40% (w/v) sucrose in 10 m*M* Tris-HCl, pH 7.5, overlaid with 8 ml of 20% sucrose in the same buffer. Centrifuge for 2 hours at 155,000 g_{av} (35,000 rpm in a Beckman SW40 Ti rotor). Recover the crude membrane preparation from the 20%–40% interface by tube puncture. It is essential not to overload the sucrose gradients, as this causes aggregation of membranes at the interface. Preparations (typically 2–3 mg/ml) can be stored in liquid nitrogen (0.1-ml aliquots).

F. Acceptor Preparations

Grow cells as for donor preparations. Wash cells (~10^8 on a 24 × 24 cm dish) four times in ice-cold PBS and incubate, with rocking, for 1 hour at 4°C within 10 μg/ml unlabeled human transferrin in 15 ml binding medium. After four further washes in ice-cold PBS, incubate cells with 15 ml binding medium containing heat-inactivated, antitransferrin serum for a further 1.5 hour at 4°C. Production of active acceptor preparations requires internalization of antibodies that possess free sites for the binding of ^{125}I-transferrin after vesicle fusion has occurred. To achieve this, apply supersaturating amounts of antibody to the surface-bound transferrin when making acceptor preparations. Use 3–5 ml rabbit antiserum, or 1 ml sheep antiserum, based on the antibody titers given before. These amounts yield ~2 molecules of rabbit antibody, or ~10 molecules of sheep antibody, bound per molecule of receptor-bound transferrin. Wash the cells three more times in PBS and incubate at 31°C in a water bath with 50 ml prewarmed binding medium. The period of incubation at 31°C should be chosen to maximize the population of internalized antibody. We have found that the rabbit antibody is endocytosed rapidly and, therefore, incubation of cells for 5 minutes is sufficient. In contrast, the sheep antitransferrin antibody, which, from our binding experiments, appears to recognize more epitopes on the receptor-bound transferrin and is therefore likely to form a larger structure on the cell surface, is internalized more slowly (results not shown). An incubation time of 15 minutes is required to reach a peak in the internal pool of antibody. It is possible that the internalization kinetics will vary according to the antibody, and so should be tested for each animal. Wash, scrape, and homogenize the cells as described for the donor preparations. Load the postnuclear supernatant onto sucrose gradients and proceed as for the donor preparations. Protein concentrations are similar to those for donor preparations. Store in liquid nitrogen.

G. Cytosol Fractions

Wash confluent monolayers of A431 cells (~5×10^8 cells on five 24×24 cm dishes) four times in KSH buffer and drain for 2 minutes. Scrape cells and homogenize as described previously. Centrifuge in an SW 50.1 rotor for 1 hour at 225,000 g_{av} (49,000 rpm). Remove the supernatant from the top using a syringe needle, taking care not to disturb the fatty layer at the surface. Apply the supernatant onto a Biogel P-6DG (Bio-Rad Laboratories) desalting column ($V_t \times 40$ ml) equilibrated with KSH buffer. Pool the protein-rich fractions from the void volume (~5 mg/ml protein). Freeze and store aliquots in liquid nitrogen.

H. Assay Conditions

For a standard assay, donor and acceptor membranes are combined in the presence of cytosol. An excess of unlabeled transferrin is also included in the incubation to quench the formation of radiolabeled immunecomplexes that would otherwise form from the interaction of ^{125}I-transferrin and antibody present in unsealed vesicles and on membrane sheets that arise during homogenization of the cells. Parallel incubations are performed in the presence of either ATP-regenerating or ATP-depleting cocktails. For a standard assay (100 μl volume), mix the following components in the order given; 10 μl transferrin (1 mg/ml in buffer A); either 10 μl of an ATP-regenerating cocktail containing (in water) 10 m*M* ATP (buffered to pH 7.0 with NaOH), 10 m*M* magnesium acetate, 50 m*M* CP, and 80 IU/ml CPK, or 10 μl of an ATP-depleting cocktail comprising hexokinase (supplied as an ammonium sulfate precipitate, pelleted and resuspended to 500 IU/ml in 50 m*M* glucose) in water. Add 4 μl cytosol, diluted in KSH buffer to give a final protein concentration of 2 mg/ml; 30 μl acceptor preparation diluted 3-fold with 10 m*M* Tris-HCl, pH 7.5, to restore the osmolarity approximately to normal; 6 μl donor preparation (usually 10,000–20,000 dpm), also diluted 3-fold with 10 m*M* Tris-HCl, pH 7.5. Incubate samples for 1.5 hour at 37°C. Then, cool the tubes on ice and add 900 μl immunoprecipitation buffer (IB; 0.1 *M* Tris-HCl, pH 8.0, 0.1 *M* NaCl, 5 m*M* $MgCl_2$, 1% w/v Triton X-100, 0.5% w/v SDS, 1% w/v sodium deoxycholate, 0.1% w/v BSA) to solubilize the membranes. Where acceptor preparations have been made using rabbit antitransferrin serum, mix fixed Staph A cells [20 μl of a 10% suspension of "Pansorbin" (Calbiochem), washed three times over 1 hour in IB] with each sample, and incubate for 1 hour at 4°C. Wash the cells by pelleting at low speed in an MSE "microcentaur" microfuge for 4 minutes at room temperature and remove the supernatant carefully using a syringe needle connected to a vacuum line. Resuspended pellets in 1 ml IB and repeat the washing procedure twice. Count the final pellet for radioactivity in a γ counter. Where the assay is performed using acceptor preparations made with sheep antitransferrin antibodies, the washed Staph A cells should be pretreated with a rabbit antisheep antibody. Incubate 1 ml Staph A with 20 μl rabbit antisheep IgG (6.0 mg IgG per milliliter; Cooper Biomedical) for 30 minutes at 4°C and wash by pelleting. Resuspended in 1 ml IB and proceed as described earlier.

III. Expected Results

The observed efficiency of fusion will depend on the immunoprecipitating activity within the acceptor preparation. It is best to titrate each preparation by incubating for the stated time at 37°C with a donor preparation in the presence of 1% (w/v) Triton X-100 and the absence of unlabeled transferrin. The amount of acceptor used in fusion assays should immunoprecipitate at least 80% of the ^{125}I-transferrin in the donor preparation under these conditions. Incubation of samples in the fusion assay for 90 minutes at 37°C should produce an ATP-dependent signal of between 10 and 20% of the total radioactivity in the donor preparation, above a background (ATP-independent) of 0.1–1%.

Having obtained an ATP-dependent signal, controls should be performed to confirm that it is the consequence of vesicle fusion. The results in Table I show that Triton X-100 abolishes the ATP-dependent formation of an immune complex, suggesting that intact vesicles are required. Furthermore, the result is seen only when both ^{125}I-transferin and antibody are within sealed vesicles. Together, these strongly suggest that the ATP-dependent signal is the result of vesicle fusion. It should also be demonstrated that the hydrolysis of ATP is required to produce the signal, as shown in Table II. Here, ATP by itself can support vesicle fusion, in contrast to the nonhydrolyzable analog, adenylyl-imidodiphosphate (AMP-PNP). However, the extent of vesicle fusion is increased greatly by the inclusion of a regenerating cocktail, which maintains the concentration of ATP >0.5 m*M* over the course of the incubation (results not shown). In the absence of ATP, the regenerating cocktail does not support fusion. Furthermore, the nonhydrolyzable analog of ATP is able

TABLE I

DONOR AND ACCEPTOR ACTIVITIES MUST RESIDE WITHIN SEALED VESICLES[a]

^{125}I-transferrin	Antibody Incubation	ATP-dependent immunoprecipitation of ^{125}I-transferrin (% of control)
Donor	Acceptor—	100 ± 12
Donor	Acceptor Triton X-100	−5 ± 2
Exogenous	Acceptor—	−6 ± 3
Exogenous	Exogenous—	9 ± 12
Donor	Exogenous—	2 ± 5

[a] Donor preparations were made as described in Section II,E, or by the addition of an equal activity of ^{125}I-transferrin to membranes prepared from mock-treated A431 cells. Acceptor preparations were prepared using the sheep antitransferrin antibody as described in Section II,F, or by adding sheep antitransferrin antibody exogenously to membranes from mock-treated cells. Also shown are the results of incubations of standard preparations in the presence of 1% (w/v) Triton X-100. Results are means of triplicate samples ± SEM from a single experiment.

TABLE II

REQUIREMENT OF ATP HYDROLYSIS FOR FUSION[a]

Addition	ATP-dependent immunoprecipitation of ^{125}I-transferrin (% of control)
None	4.3 ± 1.0
Hexokinase (50 U/ml) + glucose (5 m*M*)	2.6 ± 0.8
CP (5 m*M*) + CPK (8 U/ml)	3.5 ± 0.5
ATP (1 m*M*)	20.8 ± 0.2
AMP-PNP (1 m*M*)	3.8 ± 1.2
CP + CPK + 1 m*M* ATP	100 ± 1.3
CP + CPK + 2 m*M* ATP	117.1 ± 1.2
CP + CPK + 1 m*M* ATP + 1 m*M* AMP-PNP	64.9 ± 2.4

[a] Donor and acceptor (made with sheep antitransferrin antibody) preparations were incubated with gel-filtered cytosol as described, except that the ATP-depleting and ATP-regenerating cocktails were replaced with the components indicated. Adenine nucleotides were added equimolar with magnesium acetate. Results are expressed as the means ± SEM of triplicate samples from a single experiment.

to reduce the efficiency of fusion when mixed with the ATP-regenerating cocktail.

IV. Future Prospects

We have described an assay that measures the fusion of endocytic vesicles carrying internalized trasnferrin. The markers for the assay are carried on the ligand itself, rather than being associated with an intracellular compartment. Therefore, it should be possible, by manipulating conditions such as incubation time, to use this assay to examine other steps in the transferrin cycle, such as the recycling of transferrin in vesicles from the endosome to the plasma membrane. Furthermore, the principle of the assay can be applied to the study of the receptor-mediated endocytosis of ligands that follow different routes within the cell. All that is required is the labeled ligand and the appropriate antibody.

ACKNOWLEDGMENTS

We thank Dr. James Pryde for immunizing the rabbits and for help in the preparation of the text. This work is supported by the Wellcome Trust.

References

Balch, W. E., Dunphy, W. G., Braell, W. A., and Rothman, J. E. (1984). *Cell (Cambridge, Mass.)* **39,** 405–416.

Bleil, J. D., and Bretscher, M. S. (1982). *EMBO J.* **1,** 351–355.

Brown, M. S., Anderson, R. G. W., and Goldstein, J. L. (1983). *Cell (Cambridge, Mass.)* **32,** 633–667.

Fraker, P. J., and Speck, J. C. (1978). *Biochem. Biophys. Res. Commun.* **80,** 849–857.

Goudie, R. B., Herne, C. H. W., and Wilkinson, P. A. (1966). *Lancet* **2,** 1224.

Hopkins, C. R., and Trowbridge, I. S. (1983). *J. Cell Biol.* **97,** 508–521.

Woodman, P. G., and Warren, G. (1988). *Eur. J. Biochem.* **173,** 101–108.

Chapter 12

Purification and Biochemical Assay of Synexin and of the Homologous Calcium-Dependent Membrane-Binding Proteins, Endonexin II and Lipocortin I

HARVEY B. POLLARD,* A. LEE BURNS,* EDUARDO ROJAS,* D. D. SCHLAEPFER,† HARRY HAIGLER,† AND KEITH BROCKLEHURST*

**Laboratory of Cell Biology and Genetics*
National Institute of Diabetes, Digestive and Kidney Diseases
National Institutes of Health
Bethesda, Maryland 20892
†Department of Physiology and Biophysics
University of California
Irvine, California 92717

I. Synexin, Granule Aggregation, and Membrane Fusion

The process of exocytosis initially involves simple contact and fusion between the granule membrane and the plasma membrane of the secreting cell. However, in many cells this simple exocytotic process is followed by fusion of intact granules to the membranes of expended granules still fused to the plasma membrane. This is called compound or "piggyback" exocytosis, and is believed to spare the granules more deeply situated in the cytosol the difficulty of wending their way through the cytoskeletal meshwork to reach the plasma membrane. Whatever its purpose, however, it is believed that the processes regulating interaction between granule and plasma membranes during simple exocytosis are similar or identical to those regulating interaction of granule membranes during compound exocytosis. Calcium is believed to be involved in both membrane fusion processes, and a search for proteins in the adrenal medulla involved in mediating calcium-dependent granule membrane contact and fusion resulted in the discovery of synexin (Creutz *et al.*, 1978). Later studies have revealed that synexin is a voltage-dependent calcium channel (Rojas and Pollard, 1987; Pollard and Rojas, 1988), and that synexin is homologous with a set of proteins characterized by their ability to bind to biological membranes or purified acidic phospholipids in a calcium-dependent manner (Pollard *et al.*, 1988; Burns *et al.*, 1988). The latter proteins include endonexin II (Schlaepfer *et al.*, 1987; Kaplan *et al.*, 1988), lipocortin I (Walner *et al.*, 1986), calpactin heavy chain (AKA p36, Glenny, 1986; Huang *et al.*, 1986), calelectrin 67K (Sudhof *et al.*, 1988), and protein II (Weber *et al.*, 1987). In this section we will describe purification of synexin and methods for analyzing synexin activity. In the subsequent section we will describe techniques for purifying and assaying endonexin II and lipocortin I.

A. Initial Steps in Purification of Synexin from Bovine Liver

1. Solutions and Materials Needed

1. Obtain 500 g of fresh bovine liver from the slaughterhouse and keep on ice until use.
2. Solution A: Measure 2 liters of 0.3 *M* sucrose and keep ice cold until use.
3. Solution B: Mix 1 liter of 0.3 *M* sucrose (solution A) with 6.2 g histidine, and adjust to pH 6.0 with HCl. Keep this ice cold until use.

4. Solution C: 5 ml of 500 m*M* EGTA, adjusted to pH 7.3 with NaOH.

5. Solution D: 350 mg phenylmethylsulfonylfluoride (PMSF, Sigma no. 8-7626) dissolved in 20 ml of 95% ETOH.

6. Solution E: Prepare immediately before perfusion the following mixture. To 1 liter of solution B add 5 ml of solution C and 20 ml of solution D. Add solution D dropwise to this solution while mixing rapidly with a magnetic stirrer. Some PMSF will not dissolve; ignore it.

7. Hardware needed: one 9 × 12-in. glass dissection tray; three 600-ml plastic beakers; three ice buckets full of ice; one metal Waring blender with 1 liter volume; six GSA rotor-sized Sorvall tubes with tops; 24 size 35 Beckman ultracentrifuge tubes; four 50-ml Sorvall plastic tubes; two 15-ml Sorvall plastic tubes; one Sorvall RC5 (or equivalent) preparative centrifuge with GSA, SS24 and SS34 rotors; one Beckman L5-75 (or equivalent) ultracentrifuge with type 35 rotor; one 15-ml and one 50-ml glass-on-glass A-type ten-broek homogenizer; one 35-ml disposable plastic syringe; one-quarter horsepower 1/4-in. drill on stand with 1-in. Teflon pestle and glass 50-ml ten-broek homogenizer pestle.

2. Initial Steps in Synexin Isolation from Tissue

1. Flush liver of clotted blood by perfusing through cut veins with 400 ml of ice-cold solution E using the plastic syringe.

2. Cut the liver into 1-in. square pieces and weigh out exactly 500 g.

3. Blend 125-g aliquots of liver in buffer such that when all liver is blended the total volume is 1200 ml. Each aliquot is exposed to two 10-second bursts at top speed.

4. Homogenize the entire volume with one pass of the power drill-driven homogenizer, and filter homogenate through four thicknesses of cheesecloth.

5. Place homogenate into six GSA-sized tubes (i.e., 200 ml per tube) and centrifuge in a GSA rotor on a Sorvall centrifuge at 13,000 rpm for 30 minutes. Recover supernatant solutions by decantation through four thicknesses of cheesecloth to catch any large particles.

6. Place supernatant into type 35 rotor in appropriate screwtop tubes and centrifuge at 34,000 rpm for 1 hour. Recover supernatant and measure volume accurately.

7. The synexin activity is now precipitated with 20% ammonium sulfate by adding 10.6 g enzyme-grade ammonium sulfate to each 100 ml of step 6 supernatant. The procedure is to add small amounts (~10% of total) of ammonium sulfate to a slowly stirring solution of crude synexin, waiting until all added ammonium sulfate is dissolved before adding more.

8. Allow the ammonium sulfate precipitate to develop over at least a 1-hour period at 4°C before proceeding to the next step. This is one

convenient overnight stopping place, because overnight incubation in the cold at this point has no deleterious consequences for the preparation.

9. Place the ammonium sulfate suspension into type 35 tubes and centrifuge in a type 35 rotor in the Beckman preparative centrifuge for 1 hour at 34,000 rpm.

10. Discard supernatant solution and suspend pellets in a total of 64 ml of solution B using the large-volume ten-broek homogenizer. Divide the solution between two 50-ml Sorvall centrifuge tubes, and centrifuge at 18,000 rpm for 30 minutes.

11. After centrifugation is complete, discard pellet and readjust the supernatant solution to 20% ammonium sulfate. This is also a convenient stopping place, and the solution can be kept at 4°C overnight.

12. Sediment the ammonium sulfate suspension by dividing the solution between two 50-ml Sorvall tubes as in step 10, and centrifuging at 18,000 rpm for 30 minutes. Resuspend the resulting smaller pellets in 3.5 ml solution B and place at −20°C. The activity is stable for as long as 6 months, but we routinely go on to the next purification steps immediately.

B. Purification of Synexin by Column Chromatography and Chromatofocusing

1. Preparation of Ultragel Column

1. Buffer A: Prepare 2 liters of Tris-acetate buffer (0.02 *M*, pH 8.3) by adjusting a 0.02 *M* Tris base solution to pH 8.3 with acetic acid.

2. Suspend 300 ml of AcA34 Ultragel size-exclusion medium (LKB) in 600 ml of buffer A. Place in a 1-liter Ehrlenmeyer side-arm flask and place under vacuum for 15 minutes, swirling occasionally. This degasses the medium so that formation of bubbles in the column is less frequent.

3. Wash and mount a Pharmacia (or equivalent) column (2.5 × 100 cm) with loading funnel on top. Place in the cold room. Fill one-quarter volume of the column with buffer A and pour Ultragel slurry into the column. Open the bottom valve and keep adding new slurry as the level drops. When column and attached funnel are full, close the bottom valve and allow to settle overnight. Add more gel the next day if needed. Finally, add top valve insert, making sure that no air is included in the system.

4. Obtain a fraction collector capable of collecting 5-ml fractions from the column. We use an LKB Ultrarac 7000. Adjust the collection rate to 5 ml per tube, and prepare to collect 500 ml total volume.

5. Prepare the column to purify synexin by first passing 3.5 ml of 1% bovine serum albumin (BSA) over the column to occupy nonspecific

binding sites in the medium. Failure to do this ensures that no synexin will emerge from the column. This need only be done once. If column is not to be used for some time, equilibrate it with buffer A and 2 m*M* sodium azide to prevent microbial contamination.

2. Purification of Synexin on Ultragel Column

1. After washing column clean of either BSA or sodium azide, load 3.5 ml of 20% ammonium sulfate fraction obtained as described earlier onto column and collect 5-ml fractions. The activity will elute from the column in the range of tubes 30–40. Measure activity as described in the following.
2. Pool active fractions and freeze. Activity is stable for 2 months under these conditions if concentration is high enough and if enough contaminants are present to prevent synexin from adhering to plastic or glass surfaces. We seldom keep synexin in this condition for more than 24 hours.
3. If one wants to keep synexin in this state for lengthy periods and the purity is high, we usually lyophilize the pooled fraction and store at −20°C, where it is stable for as long as 1 year.

3. Final Purification of Synexin by Chromatofocusing

a. Solutions and Materials

1. Purchase chromatofocusing medium PBE 94 from Pharmacia. This material is a type of ion exchange resin that allows one to separate proteins from one another on the basis of pI. Separation is usually accomplished by developing a pH gradient with carrier ampholines. However, ampholines are virtually impossible to separate from synexin after purification, and we simply use two buffer steps to separate synexin from contaminants.
2. Prepare 1 liter of Tris-acetate buffer, 0.02 *M*, pH 8.3 (buffer A) and 200 ml of Tris-HCl buffer, 0.1 *M*, pH 7.5 (buffer B).
3. Mix 10 ml of PBE 94 medium with 500 ml of buffer A and degas for 15 minutes as described previously for Ultragel medium. Ensure that bubbles can be seen evolving.
4. Prepare a column to hold the chromatofocusing medium by placing a wisp of glass fiber material in the barrel of a plastic 6-ml syringe. Place a plastic three-way stopcock on the tip of the syringe. Mount the syringe on a metal column rack, and add chromatofocusing medium with a 10-ml

pipet from the bottom of the degassing flask until a 5-ml packed volume is obtained.

5. As a buffer or sample reservoir, mount a 35-ml plastic syringe 20 cm above the column, using the gasket from the 6-ml syringe to seal the top. Connect buffer reservoir to the column by means of a 20-cm length of tubing at the end of which is a 20-gauge needle. The needle is allowed to pierce the gasket sealing the top of the column. We usually use a butterfly needle assembly, commonly found in a clinical setting.

b. *Purification of Synexin*

1. This step is performed at room temperature. Immediately before adding synexin to the column, wash the column with 20 ml of buffer A. As soon as buffer A treatment is complete, add the pooled fractions from the Ultragel column to the reservoir and allow to flow through the chromatofocusing column. If Ultragel column fractions have been lyophilized, reconstitute in the initial volume with distilled water and ensure that the pH is 8.3. Alternatively, reconstitute in a smaller volume, but run through a PD 10 desalting column (Pharmacia, or equivalent G-25) equilibrated with buffer A. It is very important that the fraction be equilibrated at the ionic strength and pH described, because otherwise the protein will simply not adhere to the column. The fraction collector (LKB Redirac 2112 or equivalent) is set to collect 1-ml fractions.

2. After protein solution is applied, wash column with an additional 24 ml of buffer A. Immediately after the completion of the wash step, change the buffer to buffer B. The activity should elute within the range of fractions 30–40. Pool the active fractions and lyophilize (Scott *et al.*, 1985).

3. Many precautions have been learned with regard to handling synexin. The molecule is extremely hydrophobic, even when in its water-soluble state. Never pass synexin through a Millipore filter; never place synexin in a dialysis bag. Avoid dilute solutions of synexin ($<10\ \mu g/ml$) if one intends to recover the material; be ever vigilant regarding proteases, especially during storage of initial purification fractions. Lyophilization is the best way to preserve active protein.

C. Preparation of Chromaffin Granules for Synexin Assay

Chromaffin granules are prepared either by the equilibrium centrifugation on metrizamide, or by the differential-sedimentation method. Both give roughly equivalent granules preparations in terms of this assay; however, we routinely use the metrizamide method. Suitability of the granule preparation at any one time is defined by the ability of synexin to aggregate them in a calcium-dependent manner; this activity declines with time. In general, we use the equilibrium method to prepare granules early in the week, and then use them for just 1 week. These are stored as a highly concentrated solution in 0.3 *M* sucrose at 4°C.

1. Preparation of a Crude Granule Fraction by Differential Centrifugation

1. Cut the glands and dissect the medullas free of cortical contamination. Place the tissue in 10 volumes of ice-cold 0.3 *M* sucrose containing other additions such as buffer (e.g., 10 m*M* MES-Tris, pH 6.5), divalent cation chelator (e.g., 1 m*M* EGTA), or protease inhibitor (e.g., 1 m*M* PMSF) as required.
2. Disrupt the tissue with two 3-second bursts in a Waring blender followed by homogenization with one complete stroke of a motor-driven Teflon pestle in a loose-fitting glass homogenizer.
3. Filter the homogenate through four layers of surgical gauze.
4. Centrifuge the filtrate at 500 *g* for 10 minutes at 4°C in a Sorvall SS34 rotor to remove cell debris and nuclei.
5. Decant the supernatant and centrifuge at 20,000 *g* for 30 minutes at 4°C.
6. Discard the supernatant and add 5 ml of cold homogenization medium to the pellets. Gently swirl off the loose material found on top of the compact pinkish granule pellet and discard.
7. Resuspend the granule pellet in homogenization medium by carefully scraping it off the wall of the tube with a flat spatula (leaving behind the red blood cells) and then transferring to a glass homogenizer and applying one stroke of a glass pestle.
8. Centrifuge the granule suspension at 10,000 *g* for 30 minutes at 4°C.
9. Repeat steps 6 and 7.
10. Centrifuge the granule suspension at 7000 *g* for 30 minutes at 4°C.
11. Repeat steps 6 and 7, resuspending the final granule pellet in the desired medium.

2. Preparation of Chromaffin Granules by Density Gradient Centrifugation on Metrizamide

The starting material for this method is the pellet from the 20,000 *g* centrifugation of an adrenal medullary homogenate prepared as described in the previous section and resuspended in a 0.3 *M* sucrose medium. The procedure is as follows:

1. Prepare a stock solution of metrizamide (Accurate Chemical and Scientific Corp., Westbury, NY) in water with a density of 1.21 g/ml by dissolving 34.77 g in a final volume of ~100 ml and checking the density by weighing. The osmotic strength of such a solution is 300 mOsm. Mix the metrizamide solution with a solution of 0.3 *M* sucrose (density of 1.03 g/ml) in the following combinations to obtain solutions of density 1.10 and 1.12 g/ml:

Density of 1.10 g/ml: 3.9 ml metrizamide + 6.1 ml sucrose
Density of 1.12 g/ml: 5.1 ml metrizamide + 4.9 ml sucrose

2. Prepare a discontinuous step gradient by underlaying 10 ml of density 1.12 g/ml metrizamide–sucrose solution under 10 ml of 1.10 g/ml metrizamide–sucrose solution in a 4 × 1.5-in. nitrocellulose tube.
3. Layer a 15-ml aliquot of the resuspended granule pellet over the step gradient.
4. Centrifuge the gradient at 100,000 *g* for 1 hour at 4°C in a Beckman SW27 rotor.
5. Remove the band of material that forms at the 1.10–1.12 g/ml interface of the two metrizamide solutions and dilute 4-fold with a solution of 0.3 *M* sucrose. This material consists of epinephrine-containing granules.
6. Centrifuge at 20,000 *g* for 30 minutes at 4°C in a Sorvall SS34 rotor and resuspend the purified granules as required.

D. Assay of Synexin Activity by Chromaffin Granule Aggregation

1. Materials and Solutions

1. To measure synexin activity a recording spectrophotometer such as the Gilson 250 can be used, with which four samples can be measured at one time. In addition, a four-cuvet holder is needed, as are the cuvets. The cuvets can either be quartz or acrylic plastic, and of a size suitable for holding 1 ml total volume. The recorder should be set with a full scale of OD 0.2 at 540 nm. The instrument can be temperature statted with a circulating water pump, but routine assays are usually performed at room temperature (~22°C in our laboratory).
2. Solution A: Prepare a stock solution of 100 m*M* $CaCl_2$ in milli-Q water.

3. Solution B: This is the same as solution B (sucrose histidine buffer) described in Section I,A,1, step 3. It is prepared by mixing 1 liter of 0.3 *M* sucrose with 6.2 g of histidine and adjusting the pH to 6.0 with HCl.

4. Solution C: Solution B is equilibrated at room temperature and chromaffin granules are diluted with solution B to an OD at 540 nm of 0.6. In practice this usually involves mixing 30 to 40 μl of ice-cold concentrated chromaffin granules with 4 ml of solution B. This solution is suitable for assay for ~1 hour at room temperature. Preequilibration of the granules at the assay temperature is advisable, because temperature gradients cause changes in the light-scattering properties of granules.

5. Solution D: 5 ml of 100 m*M* EGTA (pH 6.0), adjusted with NaOH.

2. Assay of Synexin Activity

1. Preadjust the recorder with mock assay solutions missing synexin as described in Section I,D,1, step 1, so that all cuvets give optical density readings in the lower part of the graph paper.

2. In the following order add to each cuvet: 10 μl solution A, 10–100 μl synexin solution, 390–300 μl solution B (depending on the amount of synexin solution added), and 500 μl of solution C (chromaffin granules). Begin measurement of optical density immediately, and record at a rate of 1 cm/minute. The assay can be terminated after 6 minutes, or can be allowed to progress to equilibrium.

3. To calculate synexin activity, the percentage change in optical density is measured over a given time interval. Kinetic constants such as $k_{0.5}$ for calcium or synexin can be measured, using a true initial rate, a total activity over an arbitrary time interval such as 6 minutes, or activity after reaching equilibrium (8–10 minutes). The values are the same.

4. The relationship between activity and calcium or synexin concentration is sigmoidal, not linear. Therefore, a low concentration of synexin might give quite low activity, but just doubling the concentration might yield quite significant activity. Synexin at a concentration of 4 μg/ml will usually generate maximum activity under these assay conditions.

E. Fusion of Intact Chromaffin Granules with Synexin and Arachidonic Acid

1. Preparation of Stock Solution of Arachidonic Acid

1. Prepare a stock solution of arachidonic acid (10 m*M*) in ethanol, and

hold on ice. This solution can be kept at −70°C for at least 1 month. Discard if even slightly discolored.

2. Immediately before the experiment dilute the arachidonic acid 20-fold into buffer B (see Section I,A,1, step 3). It will form a slightly opalescent suspension, which should be completely mixed prior to use. This solution is good for only 2–3 hours, on ice.

2. Initiation of Granule Aggregation

1. Prepare a synexin assay mixture as described in Section I,D,2, step 2, and allow the reaction to proceed for 15 minutes or until the increase in optical density appears to be leveling off.

2. Samples of the assay mixture will appear as clumps of particles, ~2–10 μm in diameter, when viewed by phase optics at 420× under a coverslip and slide. Be sure to *wash the slide and the coverslip with ethanol and then acetone*.

3. Initiation of Fusion

1. Add 10 μl of the aqueous arachidonic acid suspension in 1.0 ml of aggregated granules. Mix by inversion using parafilm to cover the cuvet, and follow the rapid decline in optical density over time. The optical density will decline only about 20% below the original starting optical density (i.e., 0.3 OD).

2. Microscopic examination of granules as described in section I,E,2 will reveal vacuolar structures of 1–10 μm in diameter. Formation of the structures can be followed visually in real time.

3. It is crucial to wash the slide and coverslip with organic solvent prior to use. Otherwise, as first noted by Dr. Carl Creutz during development of this assay at NIH, organic wetting agents added by some manufacturers will cause aggregated granules to fuse directly upon addition to the glass slide (Creutz, 1981; Creutz and Pollard, 1982). These "wetting agents" are said to include fatty acids from abattoir sources.

F. Fusion of Chromaffin Granule Ghosts with Synexin

Fusion of chromaffin granule ghosts can be followed by first loading the ghosts with self-quenching concentrations of FITC–dextran using a freeze–thaw technique in liquid nitrogen (Stutzin *et al.*, 1987). When loaded ghosts are mixed with empty ghosts in the presence of synexin, fusion ensues and can be detected by acquisition of a fluorescence signal as FITC–dextran becomes dequenched by dilution into the empty ghost. Any possible signal from lysis into the medium is suppressed by including an antifluorescein antibody in the reaction mixture.

1. Preparation of Chromaffin Granule Ghosts

Chromaffin granule ghosts can be prepared by subjecting granules to osmotic lysis, repeated washing with hypoosmotic medium, and finally resuspending in the experimental medium. We have used a procedure based on this protocol to prepare granule membranes in order to study actin binding as described by Fowler and Pollard (1982). This procedure originally employed as starting material the classically purified granule pellet obtained after centrifugation of a crude granule fraction through hyperosmotic 1.6 *M* sucrose (Bartlett and Smith, 1974). However, granules prepared on a metrizamide gradient (see Section I,C,2) work just as well.

We have further purified the granule membranes obtained after hypoosmotic lysis of this pellet by centrifugation over a 1 *M* sucrose shelf. During this centrifugation the granule membranes accumulate as an intensely pink band on top of the 1 *M* sucrose shelf, whereas contaminating material forms a brown pellet. Granule membranes isolated in this manner consist of a fairly homogeneous preparation of closed vesicles similar in size to chromaffin granules. The procedure is as follows:

1. Prepare a purified-granule pellet by centrifugation of a crude granule preparation through a 1.6 *M* sucrose solution as described in the previous section using a homogenization buffer consisting of 0.3 *M* sucrose, 40 m*M* HEPES–NaOH (pH 7.4–7.6), 1 m*M* EDTA, 1 m*M* dithiothreitol (DTT), and 1 m*M* PMSF.
2. Lyse the granules by homogenization with a glass homogenizer and pestle in 25 volumes of ice-cold 5 m*M* HEPES–NaOH (pH 7.8–8.0), 1 m*M* EDTA, 1 m*M* DTT, and 1 m*M* PMSF.
3. Centrifuge at 48,000 *g* for 30 minutes at 4°C in the Sorvall SS34 rotor.
4. Discard the supernatant and resuspend the pellet with the same volume of lysis buffer.
5. Centrifuge at 48,000 *g* for 30 minutes at 4°C.
6. Resuspend the pellet in a suitable volume of lysis buffer.
7. Carefully layer 20 ml of the membrane suspension on top of 17 ml of an ice-cold solution containing 1 *M* sucrose, 40 m*M* HEPES–NaOH (pH 7.6–7.8), 1 m*M* EDTA, 1 m*M* DTT, and 1 m*M* PMSF in Beckman SW27 rotor tubes.

8. Centrifuge at 100,000 *g* for 60 minutes at 4°C in the SW27 rotor.
9. Carefully remove the chromaffin granule membranes accumulated as a pink band on top of the 1 *M* sucrose shelf using a Pasteur pipet and resuspend in ~20 volumes of lysis buffer.
10. Centrifuge at 48,000 *g* for 30 minutes at 4°C in the Sorvall SS34 rotor.
11. Resuspend the granule membrane pellet in a small volume of 140 m*M* KCl, 20 m*M* HEPES–KOH, and 0.1 m*M* EGTA, pH 7.2.

2. Loading of Chromaffin Granules Ghosts with FITC–Dextran

1. Resuspend an aliquot of ghosts in 140 m*M* KCl, 20 m*M* HEPES–KOH, pH 6.03 (buffer A) containing 10 m*M* FITC–dextran M_r 20,000 (Sigma). Prepare an identical mixture, but lacking FITC–dextran.
2. Submerge both samples in liquid nitrogen for 1 minute and allow to thaw at room temperature. Repeat three times.
3. Wash loaded and blank ghosts twice in buffer A by centrifugation in 5.0 ml of buffer A at 20,000 *g* for 30 minutes on a Sorvall RC2B or equivalent centrifuge.
4. Pass membranes in 0.5 ml volume through a Sephacryl S-3000 Superfine (Pharmacia) column, 10 × 0.5 cm, using buffer A as eluant, to remove excess dye. Adjust protein content of both samples to 0.3 mg/ml.

3. Measuring Synexin-Induced Membrane Fusion

1. In a 3-ml fluorescence cuvet mix 1.5 μg FITC–dextran-loaded ghosts and 7.5 μg empty ghosts (1 : 5 mass ratio) in 2.0 ml of buffer A. Add to the mixture a previously calibrated excess amount of antifluorescein antibodies to quench free FITC–dextran. Add different amounts of synexin up to a maximum of ~15 μg/ml to initiate fusion. In this system of frozen/thawed granule ghosts the calcium dependence of synexin-driven fusion is less profound. At pCa 7 (as in the experiment shown in Fig. 1), fusion can be increased by up to 40% by raising the pCa to 5. This contrasts with results with fusing liposomes, where direct fusion is entirely calcium-dependent (Hong *et al.*, 1981, 1982).
2. Fusion is estimated by following increase in fluorescence, activated by 465 nm and measured at 520 nm. Carefully follow the detailed protocol for measuring and calibrating the signal as described by Stutzin *et al.* (1987; see legend to Fig. 1). Further details may be found in Nir *et al.* (1987) and Stutzin *et al.* (1987).

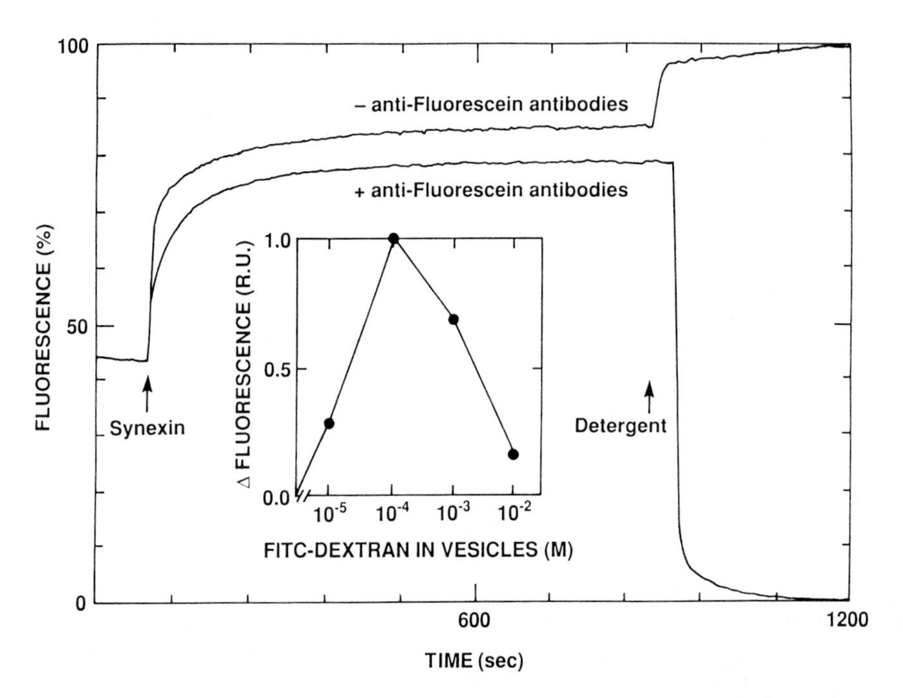

FIG. 1. Synexin-induced fusion of chromaffin granule ghosts. The increase of fluorescence due to the addition of synexin (34.08 μg, first arrow) is shown in the presence and absence of antifluorescein antibody. The medium is 2 ml of buffer A (see Section I,F,2, step 1) and has a pCa of 7 at 37°C. After addition of detergent (NP-40, second arrow), the fluorescence either increased (calibration experiment, upper trace) or decreased (fusion experiment, lower trace). Inset: Self-quenching properties of FITC–dextran trapped inside chromaffin granule ghosts. See Stutzin *et al.* (1987) for details.

II. Purification and Assay of Lipocortin I and Endonexin II

Recent research has revealed the existence of a family of structurally related proteins that undergo calcium-dependent binding to membrane phospholipids. By using the calcium-dependent binding properties as an affinity step, these related proteins have been detected in a wide range of experimental systems. Members of this family include proteins initially investigated as substrates for protein-tyrosine kinases, mediators of exocytosis, inhibitors of phospholipase A_2, inhibitors of coagulation, and components of the cytoskeleton, but their exact physiological roles are not yet known. For a description of the structural relationships of these proteins see Fig. 5 in the article by Pepinsky *et al.* (1988), and references therein. The realization that the proteins in this family were related was the unexpected result of their primary amino acid sequence elucidation. These proteins have ~50% overall sequence identity and consist of what can be thought of as a core and an amino-terminal domain. The amino-terminal domains are not required for calcium or phospholipid binding and have only limited sequence similarity, whereas the larger core domain consists of a 4-fold repeat motif containing a highly conserved 17-amino acid consensus sequence that may be involved in the calcium- and phospholipid-binding properties of these proteins.

A. Protein Isolation Procedure

The shared structural characteristics of the lipocortin family members are presumed to be the basis of their similar biochemical properties; all are calcium- and phospholipid-binding proteins. This property of a reversible calcium-dependent association with the particulate fraction from human placenta was used as the first step in the purification of these proteins (Haigler *et al.*, 1987). This protocol initially was designed for the isolation of a substrate for the epidermal growth factor (EGF) stimulated kinase that was identified in fibroblasts (Giugni *et al.*, 1985) and A431 membranes (Fava and Cohen, 1984).

The amnion, chorion, and large blood vessels were removed from fresh term human placenta obtained by cesarean section, and the remaining tissue was rinsed in four volumes (w/v) of a cold NaCl solution (0.15 *M*) and minced with a razor blade. Unless otherwise indicated, all procedures were performed at 4°C. The washed tissue was homogenized in a Waring blender for 1 minute with two volumes (w/v) of homogenization buffer with EGTA (1 m*M*): Buffer H contained HEPES (20 m*M*, pH 7.4), KCl (150 m*M*), $MgCl_2$ (2 m*M*), aprotinin (1 μg/ml), leupeptin (1 μg/ml), iodoacetic acid (4 m*M*), benzamidine (1 m*M*), and PMSF (1 m*M*). The homogenate was centrifuged at 700 *g*, the supernatant was saved, and the pellet was reextracted as before in one-half volume of buffer H with EGTA. The combined supernatants were filtered through cheesecloth. $CaCl_2$ was added to the filtrate to a final conentration of 2 m*M*. After 10 minutes, the extract was centrifuged (100,000 *g*, 20 minutes) and the supernatant was discarded. The pellets were resuspended with a dounce and pestle in 300 ml of buffer H with added $CaCl_2$ (1 m*M*) followed by centrifugation (100,000 *g*, 20 minutes) (Fig. 2, lane a). This pellet–resuspension–wash step was repeated three times or until the supernatant was clear. After the wash steps, the pellet was resuspended by dounce and pestle in 300 ml of buffer H minus iodoacetic acid containing EGTA

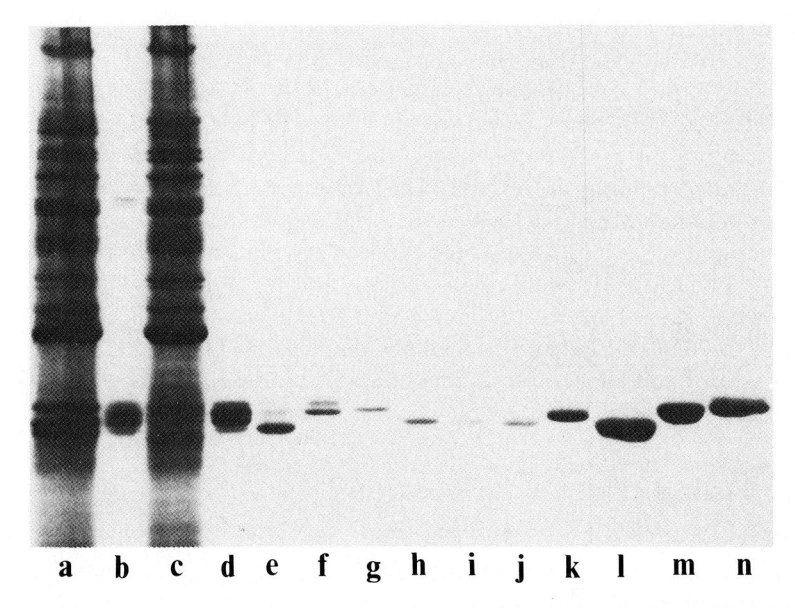

FIG. 2. PAGE and protein staining of samples from the purification steps. The indicated fractions from the purification steps of the placental proteins were analyzed by PAGE and stained with Coomassie blue. Lane a, 100 μg placental particulate fraction after washing in $CaCl_2$-containing buffer but before EGTA elution; lane b, 10 μg EGTA eluate from placental membranes; lane c, 100 μg placental membranes after EGTA elution; lane d, 5 μg pooled fractions 26–31 of Sephadex G-100 column; lanes e–k, 1 μg of peaks A–G, respectively, from the DE-52 anion exchange column; lane l, 6 μg of purified peak A from the CM-52 cellulose column; lane m, 6 μg of purified peak B from the CM-52 cellulose column; lane n, 6 μg of purified peak C from the Bio Rex 70 column.

(2 m*M*). After 10 minutes the solution was centrifuged (100,000 *g*, 20 minutes) and the supernatant was frozen at −70°C. The frozen extract was thawed, centrifuged (100,000 *g*, 20 minutes), then concentrated to 10 ml by pressure filtration through an Amicon YM-10 membrane (Fig. 2, lane b). The concentrated EGTA eluate was fractionated on a Sephadex G-100 column (2.5 × 90 cm) equilibrated with ammonium acetate (0.06 *M*, pH 8.0). The flow rate of was 30 ml/hour, and 10-ml fractions were collected. The following standard proteins eluted with peaks at the indicated fractions during the G-100 column standardization: γ-globulin (150 kDa), fraction 19; transferrin (95 kDa), fraction 1; ovalbumin (45 kDa), fraction 26; carbonic anhydrase (31 kDa), fraction 29; and cytochrome *c* (12.5 kDa), fraction 34.

All G-100 column fractions were monitored for OD_{280}, and fractions 26–31, representing the second protein absorption peak (Fig. 2, lane d), were pooled and adjusted to 5 m*M* ammonium acetate, pH 8.65, by exhaustive dialysis. The pooled sample was concentrated to 8 ml by pressure filtration and applied to a column of DE-52 cellulose (1 × 4.0 cm) equilibrated with ammonium acetate buffer (5 m*M*, pH 8.65) at room temperature (it is important that the pH and ionic strength of the equilibrated column are exact). The column was eluted with 35 ml of equilibration buffer and then with three sequential linear gradients of ammonium acetate (pH 8.65) of the following volume and concentrations: 160 ml of 5–30 m*M*; 60 ml of 30–75 m*M*, and 125 ml of 75–350 m*M*. The flow rate was 0.5 ml/minute and 2.5-ml fractions were collected. Seven peaks of OD_{280} were resolved and labeled peaks A–G (Fig. 3). Each peak was pooled and concentrated, and could be stored frozen at −70°C (Fig. 2, lane e–k).

B. Peak B: Placental Lipocortin I

Peaks A, B, and C (Fig. 3) reacted with antiserum specific for lipocortin I. These three peaks were further purified by cation exchange chromatography at room temperature as follows:

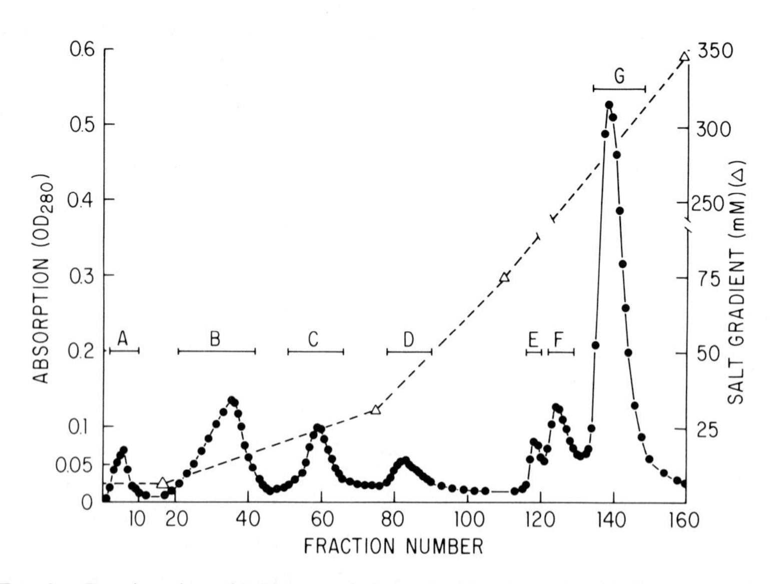

FIG. 3. Fractionation of 35-kDa peak from G-100 column by DE-52 anion exchange column chromatography. Fractions 26–31, representing the 35-kDa protein peak, from the Sephadex G-100 column were pooled, and fractionated on a DE-52 column as described in text. The indicated fractions were pooled, assigned the indicated letters, concentrated by pressure filtration (Amicon YM-10), and stored at −70°C.

1. Peak A (6 ml) was dialyzed against 10 m*M* ammonium acetate, pH 6.0, and loaded onto a column (1 × 4 cm) of CM-52 cation exchange resin equilibrated with the same buffer. The column was eluted with 30 ml of the equilibration buffer before 300 ml of a linear salt gradient (10–400 m*M* ammonium acetate, pH 6.0) was applied. Fractions (2.5 ml) were collected at a flow rate of 1 ml/minute. The major peak of OD_{280} eluted at fractions 81–95. These fractions were pooled and concentrated, and constitute the purified peak A sample, which contained a single sharp band with an apparent M_r of 32,000 when analyzed by polyacrylamide gel electrophoresis (PAGE; Fig. 2, lane l).

2. Peak B (10 ml) was processed and loaded onto a CM-52 column exactly as peak A. The column was eluted with 35 ml of equilibration buffer, and then two linear salt gradients of ammonium acetate (pH 6.0) were applied sequentially: 140 ml of 10–70 m*M*; then 60 ml of 70–175 m*M*. Fractions (2.5 ml) were collected at a flow rate of 1 ml/minute. Three OD_{280} peaks were resolved: fractions 2–10, 46–60, and 70–85; only the proteins in the major peak, fractions 46–60, were recognized by polyclonal antiserum raised against lipocortin I. These fractions were pooled and concentrated, and constitute the purified peak B sample, which contained a single sharp band with an apparent M_r of 35,000 when analyzed by PAGE (Fig. 2, lane m). Approximately 2 mg of purified peak B can be recovered from the tissue of a single placenta.

3. Peak C (10 ml) was dialyzed against 10 m*M* ammonium acetate, pH 6.0, and applied to a column (1 × 3 cm) of Bio Rex 70 cation exchange resin equilibrated with the same buffer. The column was eluted with 30 ml of the equilibration buffer, and then two linear gradients of ammonium acetate, pH 6.0, were applied sequentially: 120 ml of 10–75 m*M*; then 120 ml of 175–500 m*M*. Fractions (2.5 ml) were collected at a flow rate of 1.0 ml/minute. Fractions 70–95 formed the major peak of OD_{280}. They were pooled and concentrated, and constitute purified peak C, which contained one sharp band with an apparent M_r of 35,000 when analyzed by PAGE (Fig. 2, lane n).

The major proteins in peaks B, C, and A were identified as intact lipocortin I, lipocortin I missing the amino-terminal 12 amino acids, and lipocortin I missing the amino-terminal 26 amino acids, respectively. A 1 mg/ml solution of cation exchange-purified Peaks A, B, and C, had an OD_{280} of 0.34, 0.49, and 0.36, respectively.

C. Peak G: Placental Endonexin II

Peak G from the DE-52 column was purified further by placing the protein back over the Sephadex G-100 column to remove a higher molecular mass contaminant as described earlier. Fractions 28–33 were pooled, dialyzed against Tris-HCl (10 m*M*, pH 7.45), and loaded onto a DE-52 cellulose column (1 × 4 cm) equilibrated with the same buffer at room temperature. The column was eluted with 25 ml of equilibration buffer, and then two linear gradients of NaCl (200 ml of 0–100 and 100–500 m*M*) in Tris-HCl (10 m*M*, pH 7.45) were applied sequentially. The single major protein peak eluted at 150–200 m*M* NaCl, and it contained one sharp band with an apparent M_r of 33,000 when analyzed by PAGE. The concentration of purified peak G was determined by the method of Lowry *et al.* (1951) using BSA as the standard. This protein is very abundant in placenta tissue; ~5 mg of the purified 33-kDa protein could be recovered from a single placenta.

A partial amino acid sequence of peak G showed that it had ~50% sequence identity with lipocortin I and ~80% sequence identity with bovine endonexin (Schlaepfer *et al.*, 1987). Based on the similarity with bovine endonexin, peak G was named endonexin II. The complete amino acid sequence of endonexin II was deduced from its cDNA (Kaplan *et al.*, 1988) and found to have 43% sequence identity with lipocortin I and 63% sequence identity with the partial sequence of bovine endonexin I (Geisow *et al.*, 1986). Peak D of the DE-52 column (Fig. 3) was identified as human endonexin I (Fig. 2, lane h).

D. Phospholipid-Binding Assay for Lipocortin I and Endonexin II

The Ca^{2+}-dependent phospholipid-binding properties of these proteins can readily be studied in a simple assay that measures copelleting with phospholipid vesicles (Schlaepfer and Haigler, 1987). The large thin-walled vesicles used in the binding assays were prepared in the presence of 240 m*M* sucrose according to the method of Reeves and Dowben (1969) using phosphatidylserine (PS; bovine brain, Sigma P8518) or phosphatidylcholine (PC; egg yolk, Sigma P5885). The vesicles were harvested by centrifugation (15 minutes at 20,000 *g*) and resuspended in buffer P: 20 m*M* HEPES (pH 7.4), 100 m*M* NaCl, 2 m*M* $MgCl_2$, 2 m*M* NaN_3.

Binding of peak B (lipocortin I), peak A (des 1–26 lipocortin I), or peak G (endonexin II) to phospholipid vesicles was performed in 100 μl of buffer P containing BSA (0.1 mg/ml). The divalent cations were added as a 1.25-fold concentrated stock solution in buffer P. After a 10-minute incubation at room temperature, the protein–phospholipid solution was subjected to centrifugation (10 minutes at 30,000 *g*) and the supernatant and phospholipid pellet were analyzed separately by PAGE followed by Coomassie blue staining (Fig. 4). Lipocortin copelleted with PS vesicles

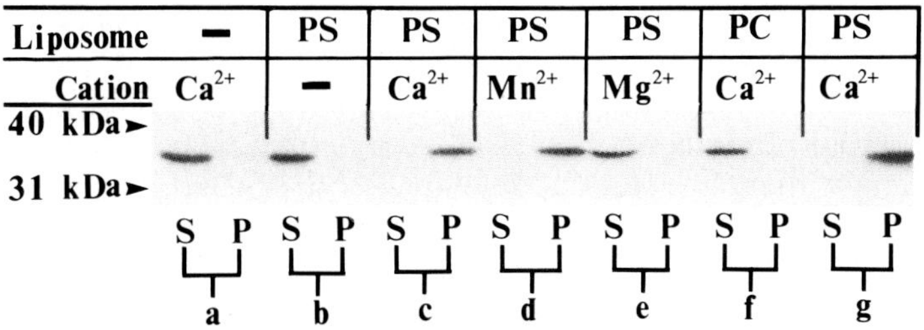

Liposome	–	PS	PS	PS	PS	PC	PS
Cation	Ca^{2+}	–	Ca^{2+}	Mn^{2+}	Mg^{2+}	Ca^{2+}	Ca^{2+}

40 kDa►

31 kDa►

S P S P S P S P S P S P S P

a b c d e f g

FIG. 4. Association of lipocortin I with phospholipid. Peak B/lipocortin I (1.0 μg, reactions a–f) or peak A/des 1–26 lipocortin I (1.0 μg, reaction g) was incubated in 100 μl of buffer P in the presence of Ca^{2+} (1.0 m*M*, reactions a and f; 0.1 m*M*, reactions c and g), Mn^{2+} (0.5 m*M*, reaction d), Mg^{2+} (20 m*M*, reaction e), or EGTA (2.0 m*M*, reaction b). The indicated reactions also contained phosphatidylserine (PS; 300 μg/ml) or phosphatidylcholine (PC; 300 μg/ml) vesicles. After a 10-minute incubation, the solutions were subjected to centrifugation (10 minutes at 30,000 *g*). The pellets (P) and supernatants (S) were analyzed by PAGE (8.5%), followed by Coomassie blue staining.

in the presence (Fig. 4, reaction c), but not the absence (Fig. 4, reaction b), of Ca^{2+}, whereas essentially all of the lipocortin I was in the supernatant in the presence of Ca^{2+} if no lipid was present (Fig. 4, reaction a). Mn^{2+} (Fig. 4, reaction d), but not Mg^{+2} (Fig. 4, reaction e), supported association with this lipid. All detectable lipocortin I was in the supernatant following incubation in Ca^{2+} and PC vesicles (Fig. 4, reaction f). It should be noted that, if lipocortin I solutions were exposed to multiple cycles of freezing and thawing or were frozen slowly at −20°C, self-aggregation appeared to occur in a Ca^{2+}- and phospholipid-independent manner.

E. Biological Function of Lipocortin I and Endonexin II

Although this family of Ca^{2+}-binding proteins is attracting intensive investigation because of their potential involvement in Ca^{2+}-mediated stimulus–response coupling, their biological functions are not known. Lipocortin I and endonexin II inhibit phospholipase A_2 *in vitro*, but this inhibition appears to be secondary to depletion of the phospholipid substrate and not due to direct interaction with the enzyme (Davidson *et al.*, 1987; Haigler *et al.*, 1987). Thus, their role as physiological inhibitors of this important enzyme remain controversial.

Endonexin II also was identified as an anticoagulant in *in vitro* assays but, as with the phospholipase A_2 assay, the inhibition was due to depletion of phospholipid in the assay and not by interaction with the enzymes involved (Funakoshi *et al.*, 1987). Additional studies are required to determine if other members of this family of Ca^{2+}-binding proteins inhibit coagulation and to determine the physiological significance of this observation.

Although the physiological function of lipocortin I is not known, indirect evidence implicates it in the mitogenic signal transduction pathway. There is compelling evidence that the tyrosine-specific protein kinase associated with the EGF receptor is involved in stimulating cell replication. This kinase phosphorylates lipocortin I in intact cells. The stoichiometry of phosphorylation of lipocortin I is $<2\%$ in quiescent cultured fibroblasts and up to 20% in rapidly growing cells (D. D. Schlaepfer and H. Haigler, unpublished results). In other indirect experiments, we found that the expression of lipocortin I was 5-fold higher in growing cultured fibroblasts than in quiescent cells (D. D. Schlaepfer and H. Haigler, unpublished results).

References

Bartlett, S. F., and Smith, A. D. (1974). *In* "Methods in Enzymology" (S. Fleischer and L. Packer, eds.), Vol. 31, pp. 379–389. Academic Press, New York.

Burns, A. L., Magendzo, K., Shirvan, A., Srivastava, M., Rojas, E., Alijani, M., and Pollard, H. B. (1988). *Proc. Natl. Acad. Sci. U.S.A.* (in press).

Creutz, C. E. (1981). *J. Cell Biol.* **91,** 247–256.

Creutz, C. E., and Pollard, H. B. (1982). *Biophys. J.* **37,** 119–120.

Creutz, C. E., Pazoles, C. J., and Pollard, H. B. (1978). *J. Biol. Chem.* **253,** 2858–2866.

Davidson, F. F., Dennis, E. A., Powell, M., and Glenney, J. R. (1987). *J. Biol. Chem.* **262,** 1698–1705.

Fava, R. A., and Cohen, S. (1984). *J. Biol. Chem.* **259,** 2636–2645.

Fowler, V. M., and Pollard, H. B. (1982) *Nature* **295**:336–339.

Frey, A. Z., and Wallner, B. P. (1988). *J. Biol. Chem.* **263,** 10799–10811.

Funakoshi, T., Heimark, R. L., Hendrickson, L. E., McMullen, B. A., and Fujikawa, K. (1987). *Biochemitry* **26,** 5572–5578.

Geisow, M. J., Fritsche, U., Hexham, J. M., Dash, B., and Johnson, T. (1986). *Nature (London)* **320,** 636–638.

Giugni, T. D., James, L. C., and Haigler, H. T. (1985). *J. Biol. Chem.* **260,** 15081–15090.

Glenny, J. R. (1986). *Proc. Natl. Acad. Sci. U.S.A.* **83,** 4258–4262.

Haigler, H. T., Schlaepfer, D. D., and Burgess, W. H. (1987). *J. Biol. Chem.* **262,** 6921–6930.

Hong, K., Duzgunes, N., and Papahadjopoulos, D. (1981). *J. Biol. Chem.* **256,** 3641–3644.

Hong, K., Duzgunes, N., and Papahadjopoulos, D. (1982). *Biophys. J.* **37,** 297–305.

Huang, K.-S., Wallner, B. P., Mattaliano, R. J., Tizard, R., Burne, C., Frey, A., Hession, C., McGray, P., Sinclair, L. K., Chow, E. P., Browning, J. L., Ramachandran, K. L., Tang, J., Smart, J. E., and Pepinsky, R. B. (1986). *Cell (Cambridge, Mass.)* **46,** 191–199.

Kaplan, R., Jaye, M., Burgess, W. H., Schlaepfer, D., and Haigler, H. T. (1988). *J. Biol. Chem.* **263,** 8037–8043.

Lowry, O. H., Rosebrougn, N. J. H., Farr, A. L., and Randall, R. J. (1951). *J. Biol. Chem.* **193,** 265–275.

Nir, S., Stutzin, A., and Pollard, H. B. (1987). *Biochim. Biophys. Acta* **903,** 309–318.

Pepinsky, R. B., Tizard, R., Mattaliano, R. J., Sinclair, L. K., Miller, G. T., Browning, J. L., Chow, E. P., Burne, C., Huang, K.-S., Pratt, D., Wachter, L., Hession, C., Frey, A. Z., and Wallner, B. P. (1988) *J. Biol. Chem.* **263,** 10799–10811.

Pollard, H. B., and Rojas, E. (1988). *Proc. Natl. Acad. Sci. U.S.A.* **85,** 2974–2978.

Pollard, H. B., Burns, A. L., and Rojas, E. (1988). *J. Exp. Biol.* **139,** 267–286.

Reeves, J. P., and Dowben, R. M. (1969). *J. Cell. Physiol.* **73,** 49–60.

Rojas, E., and Pollard, H. B. (1987). *FEBS Lett.* **217,** 25–31.

Schlaepfer, D. D., and Haigler, H. T. (1987). *J. Biol. Chem.* **262,** 6931–6937.

Schlaepfer, D. D., Mehlman, T., Burgess, W. H., and Haigler, H. T. (1987). *Proc. Natl. Acad. Sci. U.S.A.* **84,** 6078–6082.

Scott, J. H., Kelner, K. L., and Pollard, H. B. (1985). *Anal. Biochem.* **149,** 163–165.

Stutzin, A., Cabanchik, I., Lelkes, P. I., and Pollard, H. B. (1987). *Biochim. Biophys. Acta* **905,** 205–212.

Sudhof, T. C., Slaughter, C. A., Leznicki, I., Barjon, P., and Reynolds, G. A. (1988). *Proc. Natl. Acad. Sci. U.S.A.* **85,** 664–668.

Wallner, B. P., Mattaliano, R. J., Hession, C., Cate, R. L., Tizard, R., Sinclair, L. K., Foeller, C., Chow, E. P., Browning, J. L., Ramachandran, K. L., and Pepinsky, R. B. (1986). *Nature (London)* **320,** 77–81.

Weber, K., Johnsson, N., Plessman, U., Van, P. N., Soling, H.-D., Ampe, C., and Vandekerckhove, J. (1987). *EMBO J.* **6,** 1599–1604.

Chapter 13

Characterization of Coated-Vesicle Adaptors: Their Reassembly with Clathrin and with Recycling Receptors

BARBARA M. F. PEARSE

MRC Laboratory of Molecular Biology
Cambridge CB2 2QH, England

I. Introduction

Coated vesicles mediate the transfer of specific molecules from one membrane compartment to another in the cell (for reviews, see Pearse and Bretscher, 1981; Goldstein *et al.*, 1985; Pearse, 1987). At the plasma membrane, certain recycling receptors assemble into coated pits, which endocytose forming coated vesicles in the cytoplasm. The coats dissociate, freeing the vesicles, which then fuse with forming endosomes. Coated pits also form on intracellular membranes particularly in the Golgi region. These are involved in the process of forming storage vacuoles and lysosomes.

The coats of coated vesicles are composed of two major structural units (for review, see Pearse and Crowther, 1987). The unique clathrin triskelions pack together to form the outer polyhedral cage characteristic of the cytoplasmic surface of coated vesicles. The adaptors assemble with clathrin to form an inner shell of the coat, which in the coated vesicle immediately surrounds the membrane. Each type of adaptor complex, a member of a heterogeneous family, interacts with a select group of receptors and is restricted to a subset of membranes in the cell. Thus a coated pit on a given membrane will contain clathrin, a characteristic type of adaptor, recycling receptors, and their ligands.

The clathrin triskelion contains three heavy chains of 180 kDa and three associated light chains the arrangement of which has been previously reviewed (Pearse and Crowther, 1987). The adaptor complex is composed of two different polypeptides of 100 kDa associated with a type of 50-kDa polypeptide and an adaptor light chain (Pearse and Robinson, 1984; Keen, 1987; Virshup and Bennett, 1988; Ahle *et al.*, 1988). The 100-kDa components of the adaptor complexes were the first to be identified as significant components of coated vesicles (Pearse, 1978; Woodward and Roth, 1978; Schook *et al.*, 1979; Keen *et al.*, 1979), and the presence of the 50-kDa polypeptide was recognized later (Unanue *et al.*, 1981; Pearse, 1982). Keen *et al.* (1979) showed that the 100-kDa polypeptides were released from the coated vesicles, along with clathrin, on addition of 0.5 *M* Tris-Cl (pH 7.0) to the buffer. This extraction has provided the basis for the separation of these two coat structural units. Keen *et al.* (1979; also Zaremba and Keen, 1983) also made the important observation that the 100-kDa polypeptides promote the assembly of clathrin under certain conditions, and thus he called them assembly polypeptides. Further purification and analytical separation of the 100-kDa polypeptides (Manfredi and Bazari, 1987; Keen, 1987; Virshup and Bennett, 1988; Ahle *et al.*, 1988), crosslinking studies (Pearse and Robinson, 1984; Virshup and Bennett, 1988), and use of monoclonal antibodies (mAb) against individual polypeptides (Robinson, 1987; Ahle *et al.*, 1988) has led to our present, more definitive picture of the complete adaptor complexes. Their localization to particular membranes in the cell (Robinson and Pearse, 1986; Robinson, 1987; Ahle *et al.*, 1988) and their reconstitution with receptors (Pearse, 1985) and receptor cytoplasmic tails have given us insight into their properties.

Because the role of these coat 100 kDa–50 kDa complexes appears to fulfil that proposed for adaptors (Pearse and Bretscher, 1981) it seems appropriate to give them this name. Thus specific adaptors on a particular membrane bind to a group of receptors with a common recognition site and mediate their assembly with clathrin to form a coated pit. Thus the adaptors sort out those molecules that are to travel on from those that are to remain behind. Just how the adaptors are restricted to a particular membrane is not yet clear. After release of clathrin and perhaps the 100-kDa proteins in the uncoating process, some portion of the adaptor may remain to direct the vesicle to fuse with a particular organelle.

The proposed nomenclature for the components of adaptors is summarized in Table I, incorporating "100-kDa" polypeptide designations introduced by Robinson (1987) and Ahle *et al.* (1988). The name "adaptin" is suggested for members of the "100-kDa" family of coat polypeptides, and the different types of adaptins are specified as α, β, or γ.

II. Purification and Characterization of Adaptors

The strategy for purifying adaptors is described in a number of steps. The first objective is to obtain reasonably clean coated vesicles with the minimum of necessary centrifugation. The next stage requires the efficient extraction of the coat proteins—that is, clathrin and adaptors—from

TABLE I

ADAPTOR COMPONENTS[a]

HA-I Adaptors: Golgi region			HA-II Adaptors: Plasma membrane		
ntibodies		Components	Antibodies		Components
Ab 100/1	Dimer	β'-adaptin	mAb 100/1	Dimer	β-adaptin (B)
Ab 100/3		γ-adaptin	AC1-M11		α-adaptin (A1, A2, or C)
		47-kDa Protein			50-kDa Protein
		19-kDa Light chain			16-kDa Light chain

[a] Derivation of terms: HA-I and HA-II, relative elution of adaptors from hydroxylapatite; location e., Golgi or plasma membrane), determined by Ab against γ-adaptin in HA-I complex and α-adaptin HA-II complex; adaptins (i.e., "100-kDa" polypeptides), coat polypeptides in range 100–120 kDa rmed adaptins (i.e., major polypeptides of the adaptor complex); α-, β-, β'-, and γ-adaptins, signation of classes of adaptins based on recognition by mAb, peptide mapping, and sequences erived from appropriate cDNAs; specific adaptin polypeptides (A1, A2, B, and C), polypeptides riginally defined by mobility on 7.5% SDS–polyacrylamide gels, which stain with particular mAb and r which some of the sequences are known (Robinson, 1989).

the vesicles and their separation from each other. After obtaining a fraction containing adaptors, various chromatographic techniques may be employed to separate them further into discrete classes. Finally, the components of the different classes of adaptors are identified by electrophoresis on SDS–polyacrylamide gels and by immunoblotting with the available mAb to specific adaptins (adaptor 100-kDa polypeptides).

Standard buffers used in the purification of adaptors are as follows:

Buffer A: Assembly Buffer
- 0.1 *M* MES–NaOH, pH 6.5 or 7.0 as specified
- 0.2 m*M* EGTA
- 0.5 m*M* $MgCl_2$
- 0.02% NaN_3
- 0.1 m*M* phenylmethylsulfonyl fluoride (PMSF)

Buffer B: Tris Extraction Buffer
- 1 *M* Tris-Cl, pH 7.0
- 1 m*M* EDTA
- 0.1% 2-mercaptoethanol
- 0.02% NaN_3
- 0.2 m*M* PMSF

A. Coated-Vesicle Preparation

The first purification of coated vesicles was described in 1975 by Pearse. The procedure can be employed for most tissues and cells (Pearse, 1976, 1978), although bullock brains are the tissue most frequently used because they are a good source of coated vesicles and are easily obtained in bulk. The tissue (obtained as soon as possible after slaughter) is cut into small pieces and suspended in an equal volume of buffer A, pH 6.5 (or three to five volumes, if livers are used instead of brains), and homogenized in a Waring blender for 1 minute at 0°–6°C. The homogenate is centrifuged for 30 minutes at 20,000 *g*. The resulting supernatant contains, among other things, coated vesicles mixed with a plethora of noncoated membranous vesicles. The next major objective is to separate the latter from the coated vesicles. One possibility is to add 1% Triton X-100 to the supernatant, which dissolves the bulk of the membranes but leaves the coat assemblies intact for subsequent collection by centrifugation and extraction of the coat proteins (Pearse and Robinson, 1984). This method is quite efficient, but some membrane components (which later tend to contaminate the dissociated coat proteins) may be released into the medium. An equally convenient and gentler procedure is that described by Campbell *et al.* (1984). First a crude coated-vesicle pellet is collected by centrifugation of the homogenate supernatant for 1 hour at 100,000 *g*. The pellet is resuspended in a small volume of buffer A in a Sorvall omnimixer at moderate speed for 1 minute. Bubbles of air are spun out of the resulting membrane suspension at low speed in a bench centrifuge if necessary. The membrane suspension is diluted with an equal volume of a mixture of 12.5% Ficoll and 12.5% sucrose (both in buffer A) and centrifuged at 43,000 *g* for 40 minutes to remove the larger membranous structures that pellet. The supernatant is diluted with four volumes of buffer A and the coated vesicles collected by centrifugation at 100,000 *g* for 1 hour. At this stage the purity of the coated vesicles from brain is adequate for extraction and purification of the coat proteins. However, particularly when liver is the source material, a number of structures contaminate the preparation as judged by electron microscopy and analysis on SDS–polyacrylamide gels (Kedersha and Rome, 1986a). These include filaments, free ferritin, and ribonucleoprotein particles termed "vaults" (Kedersha and Rome, 1986b), contaminants that can mostly be removed by isopycnic centrifugation and gel filtration through Sephacryl S-1000 (Wiedenmann *et al.*, 1985) if highly purified coated vesicles are required. Preparative agarose gel electrophoresis has also been used to purify the coated vesicles (Rubenstein *et al.*, 1981; Kedersha and Rome, 1986a).

B. Disassembly and Separation of Clathrin and Adaptors

Coat pellets are gently resuspended in a small volume of buffer B that contains 1 *M* Tris-Cl (pH 7.0) and effectively disassembles the coat proteins and removes them from the vesicles (Keen *et al.*, 1979). The extract is clarified by centrifugation for 1 hour at 100,000 *g* and the pellet discarded.

At this stage there are two different ways to proceed to purify the adaptors. These are discussed in the following paragraphs.

1. Procedure 1

The first option is to separate the clathrin and the adaptors by gel filtration of the sample on Sepharose CL-4B essentially as first described by Keen *et al.* (1979). The clathrin from such a separation can then be used for reconstitution experiments after a cycle of polymerization and depolymerization as described previously (Crowther and Pearse, 1981; Pearse and Robinson, 1984). After the clathrin, a second peak of material elutes from the column that contains the major species of adaptors so far identified, plus some contaminants, chiefly ferritin. Different species of adaptors (namely, the HA-I and HA-II types) can then be separated by chromatography on hydroxylapatite as described by Manfredi and Bazari (1987).

Another option is to affinity-purify the adaptors by assembly onto a matrix coated with clathrin (Keen, 1987), but this is so far untried by this author.

2. Procedure 2

A second method of purifying adaptors has been introduced by Ahle *et al.* (1988). In this procedure, the coat protein extract, in buffer B containing clathrin and adaptors, is extensively dialyzed against 20 m*M* ethanolamine (pH 8.9), 2 m*M* EDTA, and 1 m*M* dithiothreitol (DTT). The extract is then chromatographed on a semipreparative Mono Q column and eluted with a linear 0–0.5 *M* NaCl gradient. This effectively separates the clathrin and adaptors and eliminates the necessity for a lengthy gel filtration step. A further similar ion exchange chromatography step on an analytical Mono Q column using a less steep salt gradient then resolves the HA-I and HA-II adaptor complexes. The separate HA-I and HA-II fractions are then applied to hydroxylapatite columns where each elutes in a characteristic region of the applied phosphate gradient (Ahle and Ungewickell, 1986). A typical preparation yields ~0.15 mg HA-I and 0.5 mg HA-II adaptors from ~50 mg coated vesicles.

C. Classes of Adaptors and Their Components

Two distinct classes of adaptors have so far been identified: those restricted to the plasma membrane coated pits (HA-II type: Robinson and Pearse, 1986; Robinson, 1987) and those restricted to coated pits in the Golgi region (HA-I type: Ahle *et al.*, 1988). An analysis of the components of these two types of adaptor complexes by SDS–PAGE is shown in Fig. 1. The HA-I complex consists of a heterodimer of "100-kDa" polypeptides, or adaptins, (apparently 115 and 105 kDa relative to marker proteins after electrophoresis on SDS gels), a 47-kDa polypeptide, and a 19-kDa light chain (Manfredi and Bazari, 1987; Keen, 1987; Ahle *et al.*, 1988). In contrast, HA-II complexes consist of a different heterodimer of adaptins (apparently in the range of 100–115 kDa) associated with a 50-kDa polypeptide and a 16-kDa light chain (Pearse and Robinson, 1984; Virshup and Bennett, 1988; Ahle *et al.*, 1988).

A number of techniques have been used to study the components of the adaptor complexes and to identify the different species of adaptins, including biochemical techniques and use of mAb.

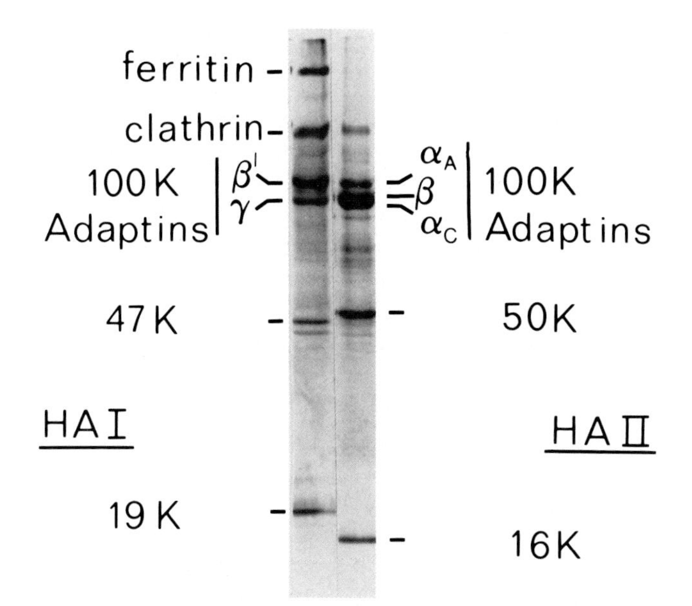

FIG. 1. Analysis by SDS gradient (7–20%) PAGE of samples of HA-I adaptors (characteristic of the Golgi region coated pits) and HA-II adaptors (restricted to plasma membrane coated pits). The two adaptor fractions were separated by chromatography on hydroxylapatite according to Manfredi and Bazari (1987), except that they were applied to the column at pH 7.0 instead of pH 8.4. The separation is better performed at the latter pH as the resolution of the adaptors from clathrin and ferritin is more efficient at pH 8.4. Samples were incubated for 5 minutes at 37°C with the addition of one-half volume of sample buffer, containing 0.2 *M* CHES (pH 9.5), 40% glycerol, 8% SDS, 4% DTT, 5% 2-mercaptoethanol, and 0.1% bromophenol blue, before application to the gel. Using this protocol, ferritin migrates as a high molecular weight band rather than as single subunits, which otherwise would run close to the HA-I adaptor light chain of 19 kDa. The gel was silver-stained. The HA-I adaptor complex consists of a heterodimer of a β'-adaptin and a γ-adaptin in combination with the 47-kDa protein and a 19-kDa light chain. The HA-II adaptor complex consists of a heterodimer of a β(B)-adaptin and an α-adaptin (either A or C) in combination with the 50-kDa protein and a 16-kDa light chain (the precise stoichiometry of the four different polypeptides, their arrangement in the complex, and the molecular weight of the total remain tentative).

1. BIOCHEMICAL TECHNIQUES

The interactions between the four components in the HA-II type adaptors have been investigated (Pearse and Robinson, 1984; Virshup and Bennett, 1988) using the crosslinking agents ethylene glycol succinimidyl succinate (EGS) and dithio-succinimidyl propionate (Lomant's reagent from Pierce Chemical Co., Rockford, IL). Two distinct adaptins (100-kDa polypeptides) interact with each other as a dimer in the adaptor complex.

After restricted crosslinking, partial complexes are observed: those of 150 kDa containing a single copy of the 50-kDa polypeptide crosslinked to a single copy of one of the 100-kDa polypeptides, and those containing a 16-kDa polypeptide linked to a single copy of one of the 100-kDa polypeptides. Immunoprecipitation of the crosslinked species should establish precisely which polypeptides are involved in these complexes. The molecular weight of the adaptor appears to be ~260,000 (although estimates vary up to ~300,000) based on the mobility on electrophoresis and hydrodynamic properties of the crosslinked species in comparison to the solution molecular weight of the native complex (calculated from the Stokes radius of 66 Å, the sedimentation coefficient of 8.65, and the partial specific volume from amino acid analysis of 0.74 cm^3/g (Pearse and Robinson, 1984; Keen, 1987; Virshup and Bennett, 1988). Sedimentation equilibrium analysis of purified adaptors should give a more definite figure for the molecular weight of the native complex. The stoichiometry of the constituent polypeptides (plausibly 1 : 1 : 1 : 1) seems to have been most reliably measured by densitometry of autoradiographs of the ^{125}I-labeled species from the native complex and the crosslinked species separated by gel electrophoresis (Virshup and Bennett, 1988).

Analysis of early preparations of the adaptins (100-kDa polypeptides) by SDS–PAGE indicated that they form a heterogeneous family of coat proteins (Pearse, 1978). Chromatography on hydroxylapatite as described in Section II,B separates the HA-I from the HA-II adaptins. Separation of the HA-II adaptins themselves is achieved by a second chromatography step on hydroxylapatite but this time after denaturation and in the presence of SDS. This allows the separation of the β(B) polypeptide, of intermediate mobility on SDS gels, from the A and C α-adaptins (Robinson and Pearse, 1986). The β(B) polypeptide is also distinguished by its lower mobility on SDS gels containing urea. Thus, on two-dimensional gels with a first dimension containing urea and a second without, the β(B) polypeptide runs off the diagonal (Ahle *et al.*, 1988). Peptide mapping and limited sequencing of the different species indicates that the A and C α-adaptins are related in amino acid sequence whereas the β(B) polypeptide has a quite different sequence (Robinson and Pearse, 1986; Robinson, 1987). Oligonucleotides based on peptide sequences from the separated polypeptides have been used in this laboratory to identify cDNA clones coding for the A, B, and C adaptins. The sequences deduced so far from the DNA confirm the similarity of the A and C polypeptides (α-adaptins) and the distinct nature of the B polypeptide (β-adaptin).

From the intensities of the adaptin bands in the adaptor complexes analyzed on SDS gels (see Figs. 1 and 3), it is likely that an HA-II complex contains a heterodimer of one α-adaptin (A or C) with one

β-adaptin (B), whereas an HA-I complex contains a heterodimer of one β'-adaptin and one γ-adaptin (Ahle *et al.*, 1988).

2. Use of mAb

Monoclonal antibodies have been raised against different members of the adaptin family. These antibodies have been used in immunofluorescence studies of cells to identify the location of the different species of 100-kDa polypeptide.

Thus the antibody AC1-M11 (mouse subclass IgG_{2a}; Robinson, 1987) recognizes the α-adaptins, A and C polypeptides, of the HA-II adaptor complexes. These adaptins appear to be restricted to plasma membrane coated pits. The mAb 100/3 (mouse subclass IgG_{2b}; Ahle *et al.*, 1988) recognizes a γ-adaptin characteristic of HA-I adaptors. In contrast to the α-adaptins, this γ-adaptin (105 kDa) appears to be restricted to coated pits in the Golgi region. A third mAb 100/1 (mouse subclass IgG_1; Ahle *et al.*, 1988) recognizes the β-adaptins including the β(B) polypeptide of the HA-II adaptor and related species in the HA-I adaptor complex. This antibody labels both plasma membrane and Golgi coated pits.

The mAb have also been exploited (Ahle *et al.*, 1988) to identify the pairs of adaptins that dimerize to form the different adaptor complexes, by immunoprecipitation. Thus, mAb 100/3, which recognizes γ-adaptin, precipitates HA-I adaptor complexes, which contain β-adaptins as well as γ-adaptin. Similarly, a monospecific anti-β-adaptin [HA-II-type β-adaptin (B)] antibody also precipitates α_C- and α_A-type polypeptides, strongly suggesting an interaction of these with the β_B-type polypeptide within the HA-II adaptor.

III. Reassembly of Adaptors with Clathrin

The strategy for the reassembly of adaptors with clathrin to form coats is as follows (Zaremba and Keen, 1983; Pearse and Robinson, 1984).

Adaptors at 0.05 mg/ml in buffer B are combined with clathrin at 0.03 mg/ml and the mixture dialyzed overnight into 100 volumes of assembly buffer A (pH 7.0) at 0.4°C. The coats are then pelleted (100,000 *g*, 1 hour at 5°C) and resuspended in a small volume of buffer A. In these conditions, neither clathrin nor the adaptors appear to sediment appreciably in the absence of the other.

There are a number of points to consider when attempting to reassemble adaptors with clathrin to form coats:

1. The clathrin, which should be freshly recycled from cages to ensure maximum assembly competence, is mixed with the adaptors in buffer B at a low concentration of $\leq$0.05 mg/ml. The level of clathrin is set below the apparent critical concentration for clathrin assembly in the absence of adaptors (Crowther and Pearse, 1981), in order to maximize the contribution of the adaptors to the formation of coats.

2. Clathrin in 20 m*M* Tris-Cl (pH 7.5), when introduced to buffer A (pH 6.5) conditions, polymerizes in seconds with a critical concentration of ~0.05 mg/ml (Crowther and Pearse, 1981). Clathrin in 1 *M* Tris-Cl (pH 7.0) will polymerize into cages with high yield at high concentrations of clathrin (>4 mg/ml) when dialyzed into buffer A (pH 6.5). At low concentrations of clathrin from these conditions, polymerization appears slow and less efficient upon dialysis into buffer A (pH 7.0). Perhaps the clathrin triskelion is in a less favorable conformation for cage assembly in 1 *M* Tris-Cl than in 20 m*M* Tris-Cl—perhaps with extended instead of bent legs (Kirchhausen *et al.*, 1986). In this case the adaptor complexes may tend to hold the clathrin triskelions together, nucleating clathrin assembly and helping to promote cage completion. The actual effect of adaptors on the critical concentration of clathrin assembly has not yet been determined. However, coats containing adaptors do seem more stable at pH 7.5 and 0.5 m*M* $MgCl_2$ than cages containing clathrin alone, which tend to disassemble at this pH (Zaremba and Keen, 1983).

3. The most dramatic effect of the adaptors on clathrin assembly is that they restrict the size range of the coats formed in comparison to the cages formed by clathrin alone. Whereas the cages are very heterogeneous (50–120 nm in diameter), their size depending on the clathrin concentration, the coats are more homogeneous in size (50–80 nm in diameter) (Zaremba and Keen, 1983).

Both the HA-I-type and the HA-II-type adaptors have this effect on clathrin assembly (Keen, 1987). Another polypeptide of 180 kDa has also been found to reassemble with clathrin with a similar effect (Ahle and Ungewickell, 1986).

4. A three-dimensional map of a reconstituted clathrin coat is shown in Fig. 2. The map is computed from electron micrographs of unstained specimens embedded in vitreous ice (Vigers *et al.*, 1986a,b). A segment has been cut away to reveal the internal structure of the coat. In addition to the two outer shells formed by the clathrin, there is a third central shell due to the adaptors. The contacts between the clathrin and the adaptors appear to be made by the shell of terminal domains of the clathrin under

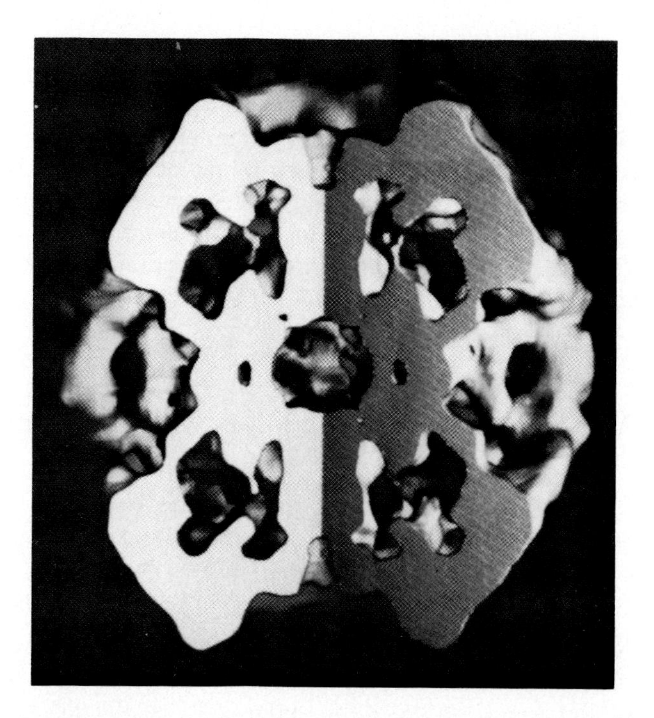

FIG. 2. Three-dimensional map of a clathrin coat computed from electron micrographs of unstained specimens of reconstituted particles embedded in vitreous ice (Vigers *et al.*, 1986a,b). A segment has been cut away to reveal the internal structure of the coat. The map shows three shells of density. The two outer shells represent the clathrin: (1) the outer lattice of the polyhedral cage of the hexagonal barrel and (2) an intermediate shell corresponding to the terminal domains of the clathrin heavy chains. (3) The third central shell is due to the adaptor complexes. The contacts between the clathrin and the adaptors appear to be made by the shell of terminal domains. The hexagonal barrel structure is too small to contain a vesicle, but in a larger coat the vesicle would be enclosed within the shell of adaptors.

the vertices. In an actual coated vesicle, which would have to have a larger coat to accommodate the vesicle, the latter would be enclosed within the shell of adaptors.

5. The stoichiometry of adaptors to clathrin triskelions in coated vesicles and in reconstituted coats is still uncertain. The preparations are heterogeneous in particle size and geometry, and in most coated-vesicle preparations, a proportion of the particles lack vesicles. Therefore, the ratio of polypeptides in individual structures is unknown. Estimates vary between one and three 100-kDa adaptin polypeptides per clathrin triskelion in the total mixture of coats, depending on the preparation (Pearse and Robinson, 1984; Keen, 1987). However, an attractive model is that

there might be three adaptor complexes (containing a heterodimer of 100-kDa adaptins) per triskelion making contact with the terminal domains of clathrin beneath a vertex of the cage. β-Adaptin polypeptides are represented in both types of adaptor complex. These are therefore obvious candidates for providing a clathrin-binding domain in the different adaptor complexes. However, some differences have been observed in the affinity of the HA-I adaptors for clathrin compared to the HA-II adaptors (Keen, 1987). The mapping of the functional domains of the proteins making up the adaptor complexes has yet to be achieved.

IV. Reassembly of Receptors with Adaptors

The interaction of receptors with adaptors has been studied in two ways: (1) reassembly of the purified mannose 6-phosphate receptor (in soluble detergent-extracted form) with adaptors (Pearse, 1985), and (2) binding of adaptors to affinity matrices of receptor cytoplasmic tails.

1. For use in reconstitution experiments, the purified mannose 6-phosphate receptor is dissolved in buffer B, dialyzed against 200 volumes of buffer B, and then centrifuged for 1 hour at 5°C at 100,000 *g* to remove any aggregated material. The concentration of the adaptors in the assembly mixture should be in the range of 50–100 μg/ml and the receptor concentration ~100 μg/ml. The assembly mixture in buffer B (1 ml) is dialyzed overnight against buffer A, pH 7.0. Complexes formed are collected by centrifugation for 1 hour at 5°C at 100,000 *g*. Typically pellets are resuspended in 20 μl of buffer A. They contain spherical structures of 30–100 nm diameter, which migrate on agarose gel electrophoresis as complexes containing roughly equivalent amounts of receptors and adaptins. When clathrin is included in the assembly mixture at 50 μg/ml, coats form that encapsulate the mannose 6-phosphate receptor.

Although this reconstitution demonstrates a direct association between a receptor and an adaptor, it is not easily applied generally to other receptors. This is because most receptors tend to aggregate by themselves in the absence of membranes, lipid, or excess detergent. Perhaps in the case of the mannose 6-phosphate receptor a minimum of residual detergent protects the relatively small hydrophobic domain of the receptor.

2. In order to study the interaction of various receptors with adaptors, the second approach is being used. The available evidence suggests that it is the cytoplasmic tail of recycling receptors that carries the signal for

assembly into coated pits (Goldstein *et al.*, 1985; Roth *et al.*, 1986; Rothenberger *et al.*, 1987; Lazarovits and Roth, 1988). For example, the tyrosine residue at position 807 in the cytoplasmic tail of the low-density lipoprotein (LDL) receptor, 18 residues from the membrane, plays a crucial role in the localization of the LDL receptor in coated pits and its subsequent endocytosis (Goldstein *et al.*, 1985).

The cDNAs coding for several receptors that enter coated pits are available and their sequences known. Expression vectors [of the type pLcII(nic$^-$); Nagai and Thøgersen, 1987] can be constructed that direct the synthesis in *Escherichia coli* of fusion proteins containing the cytoplasmic tail of a particular receptor attached to a soluble domain of a cytoplasmic protein. Thus, for instance, the protein BP2 consists of the N-terminal 31 amino acid residues of λ cII protein, 112 residues of chicken myosin light chain, and 50 residues of the human LDL-receptor cytoplasmic tail, interspersed with blood coagulation factor X_a-recognition sites. BP2 is easily purified from the inclusion bodies of *E. coli* and is soluble in both buffer A and buffer B. Therefore, the complete fusion protein is used to generate an affinity column by coupling to Affi Gel 10.

Adaptors (HA-II type) bind specifically to the BP2 column in buffer A and are eluted in buffer B (see Fig. 3). No such binding is observed under the same conditions using a column to which a similar fusion protein is attached but which lacks the LDL-receptor cytoplasmic portion and contains instead the rest of the myosin light chain. Excess unlabeled adaptors (HA-II type) compete with ^{125}I-labeled adaptors for binding to the BP2 column. Thus it is reasonable to assume that the BP2 resin is behaving as an affinity matrix of LDL-receptor cytoplasmic tails that specifically binds the adaptors of the type restricted to plasma membrane coated pits.

V. Summary

Adaptors sort out those receptors that participate in assembly of coated pits from those that are excluded.

Two distinct adaptor units have so far been identified: (1) adaptors restricted to plasma membrane coated pits (HA-II type, named according to their elution position during hydroxylapatite chromatography) and (2) adaptors restricted to Golgi region coated pits (HA-I type).

Adaptors contain a heterodimer of two 100-kDa polypeptides, a β-adaptin (possibly carrying an essentially common clathrin-binding

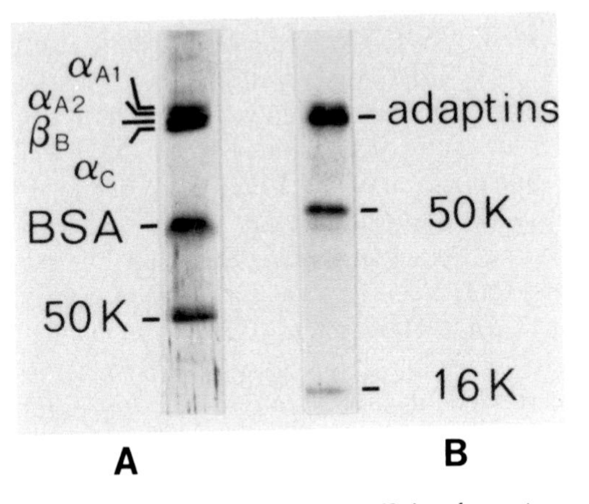

FIG. 3. ^{125}I-labeled HA-II adaptors purified by affinity chromatography on a matrix of LDL-receptor cytoplasmic portions and analyzed by SDS–PAGE. (A) A silver-stained 7.5% gel showing the α_{A1}-, α_{A2}-, β_B-, and α_C-adaptins, and the 50-kDa protein of the plasma membrane-restricted HA-II adaptor. (B) An autoradiograph of a 7–20% gradient gel showing the 100-kDa adaptins, the 50-kDa protein, and the 16-kDa light chain of the HA-II adaptors.

domain) and a distinct α- or γ-adaptin characteristic of the type of adaptor and its specific location. Each adaptor is constructed from four different polypeptides. Thus HA-II adaptors contain a β-adaptin and an α-adaptin in combination with a 50-kDa protein and a 16-kDa polypeptide. The HA-I adaptors contain a β-adaptin and a γ-adaptin in combination with a 47-kDa protein and a 19-kDa polypeptide.

Both types of adaptors and also a 180-kDa polypeptide will promote the assembly of clathrin to form coats, the size range of which appears to be relatively restricted compared to cages made from clathrin alone.

The HA-II adaptors, characteristic of plasma membrane coated pits, bind to the cytoplasmic tail of the LDL receptor. They also assemble with the mannose 6-phosphate receptor *in vitro* in the absence of membrane. When clathrin is included, the adaptors promote the assembly of coats containing bound receptor.

REFERENCES

Ahle, S., and Ungewickell, E. (1986). *EMBO J.* **5,** 3143–3149.

Ahle, S., Mann, A., Eichelsbacher, U., and Ungewickell, E. (1988). *EMBO J.* **7,** 919–929.

Campbell, C., Squicciarini, J., Shia, M., Pilch, P. F., and Fire, R. E. (1984). *Biochemsitry* **23,** 4420–4426.

Crowther, R. A., and Pearse, B. M. F. (1981). *J. Cell Biol.* **91,** 790–797.

Goldstein, J. L., Brown, M. S., Anderson, R. G. W., Russell, D. W., and Schneider, W. J. (1985). *Annu. Rev. Cell Biol.* **1,** 1–19.

Kedersha, N. L., and Rome, L. H. (1986a). *Anal. Biochem.* **156,** 161–170.
Kedersha, N. L., and Rome, L. H. (1986b). *J. Cell Biol.* **103,** 699–709.
Keen, J. H. (1987). *J. Cell Biol.* **105,** 1989–1998.
Keen, J. H., Willingham, M. C., and Pastan, I. H. (1979). *Cell (Cambridge, Mass.)* **16,** 303–312.
Kirchhausen, T., Harrison, S. C., and Heuser, J. (1986). *J. Ultrastruct. Mol. Struct. Res.* **94,** 199–208.
Lazarovits, J., and Roth, M. (1988). *Cell (Cambridge, Mass.)* **53,** 743–752.
Manfredi, J. J., and Bazari, W. L. (1987). *J. Biol. Chem.* **262,** 12182–12188.
Nagai, K., and Thøgersen, H. C. (1987). *In* "Methods in Enzymology" (R. Wu and L. Grossman, eds.), Vol. 153, pp. 461–481. San Diego, California.
Pearse, B. M. F. (1975). *J. Mol. Biol.* **97,** 93–98.
Pearse, B. M. F. (1976). *Proc. Natl. Acad. Sci. U.S.A.* **73,** 1255–1259.
Pearse, B. M. F. (1978). *J. Mol. Biol.* **126,** 803–812.
Pearse, B. M. F. (1982). *Proc. Natl. Acd. Sci. U.S.A.* **79,** 451–455.
Pearse, B. M. F. (1985). *EMBO J.* **4,** 2457–2460.
Pearse, B. M. F. (1987). *EMBO J.* **6,** 2507–2512.
Pearse, B. M. F., and Bretscher, M. S. (1981). *Annu. Rev. Biochem.* **50,** 85–101.
Pearse, B. M. F., and Crowther, R. A. (1987). *Annu. Rev. Biophys. Biophys. Chem.* **16,** 49–68.
Pearse, B. M. F., and Robinson, M. S. (1984). *EMBO J.* **3,** 1951–1957.
Robinson, M. S. (1987). *J. Cell. Biol.* **104,** 887–895.
Robinson, M. S. (1989). *J. Cell Biol.* **108,** 833–842.
Robinson, M. S., and Pearse, B. M. F. (1986). *J. Cell Biol.* **102,** 48–54.
Roth, M. G., Doyle, C., Sambrook, J., and Gething, M.-J. (1986). *J. Cell Biol.* **102,** 1271–1283.
Rothenberger, S., Iacopetta, B. J., and Kühn, L. C. (1987). *Cell (Cambridge, Mass.)* **49,** 423–431.
Rubenstein, J. L. R., Fine, R. E., Luskey, B. D., and Rothman, J. E. (1981). *J. Cell Biol.* **89,** 357–361.
Schook, W., Puszkin, S., Bloom, W., Ores, C., and Kochwa, S. (1979). *Proc. Natl. Acad. Sci. U.S.A.* **76,** 116–120.
Unanue, E. R., Ungewickell, E., and Branton, D. (1981). *Cell (Cambridge, Mass.)* **26,** 439–446.
Vigers, G. P. A., Crowther, R. A., and Pearse, B. M. F. (1986a). *EMBO J.* **5,** 529–534.
Vigers, G. P. A., Crowther, R. A., and Pearse, B. M. F. (1986b). *EMBO J.* **5,** 2079–2085.
Virshup, D. M., and Bennett, V. (1988). *J. Cell Biol.* **106,** 39–50.
Wiedenmann, B., Lawley, K., Grund, C., and Branton, D. (1985). *J. Cell Biol.* **101,** 12–18.
Woodward, M. P., and Roth, T. F. (1978). *Proc. Natl. Acad. Sci. U.S.A.* **75,** 4394–4398.
Zaremba, S., and Keen, J. H. (1983). *J. Cell Biol.* **97,** 1339–1347.

Part III. Subcellular Fractionation Procedures

"Classical" subcellular fractionation procedures for solid tissues are discussed and presented by Beaufay and Amar-Costesec (1976), de Duve (1971), and in Volume 31 of *Methods in Enzymology* (See Appendix). Several routine procedures for cells in culture are also available, although the resulting fractions are often not as thoroughly characterized as those recovered from solid tissues (Tulkens *et al.*, 1974; Knipe *et al.*, 1977; Storrie *et al.*, 1984; Schmid *et al.*, 1988).

The present emphasis in organelle fractionation is increasingly preparative. Several high-resolution fractionation methods have been developed, based on the recognition of superficial antigenic determinants of the endodomain of membranes (Morré *et al.*, 1988) or on electrical charge characteristics of vesicles (Morré *et al.*, 1988), or determined by the endocytic history of the organelle in question (Quintart *et al.*, 1984; Helmy *et al.*, 1986). Examples of the latter elegant approach have involved loading endocytic vesicles with enzyme tracers whose activity allows them to be separated from nonendocytic contaminants. A variant of such methods makes use of fluorescent tracers and flow cytometry (see Wilson and Murphy, Chapter 16 of this volume). Despite the sophistication of these several procedures, subcellular fractionation is still largely empirical. Issues such as particle aggregation, distortion, or rupture are still of major concern, and the question of whether an isolated fraction is representative of an organelle *in situ* can only be answered after rigorous extended scrutiny.

Basic references to the use of Percoll (Schmitt and Herrmann, 1977) and iodinated solutes (Rickwood *et al.*, 1982) may be of interest. A useful homogenization apparatus for cultured cells has been described (Balch and Rothman, 1985).

References

Balch, W., and Rothman, W. (1985). Characterization of protein transport between successive compartments of the Golgi apparatus: Asymmetric properties of donor and acceptor activities in a cell-free system. *Arch. Biochem. Biophys.* **240,** 413–425.

Beaufay, H., and Amar-Costesec, A. (1976). Cell fractionation techniques. *Methods Membr. Biol.* 1–100.

de Duve, C. (1971). Tissue fractionation, past and present. *J. Cell. Biol.* **50,** 20–55.

Helmy, S., Porter-Jordan, K., Dawidowiez, P., Pilch, P., Schwartz, A., and Fine, R. (1986). Separation of endocytic from exocytic coated vesicles using a novel cholinesterase-mediated density shift technique. *Cell (Cambridge, Mass.)* **44,** 497–506.

Knipe, D., Baltimore, D., and Lodish, H. (1977). Separate pathways of maturation of the major structural proteins of vesicular stomatitis virus. *J. Virol.* **21,** 1128–1139.

Morré, D., Howell, K., Cook, G., and Evans, W. (eds.). (1988). "Cell-Free Analysis of Membrane Traffic." Liss, New York.

Quintart, J., Courtoy, P., and Baudhuin, P. (1984). Receptor-mediated endocytosis in rat liver: Purification and enzymic characterization of low density organelles involved in uptake of galactose-exposing proteins. *J. Cell Biol.* **98,** 877–884.

Rickwood, D., Ford, T., and Graham, J. (1982). Nycodenz: A new nonionic iodinated gradient medium. *Anal. Biochem.* **123,** 23–31.

Schmid, S., Fuchs, R., Male, P., and Mellman, I. (1988). Two distinct subpopulations of endosomes involved in membrane recycling and transport to lysosomes. *Cell (Cambridge, Mass.)* **52,** 73–83.

Schmitt, J., and Herrmann, R. (1977). Fractionation of cell organelles in silica colloid gradients. *In* "Methods in Cell Biology" (D. M. Prescott, ed.), Vol. 15, pp. 177–200. Academic Press, Orlando, Florida.

Storrie, B., Pool, R., Sachdeva, M., Maurey, K., and Oliver, C. (1984). Evidence for both prelysosomal and lysosomal intermediates in endocytic pathways. *J. Cell Biol.* **98,** 108–115.

Tulkens, P., Beaufay, H., and Trouet, A. (1974). Analytical fractionation of homogenates from cultured rat embryo fibroblasts. *J. Cell Biol.* **63,** 383–401.

Chapter 14

Lectin–Colloidal Gold-Induced Density Perturbation of Membranes: Application to Affinity Elimination of the Plasma Membrane

DWIJENDRA GUPTA[1] AND ALAN M. TARTAKOFF

Institute of Pathology
Case Western Reserve University School of Medicine
Cleveland, Ohio 44106

I. Introduction

Several novel subcellular compartments such as the trans-Golgi network (TGN, Griffiths and Simons, 1986) and the mannose 6-phosphate

[1] Present address: Department of Biochemistry, Faculty of Sciences, University of Allahabad, Allahabad 211 002, U. P., India.

receptor-rich compartment (Griffiths *et al.,* 1988), have recently been distinguished from those that are conventionally known to play a role in vesicular transport along the secretory and endocytic paths. Because any systematic functional dissection of transport will require precise organelle characterization and reconstitution of membrane traffic events in cell-free systems, increasingly selective subcellular fractionation procedures are essential.

In an effort to document the composition of the membranes of the Golgi complex and compare and contrast this organelle with the rough endoplasmic reticulum, we have described a subcellular-fractionation procedure for rat myeloma cells that effectively eliminates rough microsomes from Golgi-enriched fractions (Gupta and Tartakoff, 1989a). The differential and isopycnic-sedimentation protocol yields Golgi-enriched fractions of considerable purity. Nevertheless, a major isopycnic overlap with plasma membrane remains. We have therefore developed a novel density perturbation procedure to eliminate these plasma membrane contaminants. This protocol makes use of the specific interaction of the plant lectin, wheat germ agglutinin (WGA), with carbohydrates on the surface of intact cells. This lectin is known to bind oligosaccharides with nonreducing terminal D-GlcNAc and clustered sialic acid residues (Bhavanandan and Katlic, 1979).

The irreversible nature of the binding of proteins such as WGA to colloidal gold, and the high electron density and physical density of gold particles make such conjugates convenient electron-microscopic cytochemical probes and potential density perturbants. We describe here the preparation and use of 40-nm colloidal gold particles conjugated to WGA to increase the density of the plasma membrane-derived vesicles and/or fragments. Analogous procedures employing other lectins or antibodies might be of general use for affinity removal–purification of intracellular organelles.

II. Preparation of Lectin–Gold Complexes

A. Colloidal Gold Solution

A 2% (w/v) stock solution of tetrachloroauric acid, $HAuCl_4$ (Sigma, St. Louis, MO, or Fisher Scientific Co.) is prepared in deionized water (pale yellow) and stored at 4°C in a glass bottle wrapped with aluminum foil to protect from light. The solution can be thus stored indefinitely at 4°C. The preparation of a colloidal gold suspension has to be carried out in siliconized or silanized glasswares. A 5% solution of dichloro-dimethyl silane (Kodak Laboratories, Rochester, NY) in chloroform was therefore used to rinse all glass prior to use. It is essential to get rid of all traces of chloroform because its presence adversely affects the reduction of Gold(ic)$^{3+}$ to Au^{2+}. For this purpose, we conveniently boil ~50 ml of deionized water in silanized Erlenmeyer flasks. The flask is then used to prepare a 0.01% aqueous solution of $HAuCl_4$ from the 2% stock. For preparing a large batch of colloid, it is *strongly advised* to make several 100-ml suspensions separately and combine them for subsequent use in lectin–gold conjugate preparation. Preparing a single large batch usually results in an undesirable *dirty-red color* with a heterogeneous range of gold particle sizes (from 5 to 40 nm), presumably because of lack of uniform heating and/or mixing.

This solution is brought to a vigorous boil in an Erlenmeyer flask capped with aluminum foil, and reduction is initiated by rapidly piercing the foil with a pipet (siliconized or plastic) and introducing freshly prepared 1% aqueous sodium tricitrate (1.5 ml/100 ml of 0.01% gold chloride solution) without interrupting the boiling. After replacing the foil cap, the solution is allowed to boil for 7–10 minutes to complete the reaction. Visually one can follow the reduction of Au^{3+}, as the color turns to black upon addition of citrate and passes from purple-blue to wine-red color (Geoghegan and Ackerman, 1977). The color is a function of particle size (Frens, 1973). If the wine-red color is not achieved, the solution should be discarded.

The colloid is cooled to room temperature and its pH measured in the presence of 1% polyethylene glycol (PEG, MW 20,000, Sigma) to protect the electrode. This colloid should originally have a pH of 4.85–5.0, which is then raised to the optimal pH for adsorption of the particular lectin. In our studies, we slowly adjusted the pH to 7.0 with successive addition of a few drops of 0.2 *M* K_2CO_3. The colloidal gold solution now can be stabilized with protein after empirical titration with the protein.

B. Wheat Germ Agglutinin–BSA

Wheat germ agglutinin was chosen for our purposes because (1) it binds essentially irreversibly to surface glycoproteins and glycolipids at 4°C, and (2) the binding is not competed by sucrose (Chang *et al.*, 1975), as, for example, in subcellular-fractionation protocols employing high concentrations of sucrose. In our experiment, a gradient of 20–45% sucrose TKM (50 m*M* Tris-HCl, pH 7.4, 100 m*M* KCl, and 5 m*M* $MgCl_2$) was employed to subfractionate microsomes. The small molecular size of WGA (36 kDa), however, is a problem inasmuch as it results in poor adsorption to the large gold particles (40 nm). It was therefore necessary to increase the molecular weight of WGA by crosslinking with bovine serum albumin (BSA; Horisberger and Rosset, 1977). Wheat germ agglutinin, 2 mg (Sigma no. L9640)[2] and BSA, 8 mg (98–99% pure fraction V, Sigma) were dissolved in a relative weight proportion of 1 : 4 in 0.5 ml 5 m*M* NaCl. *N*-Acetyl-D-glucosamine (Sigma) was added to a final concentration of 0.1 *M* to protect the active site of WGA during the crosslinking, which was started by adding 1/100 volume 20% glutaraldehyde (Sigma) and incubating at 25°C for 2 hours. Subsequently, 5 m*M* NaCl was added to adjust the WGA–BSA concentration to 2 mg/ml. This preparation was passed through a Millipore filter (0.45 μm), prewashed with 30 ml deionized water, and dialyzed overnight against two changes of 5 m*M* NaCl at 4°C. The dialyzed preparation was again passed through a 0.45-μm Millipore filter to remove any nonspecific large aggregates. This preparation can be used immediately for adsorption to colloidal gold or can be stored in presence of 0.02% NaN_3 for as long as 10 days at 4°C. We routinely used preparations that were not more than 2–3 days old. Dialysis may be used to remove the azide from the WGA–BSA solution prior to adsorption.

III. Lectin–Gold Complex

The amount of WGA–BSA required to prevent flocculation of colloidal gold was empirically determined for each preparation as follows: colloidal gold solution (1 ml) adjusted to pH 7.0 was pipeted into a series of plastic or siliconized-glass tubes. Varying amounts of the WGA–BSA preparation (0.5–5.0 μg) were added next and vortexed briefly. After a 10-minute incubation at room temperature, 100 μl of 9% (w/v) NaCl solution was added and flocculation was judged visually by change in color from wine-red to violet and finally light blue. The protein concentration that prevents change in color of the solution was taken as the stabilization point. In general, we found 0.5 mg of WGA (as WGA–BSA preparation) sufficient for 100 ml colloidal gold. Flocculation can also be evaluated spectrophotometrically by reading the absorbance at 520–580 nm against colloidal gold solution as a blank (Geoghegan and Ackerman, 1977; Horisberger and Rosset, 1977). The protein concentration corresponding to the point where the curve first appears asymptotic with the *x* axis is taken as the stabilization point.

[2] Many, but not all, batches of the lectin supplied by Sigma Chemical Co. performed well for our density perturbation experiments.

After titration, an amount of WGA–BSA just sufficient to stabilize the colloidal gold was added with gentle stirring and held at room temperature for 15 minutes. The lectin–gold complexes were then further stabilized by adding 1% PEG (MW 20,000, 7.5 ml/100 ml suspension), and centrifuged after 1 hour at 20,000 rpm for 30 minutes at 4°C (70 Ti, Beckman Instruments, Inc., Palo Alto, CA) to pellet lectin–gold particles with average diameter of 40 nm. (Higher centrifugal forces are needed to sediment smaller particles; however, excessive force reduces the yield.) The ultracentrifuge tubes were placed on a rack for 30 minutes to allow the red sediment to collect at the bottom. The clear supernatant was carefully removed (not aspirated) with a Pasteur pipet and the sediment was resuspended in the remaining 100–200 μl of fluid. A tiny deposit of metallic gold on the side of the tube is left untouched. The suspension of colloidal gold–lectin conjugate was adjusted to isotonic phosphate-buffered saline (PBS; 10 m*M* NaP_i, pH 7.4, 0.9% NaCl) or Dulbecco's modified Eagle medium (DMEM) using a 10-fold concentrated stock solution, stored at 4°C and used within 10 days. Alternately, the WGA–BSA–gold conjugate can be sterilized by Millipore filtration and/or stored at 4°C with 0.5 mg/ml sodium azide for >1 year (Horisberger and Rosset, 1977). The conjugate may contain limited amounts of free, unadsorbed WGA–BSA. This can be eliminated by ultracentrifugation of the lectin–gold suspension through a 1-ml aqueous cushion of 2% BSA (99% pure); however, resedimentation can lead to major losses.

IV. Dose Optimization of WGA–BSA–Gold

Preliminary experiments were performed to determine the maximal nonagglutinating dose of the WGA–BSA for rat myeloma cells. Samples of cell suspensions (IR202, IgM secretor, or 983F, nonsecretor), recovered from the peritoneal cavity of LOU rats and freed of red cells by NH_4Cl lysis were incubated at 2×10^7 cells/ml with increasing concentrations of the lectin–gold conjugate preparation in PBS or DMEM for 30 minutes on ice, with gentle mixing every 5 minutes. Subsequently, cells were pelleted (5 minutes at 1000 *g*, 4°C), washed twice with ice-cold DMEM, and observed under the light microscope. The cell pellet acquires a wine-red color due to the conjugate. Cell viability judged by trypan blue exclusion was slightly reduced to 81–85% relative to controls (90–92%). Modest cell agglutination was sometimes observed, but a single-cell suspension could be easily produced by gently pipeting. For radiometric determination of the amount of conjugate that was bound, ^{14}C-labeled lectin–BSA–gold conjugate was employed and the percentage radioactivity bound to the cells was calculated. In such experiments, ^{14}C-labeled BSA was prepared by reductive methylation of BSA with 0.2 m*M* ^{14}C-HCHO (40–60 mCi/mmol, New England Nuclear, Boston) and 50 m*M* $NaCNBH_3$[3] (Aldrich Chemical Co., Milwaukee, WI) in 40 m*M* NaP_i (pH 7.0), according to Haas and Rosenberry (1985). This procedure methylates ~72% of the N-terminal α-amino groups of BSA. The highly reactive ε-NH_2 groups of lysine are methylated with a stoichiometry of 7.9 residues per molecule. The radiolabel is 96% incorporated.

Figure 1 shows a representative titration curve for such radiometric evaluations.

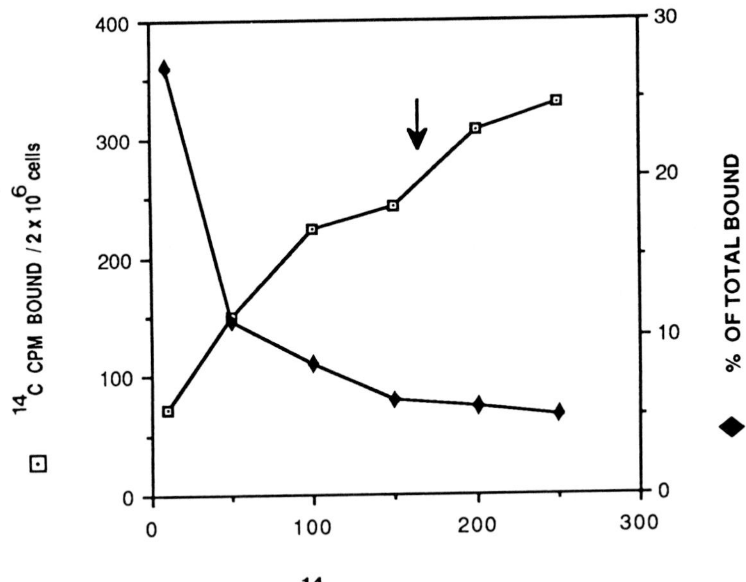

FIG. 1. Estimation of optimal WGA–gold concentration for binding to the cell surface. Myeloma cell suspensions (2 × 10^6 cells in 100 μl DMEM) were incubated on ice for 30 minutes with increasing doses of the radioactive lectin–BSA–gold preparation, where the BSA was ^{14}C-dimethylated according to Haas and Rosenberry (1985). The cells were then sedimented, washed twice, dissolved in 4 ml scintillation cocktail, and counted. The amount of lectin conjugate just below that required to cause agglutination was routinely employed for subsequent experiments and is indicated by an arrow. Generally, lectin–gold conjugate preparation from ~300 ml of colloidal gold solution was used for ~2 × 10^8 cells. In the experiment illustrated, 100 μl conjugate = 2700 cpm.

[3] $NaCNBH_3$ was recrystallized according to Jentoft and Dearborn (1979).

V. Density Perturbation of the Plasma Membrane

A. Electron-Microscopic Studies: Differential Sedimentation

Because the Golgi-enriched microsomal subfractions obtained by isopycnic sedimentation in sucrose gradients (Gupta and Tartakoff, 1989a) were contaminated with alkaline phosphatase, a marker of plasma membrane (Green and Newell, 1974), we exploited the lectin–BSA–gold conjugate bound to the cell surface to increase the density of the plasma membrane. When myeloma cells were incubated with the gold conjugate in isotonic PBS or DMEM at 4°C for 30–60 minutes, washed, and subsequently fixed and examined in the electron microscope (Fig. 2), extensive labeling of much of the cell surface was seen. No endocytosis of the lectin–gold conjugate was observed.

After binding of the conjugate, the myeloma cells were spun at 800 rpm for 5 minutes at 4°C, washed three times with ice-cold PBS to remove unbound lectin–gold conjugate, and once with cold hypotonic buffer (10 m*M* Tric-HCl, pH 7.4, 10 m*M* KCl, 1.5 m*M* $MgCl_2$) and then swollen with the same buffer (0.2 ml per 2 × 10^8 cells) for 15 minutes on ice. After homogenization in the presence of cycloheximide (10 μg/ml), DNase (10 μg/ml), phenylmethylsulfonyl fluoride (PMSF, 0.25 m*M*), and leupeptin (50 μg/ml) with a Dounce homogenizer (50 strokes), the homogenate was adjusted to 0.25 sucrose–TKM (STKM) by adding 0.1 ml of a compensating solution. During all these and subsequent manipulations, lectin–gold conjugate remains bound to the plasma membranes as judged by electron-microscopic examination of the homogenate, nuclear, mitochondrial, and microsomal fractions (not shown). This is to be anticipated from earlier studies using labeled WGA in similar protocols.

As seen in the mitochondrial pellet, for example (Fig. 3), numerous closed and open vesicles representative of plasma membrane fragments are labeled with gold particles. Sedimentation of these vesicles to the mitochondrial pellet is accompanied by removal of the alkaline phosphatase activity from the postmitochondrial supernatant. The alteration of the distribution of alkaline phosphatase is striking. Compared with controls, a considerably greater percentage of alkaline phosphatase (45–67% versus 20%) is recovered in the mitochondrial pellet (Fig. 4B). Parallel studies making use of surface-radioiodinated cells (Hubbard and Cohn, 1975) also reveal a considerable increase in the sedimentation (70–80% versus 25%) of ^{125}I versus controls (data not shown). Experi-

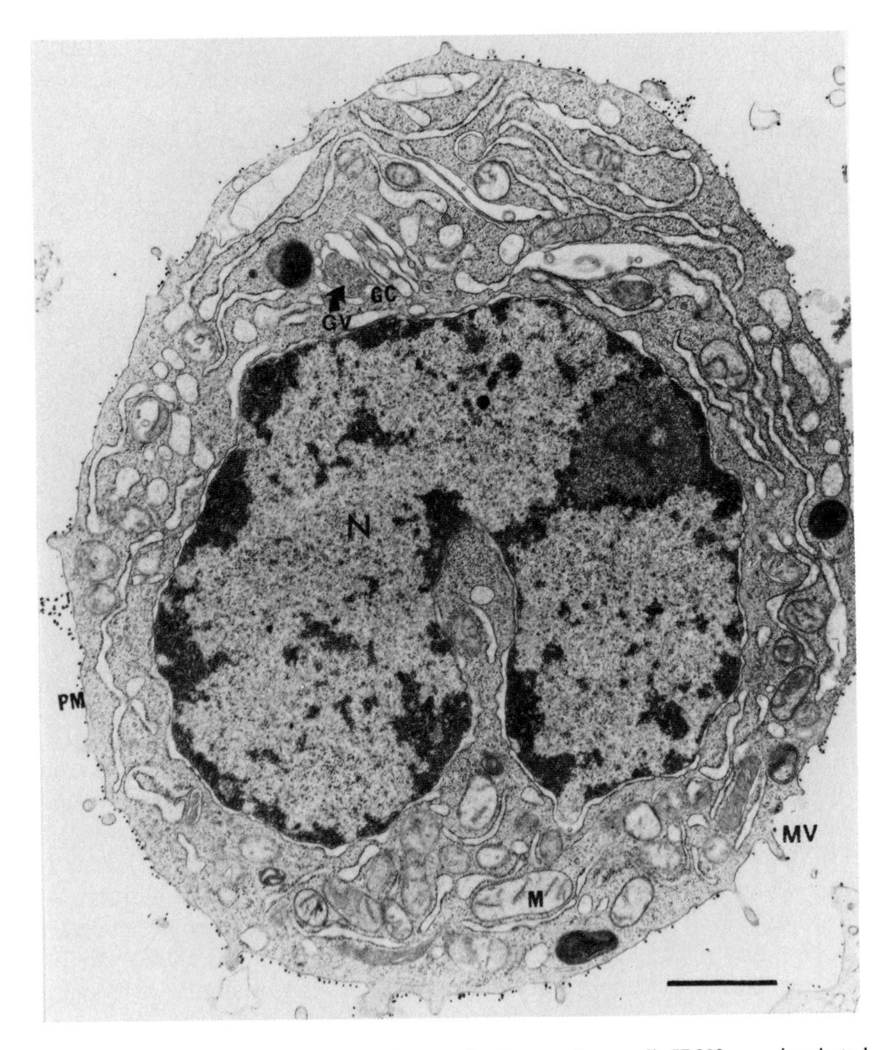

FIG. 2. WGA–gold labeling of myeloma cells. Rat myeloma cells IR202 were incubated with the optimum dose of WGA–BSA–gold (see Section IV) at 4°C for 30 minutes, centrifuged, washed with PBS, fixed in glutaraldehyde and osmium tetroxide, and processed. Thin sections were examined in the JEOL CS 100 II transmission electron microscope. Labeling of the plasma membranes (PM) with gold particles is observed. Golgi cisternae (GC) and vesicles (GV), nucleus (N), mitochondria (M), and microvilli (MV) are indicated. Bar = 1 μm.

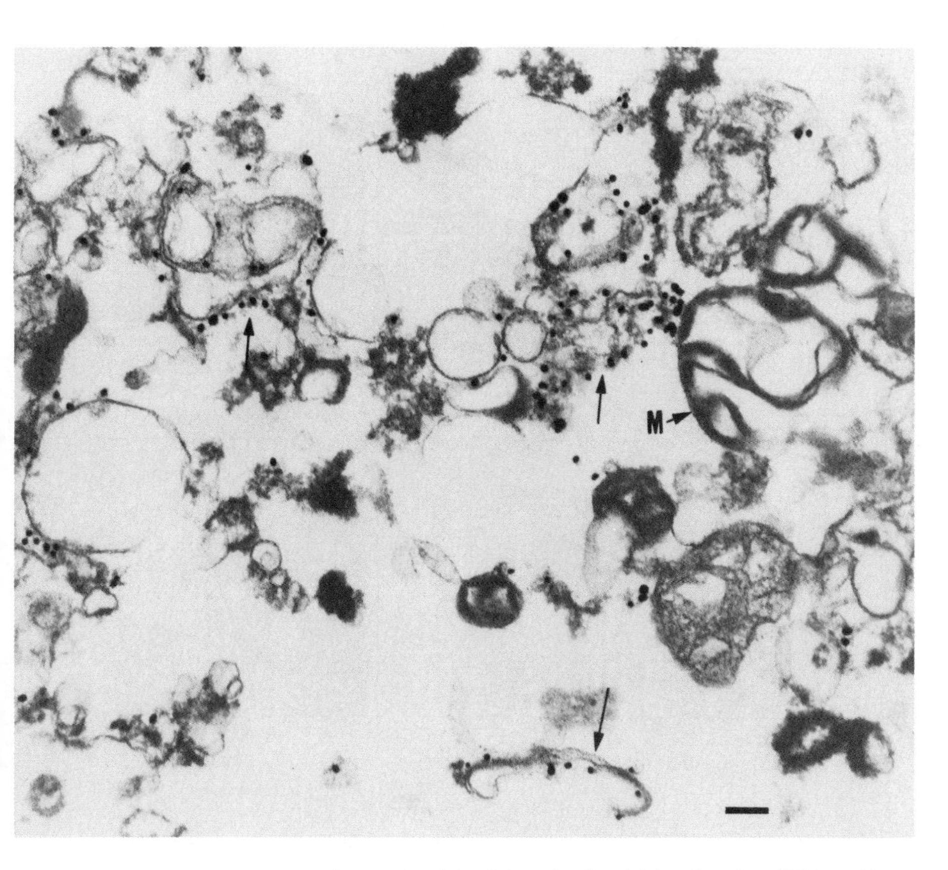

FIG. 3. Electron-microscopic examination of the mitochondrial pellet after WGA–gold binding. The mitochondrial pellet derived from lectin–gold-treated cells was fixed with 2% glutaraldehyde and osmium tetroxide and processed for electron microscopy. A thin section representative of the entire depth of the pellet is illustrated. Numerous closed and open (possibly broken during electron-microscopic processing) vesicles tagged with gold particles are seen (arrows). M, Mitochondria. Bar = 0.25 μm.

ments employing ^{14}C-labeled lectin–gold conjugate also documented the extensive sedimentation of the conjugate itself in association with membranes (Fig. 4B). By contrast, judging from the distribution of biochemical markers monitored in the presence or absence of lectin–gold labeling (Fig. 4A, C, and D), lectin–gold labeling of the surface does not influence the differential sedimentation of β-hexosaminidase, galactosyltransferase, and RNA.

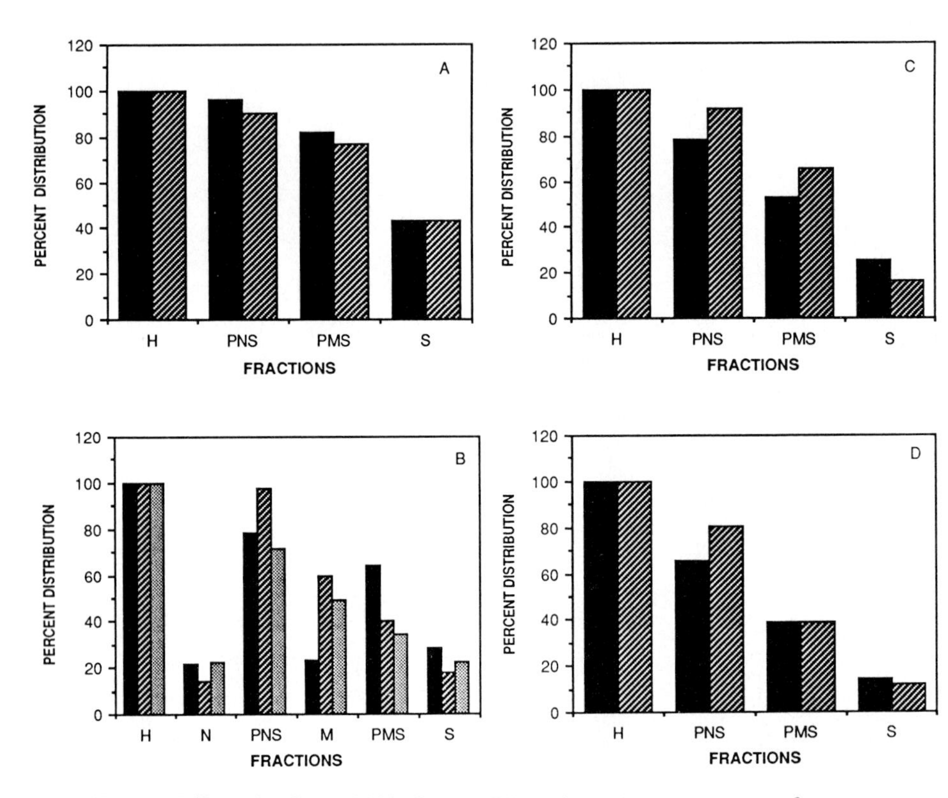

FIG. 4. Effect of WGA–gold binding on differential sedimentation. 2×10^8 cells were labeled with the lectin–gold conjugate at 4°C for 30 minutes, washed, homogenized, and subjected to differential sedimentation in parallel with controls lacking the conjugate. For (B), a preparation of ^{14}C-labeled lectin–BSA–gold was also used. In each case, the data have been normalized to 100%. The distributions shown are averages of four such experiments. Solid columns, control; hatched columns, lectin–gold labeled. (A) Galactosyltransferase. There are no differences observed as a result of lectin–gold binding to the cells. (B) Alkaline phosphatase. Note that the distribution of alkaline phosphatase in the mitochondrial pellet (M) and postmitochondrial supernatant (PMS) is drastically altered in lectin–gold-labeled cells compared to controls. Distribution of ^{14}C-BSA-labeled lectin–gold (crosshatched columns) is also affected in parallel with alkaline phosphatase. (C) β-Hexosaminidase and (D) RNA exhibit little or no change. H, Homogenate; N, nuclear pellet; PNS, postnuclear supernatant; S, microsomal fraction.

B. Isopycnic Distribution of Marker Activities in the Presence of Lectin–Gold

Following the labeling of cells with lectin–gold, the isopycnic behavior of alkaline phosphatase activity is also drastically altered. When microsomal fractions are recovered by a gel filtration procedure from the postmitochondrial supernatant and subsequently analyzed on continuous gradients of STKM (Fig. 5A), there is extensive sedimentation of this plasma membrane marker: nearly 35–40% of alkaline phosphatase is pelleted through the 40% STKM.[4] Additionally, the penultimate four dense fractions account for 20% of the enzyme activity.

The amount of residual (unperturbed) alkaline phosphatase was not reduced by doubling the dose of the conjugate. This activity may therefore not be derived from the cell surface. Considering that after binding of the conjugate only ~33–55% of alkaline phosphatase activity of the homogenate is recovered in the postmitochondrial supernatant (PMS), and that the total microsomal fraction contains only 35–40% of the alkaline phosphatase activity of the PMS, only ~18% of homogenate activity is loaded onto the isopycnic gradients. The corresponding figure for two Golgi enzyme markers (galactosyltransferase, UDPase) is ~25% (Gupta and Tartakoff, 1989a). The observations of the distribution of alkaline phosphatase activity were confirmed by immunoblots stained with anti-rat alkaline phosphatase antiserum (gift from Dr. Y. Ikehara, Fukuoka, Japan).

Figure 5B gives the isopycnic-distribution profile of galactosyltransferase in both control and density-perturbed situations. There is no effect of the lectin conjugate on this distal-Golgi marker activity. UDPase, which is essentially coisopycnic with galactosyltransferase (Yamazaki and Hayaishi, 1968), also does not undergo any perturbation (not shown). Binding of lectin–gold conjugate does not inhibit any of the enzyme activities investigated.

The distribution of acid CMPase is slightly altered (Fig. 5C), although much of its activity stays in the lighter density regions, roughly coincident with galactosyltransferase activity. There is no significant change in the distribution profile of the proximal-Golgi marker, the overosmicated material (Gupta and Tartakoff, 1989a; Novikoff *et al.*, 1971; Locke and Huie, 1983). Also, the distribution of the cation-independent mannose 6-phosphate receptor (215 kDa) and an integral Golgi protein (GCIII, Chicheportiche and Tartakoff, 1987) was unaffected in the density-perturbed situation (Gupta and Tartakoff, 1989b).

Electron-microscopic examination of an isopycnic gradient pellet from lectin–gold-treated cells shows numerous gold-tagged closed vesicular profiles (Fig. 6). Galactosyltransferase-enriched fractions from the same

[4] Similar lectin–gold labeling studies performed with the human erythroleukemic line K562 also demonstrate that ~35% of total surface ^{125}I-labeled proteins are pelleted to the bottom of Percoll gradients (Park and Snider, personal communication).

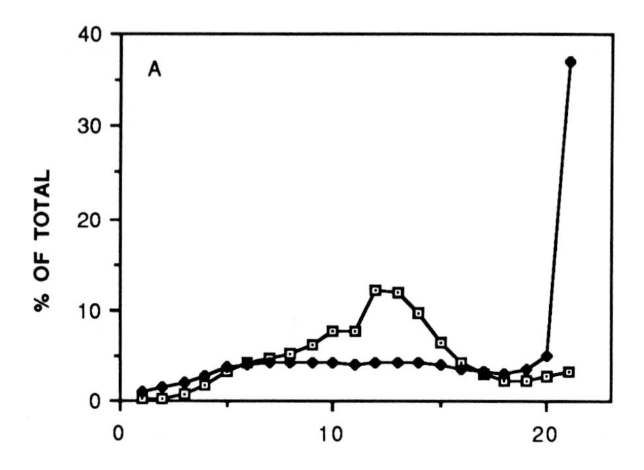

A
% OF TOTAL
40
30
20
10
0
0
10
20

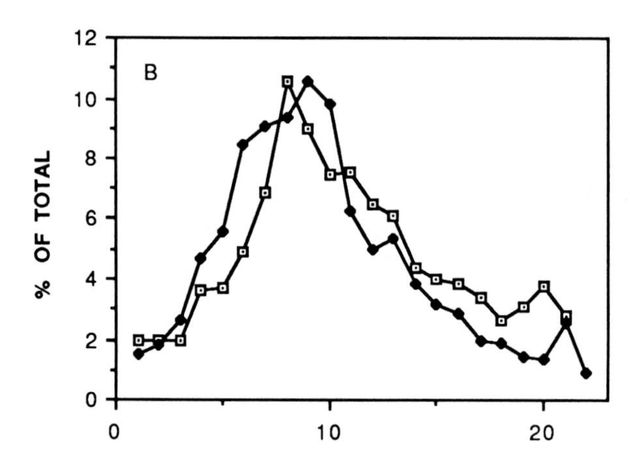

B
% OF TOTAL
12
10
8
6
4
2
0
0
10
20

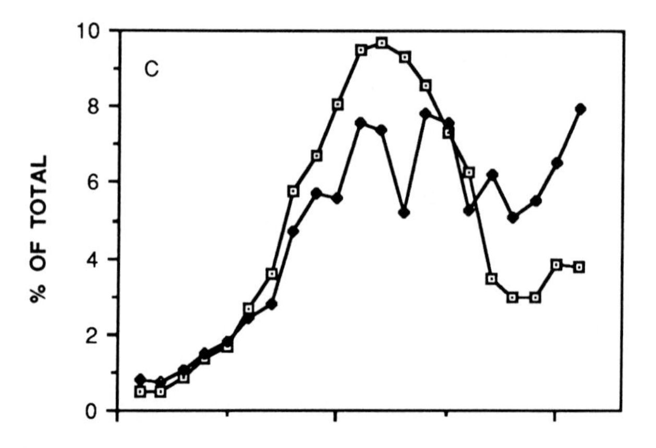

C
% OF TOTAL
10
8
6
4
2
0

gradient contain primarily cisternae and vesicles, presumably derived from the Golgi, but not gold.

It is thus clear that lectin–gold conjugate binding does not drastically alter the distribution of any of the markers of Golgi or Golgi-associated membranes and is satisfactory for removing plasma membrane contaminants.

C. Purification of Golgi Subfractions

As a result of the plasma membrane density perturbation, the ratio of galactosyltransferase to alkaline phosphatase activities in the transferase-rich part of the isopycnic gradient increased ~7 to 14-fold in several experiments (Table I). Even in the absence of the density perturbation, the transferase specific activity of these fractions is ~10× that of total microsomes. Thus, the lectin–gold binding to cells results in Golgi subfractions of strikingly high purity.

VI. Discussion

The WGA–gold conjugate was used to label the surface of myeloma cells under conditions that did not cause cell agglutination, morphological alterations as judged by electron microscopy, or endocytosis of colloidal gold.

The binding of the lectin conjugate to the cell surface makes it possible to achieve, during subcellular fractionation, extensive sedimentation of

---

FIG. 5. Effect of WGA–gold binding on the isopycnic distribution of enzymatic markers. Microsomes derived from 2×10^8 untreated (⊡, control) or lectin–gold-treated cells (◆) were fractionated on a 20–40% isopycnic STKM gradient by an overnight centrifugation at 50,000 rpm using the SW50.1 rotor (Beckman). The gradient fractions were analyzed for alkaline phosphatase (A), galactosyltransferase (B), and CMPase (C), and the data from a typical experiment plotted. In each case, the recovered enzyme activity is normalized to 100%. (A) Alkaline phosphatase. Note that the majority of the enzyme activity in lectin–gold-treated cells has shifted to denser regions of the gradient compared to controls. Because of the extensive sedimentation of alkaline phosphatase to the "mitochondrial pellet," the actual amount loaded onto the experimental gradient is only about one-third that of the control gradient. Recovery was 90–127%. (B) Galactosyltransferase. No influence is observed on the isopycnic distribution. Recovery was 90–102%. (C) CMPase. There is a small amount of sedimentation of this marker. However, the majority of activity stays within the light-density regions coincident with Golgi markers. Recovery was 115–130%. Fraction 1 is the top of the gradient.

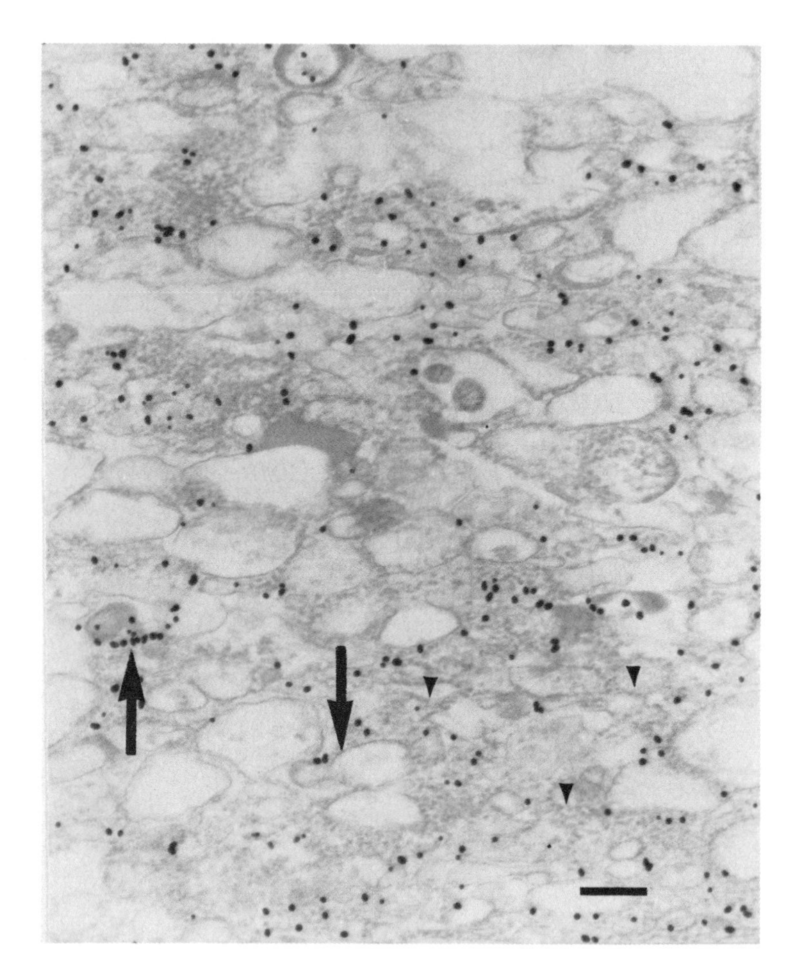

FIG. 6. Extensive pelleting of plasma membrane visualized by electron microscopy. Microsomes derived from 4×10^8 lectin–gold-labeled cells were centrifuged overnight on a 20–40% STKM gradient. The pellet was fixed *in situ* with 2.5% glutaraldehyde and osmium tetroxide and processed for electron microscopy. Many of the gold-tagged membranes have a large, vesicular appearance and may derive from the plasma membranes (arrows). The pellet is also highly enriched in ribosomes and rough microsomes (arrowheads). Note the absence of cisternal elements. Bar = 0.25 μm.

alkaline phosphatase activity, the lectin conjugate itself, and iodinated cell surface proteins to the mitochondrial pellet. For this reason, the lectin–gold binding resulted in Golgi subfractions of unusually high purity.

What is especially striking is the lack of effect of density perturbation on the distribution profile of biochemical markers characteristic of distal

TABLE I

PURIFICATION OF GOLGI MARKER (GALACTOSYLTRANSFERASE) ON ISOPYCNIC SUCROSE GRADIENTS AFTER ELIMINATION OF PLASMA MEMBRANE[a]

Fraction number	Galactosyltransferase–alkaline phosphatase ratio		Enrichment over control (-fold)
	Control	Lectin–gold treated	
8	7.7	53	7
9	5.1	55	11
10	3.1	43	14
11	2.3	21	9
12	1.5	16	11

[a] Isopycnic gradient fractions from the galactosyltransferase activity-rich region have been evaluated for enrichment in terms of this trans-Golgi marker after lectin–gold-induced sedimentation of alkaline phosphatase. On this basis, lectin–gold treatment of cells results in 7- to 14-fold purification of Golgi membranes compared to control. Data have been taken from the experiment illustrated in Fig. 5A and B.

Golgi (galactosyltransferase, UDPase), GERL or trans-Golgi reticulum (acid CMPase), and proximal Golgi (osmicatable material), as well as antigenic markers.

It is not clear why any residual alkaline phosphatase remains unperturbed. Doubling the dose of conjugate does not alter this observation. We therefore suspect that this "residual" activity could be a bona fide contribution from endomembranes, including Golgi membranes. Indeed, the "residual" alkaline phosphatase activity in the Golgi region exhibits a latency of >30% (Gupta and Tartakoff, unpublished observation). Moreover, cytochemical studies on HeLa cells (Tokumitsu and Fishman, 1983) demonstrate that this enzyme is also localized in the endoplasmic reticulum, Golgi, and lysosomes. In our electron-microscopic cytochemical studies of rat myelomas, alkaline phosphatase is principally detected at the cell surface (Gupta and Tartakoff, unpublished observations) with a sporadic minor staining of Golgi cisternae and the rough endoplasmic reticulum (not shown). It therefore appears that the sedimentation of alkaline phosphatase and thus the plasma membrane is almost quantitative.

Previous investigators have used density or size perturbation strategies to recover plasma membrane fractions from cells in suspension. Certain ones of the perturbants bind electrostatically to the cell surface (Gottlieb and Searls, 1980; Chaney and Jacobson, 1983), whereas others insert into the lipid bilayer (Beaufay *et al.*, 1974) or bind to surface saccharides (Wallach and Kamat, 1964; Kang *et al.*, 1985; Molday and Molday, 1987).

Moreover, as in studies of affinity retrieval of cytoplasmic organelles (Ito and Palade, 1978; Merisko *et al.,* 1982; Luzio and Stanley, 1983), immunoabsorbants consisting of antibodies bound to magnetic particles or fixed *Staphylococcus aureus* have been used (Roman and Hubbard, 1984; Devaney and Howell, 1985; Gruenberg and Howell, 1985).

By contrast with these latter procedures, we have used a density perturbant that is much smaller than the membrane vesicles being isolated. The conjugate is also much simpler from the compositional point of view: no intact bacteria need be added. A possible disadvantage of our procedure might be promotion of vesicle aggregation if lectin-binding sites were present on both ectodomains and endodomains of membranes; however, the absence of cosedimentation of Golgi markers argues against this. Because immunoglobulin can be directly bound to colloidal gold (or to gold-bearing protein A), the present approach might be readily complemented by positive immunoselection procedures for organelle isolation, as in recent studies of epidermoid carcinoma cells in which endocytosis of colloidal gold–antitransferrin receptor antibody complex resulted in recovery of a 200-fold purified endosomal fraction in an isopycnic sucrose gradient (Beardmore *et al.,* 1987). Such double density-perturbation schemes (first eliminate the plasma membrane, then retrieve the organelle of interest) should result in Golgi subfractions of the purity required for documentation of their membrane protein composition.

Acknowledgments

Technical assistance given by Al Haddad, Erica Colton, Anita Saha, and Thomas Massella is gratefully acknowledged. Thanks are also due to Dr. Yves Chicheportiche, Dr. William Brown, and Dr. Ikehara for antibodies, Dr. Robert Haas for advice in the preparation of ^{14}C-BSA, Dr. Neal Smith for maintenance of inbred colonies of Lou CN rats, and to Sonya Olsen for typing this manuscript. Work was supported by NIH grants DK27651 (to D.K.G.) and AI21269 (to A.M.T.).

References

Beardmore, J., Howell, K. E., Miller, K., and Hopkins, C. R. (1987). *J. Cell Sci.* **87,** 495–500.

Beaufay, H., Amar-Costesec, A., Feytans, E., Thines-Sempoux, D., Wibo, M., Robbi, M., and Berthet, J. (1974). *J. Cell Biol.* **61,** 188–200.

Bhavanandan, V., and Katlic, A. (1979). *J. Biol. Chem.* **254,** 4000–4007.

Chaney, L. K., and Jacobson, B. C. (1983). *J. Biol. Chem.* **258,** 10062–10072.

Chang, K.-J., Bennett, V., and Cuatrecasas, P. (1975). *J. Biol. Chem.* **250,** 488–500.

Chicheportiche, Y., and Tartakoff, A. M. (1987). *Eur. J. Cell Biol.* **44,** 135–143.

Devaney, E., and Howell, K. E. (1985). *EMBO J.* **4,** 3123–3130.

Frens, G. (1973). *Nature (London) Phys. Sci.* **241,** 20–22.

Geoghegan, W. D., and Ackerman, G. A. (1977). *J. Histochem. Cytochem.* **25,** 1187–1200.

Gotlieb, L. J., and Searls, D. B. (1980). *Biochim. Biophys. Acta* **602,** 207–212.
Green, A. A., and Newell, P. C. (1974). *Biochem. J.* **140,** 313–322.
Griffiths, G., and Simons, K. (1986). *Science* **234,** 438–443.
Griffiths, G., Hoflack, B., Simons, K., Mellman, I., and Kornfeld, S. (1988). *Cell (Cambridge, Mass.)* **52,** 329–341.
Gruenberg, J., and Howell, K. E. (1985). *Eur. J. Cell Biol.* **38,** 312–321.
Gupta, D. K., and Tartakoff, A. M. (1989a). *Eur. J. Cell Biol.* **48,** 52–63.
Gupta, D. K. and Tartakoff, A. M. (1989b). *Eur. J. Cell Biol.* **48,** 64–70.
Haas, R., and Rosenberry, T. L. (1985). *Anal. Biochem.* **148,** 154–162.
Horisberger, M., and Rosset, J. (1977). *J. Histochem. Cytochem.* **25,** 295–305.
Hubbard, A. L., and Cohn, Z. A. (1975). *J. Cell Biol.* **64,** 438–460.
Ito, A., and Palade, G. E. (1978). *J. Cell Biol.* **79,** 590–597.
Jentoft, N., and Dearborn, D. G. (1979). *J. Biol. Chem.* **254,** 4359–4365.
Kang, M. S., Au-Young, J., and Cabib, E. (1985). *J. Biol. Chem.* **260,** 12680–12684.
Locke, M., and Huie, P. (1983). *J. Histochem. Cytochem.* **31,** 1019–1032.
Luzio, J. P., and Stanley, K. K. (1983). *Biochem. J.* **216,** 27–36.
Merisko, E. M., Farquhar, M. G., and Palade, G. E. (1982). *J. Cell Biol.* **260,** 846–848.
Molday, R. S., and Molday, L. L. (1987). *J. Cell Biol.* **105,** 2589–2601.
Novikoff, P. M., Novikoff, A. B., Quintana, N., and Hauw, J.-J. (1971). *J. Cell Biol.* **50,** 859–886.
Roman, L. M., and Hubbard, A. L. (1984). *J. Cell Biol.* **98,** 1497–1504.
Tokumitsu, S., and Fishman, W. H. (1983). *J. Histochem. Cytochem.* **31,** 647–655.
Wallach, D. F. H., and Kamat, V. B. (1964). *Biochemistry* **52,** 721–728.
Yamazaki, M., and Hayaishi, O. (1968). *J. Biol. Chem.* **243,** 2934–2942.

Chapter 15

Immunoisolation Using Magnetic Solid Supports: Subcellular Fractionation for Cell-Free Functional Studies

KATHRYN E. HOWELL[1]

European Molecular Biology Laboratory
6900 Heidelberg, Federal Republic of Germany

RUTH SCHMID

SINTEF
7034 Trondheim, Norway

JOHN UGELSTAD

Department of Industrial Chemistry
University of Trondheim
7034 Trondheim, Norway

JEAN GRUENBERG

European Molecular Biology Laboratory
6900 Heidelberg, Federal Republic of Germany

---

[1] Present address: University of Colorado School of Medicine, Department of Cellular and Structural Biology, Denver, Colorado 80262.

I. Introduction

The functions performed at each stage of the biosynthetic and the endocytic pathways have been the focus of study in the field of membrane traffic. In several cases, the molecules or enzymes that carry out these functions have been identified and characterized. These encompass part of the machinery involved in the translation, insertion, and posttranslational modification of newly synthesized membrane and secretory proteins (for review, see Walter and Lingappa, 1986). Others include the motor proteins that mediate the transport of vesicles on microtubules (for review, see Vale, 1987), the vacuolar ATPase responsible for the luminal acidification of endosomes and lysosomes (for review, see Mellman *et al.*, 1986), and GTP-binding proteins involved in the secretory pathway (Melançon *et al.* 1987; Segev *et al.*, 1988; Gould *et al.*, 1988). However, the major functions of these organelles rely on protein sorting and membrane traffic between compartments. These mechanisms, now under intense scrutiny, still remain poorly understood.

The hypothesis of vesicular transport was first proposed by Jamieson and Palade in 1967 from their work on the secretory process of the guinea pig exocrine pancreas and is reviewed by Palade (1975). This transport is clearly selective: some molecules, both proteins and lipids, move from one subcellular compartment to another whereas others are retained. It has been proposed that constitutively secreted protein follow the "bulk flow" of membrane transport to the plasma membrane and that proteins destined to compartments along the pathway contain specific retention signals (Pfeffer and Rothman, 1987). The existence of retention signals is consistent with the finding that a specific KDEL sequence is present at the C terminus of soluble proteins that reside in the lumen of the ER (Munro and Pelham, 1987). Pelham (1988) has proposed that this sequence is recognized by an unidentified receptor, which recycles the KDEL-containing proteins that have escaped to a compartment distal to the ER.

It is now well established that specific transport signals and receptors direct the traffic of many molecules. The best example is the mannose 6-phosphate signal, which is required for the transport of newly synthesized lysosomal hydrolases to the lysosomes (for reviews, see von Figura and Hasilik, 1986; Kornfeld, 1987). Two receptors have been identified that recognize this signal, the cation-dependent and cation-independent mannose 6-phosphate receptors. They are believed to transport the hydrolases from the trans-Golgi network to a prelysosomal compartment, where dissociation occurs. The hydrolases are then packaged in the lysosomes and the receptors recycle for reutilization. The identification of other signal–receptor couples involved in protein sorting is the subject of many investigations.

The data accumulated since 1967 have provided strong support for vesicular transport in the exocytic pathway, but the hypothesis has never been proven in the most rigorous sense. Rothman and colleagues have pioneered the use of cell-free approaches to dissect the mechanisms controlling vesicular transport in the Golgi complex (for review, see Rothman, 1988). Recently, they have identified a cytosolic factor sensitive to *N*-ethylmaleimide that is important for the fusion step (Glick and Rothman, 1987; Malhotra *et al.,* 1988). In the endocytic pathway the formation of clathrin-coated vesicles from coated pits on the plasma membrane is the only well-documented example of transport vesicle formation. This process has focused attention on the role of clathrin and other coat proteins, although the mechanisms are not explained (for a review, see Pearse, 1987). A different coat protein has been visualized by electron microscopy on the surface of vesicles thought to be involved in transport between Golgi cisternae (Orci *et al.,* 1986). However, it is still unclear how the different subcellular compartments specifically interact with each other during transport and what mechanisms are involved. Moreover, several additional routes of membrane traffic between compartments or interactions between elements of the same compartment probably exist but have not been documented. In the endocytic pathway, the different steps of membrane traffic remain poorly understood and controversial.

Our experimental strategy to study membrane traffic in endocytosis has been to reconstitute a specific step of the pathway in a cell-free system using defined subcellular fractions. We have developed immunoisolation, which uses the specificity of antibodies, as a subcellular fractionation method. We use monodisperse magnetic solid supports, which efficiently and rapidly isolate the fraction of interest. The methods used in our experimental approach are described in Section II. The fraction can then be sequentially introduced into and retrieved from different reaction mixtures throughout an experimental protocol via the magnetic properties of the solid support. We have used this approach to dissect the early stages of the endocytic pathway, and this will be discussed in Sections III and IV.

II. Immunoisolation

Traditional subcellular-fractionation procedures are based on differences in the size and density of the organelles and organelle-derived vesicles after homogenization of the cells and separation by density gradient centrifugation (for review, see Beaufay and Amar-Costesec, 1976). However, many of the smooth membrane vesicles derived from the compartments of membrane traffic share common densities and are difficult to isolate by these procedures. In addition, many new approaches in cellular and molecular biology rely on the use of cells grown in culture. These cells are often more difficult to homogenize and fractionate than rat liver, the "classical" tissue that has been used for subcellular-fractionation studies. Alternative techniques—in particular, immunoisolation—have proved successful in overcoming some of the problems encountered using tissue culture cells (for review, see Howell *et al.*, 1989).

Immunoisolation relies on the presence of an antigenic site exposed on the outer surface of the compartment of interest, rather than a physical property of the organelle. This method provides a means to isolate a subcellular compartment rapidly and efficiently, no matter how small a percentage of the total cellular membrane it represents, and may be uniquely suited for the isolation of vesicular carriers. The antibody that recognizes the antigen is usually bound to a solid support, which is used to isolate the specific organelle from the other components of the input fraction (in many experiments the input fraction would be a postnuclear supernatant). The immobilization of the organelle allows its introduction and retrieval from a series of reaction mixtures, which significantly facilitates any assay.

Subcellular fractionation by immunoisolation was introduced by Luzio *et al.* (1976) and Ito and Palade (1978), and has been reviewed by a number of groups, each using different solid supports and protocols: cellulose fibers (Bailyes *et al.*, 1987; Luzio *et al.*, 1988), *Staphylococcus aureus* cells (Hubbard *et al.*, 1988), and magnetic beads (Howell *et al.*,1988b). These papers present a good overview of the application of immunoisolation in a number of different systems. In this article we will focus on the details of the method as we apply it.

A. Magnetic Solid Support

Magnetic solid supports have many advantages because separation is based on a principle other than sedimentation properties and retrieval is rapid and efficient. For these reasons, we have focused on designing an immunoisolation system for organelles using magnetic solid supports. A variety of magnetic particles have been constructed (for reviews, see Rembaum *et al.*, 1982; Kemshead and Ugelstad, 1985) and used for many purposes, including positive and negative cell selection and immunoassays (for review, see Lea *et al.*, 1988).

The solid support developed for subcellular fractionation is a nonporous, magnetic, monosized polymer particle of 4.5 μm diameter. These beads are identical in size and contain equal amounts of magnetizable material, features that allow all beads to be collected homogeneously in a magnetic field or sedimented by centrifugation. Retrieval with a magnet is rapid and requires no special equipment. Centrifugation is effective only when none of the components of the input fraction can be sedimented by the force required to sediment the solid support. Magnetic systems are also an advantage when sterile conditions are required during an experiment (Roman *et al.*, 1988).

The magnetic monosized polymer particles are produced by a technique developed by Ugelstad *et al.* (1980). A macroreticular particle is formed by an activated swelling of a polymer made from styrene divinyl benzene. Oxidative groups are introduced at the surface of the pores, and then the magnetization process is carried out *in situ* by stirring the particles in an aqueous solution of Fe^{2+} salts. Under appropriate conditions Fe^{2+} is continuously transferred to the interior of the pores, where it is oxidized and precipitates as insoluble iron hydroxide, which on heating is transformed to maghemite (γFeO_2O_3) (Ugelstad *et al.*, 1983). The fact that the maghemite is present as very fine grains throughout the particle volume (see Fig. 6) ensures that the particle is superparamagnetic. The lack of remnant magnetism of the beads is essential for their application in separation procedures (Ugelstad *et al.*, 1986, 1988).

The pores remaining after magnetization may be filled with polymeric materials. The final treatment of the beads may be selected to provide functional groups at the surface, for example, epoxy, hydroxyl, amino, hydrazide, and sulfhydryl groups (or chelating groups). These groups are used for the covalent coupling of proteins or other hydrophilic molecules, such as spacer arms, to the bead surface.

The magnetic beads are marketed under the trademark Dynabeads by Dynal A/S (Oslo, Norway and Great Neck, NY).

The beads, denoted M-450, have been used by several laboratories, especially in cell separation experiments. They have a diameter of 4.5 μm, a density of 1.5 g/cm^3, and contain 20% Fe by weight; 1 g corresponds to 1.4×10^{10} beads. The surface area accessible by a gas adsorption method is 2–4 m^2/g, whereas the outer surface calculated from $4\pi r^2$ is 1.08 m^2/g. This difference is explained by some microporosity of the bead surface. The M-450 bead exhibits a rather hydrophobic surface, allowing a relatively strong physical adsorption of protein. The surface polymer also has free hydroxyl groups, which are used to couple protein covalently. Also available are M-450 beads with a variety of linker antibodies or specific antibodies already covalently coupled to the bead surface.

There are many special demands of beads used for immunoisolation, and various new particles are being designed, produced, and tested in our experiments. Especially promising are those that have a thin shell of nonmagnetic material on the bead surface. This is achieved by repeated coating of the macroporous particles with oligomeric compounds and results in a compact bead with a thin layer of nonmagnetic hydrophilic material. The shell reduces the interactions between the beads in the magnetic field (aggregation) and limits the nonspecific binding of cellular material. Both effects help to decrease significantly the contamination of the immunoisolated fraction with other vesicles.

B. Experimental Protocol

An outline of our usual experimental protocol is presented here and diagrammed in Fig. 1. The detailed methods of specific steps will be presented following the outline.

1. A linker molecule is covalently coupled to the surface of the bead.
2. The specific antibody is bound to the linker molecule.
3. The complex—termed the immunoadsorbant (ImAd), which consists of the solid support, the linker molecule, and the specific antibody—is incubated with the input fraction for the immunoisolation. We routinely carry out this step in PBS containing 5 mg/ml BSA in 1 ml volume in an Eppendorf tube and mix on a rotator at 2 rpm for 2 hours. Because of the magnetic properties of the beads, the tube should remain insulated from ferromagnetic material to prevent the beads from aggregating on the tube walls. We have made a very simple modification of a conventional rotator by placing a plastic tube holder on the rotator spindle. This is effective in preventing the aggregation of the magnetic beads during the binding steps (Fig. 2)
4. At the end of the incubation the beads plus bound fraction are retrieved on a simple permanent magnet either hand-held (Fig. 1) or using a magnetic rack (Fig. 3). We have had a magnetic rack constructed for Eppendorf tubes, so that we can process many tubes at once. Placing the magnet at the side of the tube, rather than the bottom, obviates formation

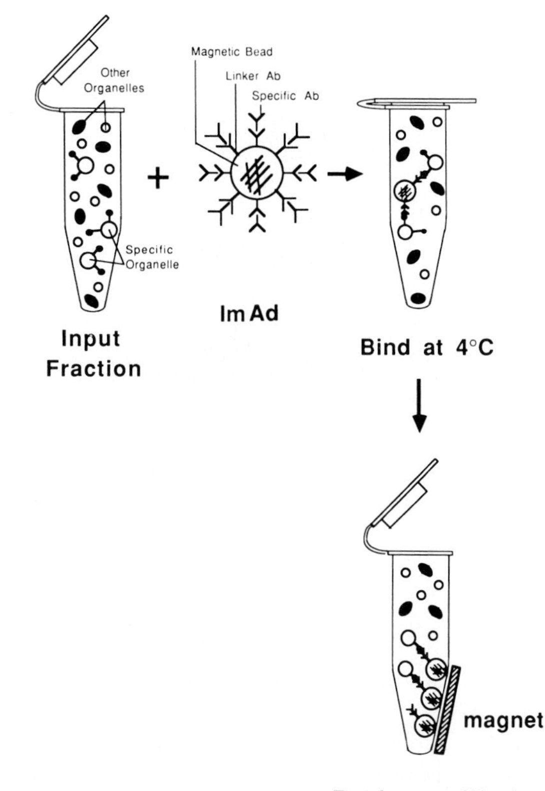

FIG. 1. Diagram of an immunoisolation experiment using ImAd prepared with magnetic solid supports. The isolation is carried out in a standard Eppendorf tube. The input to the immunoisolation is subcellular fraction (≈500 μg), usually a PNS, which is uniformly suspended in ≈0.9 ml of the following mixture: one volume of 0.25 *M* sucrose, 3 m*M* imidazole, pH 7.4 (homogenization buffer), and two volumes of PBS–5 mg/ml BSA. The input fraction contains both specific organelles (diagrammed to have two exposed antigenic sites and other organelles that do not contain the antigenic site exposed in a position available for the specific antibody to bind. The ImAd (1 mg), which has been prepared in advance and is uniformly suspended in 100 μl of the above buffer is added to the input fraction. The ImAd consists of the magnetic beads with covalently attached linker antibody to which a specific antibody against the exposed antigenic sites is immunobound. The two are mixed at 4°C for ~2 hours on a rotator like that diagrammed in Fig. 2. Then the ImAd and the organelles bound via the specific antibody are retrieved using a small hand-held magnet or in a magnetic rack as diagrammed in Fig. 3. The magnet is placed at the side of the tube rather than the bottom to avoid any clumps that would tend to sediment, and the nonbound fraction is removed with the buffer and set aside for assay. The ImAd and bound fraction can be resuspended for washing nonspecifically bound and entrapped vesicles, and retrieved again on the magnet. The final immunoisolated fraction is resuspended in the appropriate medium and volume required for the next step.

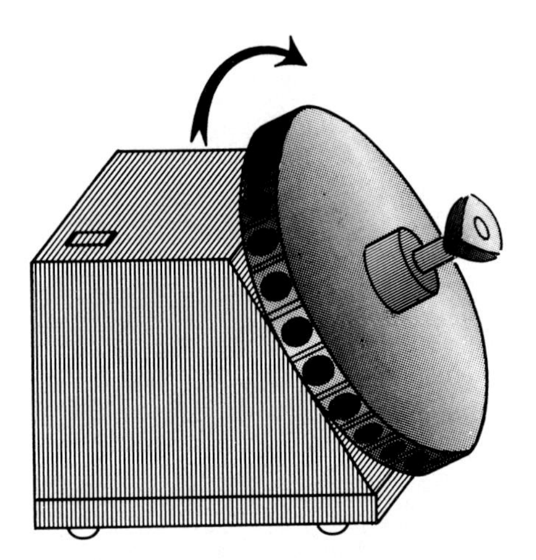

FIG. 2. Diagram of a rotator used for mixing in immunoisolation experiments. A plastic holder (≈20 cm diameter) for Eppendorf tubes was set onto the spindle of a small rotator used for electron microscopy sample preparation, which rotates at 2 rpm. The tubes are held in place with a rubber band or with tape.

of any clumps of cellular material that would normally sediment. The ImAd can be resuspended a number of times and washed. Analysis of the washes will provide an estimate of how much washing is required.

5. There are two alternative retrieval protocols we use when the fraction to be isolated is complex and/or fragile, to guarantee the cellular structure remains intact when immunoisolated on the beads. With one method the beads are brought down on a plate magnet (Fig. 4) so that they form a monolayer. This method has been used to immunoisolate a stacked Golgi fraction (Salamero *et al.*, 1989). In the second method, retrieval and resuspension of the beads and bound material is completely eliminated using a free-flow system. An instrument has been constructed in which the beads remain suspended as a dilute dispersion within a chamber, which is placed within a magnetic field generated by an electromagnet. The beads and bound vesicles are retained within the chamber by the magnetic field, and washing buffers are pumped through to remove the unbound and nonspecifically bound components of the input fraction. This system is presented in detail in Howell *et al.* (1988a,b).

6. Depending on the nature of the fraction to be isolated and the input fraction, sometimes it is useful to pass the fraction bound to the ImAd through a cushion made of sucrose or any of the other media used for density gradients. This step can reduce the number of washes required.

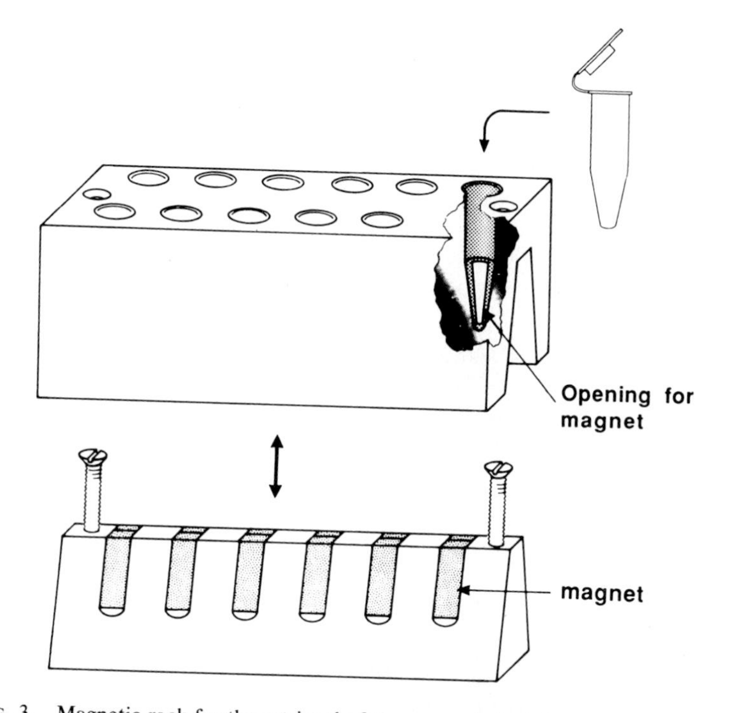

FIG. 3. Magnetic rack for the retrieval of the magnetic ImAd. A magnetic rack for the retrieval of magnetic beads in Eppendorf tubes was constructed in two pieces to allow easy setting of the magnet. When put together with the two screws at either end of the rack, the magnet is in the same orientation to the tube as the hand-held magnet shown in Fig. 1. The rack holds 12 tubes and is made of metal, so that it can be easily cooled by placing on a metal plate in an ice bucket. Constructed in the mechanical workshop of EMBL (Heidelberg, FRG).

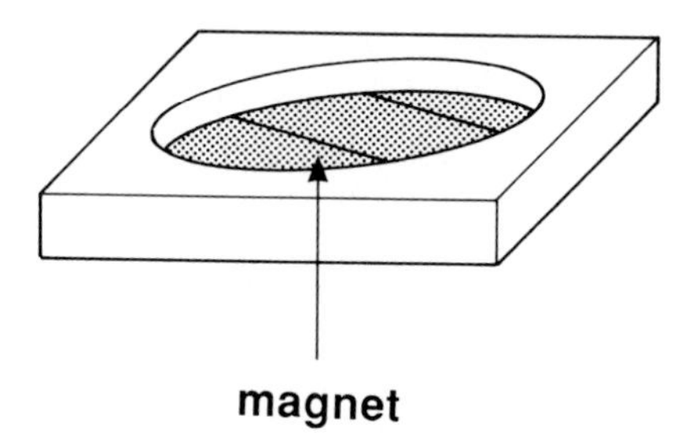

FIG. 4. Plate magnet for the retrieval of the magnetic ImAd. A small metal holder was constructed for a 5-cm-diameter tissue culture Petri dish. The entire bottom of the dish sits directly on the flat surface of the magnet. With this plate magnet the beads can be retrieved in a monolayer.

C. Methods

1. Activation of Beads

With the commercially available M-450 beads, linker molecules or specific antibodies can be coupled to the bead surface either by hydrophobic interactions or by covalent coupling. Hydrophobic interactions require no previous activation and provide a strong attachment of antibodies to the bead surface. Little, if any, of the bound antibody is lost during subsequent storage or use. However, these interactions may distort the bound molecules more than covalent coupling. We find the covalent coupling of antibodies more effective because a higher proportion of the bound antibodies retain their capacity to bind their antigen and because fewer linker molecules are required. It is preferable to prevent bead aggregation during the preparation of the ImAd. Therefore, we maintain the beads insulated from ferromagnetic material during the entire experiment, until the final step, the retrieval and washing of the desired fraction.

The hydroxyl groups of the polymer surface are activated with toluene-4-sulfonyl chloride (tosylation) as described by Nustad *et al.* (1984, 1988). The activation reaction is carried out in organic solvents in a hydrophobic environment. The solvents, acetone and pyridine, must be dried to prevent hydrolysis of the reagent; therefore, they are mixed overnight with molecular sieves (Union Carbide type 3 Å; Fluka AG, Buchs, Switzerland), which have been predried for 4 hours at 200°C. The magnetic beads (100 mg dry weight) are suspended in 10 ml 0.1 M KPO_4 buffer (pH 7.8) in a siliconized glass tube with a Teflon-lined screwcap. Before the beads have protein bound to their surface, they tend to stick to the tube walls and are difficult to handle; siliconization of the tube makes the processing easier. The beads are transferred from the water solution to 100% acetone in four sequential steps using 10 ml each of acetone–water: 3 : 7, 4 : 6, 8 : 2, and 100% acetone. The washing is carried out by pelleting the beads in a centrifuge (1500 rpm, 5 minutes) and resuspending with a siliconized Pasteur pipet. The final pellet is resuspended in 1 ml acetone and pyridine and toluene-4-sulfonyl chloride (Fluka AG, cat. no. 89730) added to final concentrations of 4.5 mM and 2.2 mM, respectively. They are rotated end-over-end at 2 rpm overnight at room temperature (RT). The next day the beads are washed three times as before with 10 ml acetone and transferred back to water with the reverse sequential washings: 10 ml acetone–water; 8 : 2, 4 : 6, 3 : 7 and three washes of water only. If sterile beads are required, the activation can be carried out under sterile conditions and/or a treatment with 70% ethanol may be added at this step. Then the beads are resuspended in 1 m*M* HCl and stored at 4°C. The activated beads are stable for longer than a year.

2. Linker Molecules

Linker molecules placed between the bead surface and the specific antibody are usually advisable. This is to provide the specific antibody with proper orientation, increased flexibility, and some distance from the surface of the solid support. In the coupling process some molecules will be inactivated or coupled in the incorrect orientation to bind a specific antibody. Linker molecules will be either a generic antibody (e.g., an antibody against mouse IgG or the Fc domain of mouse IgG) or an IgG-binding protein (e.g., protein A from *S. aureus* or protein G from *Streptococcus*). Those that bind the Fc domain of the specific antibody provide an optimal ImAd because all the specific antibodies are then in the correct orientation to bind their antigen and a high density of specific antibody is achieved. Flexibility and distance from the bead surface allows the bound specific antibody to bind its antigen more readily, which is particularly important when dealing with an antigen that is buried or close to the membrane surface.

With Mathias Uhlén (Royal Institute of Technology, Stockholm), we have been investigating a series of linker molecules, designed and engineered from the cDNAs of protein A and protein G. Uhlén and colleagues have previously cloned protein A (1984) and protein G (Olsson *et al.*, 1987), and have produced chimeric IgG-binding molecules engineered from protein A and protein G (Eliasson *et al.*, 1988). The "new" engineered linkers contain multiple mixed IgG-binding domains from both protein A and protein G, are produced in *Escherichia coli,* and are easily purified from the inclusion bodies formed in the bacteria. Our goal is to produce a universal linker to function for most of the commonly used specific antibodies, which will be effective and inexpensive.

It is not always necessary to use linker molecules; the specific antibody can be coupled directly to the bead surface. This needs to be tested directly with the specific antibody–antigen used. Kvalheim *et al.* (1987) have found that binding directly to the bead surface is advantageous when the specific antibody is a monoclonal IgM.

We raised antibodies in sheep against the Fc domain of rabbit and mouse IgG for linker molecules. These must be affinity-purified on a rabbit or mouse IgG column before coupling to the bead surface. Only a fraction (~10%) of the serum immunoglobulins recognize the immunogen, so an IgG fraction does not provide a high enough proportion of the correct antibody.

3. Binding of Linker Molecules

The binding of the linker molecules to the activated beads occurs in two steps. First, the antibody is adsorbed within 20 minutes to the hydrophobic group introduced by the tosylation reaction. The chemical coupling is completed by 16 hours (Nustad *et al.*, 1988). Our experience is that this coupling protocol is reasonably "gentle" for antibodies. However, for less stable ligands other coupling methods are possible (see Lea *et al.*, 1988).

The beads are first pelleted in the centrifuge (1500 rpm, 5 minutes) to remove the 1 m*M* HCl, and then resuspended in 0.1 *M* borate buffer (pH 9.5) for coupling. Then, 5–15 μg IgG/mg activated beads are mixed (at a concentration of 2 mg beads/ml) and rotated end-over-end at 2 rpm overnight at RT. The beads are pelleted and to estimate how much protein has bound; the OD_{280} is measured or the protein concentration of the unbound linker is assayed. The approximate number of micrograms bound per milligram beads is calculated. Then the beads are washed three times in 0.05 *M* Tris-HCl (pH 7.8), 0.1 *M* NaCl, 0.01% BSA, 0.1% Tween-20, and twice in PBS–5 mg/ml BSA, which quenches any remaining activated sites. The beads can be stored in PBS–5 mg/ml BSA at 4°C with 0.02% sodium azide for extended periods (certainly $\leq$6 months). To have a more accurate estimate of the amount of linker on the final ImAd, labeled linker molecules should be used (usually iodinated); then all fractions can be counted and a balance sheet produced, as in Table I. With the given conditions, 3–5 μg protein will be bound per milligram of beads, corresponding to ~700,000 IgG molecules bound per bead. One IgG occupies a surface area of $\approx$100 nm^2 (Bagghi and Birnbaum, 1981), so these values imply that close to 100% of the bead surface is covered with IgG (bead surface = 64 μm^2).

4. Binding of Specific Antibody

At least a 2-fold excess of specific antibody relative to the amount of bound linker molecules should be used (i.e., 6–10 μg/mg beads, again at a concentration of 2 mg beads/ml). Bind 4 hours to overnight at 4°C with end-over-end rotation, and then wash the beads three times in PBS–5 mg/ml BSA. It may be possible to store the completed ImAd at this step, but this will depend on the stability of the specific antibody. We normally add the specific antibody the evening before the experiment and bind overnight. With this protocol, 2–4 μg specific IgG are bound, demonstrating that with the tosylation protocol for coupling, $\approx$75% of the covalently coupled linker molecules have retained their binding capacity.

TABLE I

COUPLING OF THE LINKER MOLECULES TO ACTIVATED BEADS[a]

Experimental stage	Binding results		
	cpm	%	μg
Beads	31,676	34.2	5.1
Unbound	56,479	60.0	9.0
Wash 1	3,211	3.5	0.52
Wash 2	628	0.7	0.11
Wash 3	254	0.3	0.04
Quench 1	189	0.2	0.02
Quench 2	101	0.1	0.01

[a] The affinity-purified antibody raised in sheep against the Fc domain of mouse IgG (15 μg unlabeled plus ~95,000 cpm of ^{125}I-labeled antibody) was mixed and incubated with 1 mg activated beads. After both the binding of the specific antibody and the subsequent immunoisolation experiment, 29,429 cpm or 92% of the coupled linker molecules remained bound to the beads. This loss may be accounted for by a small loss of the beads themselves during the various steps.

As with linker antibodies, the specific antibody must be affinity-purified if polyclonal to provide a high enough density of the specific antibody on the ImAd surface. The efficiency of immunoisolation can be correlated with the density of the specific antibody on the ImAd surface (Gruenberg and Howell, 1985).

5. BINDING OF SPECIFIC LIGAND (SUBCELLULAR FRACTION)

In order to obtain an optimal ratio of input fraction and ImAd, it is advisable to start by making a concentration curve; each point should contain 1 mg ImAd and an increasing concentration of the input fraction, for example, 10–200 μg protein (Fig. 5). At saturation the surface of the bead should be almost fully occupied with the isolated fraction, as shown in the electron micrograph in Fig. 6. Assays for specificity of the isolation and for the amount of antigen remaining in unbound fraction are necessary to evaluate the efficiency of isolation. Here efficiency is expressed as percentage of the specific component of the input fraction isolated.

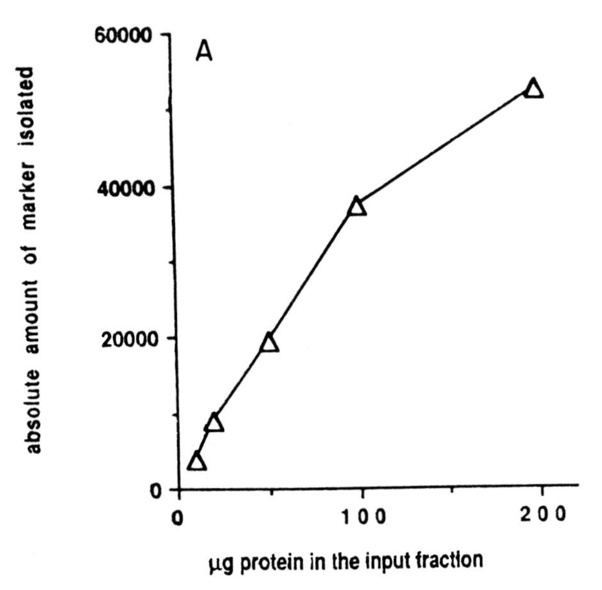
A
absolute amount of marker isolated
60000
40000
20000
0
0
100
200
μg protein in the input fraction

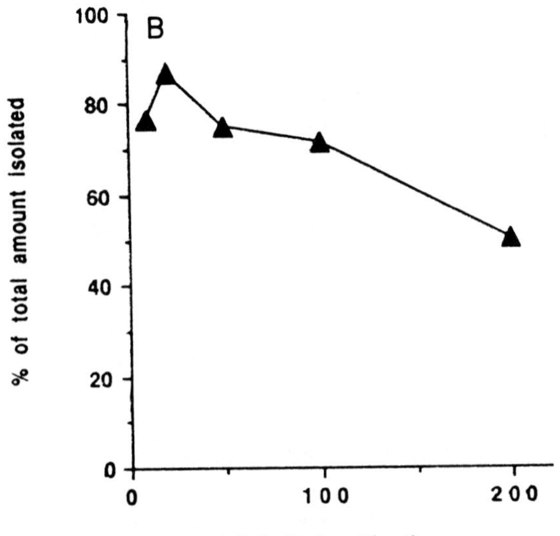
B
% of total amount isolated
100
80
60
40
20
0
0
100
200
μg protein in the input fraction

To follow the immunoisolation of a compartment, the standard criteria of subcellular fractionation should be applied (Beaufay and Amar-Costesec, 1976). In addition, the amount of nonspecific binding can easily be assessed using an ImAd prepared with a nonspecific antibody. The components that are not specifically associated with or "contaminating" the specifically isolated fraction can be identified, and the extent of the contamination can be quantitated.

To determine a possible effect of the slight microporosity of the beads (discussed earlier), we determined the contamination due to soluble components in a fractionation experiment. The cytoplasm that is often a component of the input fraction is composed of a high concentration of soluble proteins and provided an appropriate control. Cytoplasm was prepared from [^{35}S]methionine metabolically labeled cells, (the supernatant of a 140,000 *g*, 60-minute centrifugation of a cellular homogenate). When tested in immunoisolation experiments, <0.1% of the labeled cytosol was associated with the beads after three washes. This provided evidence that the microporosity of the beads did not interfere with the immunoisolation experiments.

6. Alternative Protocol

The protocol just described is preferable because of speed and efficiency in carrying out the isolation. However, with some antibody–antigen couples a better yield is achieved by first incubating the input

---

FIG. 5. Example of an immunoisolation experiment. The input fraction is a Golgi-enriched fraction isolated from rat liver, GF3 (Howell *et al.*, 1978) prepared from a rat that was metabolically labeled with [^{3}H]fucose for <5 minutes to provide a marker for newly synthesized glycoproteins. The ImAd was prepared with M-682 magnetic beads with an affinity-purified antibody against the Fc domain of mouse IgG covalently attached. The specific antibody against the cytoplasmic domain of the polymeric IgA receptor (Kühn and Kraehenbuhl, 1983) was bound to the linker antibody. Increasing concentrations (10–200 μg fraction) were mixed with 1 mg ImAd for 2 hours at 4°C, retrieved, and washed two times. (A) The absolute amount (cpm) of ^{3}H label immunoisolated. The isolation is linear up to 100 μg protein input and then falls off. (B) Percentage of the total ^{3}H label isolated. The isolation is most efficient at low concentrations of input fraction. The efficiency drops rapidly after 100 μg, when saturation is approached. The finding that the antibody against the cytoplasmic domain of the polymeric IgA receptor efficiently isolated vesicles derived from the Golgi complex was confirmed by the observation that galactosyltransferase was as efficiently isolated as the 5-minutes [^{3}H]fucose pulse. Nonspecific binding measured with a control ImAd prepared with an irrelevant antibody was <2% of the counts of the input fraction.

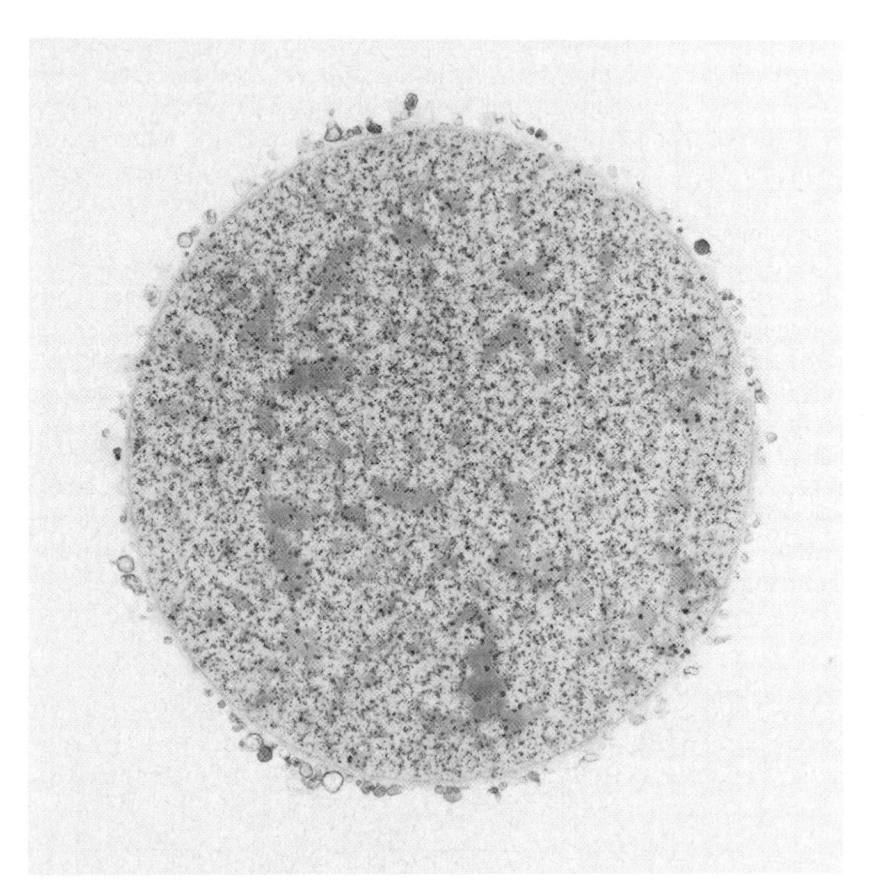

FIG. 6. Electron micrograph of magnetic bead from experiment of Fig. 5. An electron micrograph of a thin section of the magnetic beads shows the fraction immunoisolted from the experiment in Fig. 5. The 200-μg input fraction/mg ImAd sample was fixed and prepared for electron microscopy. The bead surface is almost completely covered with the specific vesicles. It is also quite easy to visualize the interior of the magnetic beads. The small black dots are the maghemite precipitates, the light-gray area is the original polymer, and the darker areas are the polymer that has been used to fill and seal the bead. The very thin layer at the surface is the nonmagnetic shell.

fraction with the specific antibody in suspension and later adding the solid support with bound linker molecules. This protocol has allowed better accessibility when the antigen is buried, for example, in coated vesicles (Pfeffer and Kelly, 1985; B. Hoflack, EMBL, Heidelberg, FRG, personal communication). Two retrieval protocols can then be followed. In the first, all unbound antibody must be removed by a washing step, entailing either flotation of the fraction on a gradient or pelleting followed by resuspension. Then the input fraction with the bound specific antibody is incubated with the solid support–linker molecule complex and the subcellular fraction is retrieved. In the second, a high-capacity solid support is used—for example, cellulose fibers or *S. aureus* cells—and all IgG both free and bound to the specific organelle are captured. This protocol can be made extremely efficient when the amount of specific antibody added is titrated experimentally to provide a stoichiometric balance between the antigen and the amount of active linker molecules on the solid support. This protocol is also advantageous if there is no way to affinity-purify the polyclonal antibody. Incubation with the appropriate fraction is a very efficient affinity purification procedure.

7. Analysis of Immunoisolated Fractions

The fraction bound to the beads, the unbound fraction, and the washes can be assayed by all the conventional techniques: chemical, radiolabel, immunoassays and enzymatic assays, and SDS–PAGE analysis. In the next section we will describe how immunoisolated fractions are used in assays of vesicle fusion.

III. Immunoisolation of the Compartments of the Endocytic Pathway

In order to provide an antigen that could be used to immunoisolate the compartments of the endocytic pathway, we selected a protocol with which we could control the subcellular localization of the antigen (Gruenberg and Howell, 1986, 1987, 1988; Gruenberg *et al.*, 1989). We used a "foreign" antigen, the glycoprotein G of vesicular stomatis virus (VSV), which is implanted into the plasma membrane of a BHK cell in its native transmembrane conformation by low pH-mediated fusion. After implantation the cells are maintained at 4°C to prevent membrane traffic, and the total amount of the antigen resides in the plasma membrane. When the cells are warmed to 37°C, membrane traffic resumes and the G protein moves as a wave through the endocytic pathway and can be stopped at any point by returning the cells to 4°C. The cells are then homogenized and a postnuclear supernatant (PNS) is prepared. An antibody raised against a synthetic peptide of the 15 carboxy-terminal amino acids of the G protein (P5D4, Kreis, 1986) is used to immunoisolate from the PNS the endocytic compartment in which the G protein resides at that specific time point.

In a typical experiment, 50 μg VSV are added to one 10-cm-diameter dish (1.3 × 10^7 cells and 2.5 mg protein). After the low-pH treatment, 80 G molecules are implanted per square micron of plasma membrane surface area, corresponding to ≈0.3% of the total plasma membrane protein. G-Protein internalization is rapid at 37°C; in 5 minutes ≈70% of the implanted G molecules have been internalized. For immunoisolation a PNS is prepared from cells with internalized G protein and used as the input fraction at 500 μg protein/mg beads. The immunoisolation is carried out for 2 hours at 4°C; then the beads are retrieved and the immunoisolated fraction is analyzed. Because endogenous markers of the endosomal compartment are not available, we have quantitated the isolation with the G protein itself labeled with [^{35}S]methionine and with horseradish peroxidase (HRP) internalized in the fluid phase. As shown in Table II, ≈70% of both markers are immunoisolated from the PNS, the yield from the homogenate is ≈35%. With these tissue culture cells, ≈50% of the markers are lost to the nuclear pellet (Howell *et al.*, 1989). The enrichment of endosomal markers is 15- to 30-fold when compared to a control ImAd prepared with nonspecific mouse IgG. A low level of contamination with compartments connected to endosomes by membrane traffic (plasma membrane, Golgi, and lysosome) was observed, accounting for 2–4% of the corresponding markers present in the PNS. We have also monitored the extent of contamination with other endosomal elements. In these experiments, a PNS was prepared from cells that had cointernalized the G protein and avidin, which was used as a marker of the endosomal content. This PNS was mixed with a PNS prepared from cells lacking the G protein that had internalized HRP for different times. The contamination of the immunoisolated fraction with other endosomal elements (monitored with HRP activity) was also 2–4% (Table II). These observations demonstrate that under our experimental conditions, the endosomal fractions are prepared with a high yield and with minimal coisolation of other compartments.

IV. Immunoisolated Endosomal Fractions in Cell-Free Assays of Vesicle Fusion

Immunoisolated fractions were used to study the *in vitro* reconstitution of endocytic-vesicle fusion. These fractions correspond to different times of G protein internalization. The G molecules move from the plasma

TABLE II

RESULTS IN EXPERIMENTAL IMMUNOISOLATION OF ENDOSOMAL FRACTION

Marker	Percentage immuno-isolated from PNS	Percentage yield from WH[a]	Percentage enrichment over control[b]
HRP co-internalized with G[c]	70	35	30
^{35}S-Labeled G protein[d]	80	38	30
^{3}H-Labeled plasma membrane[e]	≤1	—	1
β-*N*-Ac-glucosaminidase[f]	6	3	1
Galactosyltransferase[g]	5	2.5	1
HRP in cells lacking G[h]	<5	—	—

[a] WH, Whole homogenate.

[b] The control is an ImAd prepared with a nonspecific IgG.

[c] HRP, Horseradish peroxidase, enzymatic activity immunoisolated after internalization at 5 mg HRP/ml.

[d] The values are obtained by Triton X-114 extraction of the immunoisolated fractions and are expressed in percentage of the total G protein internalized. The internalized G protein corresponds to the amount of implanted molecules corrected for the G protein on the cell surface.

[e] The plasma membrane is labeled with the NaB[^{3}H]$_4$-galactose oxidase method.

[f] The β-*N*-Ac-glucosaminidase activity is used as a marker of the lysosomes.

[g] The galactosyltransferase activity is used as a marker of the Golgi.

[h] The cross-contamination of endosomal elements during immunoisolation was determined in a series of experiments. Each PNS (input) prepared from cells that had cointernalized for 5, 15, and 30 minutes G protein and avidin, as a marker of the endosomal content, was separately mixed with the PNS prepared from cells lacking the G protein that had internalized HRP for, respectively, 5, 15, and 30 minutes.

membrane to an "early", tubulovesicular endosome located at the cell periphery and then to "late" acid phosphatase-positive endosomal structures with a complex morphology in the perinuclear region (Gruenberg *et al.*, 1989). Because each fraction is prepared under conditions in which cross-contamination with other endosomal elements is minimal, the different subcompartments of the pathway can be separated and analyzed for their ability to fuse in the cell-free assay. These defined fractions immunoisolated on the solid support provide different "acceptor" fractions in the fusion reaction. The "donor" fractions are PNS prepared from cells without the G protein and provide both the vesicular partners and the potential cytosolic compounds required for the fusion reaction. After the fusion reaction, the magnetic properties of the beads are used to retrieve and wash the immunoisolated fusion product. This makes it possible to eliminate the excess donor fraction.

Three different detection systems have been used to monitor and quantitate the occurrence of fusion (Fig. 7):

1. Fusion-specific iodination of the G protein. The G protein that is used as the antigen for the immunoisolation of the acceptor fractions is absent from the donor cells and therefore provides a specific marker. Acceptor fractions immunoisolated after different times of G-protein internalization are mixed in the fusion assay with a donor prepared after continuous internalization of lactoperoxidase. Fusion delivers lactoperoxidase from the donor to the acceptor vesicles containing the G protein. After the assay, the fusion product is retrieved and an iodination reaction is carried out. The occurrence of fusion is monitored by the iodination of the G protein and analyzed by gel electrophoresis and immunoblotting and by immunoprecipitation.

2. Morphology. To visualize the occurrence of fusion by electron microscopy, it is necessary to provide a morphological marker of the acceptor that can be identified in all thin sections. After G implantation, the surface of the acceptor cells is labeled with a polyclonal antibody that recognizes several plasma membrane proteins, followed by protein A–colloidal gold. The antigen–gold complexes are rapidly internalized, as a result of the crosslinking. These cells are used to prepare the acceptor fraction after cointernalization of the G protein and the gold complexes for 5 minutes. The donor is prepared after continuous HRP internalization for 30 minutes. Colocalization in the same vesicles of the gold complexes and the HRP reaction product scores the occurrence of fusion.

3. Formation of a fusion-specific complex. To quantitate the fusion activity of the acceptor, avidin is cointernalized with the G protein in acceptor cells, and biotinylated HRP (biotHRP) is internalized in donor cells. This detection system is similar to that described by Braell (1987). After retrieval of the fusion product and detergent solubilization, the

---

FIG. 7. Outline of the three fusion detection systems. (A) and (B) Biochemical detection systems. (A) Fluid phase: X, avidin, cointernalized with the G protein in acceptor cells; ○, biotHRP internalized in donor cells. The quantitation is with an ELISA using an antiavidin antibody. (B) Membrane protein: ●, lactoperoxidase (LPO) internalized in donor cells. Detection is by a fusion-specific iodination of the G protein, analyzed by SDS–PAGE, immunoblotting with an anti-G antibody, and autoradiography (AR), or by immunoprecipitation, SDS–PAGE, AR, and counting. (C) Morphological detection system. The surface of acceptor cells is labeled with an anti-BHK cell surface antibody and protein A–colloidal gold. The complex is cointernalized with the G protein. The donor fraction is prepared from cells after HRP internalization. After fusion, the diamino-benzidine reaction was carried out to colocalize the HRP reaction product with colloidal gold in the same vesicle.

formation of the fusion-specific complex between avidin and biotHRP is quantitated with an ELISA using an antibody against avidin.

The results obtained with all three detection systems show that the cell-free fusion of endocytic vesicles depends on ATP, provided by an ATP-regenerating system, the presence of cytosol, and temperature.

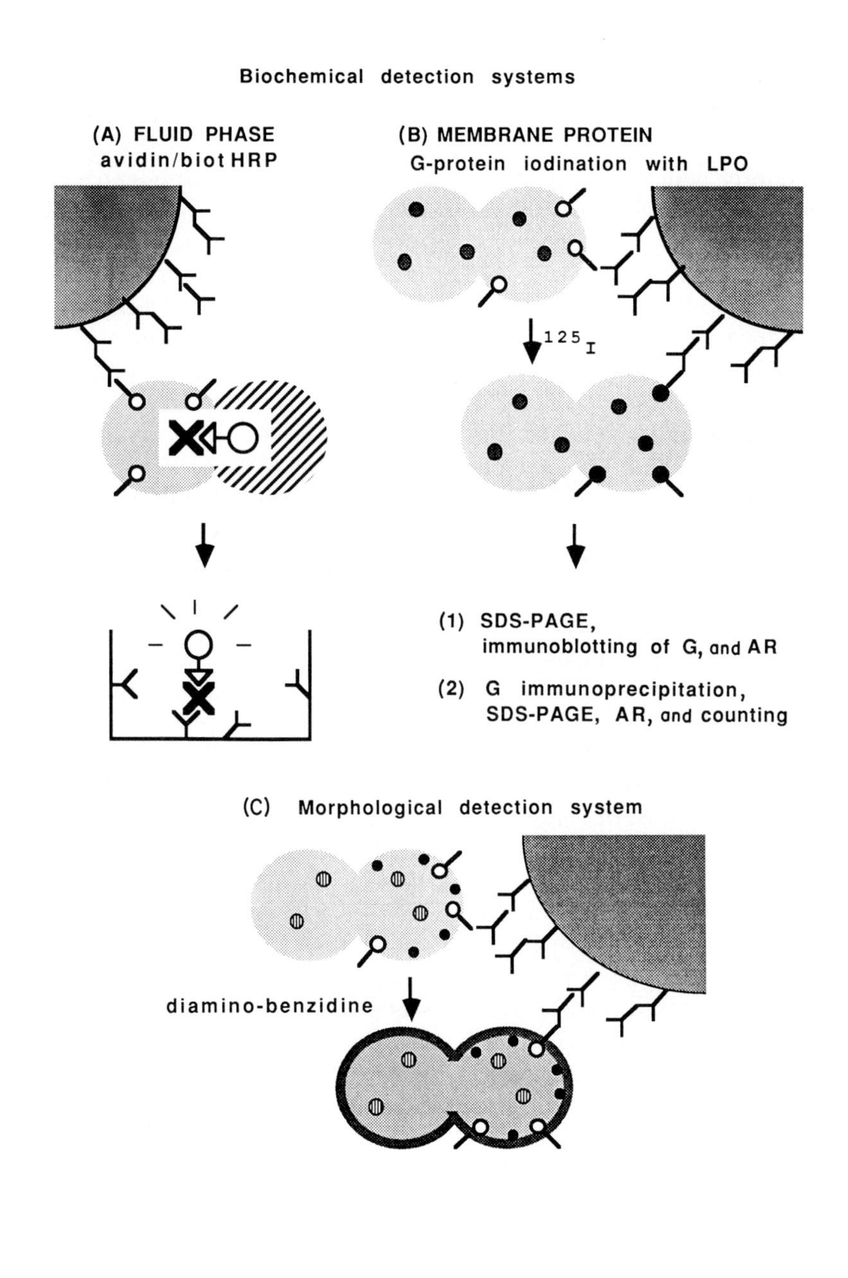

These observations agree with the findings of several groups (Davey *et al.*, 1985; Braell, 1987; Diaz *et al.*, 1988; Woodman and Warren, 1988). In the cell-free assay, the amount of fusion-specific complex formed increases linearly for ~20 minutes and reaches a maximum after 40 minutes. During the course of the assay and subsequent washes, the acceptors prepared after different times of internalization retain their integrity, because most of the fluid-phase marker present in the endocytic vesicles remains latent. This is supported by a morphological examination of the immunoisolated fraction before and after fusion. In both cases the characteristic features of the corresponding *in vivo* structures are retained.

The three detection systems show that maximal fusion activity occurs 5 minutes after G-protein internalization. The donor is prepared after continuous internalization of the markers for 30 minutes, to guarantee the detection of all fusion events that may occur. At this early time of internalization, 65% of the avidin immunoisolated in the acceptor is complexed by biotHRP during fusion. This value agrees well with our morphological examination of the fusion product, which shows that the donor HRP is delivered to ≈50% of the gold-containing acceptor vesicles during fusion. The fusion activity decreases with a $t_{1/2} \approx 5$ minutes with acceptors prepared after longer times of internalization. These kinetics are similar when followed either with the iodination of the transmembrane G protein or with the formation of the fusion-specific complex between avidin and biotHRP internalized in the fluid phase. Maximal fusion activity at an early stage of endocytosis was also reported by Braell (1987) using fluid-phase markers and by Diaz *et al.* (1988) using ligands specific for the mannose receptor.

Because the markers used in the donor cells are internalized continuously for 30 minutes, the vesicular partners of the acceptor in the fusion reaction may originate from different stages of the pathway. The same experiments repeated with donors prepared after different times of marker internalization show that maximal fusion activity is already supported by a donor prepared after 5 minutes. In addition, the signal is abolished when donors are used after a 5-minute pulse of marker internalization followed by a chase in marker-free medium. A morphological examination showed that after 5 minutes of internalization, the G protein and the markers used in the fusion reaction distribute in the network of tubulovesicular elements at the cell periphery, which correspond to the early endosomal compartment. The high activity and specificity of the fusion reaction indicates that these elements provide the vesicular partners of both the acceptor and the donor fractions in the assay, suggesting that these elements exchange membrane and content *in vivo*.

After 5–15 minutes of incubation, the fusion activity of the corresponding endosomal fractions decreases. Both the fusion markers and the G protein are no longer in the peripheral endosome and are observed in acid phosphatase-negative spherical structures of 0.5 μm diameter. These vesicles are distinct from the later, acid phosphatase-positive endosomal structures in the perinuclear region, where the markers are found after longer incubations. When the microtubules are depolymerized, the fusion markers and the G protein are exported from early endosomal elements to the spherical vesicles but not beyond. The spherical vesicles immunoisolated after microtubule depolymerization exhibited little, if any, fusion activity with themselves or with early endosomal elements.

In conclusion, these experiments indicate that we have immunoisolated and characterized (1) the early endosomal compartment, where routing of internalized molecules to recycling or degradation occurs, and (2) a putative carrier vesicle, involved in the microtubule-dependent transport to a late endosome or a lysosome. It is now possible to make use of this approach for the identification of molecular components controlling the recognition between vesicles and subsequent fusion, as well as those responsible for the interaction with microtubules.

V. Perspectives

Immunoisolation is becoming one of the most powerful techniques for isolating the different compartments of the pathways of membrane traffic. New developments in cell and molecular biology are providing methodological alternatives to obtain appropriate antibodies that recognize antigenic sites exposed on the surface of intracellular compartments. The rather limited number of such antibody–antigen couples has restricted the general application of immunoisolation until now. Golgi membranes have been immunoisolated using an antibody against NADPH–cytochrome *P*-450 reductase (Ito and Palade, 1978). Clathrin-coated vesicles have been immunoisolated using anticlathrin antibodies (Merisko *et al.*, 1982; Hanover *et al.*, 1984; Schulze-Lohoff *et al.*, 1985) and monoclonal antibodies against the cytoplasmic domain of two proteins (Pfeffer and Kelly, 1985). Monoclonal antibodies recognizing cytoplasmic epitopes were also used to immunoisolate secretory granules (Lowe *et al.*, 1988).

Endosomal elements were immunoisolated using a polyclonal antibody against the asialoglycoprotein receptor (Mueller and Hubbard, 1986).

It has proved particularly difficult to raise antibodies against cytoplasmic-exposed epitopes, using subcellular fractions as antigens. Several groups have succeeded in raising antibodies that recognize elements of the Golgi complex (Burke *et al.*, 1982; Lin and Queally, 1982; Louvard *et al.*, 1982; Smith *et al.*, 1984; Saraste *et al.*, 1987; Yuan *et al.*, 1987). However, none of these antibodies has been reported so far to function in immunoisolation experiments. The antibodies we have raised using Golgi membrane fractions as antigen and an immunization protocol similar to Louvard *et al.* (1982), all recognized epitopes facing the lumen of the compartment. Chicheportiche *et al.* (1984) and Chicheportiche and Tartakoff (1987) have produced antibodies to the cytoplasmic domain of Golgi membrane proteins or Golgi-associated proteins; their potential use in immunoisolation remains to be tested.

Other experimental alternatives to provide an appropriate antigen–antibody couple are now possible. Several membrane proteins transported along the intracellular pathways of membrane traffic have been identified. These include the two receptors responsible for the transport of lysosomal enzymes (Lobel *et al.*, 1987, 1988; Dahms *et al.*, 1987; Pohlman *et al.*, 1987), a lysosomal acid phosphatase (Pohlman *et al.*, 1988; Waheed *et al.*, 1988), and three lysosomal membrane proteins from chicken (Fambrough *et al.*, 1988), mouse (Chen *et al.*, 1988), and human (Viitala *et al.*, 1988). Because their sequence have been deduced from the cDNA, peptides of the cytoplasmic domain can be synthesized and used as antigen to raise anti-cytoplasmic domain antibodies. However, these proteins distribute in more than one compartment, a situation that may complicate some immunoisolation experiments.

Antibodies recognizing cytoplasmically exposed epitopes potentially involved in protein targeting have been generated using an antiidiotype approach (Pain *et al.*, 1988). The use of paired *in vitro* immunizations for the generation of monoclonal internal-image antiidiotype antibodies is showing considerable promise (Vaux *et al.*, 1988). Antibodies raised in this way recognize organelle-specific epitopes, which are exposed to the cytoplasm and are therefore potential candidates for immunoisolation methods (D. J. T. Vaux, personal communication). These may be also relevant to the control of interactions between organelles. Another alternative we have tried to obtain an adequate antibody–antigen couple is to use a foreign antigen. We have discussed the use of this approach to immunoisolate the compartment of the endocytic pathway with a transmembrane viral protein implanted into the plasma membrane and an antibody raised against a synthetic peptide of the cytoplasmic domain

(Gruenberg and Howell, 1986, 1987, 1988; Gruenberg *et al.*, 1989). The same antibody–antigen couple has also been used by de Curtis *et al.* (1988) for the immunoisolation of elements of the trans-Golgi network using virally infected cells.

A very promising approach has been initiated by Luzio *et al.* combining selection of clones from a cDNA library using a polyclonal antibody with molecular cloning (Luzio *et al.*, 1989). A polyclonal antibody was raised against a Triton X-114 extract of a Golgi membrane fraction and recognized at least eight membrane proteins (Howell and Sztul, 1982). This antibody was used to screen a selected rat liver library. Positive clones were identified, cloned, and sequenced. Hydrophobicity plots were used to identify the transmembrane domains and predict the cytoplasmic domain. Finally, peptides of the predicted cytoplasmic domain were synthesized and used to raise specific antibodies, which will be tested in immunoisolation experiments.

Molecular-cloning techniques will permit the identification of other transmembrane proteins, thus increasing the repertoire of cytoplasmic domains that provide antigens for raising specific antibodies. These new candidates will make it possible to isolate the compartments of the different intracellular pathways and the carrier vesicles involved in transport between these compartments. Characterization of these compartments and vesicles will offer new insights in our understanding of the molecular mechanisms controlling protein sorting and membrane traffic.

Acknowledgments

We wish to thank Hans Floesser of the EMBL Mechanical Workshop for constructing the magnetic rack, and Petra Reidinger for drawing the figures. We also wish to thank David J. T. Vaux and Angela Wandinger-Ness for critically reading the manuscript and our colleagues at EMBL for their interest in the development and use of immunoisolation.

We also thank K. Nustad, A. Berge, T. Ellingsen, P. Stentad, S. Funderund, T. Lea, and F. Vartdal for their important contributions in the development and application of the magnetic beads.

References

Bagghi, L., and Birnbaum, C. (1981). *J. Colloid Interface Sci.* **83,** 460–478.

Bailyes, E. M., Richardson, P. J., and Luzio, P. J. (1987). *In* "Biological Membranes: A Practical Approach" (J. B. C. Findlay and W. H. Evans, eds.), pp. 73–98. IRL Press, Oxford.

Beaufay, H., and Amar-Costesec, A. (1976). *Methods Membr. Biol.* **6,** 1–99.

Braell, W. A. (1987). *Proc. Natl. Acad. Sci. USA* **84,** 1137–1141.

Burke, B., Griffiths, G., Reggio, H., Louvard, D., and Warren, G. (1982). *EMBO J.* **1,** 1621–1628.

Chen, J. W., Cha, Y., Yuksel, K. U., Gracy, R. W., and August, J. T. (1988). *J. Biol. Chem.* **263,** 8754–8758.
Chicheportiche, Y., and Tartakoff, A. M. (1987). *Eur. J. Cell Biol.* **44,** 135–143.
Chicheportiche, Y., Vassalli, P., and Tartakoff, A. M. (1984). *J. Cell Biol.* **99,** 2200–2210.
Dahms, N. M., Lobel, P., Breitmeyer, J., Chirgwin, J. M., and Kornfeld, S. (1987). *Cell (Cambridge, Mass.)* **50,** 181–192.
Davey, J., Hurtley, S. M., and Warren, G. (1985). *Cell (Cambridge, Mass.)* **43,** 643–652.
de Curtis, I., Howell, K. E., and Simons, K. (1988). *Exp. Cell Res.* **175,** 248–265.
Diaz, R., Mayorga, L., and Stahl, P. (1988). *J. Biol. Chem.* **263,** 6093–6100.
Eliasson, M., Olsson, A., Palmcrantz, E., Wiberg, K., Inganas, M., Guss, B., Lindberg, M., and Uhlén, M. (1988). *J. Biol. Chem.* **263,** 4323–4327.
Fambrough, D., Takeyasu, K., Lippincott-Schwartz, J., and Siegel, N. R. (1988). *J. Cell Biol.* **106,** 61–67.
Glick, B. S., and Rothman, J. E. (1987). *Nature (London)* **326,** 309–312.
Goud, B., Salminen, A., Walworth, N. C., and Novick, P. J. (1988). *Cell (Cambridge, Mass.)* **53,** 753–768.
Gruenberg, J., and Howell, K. E. (1985). *Eur. J. Cell Biol.* **38,** 312–321.
Gruenberg, J., and Howell, K. E. (1986). *EMBO J.* **5,** 3091–3101.
Gruenberg, J., and Howell, K. E. (1987). *Proc. Natl. Acad. Sci. U.S.A.* **84,** 5758–5762.
Gruenberg, J., and Howell, K. E. (1988). *In* "Cell-Free Analysis of Membrane Traffic" (D. J. Morré, K. E. Howell, G. M. W. Cook, and W. H. Evans, eds.), pp. 317–331. Liss, New York.
Gruenberg, J., Griffiths, G., and Howell, K. E. (1989). *J. Cell Biol.* **108,** 1301–1316.
Hanover, J. A., Willingham, M. C., and Pastan, I. (1984). *Cell (Cambridge, Mass.)* **39,** 283–293.
Howell, K. E., and Sztul, E. (1982). *J. Cell Biol.* **95,** 242a.
Howell, K. E., Ito, A., and Palade, G. E. (1978). *J. Cell Biol.* **79,** 581–589.
Howell, K. E., Ansorge, W., and Gruenberg, J. (1988a). *In* "Microspheres: Medical and Biological Applications" (A. Rembaum and Z. A. Tökés, eds.), pp. 33–52. CRC Press, Boca Raton, Florida.
Howell, K. E., Gruenberg, J., Ito, A., and Palade, G. E. (1988b). *In* "Cell-Free Analysis of Membrane Traffic" (D. J. Morré, K. E. Howell, G. M. W. Cook, and W. H. Evans, eds.), pp. 77–90. Liss, New York.
Howell, K. E., Devaney, E., and Gruenberg, J. (1989). *Trends Biochem. Sci.* **14,** 44–47.
Hubbard, A. L., Dunn, W. A., Mueller, S. C., and Bartles, J. R. (1988). *In* "Cell-Free Analysis of Membrane Traffic" (D. J. Morré, K. E. Howell, G. M. W. Cook, and W. H. Evans, eds.), pp. 115–127. Liss, New York.
Ito, A., and Palade, G. E. (1978). *J. Cell Biol.* **79,** 590–597.
Kemshead, J. T., and Ugelstad, J. (1985). *Mol. Cell. Biochem.* **67,** 11–18.
Kornfeld, S. (1987). *FASEB J.* **1,** 462–468.
Kreis, T. (1986). *EMBO J.* **5,** 931–941.
Kühn, L., and Kraehenbuhl, J.-P. (1983). *Ann. N. Y. Acad. Sci.* **409,** 751–759.
Kvalheim, G., Fodstad, G., Phol, A., Nustad, K., Phare, A., Ugelstad, J., and Funderud, S. (1987). *Cancer Res.* **47,** 846–851.
Lea, T., Vartdal, F., Nustad, K., Funderud, S., Berge, A., Ellingsen, T., Schmid, R., Stenstad, P., and Ugelstad, J. (1988). *J. Mol. Recognition* **1,** 9–18.
Lin, J. J.-C., and Queally, S. A. (1982). *J. Cell Biol.* **98,** 108–112.
Lobel, P., Dahms, N. M., Breitmeyer, J., Chirgwin, J., and Kornfeld, S. (1987). *Proc. Natl. Acad. Sci. U.S.A.* **84,** 2233–2237.
Lobel, P., Dahms, N. M., and Kornfeld, S. (1988). *J. Biol. Chem.* **263,** 2563–2570.

Louvard, D., Reggio, H., and Warren, G. (1982). *J. Cell Biol.* **92,** 108–112.
Lowe, A. W., Madeddu, L., and Kelly, R. B. (1988). *J. Cell Biol.* **106,** 51–59.
Luzio, J. P., Newby, A. C., and Hales, C. N. (1976). *Biochem. J.* **154,** 11–21.
Luzio, J. P., Mullock, B. M., Branch, W. J., and Richardson, P. J. (1988). *In* "Cell-Free Analysis of Membrane Traffic" (D. J. Morré, K. E. Howell, G. M. W. Cook, and W. H. Evans, eds.), pp. 91–100. Liss, New York.
Luzio, J. P., Banting, G., Howell, K. E., Brake, B., Braghetta, P., and Stanley, K. K. (1989). Submitted for Publication.
Malhotra, V., Orci, L., Glick, B. S., Block, M. R., and Rothman, J. E. (1988). *Cell (Cambridge, Mass.)* **54,** 221–227.
Melançon, P., Glick, B. S., Malhotra, V., Weidman, P. J., Serafini, T., Gleanson, M. L., Orci, L., and Rothman, J. E. (1987). *Cell (Cambridge, Mass.)* **51,** 1053–1062.
Mellman, I., Fuchs, R., and Helenius, A. (1986). *Annu. Rev. Biochem.* **55,** 663–700.
Merisko, E. M., Farquhar, M. G., and Palade, G. E. (1982). *J. Cell Biol.* **92,** 846–857.
Mueller, S. C., and Hubbard, A. L. (1986). *J. Cell Biol.* **102,** 932–942.
Munro, S., and Pelham, H. (1987). *Cell (Cambridge, Mass.)* **48,** 899–907.
Nustad, K., Johansen, L., Ugelstad, J., Ellingsen, T., and Berge, A. (1984). *Eur. Surg. Res.* **16,** Suppl. 2, 80–87.
Nustad, K., Danielsen, H., Reith, A., Funderund, S., Lea, T., Vartdal, F., and Ugelstad, J. (1988). *In* "Microspheres: Medical and Biological Applications" (A. Rembaum and Z. A. Tökés, eds.), pp. 53–75. CRC Press, Boca Raton, Florida.
Olsson, A., Eliasson, M., Guss, B., Nilsson, B., Hellman, U., Lindberg, M., and Uhlén, M. (1987). *Eur. J. Biochem.* **168,** 319–324.
Orci, L., Glick, B. S., and Rothman, J. E. (1986). *Cell (Cambridge, Mass.)* **46,** 171–184.
Pain, D., Kanwar, Y. S., and Blobel, G. (1988). *Nature (London)* **331,** 232–237.
Palade, G. E. (1975). *Science* **198,** 347–358.
Pearse, B. M. F. (1987). *EMBO J.* **6,** 2507–2512.
Pelham, H. (1988). *EMBO J.* **7,** 913–918.
Pfeffer, S. R., and Kelly, R. B. (1985). *Cell (Cambridge, Mass.)* **40,** 949–957.
Pfeffer, S. R., and Rothman, J. E. (1987). *Annu. Rev. Biochem.* **56,** 829–852.
Pohlman, R., Nagel, G., Schmidt, B., Stein, M., Lorkowski, G., Krentler, C., Culley, J., Meyer, H. E., Grzeschnik, K.-H., Mersmann, G., Hasilik, A., and von Figura, K. (1987). *Proc. Natl. Acad. Sci. U.S.A.* **84,** 5575–5579.
Pohlman, R., Krentler, C., Schmidt, B., Schröder, W., Lorkowski, G., Culley, J., Mersmann, G., Geier, C., Waheed, A., Gottschalk, S., Grzeschnik, K.-H., Hasilik, A., and von Figura, K. (1988). *EMBO J.* **7,** 2343–2350.
Rembaum, A., Yen, R. C. K., Kempner, D., and Ugelstad, J. (1982). *J. Immunol. Methods* **52,** 341–347.
Roman, L. M., Scharm, A., and Howell, K. E. (1988). *Exp. Cell Res.* **175,** 376–387.
Rothman, J. E. (1988). *In* "Cell-Free Analysis of Membrane Traffic" (D. J. Morré, K. E. Howell, G. M. W. Cook, and W. H. Evans, eds.), pp. 311–316. Liss, New York.
Salamero, J., Sztul, E., and Howell, K. E. (1989). Submitted for publication.
Saraste, J., Palade, G. E., and Farquhar, M. G. (1987). *J. Cell Biol.* **105,** 2021–2029.
Schulze-Lohoff, E., Hasilik, A., and von Figura, K. (1985). *J. Cell Biol.* **101,** 824–829.
Segev, N., Mulholland, J., and Botstein, D. (1988). *Cell (Cambridge, Mass.)* **52,** 915–924.
Smith, A. D. J., D'Eugenio-Gumkowski, F., Yanagisawa, K., and Jamieson, J. D. (1984). *J. Cell Biol.* **98,** 2035–2046.
Ugelstad, J., Mørk, P. C., Kaggerund, K. H., Ellingsen, T., and Berge, A. (1980). *Adv. Colloid Interface Sci.* **13,** 101–140.
Ugelstad, J., Ellingsen, T., Berge, A., and Helgee, B. (1983). PCT Int. Appl. WO 83/03902.

Ugelstad, J., Berge, A., Schmid, R., Ellingsen, T., Stenstad, P., and Skjeltorp, A. (1986). *In* "Polymer Reaction Engineering" (K. H. Reichert and W. Geissler, eds.), pp. 77–93. Nijhoff Publ., Dordrecht, The Netherlands.

Ugelstad, J., Berge, A., Ellingsen, T., Aune, O., Kilaas, L., Nilsen, T. N., Schmid, R., Stenstad, P., Funderud, S., Kvalheim, G., Nustad, K., Lea, T., Vartdal, F., and Danielsen, H. (1988). *Makromol. Chem. Macromol. Symp.* **17,** 177–211.

Uhlén, M., Guss, B., Nilsson, B., Gatenbeck, S., Philipson, L., and Lindberg, M. (1984). *J. Biol. Chem.* **259,** 1695–1702.

Vale, R. D. (1987). *Annu. Rev. Cell Biol.* **3,** 347–378.

Vaux, D. J. T., Helenius, A., and Mellman, I. (1988). *Nature (London)* **336,** 36–42.

Viitala, J., Carlsson, S. R., Siebert, P. D., and Fukuda, M. (1988). *Proc. Natl. Acad. Sci. U.S.A.* **85,** 3743–3747.

von Figura, K., and Hasilik, A. (1986). *Annu. Rev. Biochem.* **55,** 167–193.

Waheed, A., Gottschalk, S., Hille, A., Krentler, C., Pohlman, R., Braulke, T., Hauser, H., Geuze, H., and von Figura, K. (1988). *EMBO J.* **8,** 2351–2358.

Walter, P., and Lingappa, V. R. (1986). *Annu. Rev. Cell Biol.* **2,** 499–516.

Woodman, P. G., and Warren, G. (1988). *Eur. J. Biochem.* **173,** 101–108.

Yuan, L., Barriocanal, J. G., Bonifacino, J. S., and Sandoval V. (1987). *J. Cell Biol.* **105,** 215–227.

Chapter 16

Flow-Cytometric Analysis of Endocytic Compartments

RUSSELL B. WILSON AND ROBERT F. MURPHY

Department of Biological Sciences and
Center for Fluorescence Research in Biomedical Sciences
Carnegie-Mellon University
Pittsburgh, Pennsylvania 15213

I. Introduction

The use of flow cytometry for the characterization of the endocytic pathway is the subject of this chapter. In particular, methods for measuring the pH of endocytic compartments in living cells, and for analyzing individual organelles, are presented here. It is our intent to provide sufficient detail about many of these techniques to allow their use without reference to previous publications. Information on topics not covered here, particularly flow-cytometric analysis of ligand binding, internalization, and degradation, may be found in several reviews (Sklar, 1987; Murphy *et al.*, 1988; Murphy, 1988, 1989).

II. Methods for Analysis of Living Cells

The most widely used applications of flow cytometry are the analysis of DNA content (cell cycle analysis) and the measurement of cell surface markers. These applications make use of a number of the features of flow cytometry, which include accurate quantitation of measured variables, lack of significant photobleaching (due to the short time for which each cell is analyzed), ability to acquire information on large numbers of cells (typically 10,000–100,000), correlated measurement of more than one variable per cell, and the potential for sorting individual cells for further analysis. Both of these applications involve *static* analysis of variables that do not change over the time course of typical experiments, and, because the measurements do not require living cells, samples are often fixed before analysis. Recently, the excellent temporal resolution of flow cytometry (the time at which individual cells are measured can be recorded with an accuracy <1 second) has been exploited for the analysis of the *kinetics* of physiological processes in living cells. Examples include continuous measurements of ligand binding and cellular responses (McNeil *et al.*, 1985), changes in intravesicular pH (Sipe and Murphy, 1987; Cain and Murphy, 1988), and degradation of endocytosed substrates (Jongkind *et al.*, 1986; Roederer *et al.*, 1987). The following section focuses on methods for measuring intravesicular pH in living cells.

A. General Methods

1. Synthesis of Fluorescent Conjugates

For the examples described here, the following ligands and labeling procedures were used. Human transferrin (Tf) was labeled with fluorescein isothiocyanate (FITC) or lissamine rhodamine sulfonyl chloride (LRSC) as previously described (Sipe and Murphy, 1987). Dye–protein ratios were determined spectrophotometrically to be 4.3 (FITC–Tf) or 5.9 (LRSC–Tf). These ratios were optimal for both specificity of binding ($>90\%$) and signal–autofluorescence ratio on cells (>3). Dextran, 70,000 D (Sigma) was labeled with FITC or substituted rhodamine isothiocyanate (XRITC) using the dibutyltin-dilaurate method (De Belder and

Granath, 1973). The labeled dextrans were purified using five cycles of ethanol precipitation to eliminate labeling artifacts due to low molecular weight contaminants (Preston *et al.*, 1987). It is especially important to purify commercial preparations of labeled dextrans. FITC and LRSC were purchased from Molecular Probes and XRITC was purchased from Research Organics.

2. Flow Cytometry

A modified FACS 440 (Becton Dickinson, Mountain View, CA) was used for all analyses reported here. For most experiments, an argon laser at 488 nm (400 mW) and a krypton laser at 568 nm (100 mW) were used for excitation. Forward scatter was collected by a photodiode. Side scatter was collected through a 488-nm bandpass filter (10-nm bandwidth) with a photomultiplier. Fluorescence from FITC was collected through a 530-nm bandpass filter (15-nm bandwidth) with a photomultiplier whereas XRITC or LRSC fluorescence was collected through a 625-nm bandpass filter (35-nm bandwidth). Because of the optical configuration of the FACS 440 dual-laser system, spillover between the two fluorescences was negligible. Data were collected in list mode using a VAXstation II/GPX computer (Digital Equipment Corp., Maynard, MA) and the Consort/VAX software package (Becton Dickinson Immunocytometry Systems, Mountain View, CA). In addition to the standard programs in this package, other utility programs were used in data analysis. These programs are available from the FACS Computer Users Group Library (c/o R.F.M.).

The methods described next should be applicable to any flow cytometer, with some restrictions. For accurate acidification kinetics, a dual-laser (either argon–krypton or argon–argon) cytometer is required. A dual-laser system is *not* required for acridine orange (AO) measurements or for many analyses of individual organelles. The cytometer should have the ability to acquire data in a time-resolved fashion, either by using an analog or digital clock and recording time as a parameter (Martin and Swartzendruber, 1980), or by recording the number of events that occur in each time interval (McNeil *et al.*, 1985).

B. Measurement of Acidification of Endocytosed Probes

Currently, there are at least three fluorescence methods used to measure ligand acidification kinetics *in vivo*. These include the dual-excitation ratio method (Ohkuma and Poole, 1978), the amine ratio method (Murphy *et al.*, 1982a,b), and the dual-fluorescence ratio method (Murphy *et al.*, 1984). These methods are based upon the decreased quantum yield of fluorescein fluorescence at acidic pH (Martin and Lindquist, 1975). To determine the pH to which the ligand is exposed, measurements of the amount of ligand present and the relative quantum yield of the fluorescein label are required. The three methods differ in how these two measurements are made. A brief description of each of the methods is given here.

1. Dual-Excitation Ratio Method

Ohkuma and Poole (1981) were the first to measure directly the pH of endocytic compartments using fluorescently labeled probes. The method is based upon the observation that the ratio of fluorescein emissions when excited at 450 and 490 nm is a monotonic function of pH. Using this ratio technique, they observed a pH of 4.5–4.8 in lysosomes of mouse peritoneal macrophages labeled with FITC–dextran for 24 hours. This technique has been widely used to assay intravesicular pH (e.g., Tycko and Maxfield, 1982; van Renswoude *et al.*, 1982). A major disadvantage of this technique is that it involves calculating the ratio of two pH-dependent emissions. Fluorescein fluorescence is highly quenched at low pH, thereby decreasing the signal–noise ratios for ligands with small numbers of receptors to levels comparable to cellular autofluorescence.

An illustration of the use of the dual-excitation method in flow cytometry is given in Section II,C on the use of membrane-permeant probes. The same general approach is applicable to fluorescein-conjugated ligands.

2. Amine Ratio Method

An alternative to the dual-excitation method is the measurement of fluorescein fluorescence (with excitation at 488 nm) before and after neutralization of acidic compartments with an equilibrating agent, such as a weak base or ionophore (Murphy *et al.*, 1982a,b). The neutralization results in the unquenching of fluorescein fluorescence and thereby allows the measurement of total cell-associated fluorophore. The intravesicular pH is then calculated from the ratio of the fluorescence emissions before and after neutralization (by comparison with a calibration curve generated *in vitro* or *in situ*). This technique has been used to demonstrate that the acidification of FITC-conjugated epidermal growth factor is biphasic, with initial rapid acidification to pH 6.3 within 3–5 minutes followed by a slower acidification to pH 5.3 over 40 minutes (Roederer and Murphy, 1986). When using this approach, care must be taken that the methods used to neutralize intravesicular pH result in reproducible equilibration to a known pH. This method is well suited for use with single-laser flow cytometers. The major disadvantage of the technique is that an average pH value is determined for a population of cells, and therefore, only discontinuous kinetics measurements can be made (i.e., by taking samples at defined time points and measuring them in the presence and absence of an equilibrating agent).

3. Dual-Fluorescence Ratio Method

The limitation of the amine ratio method can be overcome by using two different fluorescent conjugates of the same ligand, with one of the conjugates being pH-dependent (fluorescein) and the other being pH-independent (rhodamine) (Murphy *et al.,* 1984). Using this strategy, simultaneous measurements of total uptake (reflected mainly by rhodamine emission) and acidification (reflected mainly by fluorescein emission) can be made. This method has been used to make measurements of lysosomal pH (Cain and Murphy, 1986), to determine the kinetics of transferrin acidification (Sipe and Murphy, 1987), and to demonstrate the role of the Na^+,K^+-ATPase in regulation of endosomal acidification *in vivo* (Cain *et al.,* 1989).

The following protocols were used to determine the kinetics of acidification of transferrin (Sipe and Murphy, 1987), and can be used for other ligands with only minor modification. Three sets of information are required to calculate a pH value: measurements of rhodamine and fluorescein fluorescence as a function of time after internalization, measurements of the amount of ligand remaining on the cell surface as a function of time after warmup (in that surface-bound transferrin contributes to the fluorescence emission from the cell and would shift the measured pH toward neutrality), and a standard curve relating the fluorescein–rhodamine fluorescence ratio to pH.

a. Acidification Measurements. The same protocol is used for both adherent and nonadherent cells (nonadherent cells are washed by centrifugation at 800 *g* for 5 minutes, at 4°C). Chill ~1×10^6 cells in phosphate-buffered saline (PBS; 0.14 *M* NaCl, 3 m*M* KCl, 8 m*M* Na_2HPO_4, 1.5 m*M* KH_2PO_4, 0.9 m*M* $CaCl_2$, 0.5 m*M* $MgCl_2$) at 4°C for 15 minutes, wash three times with PBS, and label the cells at 4°C for 20–30 minutes in 2 ml of PBS containing 10 μg/ml FITC–Tf and 5 μg/ml LRSC–Tf. After incubation, wash the cells four times with PBS and scrape (adherent cells) or resuspend (nonadherent cells) into 1 ml of PBS at 4°C. The amounts of FITC–Tf and LRSC–Tf can be adjusted to provide approximately equal signal–noise ratios for both probes. If an energy source is desired, 2 g/liter glucose can be added to PBS (incubation and warmup to 37°C in PBS containing 2 g/liter glucose yields results identical to those obtained in the absence of glucose).

Analyze the cell suspension by flow cytometry at 4°C for ~4–6 minutes to establish initial values for green and red fluorescence at the external pH. Warm the cells to the desired temperature and record forward and side light scatter, green and red fluorescence values, and time of analysis in list mode at ~200–400 cells/second. (This may be done using the kinetics option of the ACQ8 program in the Consort/VAX software.) Of practical importance is the temporal resolution with which data are collected. Because the computer word (or byte) used to store the list mode data has finite length, the total acquisition time for an experiment is inversely proportional to the temporal accuracy of the measurements. We generally use an accuracy of 1 second, allowing a maximum acquisition time of ~5 hours.

Background values for green and red fluorescence should also be recorded using unlabeled cells. These autofluorescence values are subtracted from those obtained with labeled cells before further analysis. Calculate average values for all parameters over 10- or 20-second intervals (the KINPRO program may be used for this purpose). An example of this type of analysis is shown in Fig. 1B,C,E.

b. Determination of Surface-Bound Transferrin. For receptor-bound probes, either the percentage of ligand remaining on the surface during the warmup must be determined or such ligand must be completely removed before analysis (e.g., using an acid wash or protease treatment), because surface ligand will artificially increase the estimate of the pH of ligand-containing compartments. Because removal of surface ligand makes continuous acidification measurements impossible, the former method is preferred.

Label cells as before using 15 μg/ml LRSC–Tf and no FITC–Tf. After washing, scrape or resuspend the cells into 2.5 ml of PBS at 4°C. To determine the total amount of Tf bound, remove 200 μl of the cell suspension and add it to 3 ml of ice-cold PBS. Warm the remaining cell suspension to the desired temperature, and at various periods of time during the warmup remove 200-μl aliquots and add them to 3 ml of ice-cold PBS. It is important that the PBS be ice cold to ensure that internalization of the LRSC–Tf is stopped as quickly as possible. After all samples have been collected, pellet the cells by centrifugation at 4°C. Resuspend the cell pellet into 250 μl of PBS containing 15 μg/ml FITC-conjugated goat anti-human Tf antibody (Tago, Burlingame, CA),

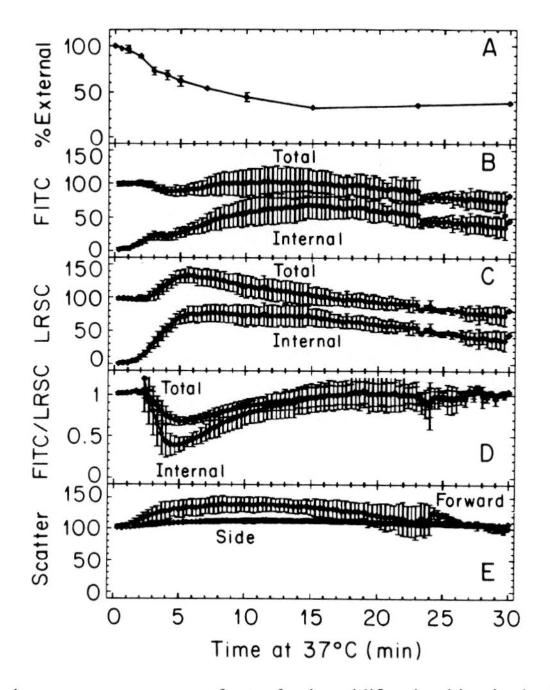

FIG. 1. Continuous measurement of transferrin acidification kinetics by flow cytometry. BALB/c 3T3 cells were allowed to bind labeled transferrin at 4°C, washed, and warmed to 37°C. Total FITC–Tf fluorescence (B) and LRSC–Tf fluorescence (C) were measured for individual cells as a function of time. The average values for six experiments, normalized to percentage of initial values, are shown. The percentages of transferrin on the surface at various times (A) were determined as described in the text and used to calculate the *internal* fluorescence (B,C). The ratio of these internal FITC–Tf and LRSC–Tf fluorescence values was calculated, and the results are shown in (D). Forward and side scatter are shown in (E). Error bars indicate 1 SD. From Sipe and Murphy (1987).

and incubate at 4°C for 20 minutes. After incubation, add 3 ml ice-cold PBS, mix, and pellet the cells by centrifugation. Resuspend the pellet in 3 ml of PBS, mix, and pellet by centrifugation. Wash three times with 3 ml of PBS, and resuspend the final cell pellet in 500 μl of PBS. Analyze ~10,000 cells at 4°C by flow cytometry. Determine the background fluorescence using unlabeled cells incubated with the FITC-conjugated antitransferrin antibody only, and subtract this nonspecific fluorescence to determine the specific transferrin fluorescence. Calculate the amount of transferrin remaining on the cell surface as a percentage of fluorescence intensity of the control sample held at 4°C. An example of this analysis is shown in Fig. 1A.

c. pH Calibration. Surface-label cells with FITC–Tf and LRSC–Tf, and wash as described before. Scrape or resuspend the cells into 1 ml of various pH buffers at 4°C. The buffers are as follows: 0.05 *M* *3-[N*-tris(hydroxymethyl)methylamino]-2-hydroxypropanesulfonic acid (HEPES), pH 7.6, 7.0, 6.6; 0.05 *M* 2-(*N*-morpholino)ethanesulfonic acid (MES), pH 6.0 and 5.6; 0.05 *M* NaOAc, pH 5.1 and 4.6; all containing 0.15 *M* NaCl. Analyze by flow cytometry at 4°C as described before. Calculate mean fluorescence values and subtract autofluorescence. An example of the resulting set of calibration curves is shown in Fig. 2.

To determine pH calibration curves for fluid-phase markers or ligands that dissociate from their receptors in response to low pH, it is necessary to generate an internal standard curve (Cain and Murphy, 1986). After internalization of the probe, incubate the cell suspension at 37°C in 200 m*M* 2-deoxyglucose and 40 m*M* sodium azide for 10 minutes to deplete

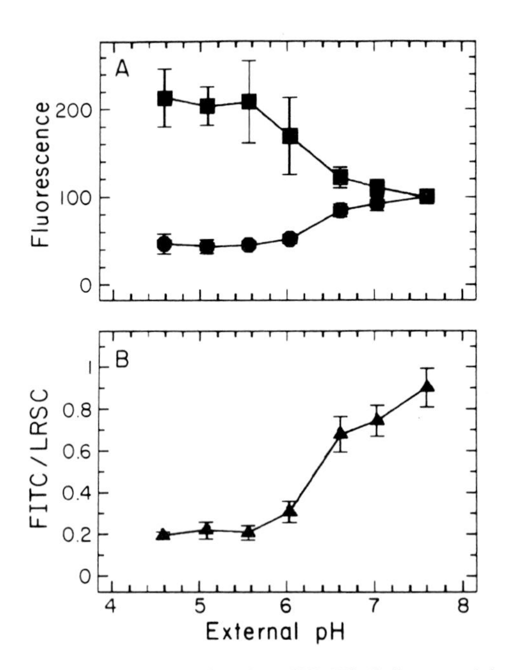

FIG. 2. pH calibration for FITC–Tf and LRSC–Tf. Cells were labeled, washed, and scraped into the appropriate buffer as described in the text. The pH dependence of fluorescence emission for surface-bound FITC–Tf (●) and LRSC–Tf (■) are shown in (A). The increase in LRSC fluorescence presumably corresponds to the conformational changes associated with the loss of iron from diferric transferrin. (B) The ratio of FITC and LRSC emissions was calculated from the data shown in (A). Note that while the fluorescence of both FITC–Tf and LRSC–Tf is pH-dependent, the fluorescence ratio is a monotonic function of pH. From Sipe and Murphy (1987).

cellular ATP pools. After incubation, dilute the suspension 1 : 1 into the pH calibration buffers used before (use at 0.1 M; final 0.05 M) and add nigericin (protonophore) to 75 μM for 10 minutes to equilibrate internal compartments to the external pH. Analyze by flow cytometry. Without the addition of the metabolic inhibitors, the cells are still able to maintain some pH gradients even in the presence of 75 μM nigericin.

d. Calculation of Internal pH. To allow comparisons between experiments, mean values for green (FITC–Tf) and red (LRSC–Tf) fluorescence for each experiment are divided by the mean initial fluorescence and averaged for all experiments. Internalization data are also averaged and interpolated over the same intervals, thereby allowing estimation of the amount of FITC–Tf and LRSC–Tf remaining on the cell surface at any time point during the experiment. After all measurements have been converted to percentages (of initial values), the percentage of external transferrin is subtracted from the mean fluorescence values for each interval. The ratio of the resulting *internal* green and red fluorescence values for each interval is then interpolated onto the normalized and averaged pH calibration curve, thereby allowing the calculation of an average pH for all compartments containing internalized transferrin (for each time interval). The calculations may be performed using the GENpH program (available through the FACS Computer Users Group Library, c/o R.F.M.). Typical results are shown in Fig. 3. For ligands that recycle, such as transferrin, acidification to approximately pH 6 is followed by alkalinization during recycling.

C. Acidification Measurements with Membrane-Permeant Probes

1. Yeast Vacuole pH Measurements with 6-Carboxyfluorescein Diacetate

As an illustration of the dual-excitation method, preliminary results on the analysis of yeast vacuole pH (Preston *et al.*, 1989) are described here. The vacuole in *Saccharomyces cerevisiae* resembles mammalian lysosomes in being an acidic organelle that contains hydrolytic enzymes. Yeast is an attractive experimental organism in that it allows facile application of classical and molecular genetic methodologies. However, it should be borne in mind that *Saccharomyces* cultures consist of free-living single cells with cell walls; this and other differences differentiate it from mammalian cells. The yeast vacuole differs from mammalian lysosomes both structurally and functionally, as does the plant vacuole. Of practical concern, *Saccharomyces* responds to drugs that are useful in

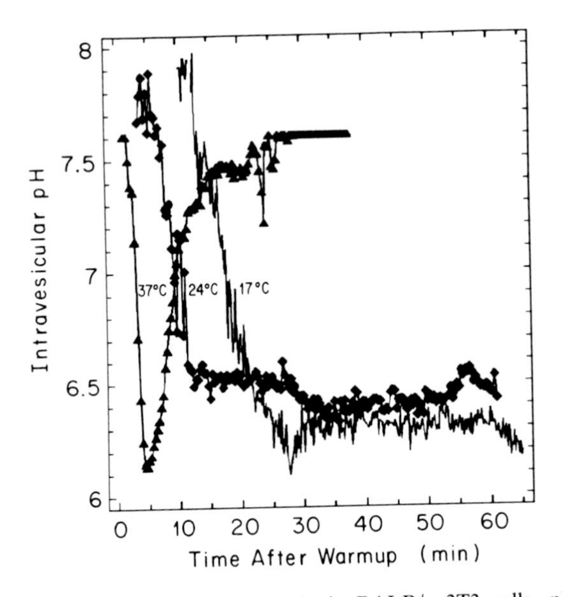

FIG. 3. Acidification kinetics of transferrin in BALB/c 3T3 cells. pH values were calculated over 20-second intervals as described in the text from the data shown in Figs. 1 and 2. Results are shown for cells warmed to 37°C (▲), 24°C (◆), and 17°C (no symbol). Note that alkalinization does not occur at the low temperatures. From Sipe and Murphy (1987).

studies of (mammalian) lysosome physiology in an idiosyncratic manner, if at all. For example, the minimal medium most widely used for yeast cultures contains nearly 0.1 *M* ammonium ion. How (and in fact whether) yeasts maintain any acidic compartment in the presence of this weak base remains to be shown. Another practical consideration involves the difficulty of labeling yeast vacuoles with macromolecular tracers, such as FITC–dextran, that have been widely used in studies of the endocytic pathway in mammalian cells (Preston *et al.*, 1987). Recent results demonstrate the feasibility of using a low molecular weight fluorescent probe to assay vacuolar pH by the dual-excitation ratio method either by fluorescence microscopy (Pringle *et al.*, Chap. 19, this volume) or by flow cytometry (Preston *et al.*, 1989).

Protocol. Yeast cultures are grown, handled, and labeled with 5 μM 6-carboxyfluorescein diacetate (6-CFDA) as described in detail in this volume (Pringle *et al.*, 19). After labeling, the cells are washed with YEPD (1% yeast extract, 2% Bacto-peptone, 2% glucose) at 0°C and kept on ice at 1×10^8 cells/ml in YEPD. The label appears to be vacuole-specific (by fluorescence microscopy) and stable for at least 2 hours under these conditions. Analysis by flow cytometry is performed as described earlier, using dual-argon lasers at 458 nm and 488 nm, and with minimal laser power consistent with useful signal–noise ratios (we use ≤40 mW at both wavelengths). For the analysis, the cells are diluted 1:100 into appropriate buffers or test media at ambient temperature (24°C). At 24°C, the rate of loss of 6-carboxyfluorescein (6-CF) is minimal (~5–10%) over this period. The pH–fluorescence ratio standard curves (Fig. 4) are constructed by analyzing cells after equilibration for 10 minutes at 24°C in analysis buffer (50 m*M* KCl, 50 m*M* NaCl, 50 m*M* HEPES, 50 m*M* MES), at various pH values, in the presence or absence of a mixture of clamping agents [0.2 *M* NH_4Ac, 10 m*M* NaN_3, 10 m*M* 2-deoxyglucose, 25 μ*M* carbonyl cyanide chlorophenylhydrazone (CCCP)]. The clamping agents do not affect 6-CF localization over the time of the analysis. Glucose, 110 m*M*, was added to the unclamped cells to prevent coalescence of the vacuole (Pringle *et al.*, Chap. 19, this volume).

An important criterion for the validity of pH measurements using fluorescence methods is the demonstration of agreement between *in vitro* and *in situ* standard curves (Cain and Murphy, 1986). These curves are

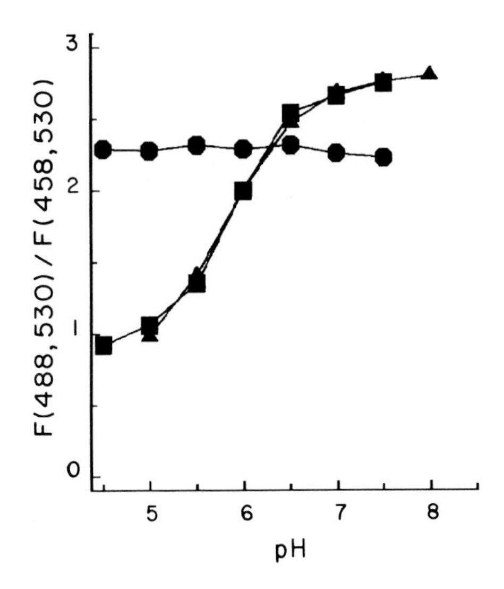

FIG. 4. Dual-emission ratio measurement of yeast vacuolar pH. Cells labeled using the protocol described in the text were diluted into analysis buffer at 24°C at the indicated pH values, in the absence (●) and presence (■) of protonophores. The fluorescence ratio was determined 10 minutes after dilutions were made. The curve obtained in the presence of protonophores constitutes the *in situ* standard curve (see text). An *in vitro* standard curve (▲) was obtained by fluorometry of solutions of hydrolyzed 6-CFDA (see text), and normalized such that the value at pH 6.0 was equal to that of the clamped curve. From Preston *et al.* (1989).

expected to be similar because the fluorescence of fluorescein derivatives is relatively insensitive to nonspecific ionic effects and molecular self-association. To construct an *in vitro* calibration curve, 6-CFDA was hydrolyzed at pH 10 and diluted to 1 μM in the buffers of different pH values described before. Fluorescence-emission curves were acquired with excitation at both 458 and 488 nm using a Gilford Fluoro IV spectrofluorometer, and the region from 515 to 545 nm was integrated. These conditions were chosen to approximate closely the detection conditions of the flow cytometer. The resulting calibration curve is in excellent agreement with the clamped *in situ* curve (Fig. 4). Comparison of the unclamped curve with the clamped calibration curve yields a value of ~6.2 for the vacuolar pH of log-phase cells, regardless of the external pH (between 5.0 and 8.0).

2. Acidification Measurements with Acridine Orange

A rapid and fairly simple method to determine whole-cell acidification activity, which we have used to screen for potential acidification mutants, is the monitoring of the pH-induced accumulation of acridine orange (AO) (Cain and Murphy, 1988). Acridine orange is a weak base that accumulates in acidic compartments in the same manner as the vacuologenic amines (i.e., chloroquine and NH_4Cl). The unprotonated form is membrane-permeant, whereas the protonated form has reduced membrane permeability. In this manner, AO becomes trapped and accumulates within acidic compartments. When the concentration of AO becomes sufficiently high, dimers and higher order multimers form, resulting in spectral shifts in absorbance and fluorescence emission from green to red. Green fluorescence (excitation 488, emission 530 nm) may be used as a measure of total cell-associated AO, whereas red fluorescence (excitation 568, emission 625 nm) may be used as a measure of the extent of stacking of AO within acidic compartments. Both green and red fluorescence can be monitored on single-laser cytometers using 488-nm excitation and separate 530-nm and 625-nm emission filters. However, significant spillover from green to red fluorescence is observed, and thus the dual-laser method is preferred. In either case, the advantages of the AO method are that it allows for the rapid screening of a large number of mutants, and that it does not rely on the endocytosis and proper delivery of probes to acidic compartments.

Protocol. Harvest cells and dilute to 1×10^6 cells/ml in PBS. Add 20

μg/ml unlabeled 7.5-μm polystyrene beads (Flow Cytometry Standards Corporation, Research Triangle Park, NC). The presence of the beads allows for the determination of background fluorescence from the AO in solution by the analysis of the apparent fluorescence of the unlabeled beads. Measure baseline autofluorescence at 37°C by analyzing the sample on the flow cytometer. Logarithmic amplifiers should be used for all fluorescence parameters in order to record accurately the large changes in fluorescence typically observed (convert data to linear scale prior to further data analysis; this is done by the KINPRO program automatically). Add AO (Kodak Laboratory and Specialty Chemicals, Eastman Kodak Co., Rochester, NY) to a concentration of 200 ng/ml (0.663 μM), and monitor AO uptake for 15 minutes. Add NH_4Cl to 100 m*M*, and continue analysis for 15 minutes to monitor loss of red fluorescence and recovery of green fluorescence as discussed later. Maintain sample temperature at 37°C and a flow rate of ~150 events per second. Upper and lower forward-scatter gates are used in the analysis of the data in order to analyze separately the fluorescence from live cells and beads.

It is important to note that the initial rate of AO accumulation in acidic compartments is dominated by existing proton gradients, and does not indicate the rate of acidification. For example, the chloroquine-resistant mutant CHL60-64, which has been shown to have a defect in the rate but not extent of lysosomal acidification, shows no difference in the initial amount of AO accumulated when compared to parental cells (Cain and Murphy, 1988). Upon addition of amines, red fluorescence is rapidly lost, indicating the loss of pH gradients within the cell. This loss of red fluorescence is stable for at least 15 minutes. Green fluorescence is also rapidly lost; however, this loss is only transient. Green fluorescence shows an exponential recovery to a value ~80% higher than the initial value. The recovery of green fluorescence represents the accumulation of AO within compartments due to continued proton pumping. The lack of recovery of red fluorescence indicates that the concentration of AO does not reach a sufficient concentration to allow multimer formation. This would occur because, in the presence of amines, pH gradients are not regenerated despite continued pumping; the compartments vacuolate as a result of osmotic pressure, thereby reducing the effective AO concentration within the compartment (de Duve *et al.*, 1974). Therefore, the recovery of green fluorescence is a measure of the total cellular rate of acidification, and, as shown in Fig. 5, the rate of recovery was almost three times higher in parental cells than in CHL60-64.

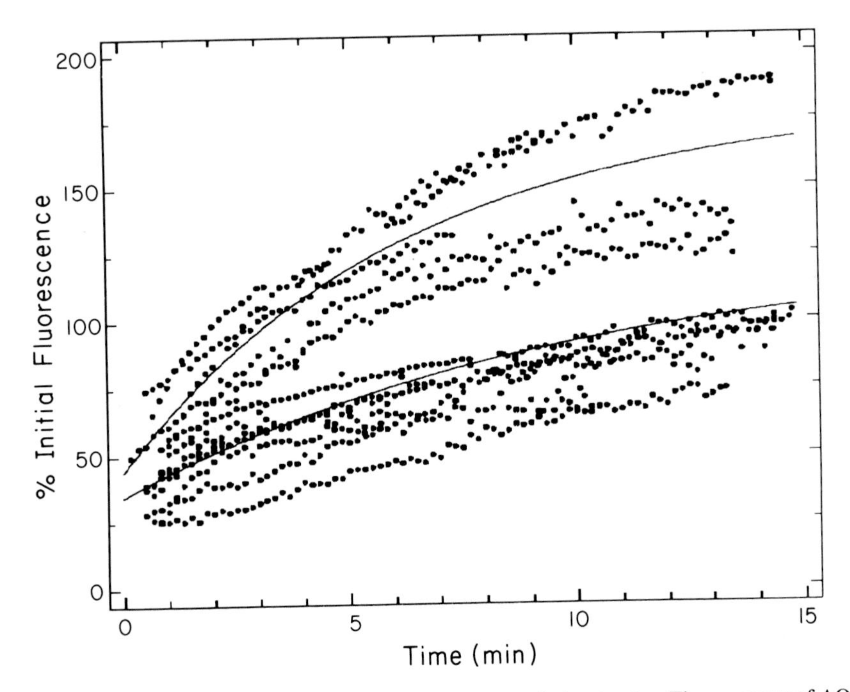

FIG. 5. Effect of ammonium chloride on AO accumulation *in vivo*. The recovery of AO green fluorescence after the addition of NH_4Cl was measured for parental 3T3 cells (upper trace) and CHL60-64 (lower trace) as described in the text. The data for multiple (five to seven) separate experiments are shown with fluorescence values normalized to percentage of initial fluorescence. The data have also been aligned so that the start of the recovery period is plotted at 0 minute. Each point represents the average of all events over a 10-second interval ~1500 cells) for a single experiment. The line is an exponential function drawn using the average values for extent of recovery, extent of loss, and the average rate. From Cain and Murphy (1988). Reproduced by copyright permission of the Rockefeller University Press.

III. Methods for Single-Organelle Flow Analysis

Endocytic compartments are extremely heterogeneous both morphologically and biophysically, making analysis and separation by standard techniques very difficult. Electron microscopy, which has been used to obtain structural and spatial information about the endocytic pathway in great detail, is limited for several reasons: only small numbers of samples can be examined easily, the information obtained is inherently static, and it is difficult to quantitate enzymatic and biochemical properties. Density gradient centrifugation, on the other hand, has been widely used to characterize the endocytic pathway because biochemical analyses may be performed on isolated fractions. This technique is also limited in that the material separated for further analysis is not homogeneous, and as a result, colocalization studies are difficult to perform and interpret. Free-flow electrophoresis has now been used to identify multiple endocytic compartments and as a preparative tool for the identification of proteins present within endosomes (Schmid *et al.*, 1988). However, trypsinization of the vesicles was required before adequate separation was achieved, and the vesicle populations isolated were again not homogeneous.

It is therefore desirable to have a method to identify compartments based on their contents (temporally defined, if necessary), rather than by morphological or biophysical criteria. The feasibility of analyzing organelles by flow cytometry has been demonstrated previously (Murphy, 1985). We have coined the term single-organelle flow analysis (SOFA) to refer to the concept of measuring the properties of individual organelles in flow (Roederer *et al.*, 1989). In addition to our studies on endocytic organelles, a similar approach has been used to analyze plasma membrane vesicles (Gorvel *et al.*, 1984; Moktari *et al.*, 1986) and mitochondria (O'Connor *et al.*, 1988; M. Roederer, D. M. Sipe, J. Halow, A. Koretsky, and R. F. Murphy, unpublished results).

The SOFA technique has several advantages over techniques currently used to investigate endocytosis.

1. A major advantage of this technique is that it can be extremely rapid; lysates can be analyzed within 2 minutes of cell disruption. At this time, pH gradients are still present in endocytic vesicles (Murphy, 1985).
2. Because of the large number of events that can be analyzed (a typical analysis rate is 3000 particles per second), meaningful statistical analyses can be performed, and low-frequency events can be detected.
3. Vesicles can be analyzed for more than one parameter at a time, including forward and side scatter (related to size and optical density) and several fluorescence parameters. Thus, colocalization studies can be performed on an individual-vesicle basis.
4. The method can be used to localize enzyme activity to single-vesicle populations, using fluorogenic substrates (Murphy, 1985).
5. Flow sorting can be used to sort vesicles based upon the fluorescence of internalized ligands, thus allowing for the isolation of highly purified populations of endocytic vesicles (Murphy, 1985).

A. General Methods

1. Flow-Cytometric Analysis

Because the small size of many organelles is at the limit of sensitivity and because of the heterogeneity in fluorescence intensity, logarithmic amplifiers, which provide 3–4 decades of intensity, should be used for all detectors. A fixed number of channels per decade should be used in all cases to facilitate conversion of data from log to linear and to allow accurate comparison between experiments.

Because the forward-scatter signal on most instruments is collected with a photodiode, which has less sensitivity than the PMT used for side scatter, a side-scatter threshold is used for all vesicle analyses. In order to set the side-scatter threshold consistently and for comparison between experiments, a mixture of 1.6-μm red and 1.333-μm green beads (Polysciences) and 0.624-μm and 0.364-μm beads (Duke Scientific) are analyzed at the beginning and end of each experiment. The threshold is set such that the 0.364-μm beads form a tight population several channels above the threshold. An example of the analysis of the bead mixture is presented in Fig. 6. At these settings, particulate matter in the normal sheath fluid gives a high background event rate; therefore, sheath fluid and all sample buffers should either be centrifuged at 27,500 g_{max} overnight or filtered through 0.1 μm Durapore membrane filters (Millipore) prior to use. In addition, light noise from extraneous sources should be reduced as much as possible by shielding the sample chamber and checking for light leaks into the optics housing. With filtered sheath fluid, there is still a "background" rate of ~40 events per second. This rate is from particulates in the sheath, PMT noise, scattering from the stream, and electronic noise. Sample flow rate or concentration should be adjusted such that event rates are at ~1000 per second to ensure a low percentage of background events. A minimum of 50,000 events should be collected for each sample.

2. Data Analysis

Frequently, only a small percentage of the events recorded are from the population of interest (e.g., containing the fluorescent probe). These labeled events often appear as a tail from the distribution of unlabeled events. Therefore, a sample prepared from unlabeled cells should always be analyzed to obtain a "background" signal that can be used in the postacquisition windowing and analysis of the labeled samples.

Experiments that detect colocalization of probes involve the use of two

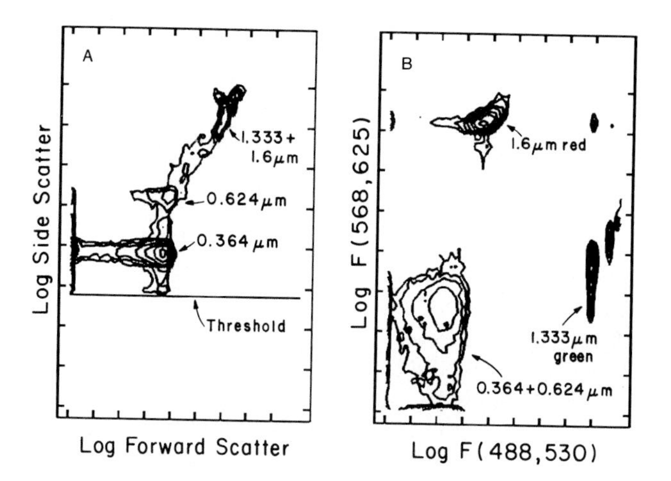

FIG. 6. Calibration standards and selection of light-scatter threshold for SOFA. A mixture of 1.6-μm red and 1.333-μm green beads (Polysciences), and 0.624-μm and 0.364-μm beads (Duke Scientific) were analyzed by flow cytometry and data for 30,000 events collected. Panels (A) and (B) show two-dimensional histograms of (A) scatter (log side scatter versus log forward scatter) and (B) fluorescence (488-nm excitation, 530-nm emission versus 568-nm excitation, 625-nm emission). The line in (A) indicates the position of the side-scatter threshold used for acquisition. Contour intervals are drawn at levels from 10 to 640 events per bin with a factor of 2 separating each contour, and tics are drawn at 0.5-log intervals.

different fluorescence signals in order to distinguish the probes. Therefore, it is often advantageous to calculate a derived parameter that represents the ratio of these two fluorescences. Because the data are collected in log mode, the log of the ratio may be obtained simply by taking the difference of the two parameters for each event (the CALC4 program may be used for this purpose). A constant offset is normally added to this log ratio to adjust the range of values obtained to within the bounds of the histogram.

B. Acridine Orange Uptake in Isolated Vesicles

As mentioned earlier, AO has been used to measure whole-cell acidification activity by flow cytometry (Cain and Murphy, 1988). In addition, AO accumulation has been used to determine the acidification activity in populations of vesicles from homogenized cells by fluorometry (e.g., Marnell *et al.*, 1984; Timchak *et al.*, 1986). We have demonstrated the feasibility of using the flow cytometer to analyze the accumulation of AO within acidic compartments on a vesicle-by-vesicle basis using the flow cytometer. A protocol describing the technique is provided here.

Protocol. Scrape or resuspend 10^7 cells into 4 ml of homogenization buffer (HB; 0.25 *M* sucrose, 2 m*M* EDTA, 10 m*M* HEPES, pH 7.4) at 4°C. Homogenize the cell suspension using 10 strokes of a tight-fitting glass–Teflon homogenizer, and then prepare a postnuclear supernatant (PNS) by centrifugation at 1000 *g* for 10 minutes. Hold the PNS on ice and perform all subsequent incubations and analyses at room temperature. Dilute an aliquot of the PNS 1 : 5 into histidine buffer (30 m*M* histidine, 130 m*M* NaCl, 20 m*M* KCl, 2 m*M* $MgCl_2$, pH 7.0), and incubate at room temperature for 1 hour to allow for dissipation of existing proton gradients. Analyze the sample on the flow cytometer as described before to obtain background autofluorescence levels. Add AO to a concentration of 7.5 μM, incubate for 10 minutes, and analyze on the flow cytometer to determine the extent of acidification remaining after vesicle isolation. Add ATP to 2 m*M*, incubate for 10 minutes, and analyze to determine acidification activity. To dissipate pH gradients, add NH_4Cl to 100 m*M*, incubate, and analyze.

The results of the analysis of AO uptake are presented in Fig. 7. From these results, it is clear that this method can be used to determine qualitatively the acidification properties of isolated vesicles.

C. Colocalization of Endocytic Markers

One powerful aspect of SOFA is the demonstration of colocalization of endocytic markers on a vesicle-by-vesicle basis. We have used this technique to determine the kinetics with which newly endocytosed material is delivered to preexisting lysosomes and to follow the segregation of XRITC–dextran and FITC–Tf during endocytosis (Roederer *et al.*, 1989).

Labeling of Lysosomes and Endosomes. Label cells with 20 mg/ml XRITC–dextran in growth medium for 1 hour at 37°C, wash cells extensively (minimum of three changes) with PBS at 4°C, and chase in growth medium for 20 minutes. Wash the cells extensively at 4°C and add prewarmed (37°C) growth medium containing 20 mg/ml FITC–dextran for 10 minutes. Wash extensively at 4°C, prepare a PNS, and analyze the PNS by flow cytometry as described earlier. It is important that the cells be thoroughly washed, because even small amounts of free labeled dextrans can lead to a dramatic increase in the fluorescence background. If there is free dextran present, the entire population of vesicles (including

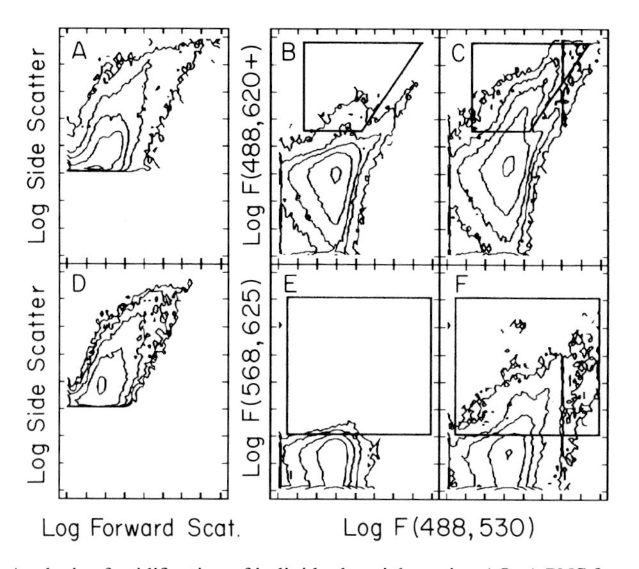

FIG. 7. Analysis of acidification of individual vesicles using AO. A PNS from Swiss 3T3 cells was prepared and assayed for AO uptake as described in the text. During analysis, 200,000 events were acquired. Contour lines in all panels are drawn to include at least 99%, 97%, 70%, and 10% of the total events. The regions drawn in the panels indicate AO-positive events. (A) A correlated histogram of side scatter versus forward scatter indicating the threshold used for acquisition. (B), (C) Two-dimensional histograms of fluorescence (488-nm excitation, ≥620-nm emission) before (B) and after (C) ATP addition. Before addition of ATP only 0.6% positive vesicles were detected compared to the unlabeled sample (data not shown). With the addition of ATP the AO-positive population increased to 11.3% of the total events recorded. This increase was due to acidification activity within the vesicles, because addition of 100 m*M* NH_4Cl decreased the frequency of positive events to 1.9% (data not shown). (D) A correlated histogram for forward scatter and side scatter for the AO-positive vesicles in (C). Because of spillover between the two emissions (see text) resulting from 488-nm excitation, the increase in AO concentration was monitored directly by exciting the higher order multimers at 568-nm light (E,F). Only 1.6% of the events were above background level before addition of ATP (E), whereas this amount increased to 10.2% after addition of ATP (F). From Roederer *et al.* (1989).

unlabeled vesicles) will be "shifted" in fluorescence as compared to an unlabeled sample.

The results of a typical analysis are shown in Fig. 8. Comparisons between an unlabeled sample (Fig. 8A) and the labeled sample (Fig. 8B) clearly demonstrate that endocytic compartments can be detected using fluorescently labeled fluid-phase markers and that endosomes (10-minute pulse) and lysosomes (1-hour pulse and 30-minute chase) can be kinetically defined with fluid-phase markers by SOFA. As discussed later, we

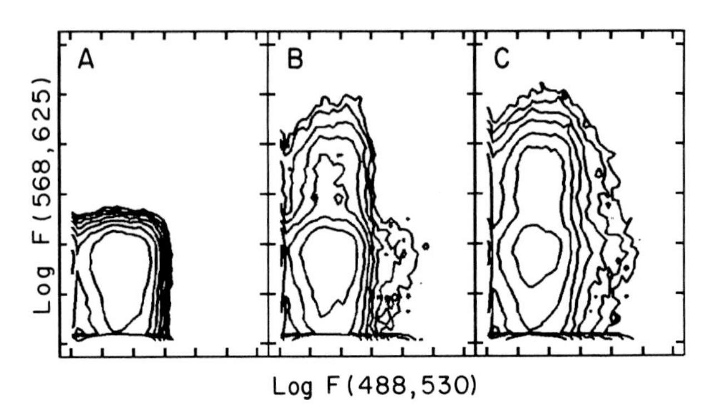

FIG. 8. Labeling of endosomes and lysosomes and the analysis of microsomal pellets. Human K562 cells were labeled and analyzed by flow cytometry for FITC– and XRITC–dextran fluorescence as described in the text. During analysis 250,000 events were recorded. Two-dimensional histograms of XRITC versus FITC fluorescences are shown for unlabeled (A) and labeled PNS (B), and a labeled microsomal pellet (C). Contour intervals are drawn at levels of 10–640 events with a factor of 2 separating each contour, and tic marks are drawn at 0.5-log intervals. Note the appearance of events positive for both red and green fluorescence after pelleting.

are currently investigating the feasibility of isolating these compartments by flow sorting for biochemical analysis. To eliminate contamination by cytoplasmic proteins present in the PNS, vesicles may be pelleted onto a sucrose cushion (see protocol in Section III,D). The effect of this purification is shown in Fig. 8C.

The SOFA technique can also be combined with density gradient centrifugation to permit correlation of fluorescence parameters with physical parameters. Fractions from density gradients (or other separation techniques) may be analyzed in sequence; if the data files from these fractions are combined in such a way that a new parameter corresponding to original fraction number is generated, the distribution of any flow-cytometer parameter versus density can be obtained (the COMBINE program from the Consort/VAX system may be used for this purpose). The analysis of a sequence of fractions is facilitated by automated sample-handling devices, such as the FACS Automate.

An illustration of this approach is shown in Fig. 9. Briefly, cells were labeled with XRITC–dextran for 1 hour, washed, chased in label-free medium for 120 minutes, and labeled with FITC–dextran for 15 minutes. Postnuclear supernatants were prepared and centrifuged on 27% Percoll gradients as described by Merion and Poretz (1981). Briefly, 2 ml of a PNS was mixed with 12 ml of a Percoll stock, to a final concentration of 1× HB and 27% Percoll. These were centrifuged in an SA600 rotor at 25,000 g_{max} for 151 minutes in an RC5C Ultracentrifuge (DuPont Instru-

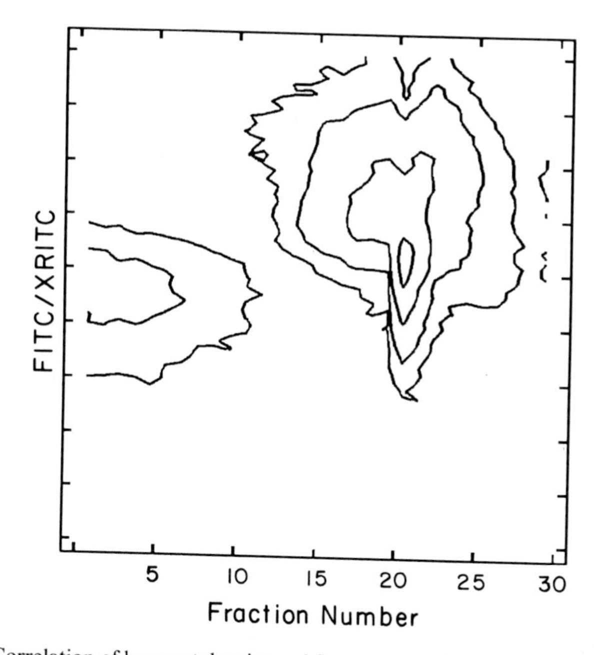

FIG. 9. Correlation of buoyant density and flow-cytometric parameters. Swiss 3T3 cells were labeled with 10 mg/ml XRITC–dextran for 60 minutes, washed at 4°C, and chased for 120 minutes. The cells were then labeled with 20 mg/ml FITC–dextran for 15 minutes. After labeling, the cells were washed at 4°C, and a PNS was prepared and fractionated on a 27% Percoll gradient as described in the text. Fractions of 0.4 ml were collected and analyzed for forward and side scatter, and FITC and XRITC fluorescence. A derived parameter, the log of the ratio of FITC–Tf and XRITC–dextran fluorescence values, is displayed versus fraction number. Because this analysis is several hours after isolation of the vesicles, pH gradients have dissipated and the fluorescein fluorescence is not quenched by low pH. Therefore, the derived parameter simply reflects the relative amounts of FITC– and XRITC–dextran in each vesicle. Contour intervals are drawn at levels of 100, 300, 1000, and 3000 events. Axis tics are drawn at 0.5-log intervals. From Roederer *et al.* (1989).

ments, Wilmington, DE). Fractions were collected from the bottom and analyzed on the flow cytometer for FITC and XRITC–dextran fluorescence. Note that the FITC–dextran positive events (high F/X ratio) are restricted to the low-density portion of the gradient, and that some intermediate-density events that are positive for FITC–dextran only are observed. We interpret the variation in density as resulting from maturation of endosomes (low density) into lysosomes (high density); the results indicate that the density change can occur without fusion with the dense XRITC–dextran-containing compartments (low F/X ratio).

D. Sorting of Endocytic Compartments

As discussed earlier, one of the advantages of SOFA is the possibility of obtaining organelle fractions of extreme purity. The procedures required to do this are no different from standard flow-sorting methods, with two exceptions. Only these exceptions will be discussed further. First, the nozzle head drive (which induces droplet formation) can induce significant noise in the scatter and fluorescence signals. This can be minimized by careful alignment of the optical system, with particular attention to the laser obscuration bars or beam stops. These bars act to prevent laser light scattered from the stream from entering the scatter and fluorescence detectors; when the droplet drive is on, scattering from the stream is increased and may be observed as an oscillation on the constant signal from the detectors and/or a dramatic increase in the event rate. Adjustment of the position of the obscuration bar to minimize this oscillation should significantly reduce this problem.

Second, because sorting of vesicles occurs in droplets, and these droplets contain volume from the original sample, contamination from cytoplasmic constituents is a potential problem. This is especially true when identification of proteins specific to endocytic compartments is desired. The solution is the removal of soluble cytoplasmic components by gel exclusion chromatography or pelleting onto a sucrose cushion.

Preparation and Analysis of Microsomal Pellets. The following protocol was adapted from Marsh *et al.* (1987). Cells (K562) were labeled as described earlier for labeling of endosomes and lysosomes. All of the following steps were performed at 0°C. Using a 3-ml syringe and 22.5-gauge needle, gently overlay 2.5 ml of HB onto 0.5 ml of cushion buffer (1 *M* sucrose in 10 m*M* HEPES, 2 m*M* EDTA pH 7.4; HB with 1 *M* sucrose as opposed to 0.25 *M*) in a 5-ml Beckman ultraclear centrifuge tube. Next, overlay the remaining PNS on top of the HB (this is possible because a small amount of PBS used in the wash steps remains during the preparation of the PNS, and therefore, the concentration of sucrose is slightly less than HB alone). Reserve 0.25 ml of the PNS for analysis. The layer of HB serves to reduce further the amount of contamination by cytoplasmic proteins. Spin at 4°C in an SW50.1 rotor at 100,000 *g* for 40 minutes. The microsomal pellet and the sucrose cushion are removed using a 3-ml syringe and 22.5-gauge needle and are added to 1.5 ml of HB without sucrose (final sucrose concentration is 0.25 *M*). Analyze both the PNS and the resuspended microsomal pellet. If the event rate is too high, dilute the samples with HB.

The results of the analysis are shown in Fig. 8C. As can be seen, there is an increase in the frequency of red and green positive events. This enrichment may be due to removal of cellular debris. In addition, a new population of vesicles containing both FITC and XRITC–dextran has appeared after preparation of the microsomal pellet. The appearance of this population could result from fusion and/or aggregation of the vesicles during the centrifugation or from the enrichment of a low-frequency population. We are currently investigating these possibilities.

Perhaps the most exciting prospect of this technique is the capability to specifically sort defined populations of vesicles for further biochemical characterization. The FACS is capable of sorting several thousand vesicles per second; preliminary vesicle-sorting experiments have been successful.

IV. Summary and Future Directions

A number of straightforward extensions of the methods described in this chapter may be envisaged. Recycling vesicles may be distinguished by coincubating with transferrin and dextran, and selecting high-transferring, low-dextran containing compartments. Endosomes may be defined as medium-transferrin, medium-dextran containing compartments. Lysosomes can be defined temporally, as late (high-density) dextran-containing (transferrin-free) compartments (Roederer *et al.*, 1989). Future experiments will utilize fluorescently tagged antibodies that are specific to the cytoplasmic tails of membrane proteins. We have obtained preliminary results on the use of antibodies to cytoplasmic determinants. Results for anticlathrin antibodies, either indirectly (D. M. Sipe and R. F. Murphy, unpublished results) or directly conjugated (F. Brodsky, unpublished results), have demonstrated that coated structures can be identified by SOFA. By the use of appropriate fluorescently labeled antibodies, in combination with endocytic markers, it will become easier to identify various compartments. Much as immunologically relevant cells have been classified on the basis of surface molecules, vesicles may be similarly identified (“vesiculotyping”).

The technique of single-organelle flow analysis is the first isolation method that does not rely on a biophysical property; rather, it relies on biological properties. Analytically, it can distinguish the contents of individual vesicles at high speed to provide statistically significant data. Preparatively, it can provide the means by which highly purified populations of vesicles can be obtained.

ACKNOWLEDGMENTS

We thank Rob Preston for his generous assistance with preparing the section on yeast analysis, and David Sipe and Cynthia Cain for helpful suggestions and critical reading of the manuscript. The original research described in this chapter was supported by National Institutes of Health grant GM 32508, training grant GM 08067, and National Science Foundation Presidential Young Investigator Award DCB-8351364, with matching funds from Becton-Dickinson Monoclonal Center, Inc.

REFERENCES

Cain, C. C., and Murphy, R. F. (1986). *J. Cell. Physiol.* **129,** 65–70.
Cain, C. C., and Murphy, R. F. (1988). *J. Cell Biol.* **106,** 269–277.
Cain, C. C., Sipe, D. M., and Murphy, R. F. (1989). *Proc. Natl. Acad. Sci. U.S.A.* **86,** 544–548.
De Belder, A. N., and Granath, K. (1973). *Carbohydr. Res.* **30,** 375–378.
de Duve, C., de Barsy, T., Poole, B., Trouet, A., Tulkens, P., and Van Hoff, F. (1974). *Biochem. Pharmacol.* **23,** 2495–2531.
Gorvel, J.-P., Mawas, C., Maroux, S., and Mishal, Z. (1984). *Biochem. J.* **221,** 453–457.
Jongkind, J. F., Verkerk, A., and Sernetz, M. (1986). *Cytometry* **7,** 463–466.
Marnell, M. H., Mathis, L. S., Stookey, M., Shia, S.-P., Stone, D. K., and Draper, R. K. (1984). *J. Cell Biol.* **99,** 1907–1916.
Marsh, M., Schmid, S., Kern, H., Harms, E., Male, P., Mellman, I., and Helenius, A. (1987). *J. Cell Biol.* **104,** 875–886.
Martin, J. C., and Swartzendruber, D. E. (1980). *Science* **207,** 199–201.
Martin, M. M., and Lindquist, L. (1975). *J. Lumin.* **10,** 381–390.
McNeil, P. L., Kennedy, A. L., Waggoner, A. S., Taylor, D. L., and Murphy, R. F. (1985). *Cytometry* **6,** 7–12.
Merion, M., and Poretz, R. D. (1981). *J. Supramol. Struct. Cell. Biochem.* **17,** 337–346.
Moktari, S., Feracci, H., Gorvel, J.-P., Mishal, Z., Rigal, A., and Maroux, S. (1986). *J. Membr. Biol.* **89,** 53–63.
Murphy, R. F. (1985). *Proc. Natl. Acad. Sci. U.S.A.* **82,** 8523–8526.
Murphy, R. F. (1988). *Adv. Cell Biol.* **2,** 159–180.
Murphy, R. F. (1989). *In* "Flow Cytometry and Sorting" (M. R. Melamed, T. Lindmo, and M. L. Mendelsohn, eds.), 2nd ed. Wiley, New York (in press).
Murphy, R. F., Jorgensen, E. D., and Cantor, C. R. (1982a). *J. Biol. Chem.* **257,** 1695–1701.
Murphy, R. F., Powers, S., Verderame, M., Cantor, C. R., and Pollack, R. (1982b). *Cytometry* **2,** 402–406.
Murphy, R. F., Powers, S., and Cantor, C. R. (1984). *J. Cell Biol.* **98,** 1757–1762.
Murphy, R. F., Roederer, M., Sipe, D. M., Cain, C. C., and Bowser, R. (1989). *In* "Flow Cytometry" (A. Yen, ed.). CRC Press, Boca Raton, Florida (in press).
O'Connor, J. E., Vargas, J. L., Kimler, B. F., Hernandez-Yago, J., and Grisolia, S. (1988). *Biochem. Biophys. Res. Commun.* **151,** 568–573.
Ohkuma, S., and Poole, B. (1978). *Proc. Natl. Acad. Sci. U.S.A.* **75,** 3327–3331.
Preston, R. A., Murphy, R. F., and Jones, E. W. (1987). *J. Cell Biol.* **105,** 1981–1987.
Preston, R. A., Murphy, R. F., and Jones, E. W. (1989). In preparation.
Roederer, M., and Murphy, R. F. (1986). *Cytometry* **7,** 558–565.

Roederer, M., Bowser, R., and Murphy, R. F. (1987). *J. Cell. Physiol.* **131,** 200–209.
Roederer, M., Cain, C. C., Sipe, D. M., and Murphy, R. F. (1989). In preparation.
Schmid, S. L., Fuchs, R., Male, P., and Mellman, I. (1988). *Cell (Cambridge, Mass.)* **52,** 73–83.
Sipe, D. M., and Murphy, R. F. (1987). *Proc. Natl. Acad. Sci. U.S.A.* **84,** 7119–7123.
Sklar, L. A. (1987). *Annu. Rev. Biophys. Biophys. Chem.* **16,** 479–506.
Timchak, L. M., Kruse, F., Marnell, M. H., and Draper, R. K. (1986). *J. Biol. Chem.* **261,** 14154–14159.
Tycko, B., and Maxfield, F. R. (1982). *Cell (Cambridge, Mass.)* **28,** 643–651.
van Renswoude, J., Bridges, K. R., Harford, J. B., and Klausner, R. D. (1982). *Proc. Natl. Acad. Sci. U.S.A.* **79,** 6186–6190.

Chapter 17

Endosome and Lysosome Purification by Free-Flow Electrophoresis

MARK MARSH

Institute of Cancer Research
Chester Beatty Laboratories
London SW3 6JB, England

I. Introduction

Endocytosis is a constitutive activity of virtually all cells. For much of the cell cycle, coated membrane-bound vesicles continually bud from the plasma membrane. In so doing, these vesicles transport cell surface receptors and other plasma membrane components, as well as ligands and small amounts of the surrounding medium, into the cell (Steinman *et al.*, 1983). The vesicles fuse with endosomes where the internalized membrane components and ligands are sorted to different sites in the cell.

Much of the membrane is recycled to the cell surface, but certain components, such as the receptors for epidermal growth factor or the Fc domains of certain immunoglobulins, are directed to lysosomes (Mellman *et al.,* 1986).

In order to gain a detailed understanding of the molecular mechanisms that underlie these sorting events, as well as other properties of the endocytic pathway, it is essential to be able to fractionate cells and isolate the relevant organelles in as pure a form as possible. However, the biochemical and physical properties of membrane-bound organelles frequently overlap, so that fractionation techniques capable of detecting small differences in density, charge, or composition are required to bring about effective separation. In addition, in both the endocytic and exocytic pathways new subcompartments are still being identified—for example, the trans-Golgi network (TGN) and prelysosomal compartment (PLC) (Griffiths and Simons, 1986; Griffiths *et al.,* 1988)—making the problem increasingly complex. Furthermore, the current emphasis on cell-free systems, requiring the isolation of organelles with at least some functional integrity, has increased the demands for effective fractionation procedures. As a consequence, novel fractionation methods must be introduced to supplement existing strategies.

Free-flow electrophoresis (FFE) has for some time been used in the separation of specific cell types (tumor cells, blood cells, spermatozoa), as well as proteins and larger multimeric complexes (Hannig, 1978; Hannig and Heidrich, 1977; Hansen and Hannig, 1984; Menashi *et al.,* 1981; Engelmann *et al.,* 1988; Crawford, 1988). In recent years, however, the technique has been applied to the fractionation of subcellular membrane-bound organelles (Harms *et al.,* 1980, 1981; Debanne *et al.,* 1982; Evans and Flint, 1985; Marsh *et al.,* 1987; Personen *et al.,* 1984). We have had particular success in using FFE, both preparatively and analytically, in our studies of endocytic organelles. In this chapter the procedures that have been adapted or developed to isolate endosomes and lysosomes are described.

II. Free-Flow Electrophoresis

A. Principle

Free-flow (or free-flow zone) electrophoretic separation occurs as a consequence of charge differences on particles or molecules.

In ionic media charged particles attract an ion cloud, the electrical boundary layer, which partially neutralizes the surface charge. The electrophoretic mobility of a particle is determined by the zeta potential, that is, the potential at the plane of shear between the electrical boundary layer and the medium surrounding the particle. In FFE, separations are made by injecting a sample (e.g. proteins, cells, or organelles) into a liquid curtain of buffer flowing between two plates. A potential is applied across the curtain and the sample will separate according to the zeta potential on the individual components.

In the case of membrane-containing organelles the glycoprotein, glycolipid, and phospholipid components that make up the membrane confer a charge on the organelles. Because the different functional properties of organelles are determined by their component polypeptides and lipids, it follows that charge differences will exist between functionally distinct compartments. As yet, detailed biochemical compositions for most intracellular membrane systems remain to be determined. At present, therefore, one is unable to predict the electrophoretic mobility of specific membrane systems and whether FFE can be used in the isolation of specific organelles. Nevertheless, charge differences can be exploited for fractionation.

B. Apparatus

Elpho Vap 11, 21, or 22 FFE machines (Bender Hobein, Munich, FRG) have been used for all protocols and experiments described here. The machines contain three basic components: (1) the free-flow separation chamber, sample delivery perfusor, and fraction collector; (2) the cooling systems through which the temperatures in the separation chamber, fraction collector, and perfusor can be regulated; and (3) the high-voltage power supply. With the Elpho Vap 11 and 21 machines these components are housed in a single cabinet, whereas the Elpho Vap 22 machine has the three components in separate units. The newer machines (Elpho Vap 21 and 22) are somewhat more versatile than their forerunners and can be easily adapted for running free-flow isoelectric focusing, free-flow isotachophoresis, free-flow field step electrophoresis, as well as free-flow zone electrophoresis.

The separation chamber (Fig. 1) consists of two flat plates separated by a gasket or spacer. The back plate is mounted within the separation chamber cabinet, and is composed of metal coated with a glass film on the front (inner) surface. Coolant from the refrigeration unit is circulated through channels in the back plate to regulate the temperature within the separation chamber. The front plate of the chamber is thick Plexiglas which is fitted with a metal frame through which it is clamped onto the

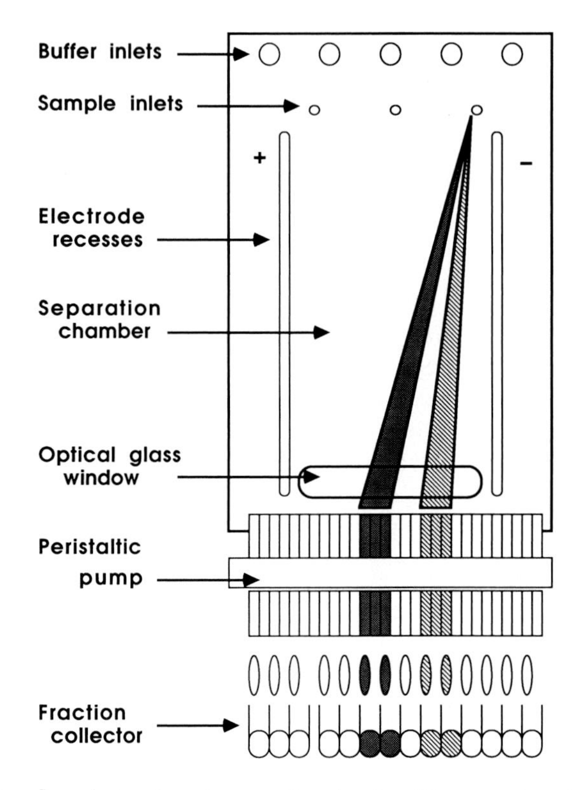

FIG. 1. Free-flow electrophoresis separation chamber. The separation chamber, viewed from the front, contains inlet ports for the separation buffer that forms the liquid curtain and the sample. The electrodes are contained in recessed compartments in the chamber front plate and are connected through ports in the front plate to the electrode buffer circuit and high-voltage supply. The 90-fold fractionator at the base of the chamber is connected to a peristaltic pump and fraction collector.

back plate. All ports into the separation chamber, bushings for the temperature sensor, electrode placements, and high-voltage connectors are accommodated in the front plate. The assembled chamber contains a space ~10 cm wide × 50 cm long (the Elpho Vap 21 chamber is shorter). Gaskets of 0.3, 0.5, or 1 mm can be used to vary the chamber depth. For endosome and lysosome separations a 0.5-mm gasket is used to separate the plates.

The electrodes are housed in recesses on either side of the front plate (Fig. 1) and are isolated from the separation chamber by ion exchange or filter membranes and leak-free sealing gaskets. Electrode buffer (see later) is pumped through the electrode compartments via ports in the front plate.

Through a series of ports at the top the separation chamber is connected, via tubings, to the separation buffer reservoir. Separation buffer (see later) is supplied to the chamber from the reservoir either by gravity feed or through a pump. A fractionator consisting of 90 ports exit the separation chamber at its base and is connected via silicon tubings to a precision 90-channel peristaltic pump. The pump draws a liquid curtain of separation buffer through the separation chamber, and the fractionator splits the curtain into 90 fractions, with minimal disturbance to the laminar flow. The fractions are delivered to a temperature-controlled fraction collector.

The front plate on the Elpho Vap 21 and 22 machines contains an optical glass window, just above the 90-channel exit port, to which a scanning device can be fitted to monitor the optical density, light scattering, and fluorescence profiles across the curtain.

The samples for fractionation are injected into the liquid curtain through one of three ports in the top of the front plate. These are positioned between the electrodes, one in the center and one each on the anodal and cathodal sides of the chamber. Each port can be connected, via tubings, to a refrigerated syringe and adjustable perfusor. Samples are placed in the refrigerated syringe and injected into the machine at a constant rate by the perfusor. The port used for injection will depend on the zeta potential of the particles; cell membranes carry a net electronegative charge and migrate to a greater or lesser extent toward the anode, therefore the port adjacent to the cathode, or in the center of the chamber must be used (for some preparative applications it is possible to use both ports at the same time).

C. Preparing the Apparatus

In preparing the machine the chamber should be first coated with bovine serum albumin (BSA). Hannig *et al.* (1975) examined the effect of a number of factors, such as electroosmosis, temperature gradients, thermal diffusion, and the geometry of the sample inlet, on the profile of the sample during transit through a free-flow separation chamber. They found that although the flow profile was influenced most significantly by electroosmotic effects in the liquid curtain, adjustment of the zeta potential on the chamber walls to that of the material to be separated considerably reduced band broadening.

The zeta potential on the cytoplasmic aspect of endosome and lysosome membranes is not currently known; however, the separation profiles are improved when the chamber is precoated with BSA (Harms *et al.*, 1980). To precoat the chamber, a 0.2% solution of BSA in distilled water (dH_2O) is injected into the chamber through a connector in tubing 1 of the 90-fold fractionator at the base of the separation chamber. Care should be taken to ensure that bubbles are not injected into the separation chamber and that the solution coats the entire surface of the chamber. When filled, the injection tubing is clamped and the BSA left to stand in the chamber for 30 minutes.

Subsequently the chamber is connected to the upper reservoir containing the separation buffer. All air must be removed from the chamber and tubings using the bubble traps above the separation chamber. The BSA solution is removed and the chamber rinsed thoroughly with separation buffer by opening, or running, the 90-fold peristaltic pump. At this time the high-voltage and cooling systems can be turned on and adjusted to the run conditions. The heat generated within the chamber by the electric field is removed through the refrigerated back plate. The cooling system in the Elpho Vap 21 and 22 machines is considerably improved on that of earlier models, making the machines more suited to operating at the low temperatures required for subcellular fractionation. The additional cooling capacity has required the incorporation of auxiliary heaters to prevent the chamber from freezing. However, care should be taken while setting up the machine to ensure that the electric field and cooling capacity are balanced, and that the chamber does not freeze. The machine can be prepared and equilibrated while the cell lysates are prepared (see below).

Following completion of the separation the machine should be shut down and washed thoroughly. Again to avoid freezing, the high-voltage power supply should not be immediately shut down. The cooling circuits can be turned off initially and the voltage gradually reduced as the temperature in the chamber increases to ambient. The separation chamber, all tubings, and the electrode buffer circuits should be flushed through with dH_2O. Precautions should be taken to remove all sucrose and to prevent algal growth in the chambers and tubings while the machine is not in use. Sucrose must not be spilt on the fragile tubings of the 90-fold peristaltic pump. Regular dusting of these tubings with French chalk will help to prevent damage. Blockages within the separation chamber or tubings can sometimes be removed by perfusing the system with aqueous solutions containing trypsin.

When not in use the machine should be stored according to the manufacturer's recommendation. For short periods the electrode compartments and chamber can be left filled with H_2O. For longer periods of nonuse (i.e., 2–3 days), the separation chamber should be emptied, however, the electrode chamber should be left containing dH_2O.

D. Buffer Systems and Separation Conditions

Various buffers have been used for biological free-flow electrophoretic separations, and the merits of different ion systems are discussed by Zeiller *et al.* (1975). For cells and organelles it is usually essential to have osmotic expanders present to maintain isotonicity. For the separation of endosomes and lysosomes we have routinely used the buffer system described by Harms *et al.* (1980). The separation buffer is 0.25 *M* sucrose, 10 m*M* triethanolamine, 10 m*M* acetic acid, and 1 m*M* EDTA adjusted to pH 7.45 with 1 *M* NaOH; the electrode buffer is 0.25 *M* sucrose, 100 m*M* triethanolamine, 100 m*M* acetic acid, and 10 m*M* EDTA adjusted to pH 7.45 with 1 *M* NaOH. Several liters of each buffer should be prepared before running the machine. In addition to providing a suitable medium for electrophoretic separation, this separation buffer also has advantages in the preparation of cell lysates (see later). A comparison of free-flow separation using a HEPES buffer and the triethanolamine buffer is given in Fig. 2. In the HEPES buffer there is little separation of endosomes and lysosomes from the major protein peak.

Conditions must be established to obtain optimal separation of the relevant components. Parameters affecting the efficiency of separation include perfusor speed, voltage, and chamber temperature, as well as the flow rate (velocity of the curtain), which in turn determines the transit time (the time an individual particle takes to pass through the chamber). Furthermore, different flow rates may be used according to whether analytical separations of small samples or larger scale preparative separations are to be undertaken. Slow sample delivery (approximately two-thirds the flow rate) can reduce the band-broadening effects and increase resolution. The conditions used for endosomes and lysosomes separation are given in Table I (Harms *et al.*, 1980; Marsh *et al.*, 1987, 1988).

III. Sample Preparation

Samples are prepared for electrophoresis using procedures that give both effective cell lysis and good recovery of intact organelles. The following protocol has worked well for a number of tissue culture cell lines including Chinese hamster ovary (CHO), baby hamster kidney (BHK), normal rat kidney (NRK), human epidermoid carcinoma (HEP-2), and human fibroblast (AF2). Unique biochemical markers for endosomes have not, to date, been described. It is therefore necessary to

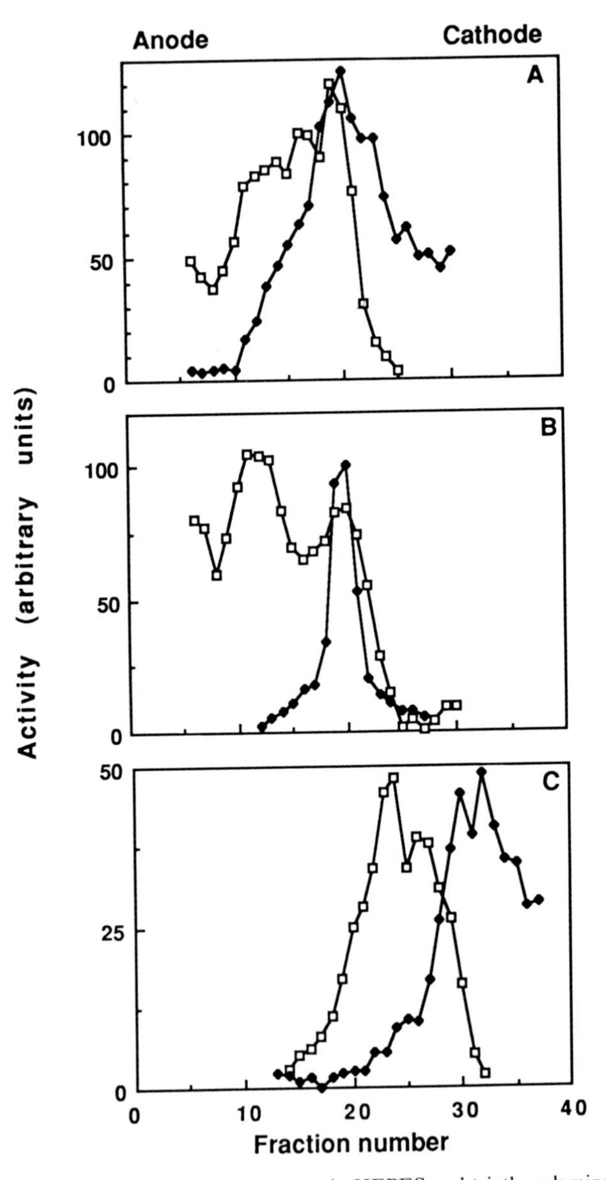

FIG. 2. Separation of CHO cell microsomes in HEPES and triethanolamine separation buffers. Monolayers of CHO cells were labeled for 10 minutes at 37°C with 10 mg/ml horseradish peroxidase (HRP) in complete medium (Marsh *et al.*, 1987). The cells were washed thoroughly and scraped from the dish in PBS. After pooling and one wash in separation buffer, the cells were lysed by five passes in a stainless-steel homogenizer in 0.25 *M* sucrose, 10 m*M* HEPES, 1 m*M* EDTA (pH 7.4). The postnuclear supernatant was treated with 10 μg/ml TPCK–trypsin for 5 minutes at 37°C and then applied to the free-flow

TABLE I

FREE-FLOW ELECTROPHORESIS SEPARATION CONDITIONS FOR PREPARING ENDOSOMES AND LYSOSOMES

Parameter	Machine type		
	Elpho Vap 11	Elpho Vap 21	Elpho Vap 22
Chamber temperature (°C)	4–5	4–5	4–5
Voltage	1400	1700	1450
Current (mA)	220	130	200
Buffer flow rate (ml/fraction/hour)	5	3.6	5
Sample flow rate (ml/hour)	3.6	2.5	2.5

label the organelles prior to preparative and analytical preparation by internalization of endocytic markers. Endosomes can be labeled by incubating the cells, prior to cell lysis, in medium containing fluid-phase markers [10 mg/ml dialyzed FITC–dextran or horseradish peroxidase (HRP), 10 minutes at 37°C or 2 hours at 20°C (Marsh *et al.*, 1987)]. Alternatively, receptor-mediated endocytosis of prebound ligands, such as radiolabeled transferrin or Semliki Forest virus, also provide effective markers (Marsh *et al.*, 1987; Schmid *et al.*, 1988). We have found it useful to include an independent marker for the plasma membrane. For this purpose cells are labeled at 0°C with ^{125}I using the lactoperoxidase method (Marsh *et al.*, 1987).

The labeled cells are placed on ice, washed with several changes of ice-cold phosphate-buffered saline (PBS), and scraped from the dish using a silicon rubber cell scraper. The cells are pooled, centrifuged (700 *g*, 5 minutes, 4°C), and washed once in 20 volumes of ice-cold separation buffer. To prevent premature lysis of the cells, care should be taken not to resuspend the cells too vigorously; gentle shaking is usually all that is required. Subsequently, the cells are centrifuged once more (800 *g*, 5

---

machine. The separation buffer was 0.25 *M* sucrose, 10 m*M* HEPES, 1 m*M* EDTA (pH 7.4). (A) and (B) Distribution of enzyme markers in the liquid curtain. (A) ◆, Total protein; □, HRP (endosomes). (B) ◆, succinate 3-(4-iodophenyl)-2-(4-nitrophenyl)-5-phenyl-2H-tetrazolium chloride (INT) oxido-reductase (mitochondria); □, β-hexosaminidase (lysosomes). (C) For comparison, the separation of endosomes in triethanolamine buffer is shown. These are the same data as described in Fig. 3A.

minutes, 4°C) and resuspended in four volumes of ice-cold separation buffer. Many cell lines lyse readily in the triethanolamine separation buffer provided the pH is maintained above 7.2. BHK and AF2 cells can be lysed by taking the cell suspension up and down (10 times) into a 10-ml glass pipet (Harms *et al.*, 1980; Marsh *et al.*, 1982). CHO cells may be broken by gentle douncing or using a ball-bearing device (Balch and Rothman, 1985; Marsh *et al.*, 1987). Between 80 and 90% of the cells are usually broken using these protocols, while most of the nuclei and other membrane-bound organelles remain intact. From the lysate a postnuclear supernatant (PNS) is derived by centrifugation at 800 *g* for 10 minutes at 4°C.

Depending on the cell type and the fractionation required, FFE can be carried out using the unfractionated PNS, microsomes, or an enriched endosome–Golgi preparation. Microsomes are prepared by centrifuging the PNS (100,000 *g* for 35 minutes in a Beckman SW55 rotor). To reduce damage during centrifugation, the organelles are centrifuged onto a 0.5-ml cushion of 1 *M* sucrose in separation buffer, and are resuspended together with the cushion by addition of 1.5 ml separation buffer without sucrose. Further fractionation of the microsomes can be carried out by flotation gradient centrifugation to provide enriched endosome–Golgi samples (Marsh *et al.*, 1987; Schmid *et al.*, 1988).

A. Trypsinization

With all the cell lines tested, we have found that the separation of endosomes and lysosomes is facilitated by treating the samples with trypsin prior to electrophoresis (see Marsh *et al.*, 1987, 1988; Schmid and Mellman, 1988). For BHK and CHO cell PNS and microsome fractions, treatment with 10 μg/ml *n*-tosyl-L-phenylalanine chloromethyl ketone (TPCK)–trypsin (Worthington Biochemical Corp.) for 5 minutes at 37°C is sufficient to enhance separation. The digestion is stopped by returning the samples to 0°C and adding a 5-fold excess of soybean trypsin inhibitor (Sigma). If endosome–Golgi-enriched fractions are used, the trypsin can be reduced to 1–2 μg/ml (Schmid and Mellman, 1988; Schmid *et al.*, 1988).

The effects of trypsin on endosome and lysosome membranes that result in the increased electrophoretic mobility is unclear. There is no change in (1) the latency of internal markers, (2) the buoyant densities of the organelles, or (3) the ATP-dependent acidification activity (Marsh *et al.*, 1987; Schmid *et al.*, 1988). Nor, at the concentrations of trypsin described, has any change in the SDS–PAGE electrophoretic pattern of proteins that fractionate into the detergent phase of a Triton X-114 lysate been detected (Marsh *et al.*, 1987; Schmid *et al.*, 1988). The digestion may reduce aggregation of the organelles in a lysate or remove cytoskeletal components present in the preparations (C. R. Hopkins, personal communication).

B. Free-Flow Electrophoresis

Following trypsin inactivation, the samples are either centrifuged (800 *g*, 10 minutes, 4°C) or passed through a 45-mm filter holder containing two layers of prewashed filter paper [Schleiches and Schull, no. 589/3 (Harms *et al.*, 1980)] to remove any large aggregates. The sample is loaded into a 10-ml precooled glass gas-tight syringe, which is placed into a refrigerated reciprocal rotator and connected to both a perfusor unit and the prepared separation chamber. The concentration of sample to be injected can be varied. A reasonable working protein concentration is 1 mg/ml. When the concentration of protein is too high, aggregates will appear in the separation chamber. In such cases the samples should be removed from the syringe, diluted with separation buffer, and reapplied to the machine.

The sample is injected into the separation chamber when the perfusor is turned on. It is often useful to inject a small aliquot first to ensure that the sample enters the liquid curtain with minimal turbulence and that separation is occurring. Turbulence at the injection port can result in considerable band broadening and loss of resolution. The flow rate of both the perfusor and the 90-fold peristaltic pump (i.e., the flow rate of the liquid curtain) can be adjusted to minimize turbulence (note that after changing the flow rate the field and chamber temperature should be allowed to reequilibrate). The turbid cell membrane preparation can usually be seen in the separation chamber, and fractionation can be checked visually by observing the initial stream of organelles separate into two or more streams. When satisfactory separation is ensured, the fraction collector can be inserted and the remainder of the sample run through the machine.

Following fractionation, the distribution of the organelles and protein across the liquid curtain can be determined using appropriate enzyme and activity assays. Using the protocol described, we have found that endosomes and lysosomes have a higher electrophoretic mobility than other components of the sample and migrate as a distinct band toward the anode. Virtually all of the components of a sample are electronegative and shift toward the anode; however, the plasma membrane, endoplasmic reticulum, Golgi apparatus mitochondria, and cytosolic proteins are less charged and display a less pronounced shift (Figs. 2, 3, and 4). Following electrophoresis, fractions containing the majority of the endosomes and lysosomes can be pooled, with minimal contamination by other organelles (Fig. 3). The organelles obtained after FFE are still latent, capable of ATP-dependent acidification (Marsh *et al.*, 1987, 1988; Schmid and Mellman, 1988), and can be further fractionated to give highly purified preparations of both endosomes and lysosomes (Fig. 3; Marsh *et al.*, 1987, 1988).

IV. Subfractionation of Endosomes by FFE

Experimental evidence indicates that the endosome compartment is highly heterogeneous and contains organelles that differ in their biochemical composition (Schmid *et al.*, 1988), their internal pH (Kielian *et al.*, 1986), their location in the cell, and their age (see Helenius *et al.*, 1983). The effect of trypsin can be titrated to modulate the electrophoretic mobility of the organelles in a cell lysate, and can be used to subfractionate the endosome compartment (Schmid and Mellman, 1988; Schmid *et al.*, 1988). In the absence of trypsin, the electrophoretic mobility of the lysosomes is slightly faster than the other organelles. The mobility is increased when trypsin at concentrations up to 0.2% (w/w protein) are used. On digestion with 1–4% trypsin the electrophoretic mobility of endosomes increases and the organelles comigrate with lysosomes. Schmid and Mellman (1988) have shown that after treatment with 2–2.5% (w/w) trypsin, CHO cell endosomes exhibit a bimodal distribution in the liquid curtain. The faster migrating species of endosomes apparently represent late endosomes, whereas the slower species is primarily early endosomes. A similar bimodal distribution of endosomes from HEP-2 cells is demonstrated in Fig. 4; however, the markers used in this experiment do not allow early and late endosomes to be distinguished.

V. Conclusions and Prospects

Free-flow electrophoresis has been successfully employed in the purification or isolation of a variety of biological materials including macromolecules (proteins and chromosomes), macromolecular complexes (viruses), and cells (parasites, tumor cells, blood cells, spermatozoa). The technique has also been applied, in both a preparative and analytical capacity, to separate cellular organelles. In combination with Percoll density gradient centrifugation, FFE has been used to prepare endosomes, endosome subpopulations, and lysosomes from tissues and cul-

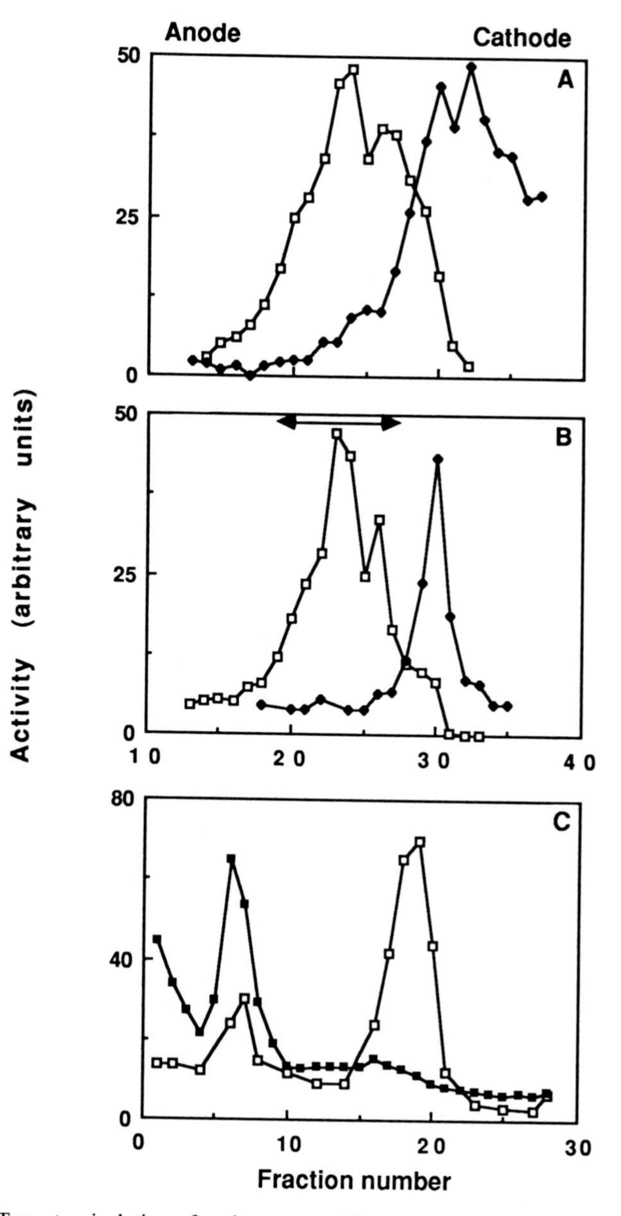

FIG. 3. Two-step isolation of endosomes and lysosomes from CHO cells, labeled as described in Fig. 2. The cells were lysed by five passes through a ball-bearing homogenizer (Balch and Rothman, 1985; Marsh *et al.*, 1987). The postnuclear supernatant was treated with 10 μg/ml TPCK–trypsin and subjected to FFE in a Bender Hobein Elpho Vap 21 electrophoresis machine. (A) and (B) Distribution of markers across the liquid curtain following electrophoretic separation. (A) □, HRP (endosomes); ◆, total protein. (B) □, β-hexosaminidase (lysosomes); ◆, INT-succinate reductase (mitochondria). A pool containing the peak of the endosome–lysosome fraction was taken as illustrated in (B) (↔), and centrifuged in 17.5% Percoll. (C) Distribution of marker enzymes in the Percoll gradient: ■, β-hexosaminidase (lysosomes); □, HRP (endosomes). The gradient of Percoll is from fraction 1 (heavy) to 28 (light).

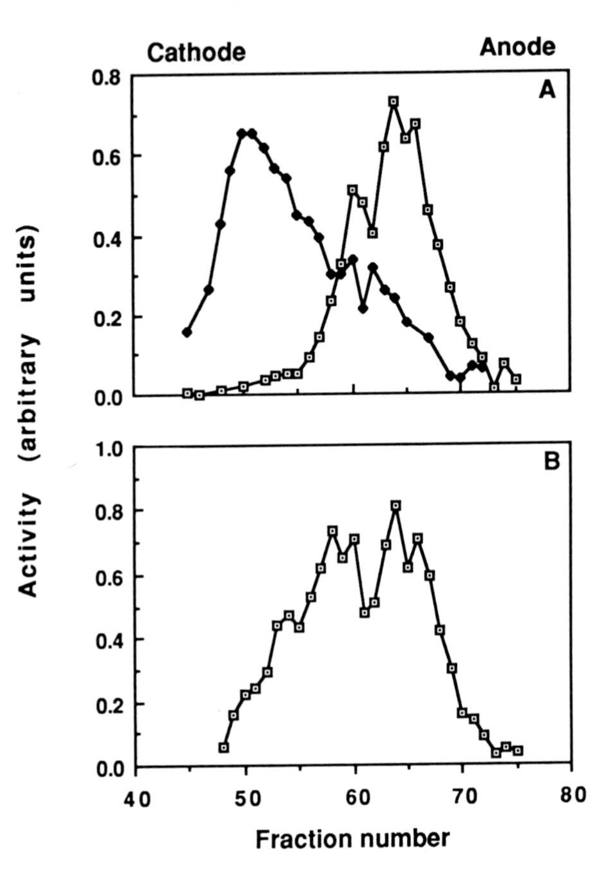

FIG. 4. Free-flow electrophoretic separation of endosomes and lysosomes from human epidermoid carcinoma (HEP-2). Monolayers of the human epidermoid carcinoma cell line (HEP-2) were labeled with HRP as described for CHO cells (Fig. 2). The cells were washed thoroughly and scraped from the dish in PBS. After pooling and one wash in separation buffer, the cells were lysed by five passes through a ball-bearing homogenizer. The postnuclear supernatant was treated with 10 μg/ml TPCK–trypsin and subjected to FFE in a Bender Hobein Elpho Vap 22 electrophoresis machine. (A) Distribution of protein (◆, determined using the Coomassie blue-binding assay; Bradford, 1976) and lysosomes (□, determined by β-hexosaminidase as described by Marsh *et al.*, 1987) following fractionation of the liquid curtain. The major protein peak contains markers for mitochondria and plasma membrane. (B) Distribution of the endosomal marker, HRP (□, determined using σ-dianisidine as described by Marsh *et al.*, 1987), is seen to overlap the lysosome profile and to be separated from the major protein peak. The endosomal profile exhibits two peaks that represent subpopulations of HEP-2 endosomes (see Schmid *et al.*, 1988). Note that the electrodes have been reversed from their positions in Figs. 1–3.

tured cells. The endosomes and lysosomes isolated by FFE and Percoll density gradient centrifugation have been enriched up to 70-fold relative to the initial homogenate. The organelles retain internalized fluid-phase markers and are capable of ATP-dependent acidification. The purification procedure can be completed within 5–6 hours and can yield amounts of endosomes (150–200 μg protein) sufficient for biochemical, immunological, and functional analysis.

It is unlikely that FFE will become a one-step technique for the purification of subcellular organelles, but in combination with other fractionation procedures, it hs a very powerful addition to the approaches that can be applied to cell fractionation. Significantly, FFE is likely to put less stress on organelles during fractionation and increase the yield of intact organelles, because the forces imposed on the organelles during brief transit through the electric field are likely to be considerably less than those imposed by centrifugation. The full potential of the technique remains, however, to be realized. Various methods for modulating the surface charges on specific organelles can be envisaged using organelle-specific antibodies or lectins, and the controlled use of proteases, or other enzymes, may permit subtle charge changes to facilitate the fractionation of other organelles. Future experiments with well-characterized systems may enable some of these potential applications to be investigated.

Acknowledgments

I thank Eric Harms and Hildegard Kern for introducing me to free-flow electrophoresis and passing on some of their considerable expertise, Professor Crawford for making available the free-flow machines at the Royal College of Surgeons, Victoria Anne Lewis for reading the manuscript, and Mrs. M. Callahan for typing. The Institute of Cancer Research is supported by the Cancer Research Campaign and Medical Research Council.

References

Balch, W. E., and Rothman, J. E. (1985). *Arch. Biochem. Biophys.* **240,** 413–425.

Bradford, M. (1976). *Anal. Biochem.* **72,** 248–254.

Crawford, N. (1988). *In* "Cell-Free Analysis of Membrane Traffic" (D. J. Morré, K. E. Howell, G. M. W. Cook, and W. H. Evans, eds.). pp. 51–67. Liss, New York.

Debanne, M. T., Evans, W. H., Flint, N., and Regoeczi, E. (1982). *Nature (London)* **298,** 398–400.

Engelmann, U., Krassnigg, F., Schatz, H., and Schill, W.-B. (1988). *Gamete Res.* **19,** 151–159.

Evans, W. H., and Flint, N. (1985). *Biochem. J.* **232,** 25–32.

Griffiths, G., and Simons, K. (1986). *Science* **234,** 438–442.

Griffiths, G., Hoflack, B., Simons, K., Mellman, I., and Kornfeld, S. (1988). *Cell (Cambridge, Mass.)* **52,** 329–341.

Hannig, K. (1978). *J. Chromatogr.* **159,** 183–191.
Hannig, K., and Heidrich, H. G. (1977). *In* "Cell Separation Methods" (H. Bloemendal, ed.) Part IV, pp. 95–116. Elsevier/North Holland, New York.
Hannig, K., Wirth, H., Meyer, B.-H., and Zeiller, K. (1975). *Hoppe-Seyler's Z. Physiol. Chem.* **356,** 1225–1244.
Hansen, E., and Hannig, K. (1984). *In* "Methods in Enzymology" (G. Di Sabato, J. J. Langone, and H. Van Vunakis, eds.), Vol. 108, pp. 180–197. Academic Press, Orlando, Florida.
Harms, E., Kern, H., and Schneider, J. A. (1980). *Proc. Natl. Acad. Sci. U.S.A.* **77,** 6139–6143.
Harms, E., Kartenbeck, J., Darai, G., and Schneider, J. (1981). *Exp. Cell Res.* **131,** 251–266.
Helenius, A., Mellman, I., Wall, D., and Hubbard, A. (1983). *Trends Biochem. Sci.* **8,** 245–250.
Kielian, M. C., Marsh, M., and Helenius, A. (1986). *EMBO J.* **5,** 3103–3109.
Marsh, M., Wellstead, J., Kern, H., Harms, E., and Helenius, A. (1982). *Proc. Natl. Acad. Sci. U.S.A.* **79,** 5297–5301.
Marsh, M., Schmid, S., Kern, H., Harms, E., Male, P., Mellman, I., and Helenius, A. (1987). *J. Cell Biol.* **104,** 875–886.
Marsh, M., Kern, H., Harms, E., Schmid, S., Mellman, J., and Helenius, A. (1988). *In* "Cell-Free Analysis of Membrane Traffic" (D. J. Morré, K. E. Howell, G. M. W. Cook, and W. H. Evans, eds.), pp. 21–33. Liss, New York.
Mellman, I., Fuchs, R., and Helenius, A. (1986). *Annu. Rev. Biochem.* **55,** 663–671.
Menashi, S., Weintraub, H., and Crawford, N. (1981). *J. Biol. Chem.* **256,** 4095–4101.
Personen, M., Ansorge, W., and Simons, K. (1984). *J. Cell Biol.* **99,** 796–802.
Schmid, S. L., and Mellman, I. (1988). *In* "Cell-Free Analysis of Membrane Transport" (D. J. Morré, K. E. Howell, G. M. W. Cook, and W. H. Evans, eds.), pp. 35–49. Liss, New York.
Schmid, S. L., Fuchs, R., Male, P., and Mellman, I. (1988). *Cell (Cambridge, Mass.)* **52,** 73–83.
Steinman, R. M., Mellman, I., Muller, W. A., and Cohn, Z. A. (1983). *J. Cell Biol.* **96,** 1–27.
Zeiller, K., Loser, R., Pascher, G., and Hannig, K. (1975). *Hoppe-Seyler's Z. Physiol. Chem.* **356,** 1225–1244.

Chapter 18

Fractionation of Yeast Organelles

NANCY C. WALWORTH, BRUNO GOUD,[1] HANNELE RUOHOLA, AND PETER J. NOVICK

Department of Cell Biology
Yale University School of Medicine
New Haven, Connecticut 06510

I. Introduction

The yeast, *Saccharomyces cerevisiae,* serves as an excellent model system for the study of many basic cellular processes common to eukaryotic cells. The well-defined classical and molecular genetics of this organism make it an attractive system in which the genes and gene

[1] Present address: Department of Immunology, Institut Pasteur, Paris, France.

products responsible for carrying out a cellular function can be identified through the isolation of mutants defective in that process. In addition, upon identification of a protein implicated in a process, the ability to disrupt or to mutagenize and replace the gene encoding that protein and to study the phenotype of the resulting mutant strain offers a powerful tool for the analysis of the function of the protein (Struhl, 1983; Botstein and Fink, 1988).

Through the isolation of a family of mutants defective in the transport of secretory and plasma membrane proteins from the endoplasmic reticulum (ER) to the cell surface of yeast (*sec,bet* mutants), the secretory pathway of *S. cerevisiae* has been defined (Novick and Schekman, 1979; Novick *et al.,* 1980; Newman and Ferro-Novick, 1987). Because secretion is an essential process required for cell growth, the *sec* mutants are, by necessity, conditional lethal mutants. Biochemical and morphological analysis of these conditional mutants has demonstrated that the machinery responsible for the transport of proteins through the secretory pathway is similar to the machinery in other eukaryotic cells (Novick *et al.,* 1981). Specifically, as in mammalian cells, proteins destined for transport to the cell surface are synthesized on free or membrane-bound ribosomes, translocated through the ER where core glycosylation of glycoproteins takes place, transported to the Golgi complex where modification of attached oligosaccharides occurs, and transported to the cell surface via membrane-enclosed secretory vesicles (Pfeffer and Rothman, 1987). Thus, the secretory pathway in yeast is a model for the secretory pathway in other eukaryotes. The availability of mutants defective in secretion and the well-defined classical and molecular genetics available make *S. cerevisiae* an attractive system in which to combine a genetic and cell biological approach to dissect the molecular machinery involved in secretion.

Classical cell biological approaches toward the study of the secretory pathway in eukaryotic cells include morphological techniques and subcellular fractionation to purify the organelles thought to be responsible for carrying out specific functions (Palade, 1975). A technique such as immunofluorescence is useful for associating a particular protein with a specific organelle. However, localization using this technique in yeast has often proved to be difficult because of the small size of the organism and, in the case of secretory organelles, their low abundance in wild-type yeast cells. A procedure for immunolocalization at the electron-microscope level has been described by von Tuinen and Riezman (1987) and may prove to be useful for assigning particular proteins to particular components of the secretory machinery. Clearly, however, subcellular fractionation techniques are necessary in order to purify an organelle and identify the components associated with it. The advantage of identifying such components in yeast is that once such a protein is isolated, antibody can be raised and the gene can be identified by, for example, screening an expression library. The cloned gene can then be manipulated *in vitro* and reintroduced into cells such that the function of the protein can be studied *in vivo*. Additionally, proteins identified through classical genetic techniques that are known to function on the secretory pathway can be assigned to particular organelles as a first step toward elucidating their function *in vivo*.

We present here two subcellular-fractionation schemes. One, based on a protocol described by Ruohola and Ferro-Novick (1987), is an organelle separation procedure that gives sufficient resolution to allow the localization of proteins to particular organelles of the secretory pathway (Goud *et al.*, 1988). The other scheme is designed for the purification of post-Golgi secretory vesicles from *S. cerevisiae* (Walworth and Novick, 1987). Through the characterization of proteins associated with the vesicles we hope to further our understanding of the processes by which the targeting, transport, and fusion of secretory vesicles to the plasma membrane take place.

Yeast presents two unique challenges to the cell biologist undertaking cell fractionation: the presence of a cell wall and the low abundance of organelles that constitute the secretory pathway. Yeast cells are surrounded by a rigid cell wall that controls the environment immediately around the cell and protects the plasma membrane from osmotic lysis (Ballou, 1982). The first step in any fractionation procedure therefore is removal of this wall. The yeast cell wall is easily removed enzymatically, and the integrity of the plasma membrane of the resulting spheroplasts is maintained by providing osmotic support in the spheroplast medium.

Components of the secretory machinery are present in low abundance in wild-type cells because secretion is rapid and constitutive. As a result, membrane components of the secretory pathway are not prominent by electron-microscopic analysis of wild-type yeast cells (Fig. 1A). Similarly, vesicles that deliver constitutively secreted and plasma membrane proteins to the cell surface in mammalian cells are not abundant in the cytoplasm. In contrast, cells that also have a regulated secretory pathway, such as pancreatic acinar cells or neurons, accumulate secretory granules that are released upon stimulation by a signal (Kelly, 1985). Because granules accumulate in unstimulated cells, they can be isolated and subjected to biochemical and morphological analysis (Cameron *et al.*, 1986; Nagy *et al.*, 1976). Constitutive secretory vesicles are transient and present in low abundance, and have therefore eluded biochemical analysis. However, the availability of yeast mutants that accumulate post-Golgi vesicles as a result of a transport block facilitates the isolation of this organelle.

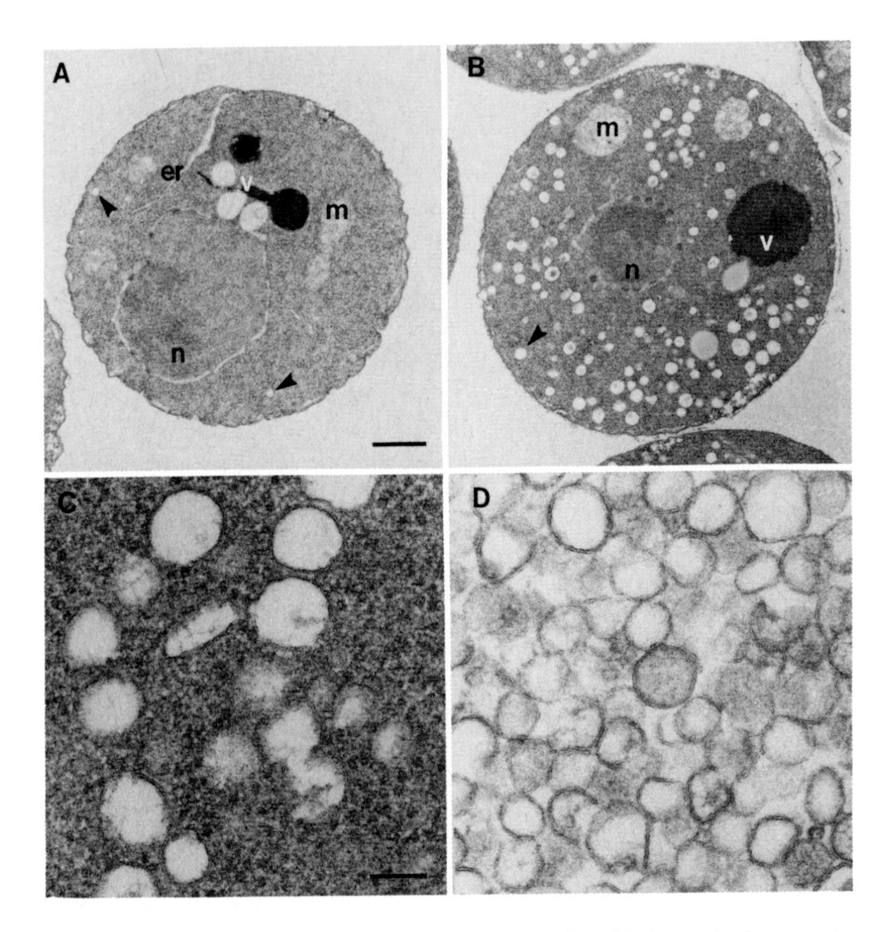

FIG. 1. Electron micrographs of intact cells and of purified constitutive secretory vesicles. (A–C) Wild-type (NY13) and *sec 6-4* (NY17) cells were shifted to 37°C in YP + 0.2% glucose for 2 hours, then fixed for electron microscopy. (A) Wild-type cells. Arrows point to secretory vesicles. **er**, Endoplasmic reticulum; **v**, vacuole; **m**, mitochondria; **n**, nucleus. Bar = 500 nm. (B) *sec 6-4* cells (NY17) at the same magnification as in (A). (C) Portion of a *sec 6-4* cell in (B), at higher magnification. Bar = 100 nm. (D) Vesicles were pooled following purification on the Sephacryl column, then fixed for electron microscopy. Representative micrograph of the vesicle pool from the Sephacryl column shown at the same magnification as the micrograph in (C). Reproduced from *The Journal of Cell Biology*, 1987, vol. 105, pp. 163–174, by copyright permission of the Rockefeller University Press.

II. Preparation of Lysate

We have used two cell fractionation schemes for the study of particular components of the secretory pathway. The first method we will describe gives sufficient separation of organelles (Ruohola and Ferro-Novick,

1987) such that each can be at least partially resolved from the others and a particular protein can be localized to a particular organelle of the secretory pathway (Goud *et al.*, 1988). The second procedure that we will discuss has been specifically designed to give a good yield of highly purified intact secretory vesicles (Walworth and Novick, 1987). The two procedures are very similar in terms of strategy, use of a reversible secretory mutant, growth conditions, and cell lysis. They differ primarily in the method by which the final separation of the membrane fractions is achieved. These fractionation protocols are shown schematically in Fig. 2. Please note, as will be described later, that the 450 *g* centrifugation is not routinely done in the vesicle purification protocol.

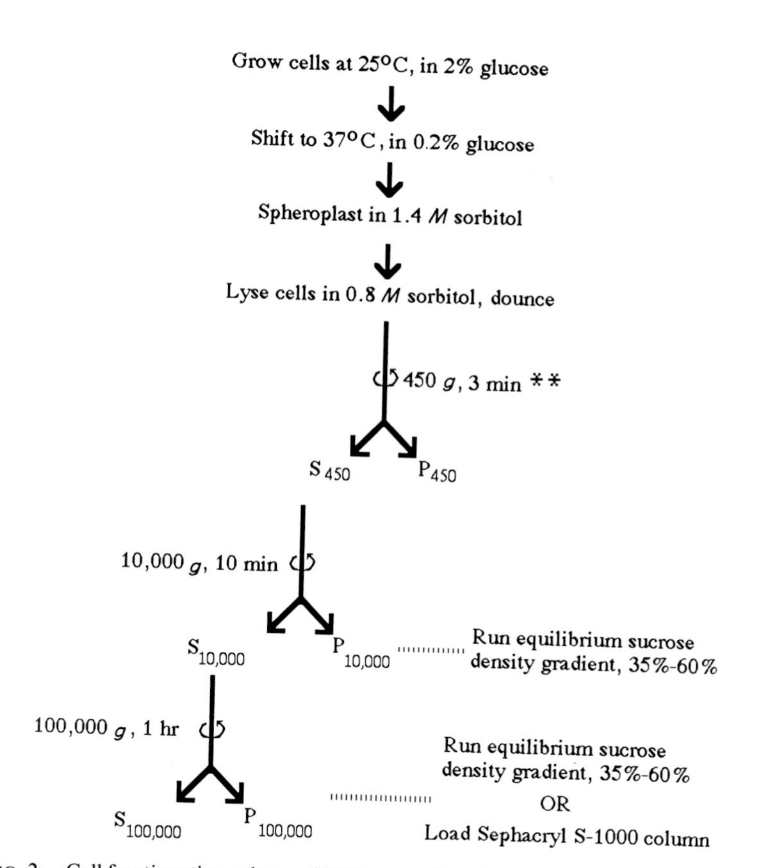

FIG. 2. Cell fractionation scheme. **450 *g* centrifugation is only necessary for organelle separation protocol using sucrose gradients. It is not necessary for vesicle purification protocol.

A. Cell Growth

For most of these studies we chose to use *sec 6-4* (NY17), a mutant that accumulates secretory vesicles in a reversible fashion (Fig. 1B,C). This strain is a conditional secretory mutant; that is, at the permissive temperature of 25°C the cells grow and secrete at the same rate as wild-type cells. However, when the strain is shifted to the restrictive temperature of 37°C the cells are blocked in the transport of secretory and plasma membrane proteins to the cell surface and they accumulate secretory vesicles (Novick *et al.*, 1980). This phenotype, however, is reversible. If cells that had been grown at 37°C are shifted back to 25°C in the absence of protein synthesis, the cells are able to transport at least 50% of those vesicles to the cell surface and release the vesicle content. This result implies that the accumulated vesicles are functional intermediates of the secretory pathway. Importantly, the vesicles that accumulate in these cells are homogeneous in size. Therefore, a purification scheme that separates particles on the basis of size should be effective in purifying these vesicles away from other organelles.

For any organelle purification scheme a marker is needed in order to follow the fate of the organelle throughout the procedure. Because the strain we use accumulates secretory products in the accumulated vesicles, we can use any one of several secreted enzyme markers to follow the vesicles throughout purification. We chose to use the secreted enzyme invertase for several reasons. The secretory form of invertase is under hexose repression such that when the level of the carbon source glucose in the medium is lowered, the expression of invertase is stimulated. This increase in expression of the protein will occur in wild-type cells in response to glucose starvation without affecting cell growth and does not require the addition of any exogenous agent to stimulate its production. Therefore, cells are first grown with 2% glucose as the carbon source and then shifted to medium containing 0.2% glucose to increase the production of invertase. Invertase can be easily assayed in a colorimetric assay by measuring the production of glucose using added sucrose as the substrate (Goldstein and Lampen, 1975). Because it is a luminal enzyme, invertase can serve as a marker for the integrity of the vesicle membrane.

B. Spheroplast Formation

The presence of the yeast cell wall protects the cells from osmotic lysis and provides them with a defined immediate extracellular space. The cell wall consists of glucans, a polymer of glucose, and mannans, a collection of glycoproteins containing a high percentage of mannose (Ballou, 1982). It can be removed by enzymatic digestion of the glucan by a crude glucanase such as the commercially available Zymolyase 100T (available from ICN Immuno-Biologicals, Lisle, IL). Upon removal of the cell wall the cells are osmotically sensitive and, therefore, during incubation with the enzyme, osmotic support in the form of sorbitol is required to retain the integrity of the plasma membrane. The buffer in which spheroplasts are generated contains 1.4 *M* sorbitol, 50 m*M* KP_i (pH 7.5), 10 m*M* NaN_3, 40 m*M* β-mercaptoethanol, and Zymolyase (amounts specified later). The Zymolyase solution is prepared in advance, allowed to stand at room temperature for 30 minutes, and then centrifuged for 15 minutes at top speed in a tabletop centrifuge before use to remove particulates. Spheroplasts are formed by incubating at 37°C for 45 minutes with occasional gentle mixing.

C. Lysis

In order to isolate intact vesicles we used the gentlest method possible to lyse the plasma membrane. Osmotic lysis of spheroplasts can be achieved in 0.8 *M* sorbitol if the buffer is triethanolamine (Makarow, 1985). The high osmolarity of this lysis buffer helps to stabilize intracellular organelles such as the vacuole (Makarow, 1985) and secretory vesicles (Walworth and Novick, 1987). As described in more detail in the sections that follow, the spheroplasts are suspended in buffer containing 0.8 *M* sorbitol, 10 m*M* triethanolamine (pH 7.2 titrated with acetic acid), and 1 m*M* EDTA (0.8 *M* sorbitol–TEA). Presumably the reduction in osmotic support from 1.4 *M* sorbitol in the spheroplasting medium to 0.8 *M* is sufficient to lyse the cells; however, because cell clumps are common, the lysate is homogenized to promote efficient lysis. Yeast cells contain a significant amount of proteases, which must be kept inactive. For the organelle separation technique all fractionation steps are performed at 4°C or on ice, and we include a protease inhibitor cocktail, which will be defined later. For the purification of vesicles we do not use protease inhibitors, but we are very careful to perform every step of the fractionation at 4°C or on ice, and all containers, including the homogenizer for example, are prechilled.

III. Differential Centrifugation

A. Organelle Marker Assays

Because distinct biochemical functions are associated with particular organelles, enzyme assays can be used to follow the fate of an organelle through a fractionation procedure. The assays we have used are described in detail elsewhere, and we will list them here. The ER is marked by NADPH–cytochrome *c* reductase, which can be assayed colorimetrically by following the reduction of cytochrome *c* using a spectrophotometer equipped with a chart recorder (Kubota *et al.*, 1977; Kreibich *et al.*, 1973). Cytochrome-*c* oxidase marks the inner membrane of the mitochondria and is assayed in a similar fashion to the reductase assay (Mason *et al.*, 1973). The yeast vacuole contains the enzyme α-mannosidase, which is assayed in a colorimetric assay measuring the breakdown of *p*-nitrophenyl-α-mannopyranoside (Tulsiani *et al.*, 1977; Opheim, 1978). The vanadate-sensitive Mg^{2+}-ATPase is an integral membrane protein of the plasma membrane and can be assayed by measuring the production of inorganic phosphate (Bowman and Slayman, 1979; Willsky, 1979). Invertase serves as the marker for the secretory vesicles and is assayed by measuring the production of glucose from added sucrose (Goldstein and Lampen, 1975). Recently an assay has been established in the laboratory of Phillips Robbins for measuring Ca^{2+}-dependent GDPase, which is thought to mark the Golgi complex of yeast (P. Robbins, personal communication). Protein was assayed using fluorescamine as described by Udenfriend *et al.* (1972). For all assays performed in our laboratory we have obtained reagents from Sigma Chemical Company (St. Louis, MO), with the exception of glucose oxidase for the invertase assay, which was obtained from Boehringer Mannheim Biochemicals (Indianapolis, IN).

B. Organelle Distribution

Differential centrifugation is used for the first steps in cell fractionation because it rapidly separates large membrane organelles from smaller ones. The distributions of organelle markers following differential centrifugation are shown in Table I as the percentage of activity in the total lysate present in each fraction. Under these conditions, most of the mitochondria (65–70%) and most of the ER (55–61%) that remain following the 450 *g* spin are pelleted in the 10,000 *g* spin. About one-fourth of the vanadate-sensitive ATPase is found in the 10,000 *g* pellet, and very little is found in the 100,000 *g* pellet. Although ~50% of the plasma membrane marker appears to be soluble, we feel that this probably reflects nonspecific soluble ATPases or phosphatases that interfere with the assay because the plasma membrane ATPase is an integral membrane protein that crosses the lipid bilayer several times. When *sec 6-4* cells are fractionated in this manner, 60% of the invertase remains in the supernatant after the 10,000 *g* spin. A more forceful centrifugation of 100,000 *g* for 1 hour pellets 30% of the total invertase. Thus, a simple two-step centrifugation effectively serves to enrich the secretory vesicle popula-

TABLE I

DISTRIBUTION OF MARKER ENZYMES IN CELL FRACTIONATION OF WILD-TYPE (WT) AND *sec 6-4* CELLS[a]

	Percentage of total lysate									
	Invertase		PM ATPase		NADPH–cyt *c* red		Cyt-*c* oxidase		α-Mannosidase	
Fraction	WT	*sec 6-4*	WT	*sec 6-4*	WT	*sec 6-4*	WT	*sec 6-4*	WT	*sec 6-4*
P_{450}	ND[b]	15 ± 1	15 ± 5	16 ± 3	30 ± 6	28 ± 4	30 ± 6	25 ± 3	6 ± 4	8 ± 2
$P_{10,000}$	ND	10 ± 2	24 ± 5	26 ± 3	61 ± 4	55 ± 6	70 ± 2	65 ± 6	25 ± 4	21 ± 6
$S_{100,000}$	ND	30 ± 6	51 ± 6	49 ± 3	4 ± 2	6 ± 3	4 ± 2	6 ± 3	40 ± 3	35 ± 3
$P_{100,000}$	ND	30 ± 3	1 ± 1	4 ± 2	4 ± 3	7 ± 2	3 ± 1	4 ± 2	15 ± 5	17 ± 4

[a] Values are expressed for each fraction as percentage of activity in the total lysate. These values represent the mean of three experiments.

[b] ND, Not determined.

tion in the 100,000 *g* pellet because other organelles are largely pelleted in the 10,000 *g* spin.

IV. Subfractionation of Membrane Pellets

A. Sucrose Gradient Centrifugation

A sucrose gradient separates organelles on the basis of their equilibrium density. In our hands, this technique has proved to be very useful for sufficiently separating organelles from each other such that the localization of a particular protein to a particular compartment can be determined with confidence. Cells are grown at 25°C to midlogarithmic phase (2×10^7 cells/ml) in rich medium (YP, 1% Bacto yeast extract, 2% Bacto peptone from DIFCO Laboratories, Detroit MI) plus 2% glucose as the carbon source. The cells are then harvested by centrifugation at top speed in a tabletop centrifuge (~1500 *g*) for 5 minutes and resuspended in YP containing 0.2% glucose prewarmed to 37°C. The low glucose concentration derepresses the synthesis of invertase and at the restrictive temperature of 37°C *sec 6-4* cells accumulate secretory vesicles containing invertase. For this procedure we start with 1×10^9 cells.

To prepare lysates the cells are converted to spheroplasts in spheroplast buffer (described earlier) in a volume of 6 ml with 1.5 mg of Zymolyase 100T. The spheroplasts are lysed in 1.5 ml of 0.8 *M* sorbitol–TEA lysis buffer that also contains 10 m*M* NaN_3, and homogenized with 20 strokes of a 2-ml tissue grinder (Wheaton Scientific, Milville, NJ). The lysis buffer also contains, as protease inhibitors, phenylmethylsulfonyl fluoride (PMSF), leupeptin, chymostatin, pepstatin, and antipain, all of which are present at 1 m*M* (obtained from Sigma). To remove unlysed cells and large debris from the homogenate, the cells are centrifuged at 450 *g* (1700 rpm) for 3 minutes in a Sorvall RT6000 centrifuge at 4°C. The resulting pellet P_{450} is washed once in lysis buffer. The S_{450} supernatants are pooled in Beckman Ultra-Clear 13 × 51 mm tubes and spun at 10,000 *g* (12,000 rpm) for 10 minutes in a Beckman SW50.1 rotor, yielding $P_{10,000}$. The resulting supernatant is then transferred to fresh tubes of the same type and spun at 100,000 *g* (33,000 rpm) in the same rotor to pellet the remaining membranes. The $P_{10,000}$ and $P_{100,000}$ pellets are resuspended in 60% sucrose (Sigma, grade I) and homogenized with six strokes of the tissue grinder. The suspended membrane fraction is placed in the bottom of a Beckman Ultra-Clear 14 × 89-mm tube and overlaid with sucrose solutions. For the $P_{10,000}$ gradients the pellet is overlaid with 1 ml 55% sucrose, 1.5 ml each of 50%, 45%, 42.5%, 40%, 37.5% sucrose, and 1 ml of 35% sucrose. For the $P_{100,000}$ gradients the fractions of interest are more dense and greater separation is needed in the denser portion of the gradient. Therefore, the resuspended $P_{100,000}$ is overlaid with 1.5 ml each of 55%, 52.5%, 50%, 45%, 40% sucrose and 2 ml of 35% sucrose. The gradients are centrifuged in a Beckman SW41 rotor at 170,000 *g* (36,000 rpm) for 14 hours at 4°C.

Fractions from the gradients are collected from the top in 550-μl aliquots using an Autodensi-Flow apparatus (Buchler Instruments Division, Nuclear-Chicago, Fort Lee, NJ) and assayed for marker enzyme activities as described earlier. The residual pellet is resuspended in the same volume of 60% sucrose. The profiles for various marker enzyme activities from fractionation of the $P_{10,000}$ from wild-type cells (NY13) are shown in Fig. 3. The peaks of activity for the mitochondrial marker (fraction 10), the vacuole marker (fraction 19), and the plasma membrane marker (fraction 12) are distinct. The ER marker elutes as two peaks, one at fraction 10 and one at fraction 17. Thus, it is possible to localize a protein to a particular organelle by determining the peak marker activity with which the protein cofractionates. Importantly, the distribution of these markers is identical whether this fractionation is performed on wild-type cells or on *sec 6-4* cells. We have used this technique to demonstrate that the product of the *SEC4* gene (Sec4p) is associated with the plasma membrane in $P_{10,000}$ (Fig. 3 and Goud *et al.*, 1988). The vesicle marker invertase is present in very low amounts in the 10,000 *g* pellet from either wild-type or *sec 6-4* cells. Fractionation of the 100,000 *g* pellet from *sec 6-4* cells on the sucrose gradient results in a single major peak of invertase activity, indicating that the vesicles are homogeneous in density as well as in size. Vesicles from wild-type cells are in very low abundance, and membrane-associated invertase activity is not detectable above background. However, we have shown that Sec4p, which has been localized to the cytoplasmic surface of secretory vesicles using the Sephacryl purification scheme described in the next section, fractionates in the same position on a sucrose gradient of the 100,000 *g* pellet from wild-type cells as it does on a gradient of the same fraction from *sec 6-4* cells (Goud *et al.*, 1988). This observation suggests that the few vesicles present in wild-type cells are of the same density and fractionate in the same position as the more abundant vesicles from *sec 6-4* cells. This method of fractionation provides sufficient resolution on a small scale to distinguish the organelle on which a protein might reside.

B. Sephacryl S-1000 Column

To examine secretory vesicles in more detail we have designed a protocol for their isolation that produces a more pure population and higher yield then can be obtained from the sucrose gradient. The technique we describe for the purification of secretory vesicles was designed based on the observation by thin-section electron microscopy that the post-Golgi vesicles accumulated in *sec 6-4* cells are homogeneous in size. As we have demonstrated, a gel filtration column that separates particles on the basis of size effectively separates secretory vesicles from other organelles present in the 100,000 *g* pellet. In order to isolate a significant amount of purified material, we start with 1.6×10^{10} cells. Because this protocol was designed to purify only the secretory vesicles, the first spin of the differential centrifugation is done at 10,000 *g* and results in removing unlysed whole cells as well as the larger organelles. The high-speed pellet is resuspended in a small volume of 0.8 *M* sorbitol–TEA and loaded onto a Sephacryl S-1000 column. Sephacryl was obtained from Pharmacia Fine Chemicals (Piscataway, NJ), and the 1.5 × 90 cm column was poured at room temperature and then moved to 4°C. More details about this procedure follow.

Cells are grown at 25°C to midlogarithmic phase (2×10^7 cells/ml) in 800 ml of YP plus 2% glucose as the carbon source. The cells are then harvested by centrifugation at top speed in a tabletop centrifuge (~1500 *g*) for 5 minutes and resuspended in 800 ml of YP containing 0.2% glucose

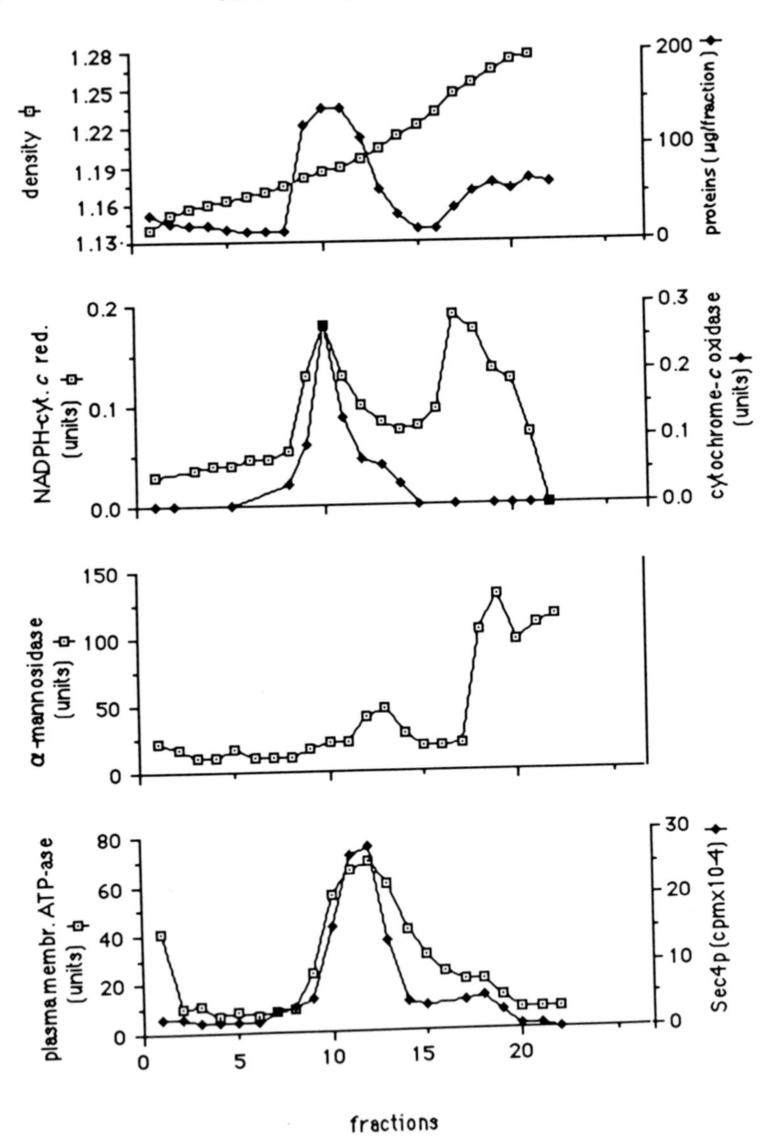

FIG. 3. Elution profile of organelle marker enzymes from sucrose gradient fractionation of $P_{10,000}$ from wild-type cells (NY13). The pellet from the 10,000 *g* spin was generated by differential centrifugation and loaded in 60% sucrose (2 ml) on the bottom of a 35%–60% sucrose gradient. After centrifugation to equilibrium, the gradient was fractionated from the top (fraction 1) in 21 fractions (0.55 ml) and the residual pellet was resuspended in 0.55 ml of 60% sucrose (fraction 22). In each fraction, various marker enzymes were assayed (see text) and the amount of Sec4p (a protein located on the cytoplasmic surface of secretory vesicles

prewarmed to 37°C. The low glucose concentration derepresses the synthesis of invertase and at the restrictive temperature of 37°C the cells accumulate secretory vesicles containing invertase.

To prepare cell lysates the cells are harvested after 2 hours of growth at 37°C by centrifugation. The cells are washed once by resuspending them in 30 ml of ice-cold 10 m*M* NaN_3, transferring them to a 35-ml tube, and spinning at 5000 rpm in a Sorvall SS34 rotor at 4°C for 4 minutes. This washed cell pellet is then resuspended in 15 ml of 10 m*M* NaN_3, and 15 ml of 2× spheroplast buffer is added. The 2× spheroplast buffer contains 2.8 *M* sorbitol, 0.1 *M* KP_i (pH 7.5), 10 m*M* NaN_3, 5 mg Zymolyase 100T, and 58 μl of β-mercaptoethanol and is prepared as described earlier. Cells are incubated at 37°C for 45 minutes and mixed by gently inverting the tube.

The spheroplasts are pelleted by spinning at top speed in a clinical centrifuge (~750 *g*) for 5 minutes. Spheroplasts are then resuspended by pipeting with a 25-ml glass pipet in 25 ml of ice-cold lysis buffer, 0.8 *M* sorbitol–TEA, which is made by combining a 10× stock buffer of TEA (pH 7.2) with 4 *M* sorbitol and water. The lysate is then transferred to a 40-ml Wheaton glass homogenizer and dounced 20 times with the A pestle keeping the homogenizer buried in ice. This material (TL, total lysate) is then transferred to a 35-ml tube and centrifuged in a Sorvall SS34 rotor for 10 minutes at 10,000 *g* (11,000 rpm). The resulting supernatant, $S_{10,000}$, is poured into a Beckman 70Ti tube, mixed by inverting and a portion of $S_{10,000}$ saved for enzyme assays. The pellet, $P_{10,000}$, from this first spin is resuspended in 23 ml of 0.8 *M* sorbitol–TEA and saved for enzyme assays. The $S_{10,000}$ is centrifuged in a Beckman 70Ti rotor at 100,000 *g* (34,000 rpm) for 60 minutes at 4°C. The high-speed supernatant ($S_{100,000}$) is taken off with a Pasteur pipet. A film of lipid is found at the top of the liquid and is included with $S_{100,000}$, and a loose viscous layer is usually seen on top of the pellet. This material contains little invertase activity and therefore is allowed to run down the side of the tube and is also collected with $S_{100,000}$. The $P_{100,000}$ is resuspended in 600 μl of 0.8 *M* sorbitol–TEA with repeated pipet using a Pipetman P-1000 (Rainin Instrument Co., Woburn, MA). The suspension is spun for 30 seconds in a microcentrifuge at 4°C to remove any aggregates and is loaded onto a 1.5

---

and the plasma membrane of yeast; see Goud *et al.*, 1988) was determined by quantitative Western blot analysis. Results are expressed as the total activity per fraction. Density (g/ml); proteins (μg); NADPH–cytochrome *c* reductase (μmol cytochrome *c* reduced per minute); cytochrome-*c* oxidase (μmol cytochrome *c* oxidized per minute); α-mannosidase (nmol of *p*-nitrophenyl-α-D-mannopyranoside hydrolyzed per hour); vanadate-sensitive Mg^{2+}-ATPase (nmol of phosphate produced per minute); Sec4p (total cpm). Reproduced from *Cell,* 1988, vol. 53, pp. 753–768, by copyright permission of the MIT press.

× 90-cm Sephacryl S-1000 gel filtration column, which is kept in a 4°C cold room. Material is eluted from the column over a period of ~15 hours in 0.8 *M* sorbitol–TEA at a flow rate of 9.2 ml/hour and collected as 80-drop (4-ml) fractions.

By performing the organelle enzyme marker assays described, the distribution of the various organelles across the column elution profile can be determined. Figure 4 shows the elution profiles for the various enzyme markers present in $P_{100,000}$. The elution profile for the vesicle marker invertase demonstrates that the vesicles elute as a homogeneous peak. Invertase activity eluting later from the column (around fraction 30) can be measured in the absence of Triton X-100, indicating that the invertase found there is not contained within sealed vesicles. We should point out that when a 100,000 *g* pellet from wild-type cells is fractionated on the column, the small amount of invertase that can be measured elutes about three fractions later than the peak when *sec 6-4* is fractionated. This result is consistent with the observation that vesicles in *sec 6-4* are slightly larger (100 nm) than vesicles in wild-type cells (80 nm). In addition to the secretory vesicles in the 100,000 *g* pellet that is loaded onto the column, small amounts of the total vacuolar α-mannosidase activity (7.3%) and of the plasma membrane ATPase activity (15%) are found. The α-mannosidase found in the 100,000 *g* pellet elutes away from the invertase in two peaks, one before and one after the vesicle peak. The elution profile for the vanadate-sensitive ATPase demonstrates that a peak of activity elutes prior to the peak of invertase activity and an additional peak coelutes with invertase activity. When wild-type cells are fractionated in the same way, a very small amount of invertase can be detected in the vesicle region on the column. However, as the elution profile for $P_{100,000}$ from wild-type cells shows (Fig. 4C), the first peak of ATPase is present, but no plasma membrane ATPase can be detected in the vesicle region, presumably because it is present in such low amounts. We postulated that the ATPase present in the fraction that coelutes with invertase represents ATPase destined for the plasma membrane that is present on the invertase-containing secretory vesicles (Walworth and Novick, 1987). Using a different vesicle isolation procedure, Holcomb *et al.* (1988) have demonstrated that another secretory enzyme, acid phosphatase, is present in the same transport vesicles as the plasma membrane ATPase.

In summary, the Sephacryl column effectively separates secretory vesicles from contaminating organelles present in the high-speed membrane fraction. To study this population of secretory vesicles, we have pooled the fractions containing the highest invertase-specific activity and studied the vesicles biochemically and morphologically.

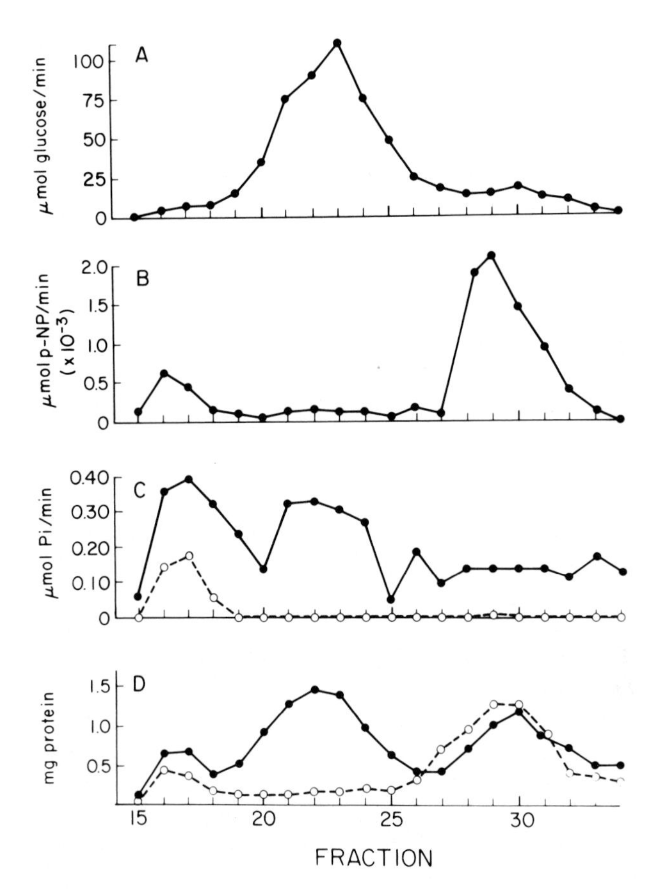

FIG. 4. Distribution of organelle marker enzyme activities across column fractions. The high-speed pellet from a mutant (*sec 6-4*) cell fractionation was applied to a Sephacryl S-1000 column. Aliquots of each column fraction were assayed for (A) invertase, (B) α-mannosidase, (C) plasma membrane ATPase, and (D) protein. In a separate experiment, $P_{100,000}$ from a wild-type fractionation was applied to the column, and aliquots were assayed for plasma membrane ATPase and protein. (A) Elution profile for invertase, the vesicle marker, from a *sec 6-4* preparation. From each column fraction, 0.5 μl was assayed for 3 minutes at 37°C. (B) Elution profile for the vacuole marker, α-mannosidase. From each fraction, 100 μl were assayed for 1 hour at 37°C. (C) Elution profile for plasma membrane ATPase from both *sec 6-4* (●) and wild-type (○) membranes. The production of inorganic phosphate (P_i) by the vanadate-sensitive plasma membrane ATPase was assayed for 10 minutes at 37°C using 50 μl of each fraction. (D) Protein elution profile for both *sec 6-4* (●) and wild-type (○) membranes. Protein was assayed from 20 μl of each fraction with fluorescamine in 0.1 *M* sodium borate (pH 9.2), with 2% SDS using BSA as standard. Reproduced from *The Journal of Cell Biology,* 1987, vol. 105, pp. 163–174, by copyright permission of the Rockefeller University Press.

V. Criteria for Purity

A. Biochemical Criteria

In order to demonstrate that purification of an organelle has been achieved, both biochemical and morphological criteria can be used. During the purification scheme the enzyme marker specific for the organelle in question should be enriched at each step. That is, the specific activity of that marker should increase such that compared to other proteins the relative abundance of the marker increases. At the same time the abundance of other organelles in the fraction should decrease. As discussed earlier, the availability of distinct biochemical activities associated with each organelle make it possible to establish the enrichment of the organelle of interest at each stage of the fractionation procedure. For example, in the vesicle purification scheme we have shown that the vesicle marker invertase is enriched 16-fold in the 100,000 *g* pellet, and in the vesicle pool from the Sephacryl S-1000 column, invertase is enriched 36-fold over the specific activity of invertase in the total lysate. In addition, the reduction of activities marking other organelles from the secretory-vesicle pool demonstrates that the pool is relatively free of contaminating organelles.

An additional biochemical technique for demonstrating that unwanted proteins have been eliminated from particular fractions is to show that a unique set of proteins increases in relative abundance throughout the purification scheme. That is, as increasingly pure fractions are isolated, proteins present in the initial total lysate should be eliminated, leaving behind those that are associated with the organelle of interest. Using a technique such as gel filtration chromatography, it is possible to determine if a particular set of proteins comigrates with the marker enzyme activity. In our procedure a set of polypeptides is enriched in the high-speed pellet fraction compared to the total lysate or $S_{10,000}$. Furthermore, as shown in Fig. 5, a subset of the proteins present in $P_{100,000}$ (lane P2) coelutes exactly with the peak (fractions 22–27) of the vesicle marker invertase when individual column fractions are assayed and aliquots of each are separated on a polyacrylamide gel. These results suggest that these particular proteins are associated in some fashion with the secretory vesicles.

In wild-type cells, vesicles are present in low abundance and, therefore, vesicle associated proteins should also be present in low abundance. By fractionating $P_{100,000}$ from these cells on the column, we can identify proteins present in the fractions where vesicles would elute, which are probably contaminants rather than vesicle-associated proteins. The major

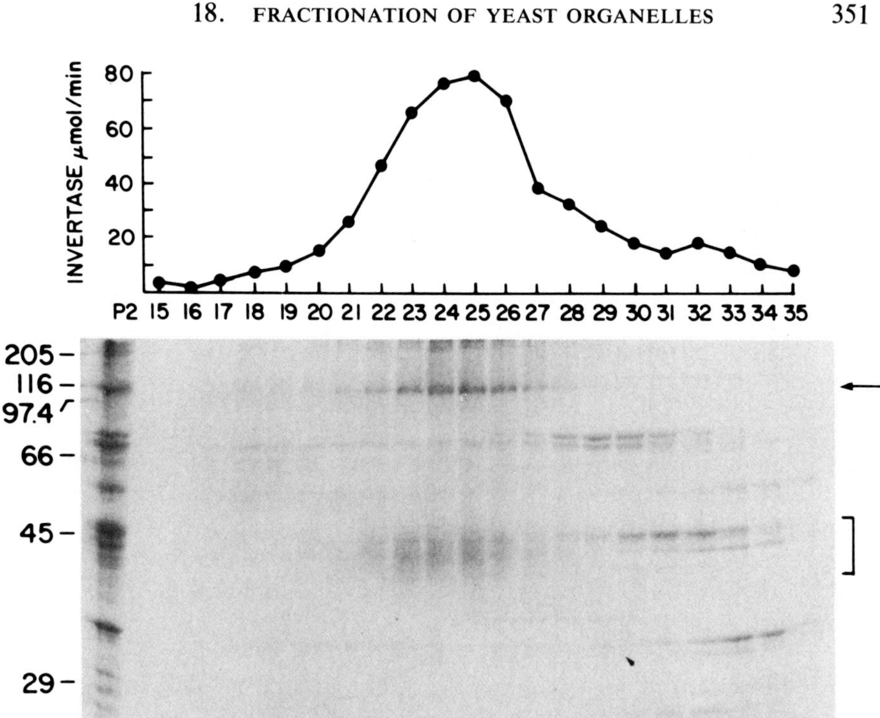

FIG. 5. Coomassie-stained protein pattern for column fractions of a *sec 6-4* cell fractionation. Aliquots (140 μl) of each fraction from the Sephacryl S-1000 column were diluted with sample buffer (60 μl) and heated at 100°C for 5 minutes. P2 is an aliquot of the high-speed (100,000 *g*) pellet, which was loaded onto the column and run on the gel at approximately the same dilution as the column fractions. On an SDS–polyacrylamide (12%) gel, 75 μl of each sample were analyzed. The Coomassie blue-stained gel is shown. Above the *sec 6-4* column protein profile, the elution profile for invertase activity is included. Proteins designated as 110, 40–45, and 18 kDa, which cofractionate with invertase are marked with arrows and bracket on the right side of the figure. Reproduced from *The Journal of Cell Biology,* 1987, vol. 105, pp. 163–174, by copyright permission of the Rockefeller University Press.

proteins that cofractionate with invertase when $P_{100,000}$ from the *sec 6-4* strain is fractionated on the Sephacryl column are conspicuously absent from the corresponding fractions when $P_{100,000}$ from wild-type cells is fractionated. The proteins that remain are thought to be contaminants. We estimate that these putative contaminants represent ~20% of the protein in the pooled-vesicle preparation from *sec 6-4* cells (Walworth and Novick, 1987).

B. Morphological Criteria

Electron microscopy of the fraction thought to contain the organelle of interest allows one to determine if the organelle is indeed present and free of contaminating organelles. As can be seen in Fig. 1D, thin-section analysis of the vesicle pool demonstrates that the fraction contains a homogeneous population of mostly closed vesicles that are 95–120 nm in diameter. Very few contaminants are seen. At low power when a wide field is viewed, a few electron-dense, membrane-enclosed particles are seen; their presence in the vesicle pool is probably due to their similar size. Although we do not know the nature of this contaminant, it represents a very minor portion of the total population. Additionally the vesicles resemble those seen inside the cells in Fig. 1A, B, and C, when thin-section analysis of whole cells is performed. The vesicles as measured in *sec 6-4* cells are slightly smaller in diameter (~100 nm) compared to those measured in the pool. This difference could be due to some osmotic swelling of the vesicles once the cells are lysed.

VI. Summary

In summary, organelles of the secretory pathway can be effectively separated from one another using differential centrifugation followed by sucrose density gradient fractionation of wild-type or vesicle-accumulating mutant yeast cells. Up to 10-fold enrichment of the plasma membrane fraction is obtained, and resolution of the peak fractions of several organelles allows one to localize specific proteins to particular components of the pathway. Additionally, a highly purified population of constitutive secretory vesicles can be isolated from the 100,000 *g* membrane fraction of *sec 6-4* cells on a Sephacryl S-1000 column. The success of this procedure is due to the homogeneous size of the vesicles and the high concentration of vesicles accumulated in the *sec 6-4* cells. From other laboratories, methods have been described for the isolation of other organelles including the vacuole (Wiemken, 1975), plasma membrane (Tschopp and Schekman, 1983), and nuclei (Mann and Mecke, 1980), as well as an alternative procedure for the purification of secretory vesicles from yeast (Holcomb *et al.*, 1987).

For the localization of proteins to particular organelles the ability to lyse cells osmotically is an important improvement over the glass bead lysis procedure. The shear forces generated during glass bead lysis could potentially remove proteins from the surface of organelles that otherwise would be membrane-attached, causing them to appear soluble. Similarly, because the conditions required for stabilizing the association of a protein with a membrane can be quite variable depending on the lysis buffer, confirmation of localization using alternative schemes is prudent. With the advent of such techniques as confocal immunofluorescent microscopy and immunoelectron microscopy, effective methods for confirming localizations are becoming available.

Acknowledgments

This work was supported by grant GM35370 to P.N. from the National Institutes of Health. B.G. was supported by a grant from the National Institutes of Health–French CNRS program no. 9 and by a Swebilius Cancer Research Award. N.C.W. was supported by training grant GM07223.

References

Ballou, C. E. (1982). *In* "The Molecular Biology of the Yeast *Saccharomyces*: Metabolism and Gene Expression" (J. N. Strathern, E. W. Jones, and J. R. Broach, eds.), pp. 335–360. Cold Spring Harbor Lab., Cold Spring Harbor, New York.

Botstein, D., and Fink, G. R. (1988). *Science* **240,** 1439–1443.

Bowman, B. J., and Slayman, C. W. (1979). *J. Biol. Chem.* **254,** 2928–2934.

Cameron, R. S., Cameron, P. L., and Castle, J. D. (1986). *J. Cell Biol.* **103,** 1299–1313.

Goldstein, A., and Lampen, J. O. (1975). *In* "Methods in Enzymology" (W. A. Wood, ed.), Vol. 42, pp. 504–511. Academic Press, New York.

Goud, B., Salminen, A., Walworth, N. C., and Novick, P. J. (1988). *Cell (Cambridge, Mass.)* **53,** 753–768.

Holcomb, C. L., Etcheverry, T., and Schekman, R. (1987). *Anal. Biochem.* **166,** 328–334.

Holcomb, C. L., Hansen, W. J., Etcheverry, T., and Schekman, R. (1988). *J. Cell Biol.* **106,** 641–648.

Kelly, R. B. (1985). *Science* **230,** 25–32.

Kreibich, G., Debey, P., and Sabatini, D. D. (1973). *J. Cell Biol.* **58,** 436–462.

Kubota, S., Yoshida, Y., Kumakoa, H., and Furumichi, A. (1977). *J. Biochem. (Tokyo)* **81,** 197–205.

Makarow, M. (1985). *EMBO J.* **4,** 1855–1860.

Mann, K., and Mecke, D. (1980). *FEBS Lett.* **122,** 95–99.

Mason, T. L., Poyton, R. O., Wharton, D. C., and Schatz, G. (1973). *J. Biol. Chem.* **248,** 1346–1354.

Nagy, A., Baker, R. R., Morris, S. J., and Whittaker, V. P. (1976). *Brain Res.* **109,** 285–309.
Newman, A. P., and Ferro-Novick, S. (1987). *J. Cell Biol.* **105,** 1587–1594.
Novick, P., and Schekman, R. (1979). *Proc. Natl. Acad. Sci. U.S.A.* **76,** 1858–1862.
Novick, P., Field, C., and Schekman, R. (1980). *Cell (Cambridge, Mass.)* **21,** 205–215.
Novick, P., Ferro, S., and Schekman, R. (1981). *Cell (Cambridge, Mass.)* **25,** 461–469.
Opheim, D. J. (1978). *Biochim. Biophys. Acta* **524,** 121–130.
Palade, G. (1975). *Science* **189,** 347–358.
Pfeffer, S. R., and Rothman, J. E. (1987). *Annu. Rev. Biochem.* **56,** 829–852.
Ruohola, H., and Ferro-Novick, S. (1987). *Proc. Natl. Acad. Sci. U.S.A.* **84,** 8468–8472.
Struhl, K. (1983). *Nature (London)* **305,** 391–397.
Tschopp, J., and Schekman, R. (1983). *J. Bacteriol.* **156,** 222–229.
Tulsiani, D. R. P., Opheim, D. J., and Touster, O. (1977). *J. Biol. Chem.* **252,** 3227–3233.
Udenfriend, S., Stein, S., Bohlen, P., Dairman, W., Leimgruber, W., and Weigele, M. (1972). *Science* **178,** 871–872.
von Tuinen, E., and Riezman, H. (1987). *J. Histochem. Cytochem.* **35,** 327–333.
Walworth, N. C., and Novick, P. J. (1987). *J. Cell Biol.* **105,** 163–174.
Wiemken, A. (1975). *In* "Methods in Cell Biology" (D. M. Prescott, ed.), Vol. 12, pp. 99–109. Academic Press, New York.
Willsky, G. R. (1979). *J. Biol. Chem.* **254,** 3326–3332.

Part IV. Morphological Procedures

Electron-microscopic immunocytochemical procedures appear to be reaching their maturity. Nevertheless, as Wright and Rine point out (Chap. 23, this volume), no one procedure is suitable for all antigen–antibody pairs. Despite the anatomic precision of methods using colloidal gold, the stoichiometry of antigen detection is certainly not quantitative. "Preembedding" procedures with peroxidase amplification, though qualitative, appear to be the most sensitive. In addition to the procedures described in the following chapters, frozen sucrose embedding procedures (Griffiths *et al.,* 1983; Keller *et al.,* 1984; see also the references cited by Bergmann, Part I, "Methods in Cell Biology," Vol. 32) and procedures based on the LR hydrophilic resins of Polysciences (Newman *et al.,* 1983) have had striking success. Protocols for "preembedding" immunocytochemistry (and immunofluorescence) are also described in Chapter 10 by Kuismanen and Saraste in Volume 32 of this series.

Other important electron-microscopic procedures are those concerned with the representative sampling of isolated subcellular fractions either for anatomic study or immunocytochemical work (Baudhuin *et al.,* 1967; Tartakoff and Jamieson, 1974; DeCamilli *et al.,* 1983; Palade *et al.,* 1983), procedures for rotary replication of etched samples (Heuser, 1981), and three-dimensional imaging (Turner, 1981).

What is equally impressive is the increased sophistication of light and fluorescence microscopy, as exemplified in Volumes 29 and 30 of this series, which describe fluorescent procedures. (See Contents of Recent Volumes for their Tables of Contents.) An outstanding example of the use of fluorescence microscopy is for intracellular pH determination, an issue corresponding to Chapter 22 on electron-microscopic immunocytochemistry by Anderson, and related to the chapters of Wilson and Murphy (Chap. 16, this volume) and Schindler *et al.* (Chap. 18, Vol. 32). Chapter 20 by Bacallao and Stelzer on confocal fluorescence microscopy illustrates a particularly dramatic technical advance that is still under development. Chapter 19 by Pringle *et al.* contains a wealth of practical advice of importance for the study of any cell type—not only yeast.

References

Baudhuin, P., Evrard, P., and Berthet, J. (1967). Electron microscopic examination of subcellular fractions. *J. Cell Biol.* **32,** 181–191.

DeCamilli, P., Harris, S., Huttner, W., and Greengard, P. (1983). Synapsin I, a nerve terminal-specific phosphoprotein II: Its specific association with synaptic vesicles demonstrated by immunocytochemistry in agarose-embedded synaptosomes. *J. Cell Biol.* **96,** 1355–1373.

Griffiths, G., Simons, K., Warren, G., and Tokuyasu, K. (1983). Immunoelectron microscopy using thin, frozen sections: Application to studies of the intracellular transport of semliki forest virus spike glycoproteins. *In* "Methods in Enzymology" (S. Fleischer and B. Fleischer, eds.), Vol. 96, pp. 466–484. Academic Press, New York.

Heuser, J. (1981). Preparing biological samples for stereomicroscopy by the quick-freeze, deep-etch, rotary-replication technique. *In* "Methods in Cell Biology" (J. N. Turner, ed.), Vol. 22, pp. 97–122. Academic Press, New York.

Keller, G.-A., Tokuyasu, K., Dutton, A., and Singer, S. J. (1984). An improved procedure for immunoelectron microscopy: Ultrathin plastic embedding of immunolabeled ultrathin frozen sections. *Proc. Natl. Acad. Sci. U.S.A.* **81,** 5744–5747.

Newman, G., Jasani, B., and Williams, E. D. (1983). A simple post-bedding system for a rapid demonstration of tissue antigens under the electron microscope. *Histochem. J.* **15,** 543.

Palade, P., Saito, A., Mitchell, R., and Fleischer, S. (1983). Preparation of representative samples of subcellular fractions for electron microscopy by filtration with dextran. *J. Histochem. Cytochem.* **31,** 971–974.

Tartakoff, A., and Jamieson, J. (1974). Subcellular fractionation of the pancreas. *In* "Methods in Enzymology" (S. Fleischer and L. Packer, eds.), Vol. 31, pp. 41–59. Academic Press, New York.

Turner, J. N. (1981). "Methods in Cell Biology," Vol. 22. Academic Press, New York.

Chapter 19

Fluorescence Microscopy Methods for Yeast

JOHN R. PRINGLE

Department of Biology
The University of Michigan
Ann Arbor, Michigan 48109

ROBERT A. PRESTON

Department of Biological Sciences
Carnegie Mellon University
Pittsburgh, Pennsylvania 15213

ALISON E. M. ADAMS, TIM STEARNS, AND
DAVID G. DRUBIN[1]

Department of Biology
Massachusetts Institute of Technology
Cambridge, Massachusetts 02139

BRIAN K. HAARER

Department of Anatomy and Cell Biology
The University of Michigan
Ann Arbor, Michigan 48109

ELIZABETH W. JONES

Department of Biological Sciences
Carnegie Mellon University
Pittsburgh, Pennsylvania 15213

---

[1] Present address: Department of Molecular and Cell Biology, University of California, Berkeley, California 94720.

I. Introduction

The yeast *Saccharomyces cerevisiae* has emerged as an important model system for study of a wide variety of problems in cell biology, including the vesicular transport associated with secretion, cell surface growth, organelle biogenesis, and endocytosis (Schekman, 1985; Riezman, 1985; Tague and Chrispeels, 1987; Payne *et al.*, 1988; Bourne, 1988). The major attraction of yeast as an experimental organism is clearly its susceptibility to the methods of classical and molecular genetics (Huffaker *et al.*, 1987; Botstein and Fink, 1988). These methods allow previously unknown components of the cellular systems to be identified (Pringle *et al.*, 1986); they also allow hypotheses about function that are based on morphological and *in vitro* biochemical studies to be tested critically *in vivo* (Payne and Schekman, 1985; Bulawa *et al.*, 1986; Lemmon and Jones, 1987). However, it is also clear that effective use of yeast for study of cell biological problems requires that satisfactory morphological (both light and electron-microscopic) and biochemical (including cell fractionation) methods also be available.

It is the purpose of this chapter to review and provide detailed protocols for the application of immunofluorescence and other fluorescence-microscopic procedures to yeast. It should be stressed that these procedures play a role that is separate from but equal to the role of electron microscopy. Although in some situations the greater resolving power of the electron microscope is clearly essential to obtain the needed structural information (see Wright and Rine, Chap. 23, this volume, for a discussion of modern electron microscopy methods for yeast), in other situations the necessary information can be obtained more easily, more reliably, or both, by light (including fluorescence) microscopy. The potential advantages of light-microscopic approaches derive from the facts (i) that they can be applied to lightly processed or (in some cases) living cells, (ii) that much larger numbers of cells can be examined than by electron microscopy (note especially the great labor involved in visualizing the structure of whole cells by serial-section methods), and (iii) that some structures (e.g., the cytoplasmic microtubules—see Adams and Pringle, 1984; Kilmartin and Adams, 1984; Jacobs *et al.*, 1988) have simply been easier to see by light microscopy than by electron microscopy. Much useful background information about fluorescence methods can be found in the volumes edited by Taylor *et al.* (1986) and by Taylor and Wang (1988, 1989). Although this chapter deals specifically with methods for studies of *S. cerevisiae,* it should be noted that related methods are effective with other yeasts such as *Schizosaccharomyces pombe* and *Candida albicans* (for examples, see Elorza *et al.*, 1983; Chaffin, 1984; Streiblová *et al.*, 1984; Mitchison and Nurse, 1985; Marks and Hyams, 1985; Anderson and Soll, 1986; Hirano *et al.*, 1986, 1988; May and Mitchison, 1986; Yanagida *et al.*, 1986; Uemura *et al.*, 1987; Hagan and Hyams, 1988).

II. General Remarks

A. The Need for a Flexible, Empirical Approach

The protocols collected here have worked for us and others in a variety of specific applications. However, there is no guarantee that they will work without complications in all other specific applications. Indeed, experience to date both with vital staining and with immunofluorescence (see later) makes clear that the procedures that are necessary and sufficient to get good results vary from case to case. We have noted here such complications and useful alternative protocols of which we are aware, but others seem certain to emerge with further experience. Thus, a flexible, empirical approach is required; this approach must incorporate controls sufficient to establish the specificity of staining and the biological meaningfulness of the results in each particular case.

B. Growing the Cells

Choices of strain and growth conditions are obviously constrained by the biological problem to be studied. Investigators should be aware, however, that these choices may affect the results obtained with the fluorescence methods discussed here. The most obvious and general problem is that cell wall removal and permeabilization for immunofluorescence are usually more difficult with nonexponential-phase cells (e.g., cells approaching, in, or just leaving stationary phase; sporulating cells; arrested cell-cycle mutants), as discussed further in Section X,D,2,b. However, more subtle effects have also been noted. For example, immunofluorescence localization of the *SAC6* 67-kDa actin-binding protein (Drubin *et al.*, 1988) was more satisfactory when the cells had been grown in defined medium (SD or SC[2]) than in rich medium (YPD). Also, the patterns of vital staining with lucifer yellow (Section VII) are different for cells grown in YPD and cells grown in YP+D (the same medium but with the glucose autoclaved separately).

C. Vital Staining

Fluorescent vital stains (Sections II,E and IV–VIII) provide an important complement to methods (notably immunofluorescence) that require

[2] Abbreviations used in this chapter: BSA, bovine serum albumin; CDCF, carboxydichlorofluorescein; CDCFDA, carboxydichlorofluorescein diacetate; CF, carboxyfluorescein; CFDA, carboxyfluorescein diacetate; Con A, concanavalin A; DAPI, diamidino phenylindole; DIC, differential interference contrast; $DiOC_6(3)$, dihexyloxacarbocyanine iodide; ER, endoplasmic reticulum; FITC, fluorescein isothiocyanate; GlcNAc, *N*-acetyl-D-glucosamine; LY-CH, lucifer yellow carbohydrazide; NA, numerical aperture; PBS, phosphate-buffered saline (made according to the formula in Section X,D,1); SC, synthetic complete medium (Sherman *et al.*, 1986); SD, synthetic minimal medium (Sherman *et al.*, 1986); TCA, trichloroacetic acid; TLC, thin-layer chromatography; TRITC, tetramethylrhodamine isothiocyanate; WGA, wheat germ agglutinin; YPD, yeast extract–peptone–dextrose medium (Sherman *et al.*, 1986); YP+D, same as YPD except that the glucose is autoclaved separately.

the use of fixed cells; clearly, some applications are greatly facilitated or only possible if the behavior of living cells can be monitored after staining. The use of vital stains also poses some special challenges to the experimenter, as follows.

1. Establishing the Specificity of Vital Stains

a. Chemical and Physiological Bases of Specific Staining. Fluorescent vital stains constitute an expanding collection of more-or-less exotic organic chemicals. We indicate later what is known of the mechanisms governing the putative macromolecule or organelle specificity of each stain discussed. However, it should be clearly noted that in many cases these mechanisms are not known in detail, and the available evidence may be limited to light-microscopic observations. Moreover, the degree of specificity may be subject to physiological variation. Thus, it is incumbent on the experimenter to establish *in each application* (including the genetic background of the strain and the physiological conditions) whether the behavior of the stain is sufficiently specific for the task at hand. Accumulation of such data will help to delineate further the degree and mechanisms of staining specificity.

b. Staining Artifacts Due to Impurities. Available preparations of vital stains cannot be assumed to be free of impurities (Preston *et al.*, 1987). Staining artifacts can be especially troublesome when using sensitive fluorescence methodology, as very low levels of impurities may prove to be highly interfering. This problem may be exacerbated by the not infrequent need to use the stains themselves at high concentrations. Thus, in many applications, it is critical that use of a vital stain be preceded or accompanied by suitable controls that address the questions: (i) Is the stain reasonably homogeneous by chromatographic assay? (ii) Is the fluorescent material in extracts of stained cells chromatographically identical to the supposed labeling agent? Such analyses can usually be conducted easily by thin-layer chromatography (TLC). We generally use reverse-phase TLC on RPS Uniplates (Analtech; see Table I for particulars on this and other suppliers). Different proportions of methanol in water seem to provide a range of polarities suitable for most dyes (and contaminants!). Parallel development in different mixtures may be required to resolve all components in some dyes. For example, 8 : 92 (v/v) methanol–water resolves 5-carboxyfluorescein and 6-carboxyfluorescein (Section VII) from each other and from monoacetate intermediates of carboxyfluorescein diacetate (CFDA) hydrolysis, but does not resolve CFDA and trace contaminants that remain at the origin. The latter are finally resolved by development with 50 : 50 (v/v) methanol–water. Very low levels of fluorescent impurities can be detected on TLC plates with

TABLE I

SUPPLIERS OF REAGENTS AND OTHER ITEMS MENTIONED IN THIS CHAPTER

Name of company	Address	Telephone number
Accurate Chemical & Scientific Corp.	300 Shames Dr., Westbury, NY 11590	800/645-6264 *or* 516/433-4900
Acufine, Inc.	5441 N. Kedzie Ave., Chicago, IL 60625	312/539-8700
Analtech, Inc.	P.O. Box 7558, Newark, DE 19714	800/441-7540
Bio Rad Laboratories	1414 Harbour Way South, Richmond, CA 94804	800/227-5589 *or* 800/227-3259
BRL (Bethesda Research Laboratories)	P.O. Box 6009, Gaithersburg, MD 20877	800/638-8992 *or* 800/638-4045
Cappel (see Organon Teknika-Cappel)		
Cel-Line Associates, Inc.	P.O. Box 35, Newfield, NJ 08344	609/697-4590
DuPont NEN Products	549 Albany St., Boston, MA 02118	800/551-2121
E·Y Laboratories, Inc.	P.O. Box 1787, San Mateo, CA 94401	800/821-0044 *or* 415/342-3296
Flow Laboratories, Inc.	7655 Old Springhouse Rd., McLean, VA 22102	800/368-3569
ICN Immunobiologicals	P.O. Box 1200, Lisle, IL 60532	800/348-7465
Leitz (see Wild Leitz)		
Lumicon Corp.	2111 Research Dr., no. 5, Livermore, CA 94550	415/447-9570
Molecular Probes, Inc.	4849 Pitchford Ave., Eugene, OR 97402	503/344-3007
Omega Optical, Inc.	3 Grove St., P.O. Box 573, Brattleboro, VT 05301	802/254-2690
Organon Teknika-Cappel	1230 Wilson Dr., West Chester, PA 19380	800/523-7620 *or* 800/622-2440
Pharmacia LKB Biotechnology, Inc.	800 Centennial Ave., P.O. Box 1327, Piscataway, NJ 08855–1327	800/922-0318
Polysciences, Inc.	400 Valley Rd., Warrington, PA 18976–2590	800/523-2575
Savant Instruments, Inc.	110–103 Bi-County Blvd., Farmingdale, NY 11735	800/634-8886
Sigma Chemical Co.	P.O. Box 14508, St. Louis, MO 63178	800/325-3010
Vector Labs, Inc.	30 Ingold Rd., Burlingame, CA 94010	800/227-6666 *or* 415/697-3600
Wild Leitz USA, Inc.	24 Link Dr., Rockleigh, NJ 07647	201/767-1100 *or* 800/654-4488
Zeiss (Carl Zeiss, Inc.)	One Zeiss Drive, Thornwood, NY 10594	914/747-1800
Zymed Laboratories, Inc.	52 S. Linden Ave., South San Francisco, CA 94080	800/874-4494 *or* 415/871-4494

near-UV illumination. The large number of spots resolved with some commercial dyes is an eye-opening experience for those who believe labels reading: "isomer-free" or "99+%."

2. Potential Toxic Effects of Stains or Impurities

An obvious concern in using vital stains is that either the stain itself or associated impurities (fluorescent or nonfluorescent) may perturb the behavior of the cells and thus give rise to misleading results. As in the case of staining artifacts, these problems may be exacerbated by the need to use some stains at high concentrations. Clearly, it is desirable (i) to determine the minimum exposure sufficient to give adequate staining for the application at hand [in this regard, the use of low-light-level video systems (Section X,E,1) may be crucial] and (ii) to monitor in all appropriate ways whether the cells have been affected by the staining regimen. These considerations and those of the preceding section also emphasize the need to validate the results of vital staining using an independent procedure whenever possible.

3. The Need for Fast and Effective Filtration

Vital stains used at high concentrations must be efficiently washed away from labeled cells to avoid high backgrounds, at least in some applications. At the same time, it is desirable to minimize the cells' exposure to the labeling regimen (see earlier). These twin requirements are addressed by quickly harvesting and washing the labeled cells on membrane filters, at 0°C when this is compatible with the needs of the experiment. Small volumes of labeled cells (1–2 ml) can be quickly processed on 13-mm-diameter membrane filters supported on the bottom half of a plastic filter holder mounted on a syringe needle that pierces a rubber stopper on a side-arm suction flask. The suction is sufficient to hold the filter without the use of any chimney. This avoids unnecessary retention of dye by parts of the filtration apparatus, and it facilitates rapid filter changes and cell resuspensions.

4. Anaerobiosis of Slide Preparations

At $\geq 10^8$ cells/ml, viable yeast cells under a coverslip on a slide become anaerobic within a few minutes, as oxygen transfer is negligible. If anaerobiosis adversely affects the images obtained in a particular application, the cell concentration can be reduced, the period of observation can be shortened, or both. However, if anaerobiosis can be tolerated, it can be used to advantage to retard the photobleaching of many stains (see Section II,F,3).

5. Immobilization of Cells for Photomicroscopy

For obvious reasons, some of the immobilization methods to be discussed (Section II,F,2) cannot be used on viable cells without drastically affecting their properties. However, we have had good success with concanavalin A (Con A) immobilization in staining of the vacuole (Section VII) and with low melting point agarose in staining of mitochondria (Section VI).

D. Preparation of Fixed Cells

1. Standard Fixation Protocol

For most purposes (some exceptions and caveats are noted later), we recommend the following procedure.

1. Add concentrated formaldehyde solution directly to the cells in growth medium to a final concentration of 3.7–5% (w/v) formaldehyde.
2. After 1–30 minutes at room temperature, recover the cells by centrifugation and resuspend in phosphate-buffered formaldehyde [40–100 m*M* potassium phosphate (pH 6.5), containing 0.5 m*M* $MgCl_2$ and 3.7–5% (w/v) formaldehyde].
3. After 2–4 hours at room temperature, wash the cells free of formaldehyde with the appropriate solution, and proceed to stain.

2. Comments on Fixation Protocol

a. Importance of Rapid Fixation. Some features of cell structure can change surprisingly rapidly when cells are subjected to the stress of harvesting. For example, the asymmetric distribution of actin in budding cells (Adams and Pringle, 1984; Kilmartin and Adams, 1984; Novick and Botstein, 1985) is much more evident when cells are fixed by addition of fixative directly to the culture medium than when cells are first harvested by centrifugation and then resuspended in fixative (Fig. 1; Drubin *et al.*, 1988). Clues to the basis for this difference come from the observations (i) that the actin distribution changes markedly within 2 minutes when the cells are treated with an energy poison such as 3 m*M* sodium azide (Novick *et al.*, 1989; D. G. Drubin, unpublished results), and (ii) that addition of 2% glucose to the wash buffer used in an experiment like that

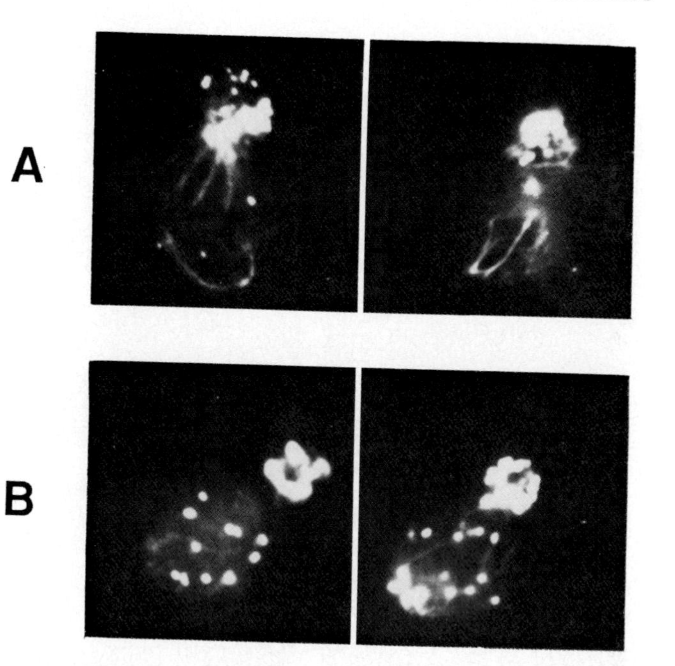

FIG. 1. The importance of rapid fixation for observation of the asymmetric actin distribution in budding yeast cells. The actin distribution in diploid *S. cerevisiae* cells (strain DBY4866) growing exponentially in YPD medium (~2×10^6 cells/ml) was revealed by antiactin immunofluorescence (Section X). (A) Cells were fixed by adding 0.67 ml of 37% formaldehyde directly to 5 ml of culture. The asymmetric distribution of actin is evident: cortical actin patches are concentrated in the buds, and actin cables are aligned with the growth axis in the mother cells. (B) Cells from the same culture were centrifuged and washed twice at room temperature in 0.1 *M* potassium phosphate (pH 6.5) over a 10-minute period before being fixed by resuspension in the same buffer containing 4% formaldehyde. Cortical actin patches are now seen in the mother cell as well as the bud, and the cables may be less prominent.

of Fig. 1B largely restores the type of pattern seen in Fig. 1A (D. G. Drubin, unpublished results). Rapid changes in mitochondrial and vacuolar morphology have also been observed (Fig. 2; Sections VI and VII). Thus, we recommend that cells ordinarily be fixed initially by adding fixative directly to the growth medium. At least with formaldehyde, this appears to stop life processes very rapidly even in rich medium (Pringle and Mor, 1975).

b. Choice of Fixative. Except for nuclear staining (Section V), nearly all fluorescence microscopy done to date on fixed yeast cells has been done using formaldehyde-fixed cells. As formaldehyde has generally given satisfactory results, and as it is also *relatively* safe and easy to use,

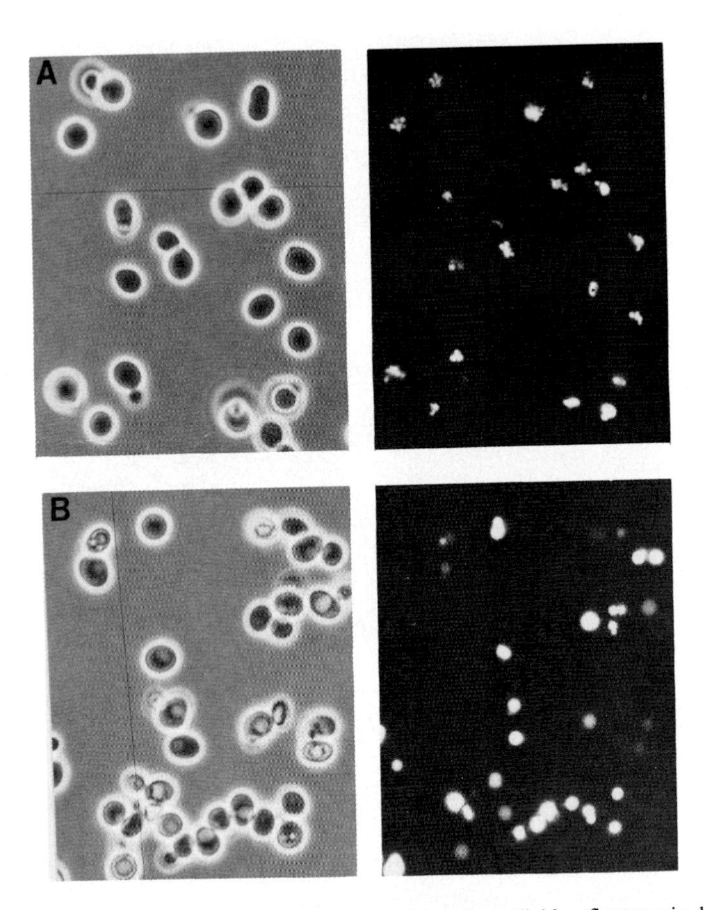

FIG. 2. Vacuoles in living yeast cells stained with carboxydichlorofluorescein diacetate (CDCFDA). Cells of strain X2180-1B growing exponentially in YP+D medium were stained with CDCFDA as described in the text (Section VII). Panels to the left show phase-contrast images; panels to the right show the fluorescence images of the same cells. Photomicrographs were taken using a Zeiss Axiophot epifluorescence microscope, 100× Plan Neofluar objective, and fluorescein filter set no. 487909. Exposures for the fluorescence micrographs were 0.125 second using Kodak Tri-X film with development in Diafine (see Section II,F,4,a). (A) After staining, cells were harvested, washed, and resuspended at 0°C using buffer containing 2% glucose. Note that the fluorescence photomicrograph represents an optical section: additional small, fluorescent vesicles that were out of focus in this section are not visible in the micrograph. (B) An aliquot of the same stained cells used in (A) was exposed to glucose-free wash buffer for 1 minute at 24°C, then resuspended in this same buffer at 0°C and photographed immediately. Note that the numerous small vesicles seen in (A) have been replaced by one or two large vacuoles. On some occasions, this coalescence of vacuoles has occurred more slowly (~10 minutes at 30°C), suggesting that additional, as yet unidentified factors also affect vacuolar morphology.

it appears ordinarily to be the fixative of choice for light microscopy. However, it should be noted that there will almost certainly be immunofluorescence (and perhaps other) applications in which formaldehyde is not the optimal fixative. First, formaldehyde clearly gives insufficient preservation of cellular fine structure to be adequate for electron microscopy; there may be applications in which this limitation is also a factor in light microscopy. In this regard, it should be noted that a protocol involving fixation with paraformaldehyde plus glutaraldehyde (and incorporating $NaBH_4$ treatments to reduce the otherwise intolerable background fluorescence) has been used successfully with yeast (Kilmartin and Adams, 1984; cf. also Chen *et al.*, 1985; Schulze and Kirschner, 1987), although the results in this particular case were not actually better than those obtained with formaldehyde. Second, some antigens may be sensitive to fixation by formaldehyde. Whether for this or other reasons, the fixation conditions yielding optimal immunofluorescence results with different antigen–antibody combinations can vary widely, as shown clearly by work on animal cells (Pruss *et al.*, 1981; Gown and Vogel, 1982; Greenwalt and Mather, 1985; Chen *et al.*, 1985). Indeed, Chen *et al.* (1985) recommend screening each new monoclonal antibody (mAb) on cells fixed by each of 10 different procedures! Third, there is at least one example in yeast work in which formaldehyde fixation had to be followed by additional treatments with methanol and acetone in order to obtain good immunofluorescence (Drubin *et al.*, 1988; see Section X,D,2,d); possibly an alternative approach to the initial fixation would also have yielded good results.

c. Deterioration of Formaldehyde Stocks. Concentrated formaldehyde stocks deteriorate with time, as evidenced by the accumulation of precipitate (polymerization product) at the bottom of the bottle. Although the presence of some of this precipitate does not appear to be a problem, we are uncertain how to gauge exactly when a stock should be replaced; thus, we recommend buying small bottles of 37% formaldehyde, removing solution as needed from the top of the undisturbed bottle, and replacing the bottle as soon as polymerization becomes extensive. Better results may be obtained with "ultrapure E.M. Grade" formaldehyde (Polysciences cat. no. 4018), but we are not certain that this really helps.

d. Resuspension in Phosphate-Buffered Formaldehyde. This step was incorporated into our standard fixation protocol based in part on general principles (the idea that cell structure would be preserved better if the pH stayed near physiological values despite the chemical action of the fixative) and in part on the routine buffering of fixatives by electron microscopists (e.g., Wright and Rine, Chap. 23, this volume). However, we are not certain how often the buffering really matters or whether any of the parameters of our particular buffered fixative are really critical. Indeed, we have often omitted this step (conducting the entire fixation right in growth medium) without obvious detriment to the results.

e. Duration of Fixation. Depending on the application, we have fixed cells for as little as a few minutes or as long as several days without detectable detriment to the results. For immunofluorescence, the duration of fixation may be more critical, although our information is anecdotal. In at least one application, 2–3 hours of fixation gave more satisfactory results in antitubulin immunofluorescence than did $\leq$1 hour of fixation (P. Schatz, personal communication). At the other extreme, we have routinely had satisfactory results after storing cells as long as 12 hours at 4°C in formaldehyde-containing growth medium or in phosphate-buffered formaldehyde after ~2 hours of fixation at room temperature. In addition, we have successfully used cells fixed for 2 hours at room temperature, then washed into solution A (Section X,D,1) and stored 24 hours at 4°C before removal of cell walls (Section X,D,2,b), as well as cells stored $\geq$2 days at 4°C after the cell wall-removal step. However, we do not know if these variations invariably yield satisfactory results.

E. Fluorescent Stains and Fluorochromes

A wide variety of fluorescent probes is now available.[3] Some of these are used directly as stains because they are specific for particular macromolecules or organelles or because their fluorescence is responsive to intracellular pH, membrane potential, or the concentration of specific ions such as Ca^{2+} (Tsien *et al.*, 1985; Waggoner, 1979, 1986; Taylor *et al.*, 1986; Paradiso *et al.*, 1987; Wang and Taylor, 1989). Some of these stains have already been used successfully in yeast work and are discussed later; many others have proved useful in work with other types of cells and are likely to find application in yeast work as more investigators appreciate the value of this approach. Other fluorochromes are of value because they can be coupled covalently to macromolecules (Haugland, 1983; Waggoner, 1986; Wang and Taylor, 1989); some of these, such as fluorescein and various rhodamine derivatives, are of especially broad utility because of their spectral properties and ease of coupling to useful reagents such as antibodies and lectins. Some useful information about the fluorescent probes discussed in this chapter is collected in Table II. A

[3] Although there are other satisfactory sources for various fluorescent probes, special note should be made of Molecular Probes. We have generally been satisfied by the purity of the reagents and the support functions provided by this company. Its "Handbook of Fluorescent Probes and Research Chemicals" and its periodical "Bioprobes" are also useful resources.

point to be noted is that the possibility of using two or more probes in double-label or multiple-label experiments depends on their spectral resolution at the level of excitatory wavelengths, emission wavelengths, or (preferably) both; inadequate spectral resolution can lead to troublesome "crossover" fluorescence (see also Sections II,F,1,d and X,F). Investigators should also be aware that work continues on developing additional fluorescent probes that are easy to use and have improved spectral properties [i.e., molar absorption and quantum yield (hence brightness), resistance to photobleaching, and spectral resolution; see Waggoner, 1986; Taylor *et al.,* 1986; Wang and Taylor, 1989; and the Molecular Probes Handbook, 1988 edition]. Particularly promising are the intensely fluorescent phycobiliproteins (Oi *et al.,* 1982; Hardy *et al.,* 1983; Glazer and Stryer, 1984; Parks *et al.,* 1984; Lanier and Loken, 1984).

F. Microscopy and Photomicroscopy

1. Microscopes and Objectives

a. Conventional Microscopes. In older fluorescence microscopes (some of which are still in service), the excitatory illumination passes from a condenser through the object and into the objective lens, very much as in ordinary light microscopy. A special "suppression filter" must then be inserted between the objective and eyepiece to absorb the excitatory illumination (but not the emitted fluorescence) so as to protect the eyes and allow the longer-wavelength fluorescence to be seen. In contrast, in modern epifluorescence microscopes, the excitatory illumination is directed onto the specimen through the objective lens. The backscattered excitatory illumination that reenters the objective is then absorbed by a supression filter that is an integral part of the packet that also supplies the excitatory illumination. [The small books "Fluorescence Microscopy" (by E. Becker; distributed by Leitz) and "Worthwhile Facts about Fluorescence Microscopy" (by H. M. Holz; distributed by Zeiss) provide helpful, simple introductions to this and other aspects of the hardware for fluorescence microscopy.] Epifluorescence microscopes thus offer major advantages in terms of simplicity of operation (there is no need to worry about alignment, focusing, and eventual deoiling of a separate condenser, and all the filters are put into position by a single manipulation), safety (there is no danger of accidentally leaving out the suppression filter and thus searing the eyes with UV excitatory illumination), brightness of object fluorescence (the excitatory illumination is focused intensely on the area of the object that is being observed), increased spectral resolution (the more intense excitatory illumination

TABLE II

SOME FLUORESCENT PROBES USED IN FLUORESCENCE MICROSCOPY OF YEAST[a]

Fluorochrome	Excitation wavelength[b] (nm)	Emission wavelength[b] (nm)	Appropriate filter sets[c]	
			Leitz	Zeiss
FITC[d]	490–495	525	I2/3 or L3	487709 or 487710
TRITC[e]	540–552	570	N2 or M2	487714 or 487715
Texas red[f]	590–596	615–620	N2	487700 or 487714
Rhodamine (on phalloidin)[g]	540	580	N2	487714 or 487715
B-phycoerythrin[h]	540–565	576	N2 or M2	487714 or 487715
C-phycocyanine[h]	605–620	645–650	N2?	487700
Calcofluor[i]	340–360	400–440	A	487702
Primulin[i]	410	550	B2, D, or A	487705 or 487718
DAPI[j]	340–365	450–488	A	487701 or 487702
Mithramycin[k]	420	575	E3, H3, or D	487705 or 487707
Acridine orange[k]	470–503	523–650	I2/3 or K3	487709 or 487711
Hoechst 33258[k]	365–374	472–480	A or B2	487702 or 487718
Propidium iodide[k]	510–520	610	N2 or N2.1	487714 or 487715
$DiOC_6(3)$[l]	478	496	I2/3 or K3	487709 or 487717
$DiIC_5(3)$[m]	540	556	N2	487714 or 487715
LY-CH[n,o]	428	540	E3	487705
Pyranine[o,p,q]			D or G	487705
Acidic form	400	515	—[q]	—[q]
Basic form	450	515	—[q]	—[q]
CDCF (from CDCFDA)[r]	505	530	L3.1 or I2/3	487709 or 487716
CF (from CFDA)[q,s]			I2/3	487709 or 487710
Acidic form	450	520	—[q]	—[q]
Basic form	490	520	—[q]	—[q]

[a] Additional information about many of these probes can be found in the Molecular Probes Handbook.

[b] The approximate maxima for excitation and emission are given. However, it should be noted that both processes actually occur over a more-or-less broad band of wavelengths, and the positions of the maxima may vary somewhat with the chemical environment.

[c] This information is provided simply for general guidance. In some applications, other standard or nonstandard filter combinations will be equally or more appropriate for use with Zeiss or Leitz microscopes, and other types of fluorescence microscopes will have their own versions of appropriate filter sets. [See also Sections II,F,1,d; VII,A,2,c(4); and X,F.] Note that the Zeiss filter numbers given are for the older-series microscopes; filters for the new Axio line are the same except for a 9 instead of a 7 in the fourth place (e.g., 487909 or 487910 for FITC).

[d] Fluorescein 5-isothiocyanate, generally used after conjugation to lectins, antibodies, or other molecules (see Sections IV, VII, IX, and X). The spectral data shown refer to conjugates at pH ≥ 8 (see the Molecular Probes Handbook).

[e] Tetramethylrhodamine 5(and 6)-isothiocyanate, generally used after conjugation to lectins, antibodies, or other molecules (see Sections IV, IX, and X). The effect of pH on the spectral properties is not clear.

[f] Trademark name of Molecular Probes, Inc., for a rhodamine X sulfonyl chloride, generally used after conjugation to lectins or antibodies (see Sections IV and X). Spectral data refer to conjugates and are reportedly pH independent near pH 7.

[g] A synthetic derivative of rhodamine conjugated to the mushroom toxin phalloidin, used in the staining of actin (Section IX).

[h] Two of a number of phycobiliproteins that have great promise as fluorescent labels because of their spectral properties. See the Molecular Probes Handbook and the references given in the text.

[i] Used in the staining of cell wall chitin (Section III).

[j] 4′,6-Diamidino-2-phenylindole · 2HCl, used in the staining of nuclear and mitochondrial DNA (Sections V,A,1 and VI).

[k] Other stains for nuclear DNA (see Section V,A,2). Note that acridine orange stains RNA heterochromatically (Dresser and Giroux, 1988), accounting for its wide range of emission wavelengths.

[l] 3,3′-Dihexyloxacarbocyanine iodide, used in the vital staining of mitochondria (Section VI,A) and perhaps ER (Section VIII).

[m] 1,1′-Dipentyl-3,3,3′,3′-tetramethylindocarbocyanine iodide, potentially useful in the vital staining of mitochondria (Section VI,A,2).

[n] Lucifer yellow carbohydrazide, used in staining vacuoles (Section VII,A,1).

[o] LY-CH and Pyranine can also be visualized, though not optimally, using standard fluorescein filter sets (e.g., Leitz I2/3 or Zeiss 487709).

[p] Used for staining vacuoles (Section VII,A,1).

[q] Ionization of weak acid groups on Pyranine (pK_a 7) and CF (pK_a 6.4) changes the excitation spectra of these dyes. The pairs of excitation wavelengths shown for the acidic and basic forms are not necessarily excitation maxima, but are appropriate for making pH measurements by ratio methods [see Sections VII,A,1,c(5), and VII,A,2,c(4)]. If the dyes are used simply for visualizing vacuoles, a wide-bandpass excitation filter spanning both wavelengths listed for each dye (e.g., those listed in the table) would give the brightest emission (albeit also the most rapid photobleaching).

[r] Carboxydichlorofluorescein, used for staining vacuoles via CDCFDA (carboxydichlorofluorescein diacetate).

[s] Carboxyfluorescein, used for staining vacuoles via CFDA (carboxyfluorescein diacetate).

means that a narrower band of excitatory wavelengths can be used), and longevity of the sample (areas of the object that are not being examined are not illuminated, and hence do not suffer photobleaching), and such microscopes should be used whenever possible.

Excellent results can be obtained with any of the high-quality epifluorescence microscopes presently on the market. It is important to realize that the vast majority of applications in yeast cell biology do not require a top-of-the-line microscope with its costly features designed to provide maximum stability, versatility, and semiautomatic operation. In particular, the same quality optics are available in the less costly versions of the research microscope lines of the major manufacturers. Nor are large numbers of objectives required. As a result, although it is not hard to spend more than $50,000 on a microscope system, it is possible to do very well while spending much less. For example, although we have been very pleased by the performance of our (rather expensive!) Leitz Orthoplan microscope, we have also been very satisfied by the performance of a Zeiss Axioscop (the lowest of three grades in Zeiss's new Axio line), which was set up for bright-field, differential interference contrast (DIC, or Nomarski), and fluorescence work on yeast for about $21,000 in 1987. Comparable results may well be obtainable at still lower cost using instruments from Leitz (Laborlux or Diaplan series microscopes), Nikon, or Olympus.

Given the high quality of the microscopes available from the various suppliers, an important consideration in purchasing a microscope is the competence and attitude of the local sales and service personnel. A helpful sales representative (such as Leitz's representative in Michigan) can provide invaluable instruction of laboratory personnel and large savings in dealing with minor service problems.

b. Confocal Scanning Microscopes. These microscopes are expensive to set up and not yet widely available, and they are certainly not essential (or even appropriate) for all applications. Nonetheless, investigators should be aware of the great potential benefits offered by this latest major advance in light microscope technology (Brakenhoff *et al.*, 1985, 1986; Boyde, 1985; White *et al.*, 1987). In confocal scanning microscopes, the object is viewed not as a whole, but as a series of points on which both illumination and detection are tightly focused. A two-dimensional scan of the object thus produces an optical section, whereas a series of such scans allows a three-dimensional image to be constructed and presented on a video screen or as stereo-pair micrographs. These systems offer an increase in potential resolution, and the digitally stored information is readily available for further computer analysis (see Section X,E,1). However, the greatest advantage of the confocal scanning systems appears to be in fluorescence applications, where a major problem is often the swamping of the in-focus fluorescence signal by noise in the form of fluorescence from out-of-focus portions of the object. By taking a thin (~0.5 μm) optical section, the confocal scanning microscope can sharply reduce the contribution of out-of-focus fluorescence, producing in some applications a dramatic increase in the sharpness and contrast of the ultimate images (White *et al.*, 1987). This feature is especially important in quantitative applications. One potential liability of confocal microscopy is that the intensely focused illumination can lead to especially rapid photobleaching, producing misleading images under some circumstances. In addition, current limitations on scanning rates limit the usefulness of these microscopes for examining vitally stained (and therefore not static) cells. Confocal microscopes have been applied successfully in fluorescence work with yeast (J. Allen and M. Douglas, personal communication).

c. Lamps. Optimal illumination (hence proper alignment of the light source) is critical to high-quality fluorescence microscopy. Another consideration is that the brightness and quality of the fluorescence image deteriorate as the mercury lamp begins to fail, especially for fluorochromes requiring UV excitation. In addition, explosions of these lamps (which can occur as the quartz gets brittle with age) are reputedly unpleasant to experience, even if successfully contained by the lamp housings (which are designed to meet this challenge). Thus, although the mercury lamps are expensive, it is generally a false economy to continue using them after they have reached their stated lifetime or after their brightness begins to decrease or fluctuate noticeably.

d. Filters. The microscope manufacturers provide a choice of prepackaged sets of excitation and suppression filters that are appropriate for use with the most commonly used fluorochromes (see Table II and the Becker and Holz books, mentioned earlier). Alternatively, special filter sets can be assembled for particular purposes—for example, certain ratio fluorescence [cf. Section VII,A,2,c(iv)] or double-label (Section X,F) experiments—using the filters available from the microscope manufacturers or from other suppliers (e.g., Omega Optical).

e. Field Diaphragm. Epifluorescence microscopes are equipped with a field diaphragm (iris) that controls the area of the object that is illuminated. To reduce background fluorescence and photobleaching of parts of the specimen that have not yet been examined, the field diaphragm should always be adjusted so that the area being illuminated is no larger than the field of view. Indeed, following the same arguments, it is sometimes helpful to stop this diaphragm down further so that only the cell(s) of immediate interest are illuminated.

f. Objectives. The objective lens chosen must obviously provide an appropriate magnification (for yeast, almost always 40–100×). However, several additional important considerations also influence the choice of an appropriate objective for fluorescence work.

1. In epifluorescence microscopy, both the intensity of the excitatory illumination and the efficiency of collection of the emitted fluorescence are proportional to the square of the numerical aperture (NA) of the objective. Thus, for objectives of a given magnification, the brightness of the fluorescence image is proportional to the fourth power of the NA. In photomicroscopy, the effect is even greater, as a brighter image allows shorter exposure times and thus less photobleaching. As the optical resolving power of an objective also increases with the NA, it is clear that there are strong incentives to use high-NA (hence oil immersion) objectives. On the other hand, it is not uncommon to find that optical resolution is not limiting and that an object is actually too brightly stained. In this situation, it can facilitate both observation (by reducing the brightness to a comfortable level and reducing the rate of photobleaching) and photography (by allowing more flexibility in the setting of exposure times) to use an objective of reduced NA. This is most conveniently accomplished by using an objective in which the NA is actually adjustable by means of an iris diaphragm. Moreover, we have the strong impression that stopping down the iris in such an objective can actually increase contrast, although we are unsure whether this visual effect is reproduced in photomicrographs.
2. Other things being equal, the brightness of the fluorescence image is inversely proportional to the square of the total magnification (objective magnification × eyepiece magnification). Thus, if image brightness is a limiting factor for observation or photography, use of a lower magnification should be considered.
3. In phase-contrast and dedicated DIC objectives, there are unavoidable losses of both excitatory and emitted light in the objectives because of the phase rings or built-in interference-contrast prisms, resulting in less bright fluorescence images. (This problem should be less severe with DIC than with phase-contrast objectives.) Thus, in choosing an objective, one must weigh the advantages of obtaining a maximally bright fluorescence image with a bright-field objective against the advantages (often considerable) of obtaining companion phase-contrast or DIC images without having to change the objective. In some new microscopes, this problem has been solved for DIC by placing the necessary prisms in a special slider (so they can be moved out of position when not in use) and using ordinary bright-field objectives.

4. The expense of objectives is determined in part by the width of the flat field that they provide. Although the uniformity of focus provided by a wide flat field is sometimes convenient for photography, it is rarely if ever essential in yeast work. Comparably good correction for chromatic and spherical aberrations and equally good fluorescence properties are available in objectives that are substantially cheaper than those offering the largest flat fields.

2. Slides and Immobilization of Cells

Although some fluorescence microscopy is done on cell sections or isolated organelles, most is done on whole cells, which must be mounted in such a way as to allow efficient observation and photography.

a. Multiwell Slides; Immobilization with Polylysine. Immunofluorescence and some other applications are best carried out on cells attached by polylysine to multiwell slides (see Sections X,D,2,c–d for a detailed protocol). This approach allows efficient staining and washing using small volumes of solutions, and the cells are immobilized for eventual photomicroscopy. We have mostly used slides either from Flow Laboratories (cat. no. 60-408-05) or from Polysciences (cat. no. 18357). We have observed some lot-to-lot variability in the wettability of the glass wells of these slides; on several occasions, the polylysine solution has beaded up rather than spreading properly. We are uncertain whether either source is routinely more reliable, and we note that a variety of similar slides (with which we have no personal experience) are also available from Cel-Line and other suppliers.

b. Ordinary Slides and Wet Mounts; Immobilization with Con A or Low Melting Point Agarose. Despite the advantages of multiwell slides, it should be noted that equally satisfactory results can sometimes be obtained more easily and cheaply using ordinary slides and coverslips. Indeed, we have sometimes had superior results with this approach. For example, for unknown reasons we have frequently had better success in staining actin with phalloidin conjugates when we have carried out the staining on cell suspensions and ultimately viewed the cells without sticking them to the slide (see detailed protocol in Section IX). Also, in using Calcofluor or DAPI staining (Sections III and V) to evaluate the terminal phenotypes of cell cycle mutants, we have frequently found it useful to have the cells still moving a little as they are viewed; this facilitates determining which cell bodies are attached to which as well as allowing the same cell to be viewed from several different angles. The cell movement is of course a problem for photomicroscopy, but for many purposes this is easily solved (for cells mounted in water or buffer) by allowing the slide to dry enough that the cells are gently squeezed between the slide and coverslip. This has the further advantage (particularly helpful for mutants with complex morphologies) of squeezing the cells more nearly into one focal plane, thus allowing more of the structure to be captured in a single photograph.

An alternative method of immobilizing viable or fixed cells suspended in water, buffer, or synthetic medium on ordinary slides employs Con A according to the following protocol. Twenty microliters of 1 mg/ml Con A in water are spread across the entire surface of a 1 × 3 in. slide with the side of a micropipet tip and allowed to air-dry. Then, 6–7 μl of cell suspension (1–2 × 10^8 cells/ml) are applied to a portion of the slide and an 18-mm^2 coverslip is added gently. There should be regions of the slide where cells are densely spaced in the same focal plane, but not clumped together. Cells that insist on clumping can sometimes be coaxed into a single focal plane by moving the coverslip back and forth slightly on the slide. This method does not work well with cells suspended in rich growth medium (YPD or YP+D), which seems to block the immobilization.

Yet another method that has proved valuable in the vital staining of mitochondria and should have other applications involves immobilization in low melting point agarose, as described in Section VI.

3. The Bleaching Problem

All fluorochromes undergo photobleaching to some extent, although the rates vary considerably with the fluorochrome, the chemical environment, and the wavelength and intensity of the excitatory illumination. Although intentional localized photobleaching can be a useful experimental tool in studies of dynamic systems (Taylor *et al.*, 1986; Petersen *et al.*, 1986; Wang and Taylor, 1989; Taylor and Wang, 1989), mostly it is a nuisance both for direct observation and for photomicroscopy. Useful countermeasures include the following.

a. Keeping Preparations Dark. We routinely perform all manipulations involving fluorochromes under minimal illumination and keep the samples in the dark (aluminum foil is convenient) during incubations and storage.

b. Anaerobiosis. Under certain conditions in vital staining applications, cells rapidly become anaerobic (Section II,C,4). Although this may cause other problems for the investigator, it can be very useful in that it greatly retards the rate of photobleaching of many fluorophores (e.g., fluorescein, Pyranine, porphyrins).

c. Chemical Antibleaching Agents. A very useful method for combating the photobleaching problem is inclusion of *p*-phenylenediamine in the mounting medium; this markedly retards the photobleaching of several important fluorochromes (notably fluorescein and rhodamine and their derivatives) and may even intensify their initial fluorescence (Johnson and Nogueira Araujo, 1981; G. D. Johnson *et al.*, 1982; Platt and Michael, 1983). A recipe for such a mounting medium is given in Section X,D,1. It has also been reported that *n*-propyl gallate greatly retards the photobleaching of both fluorescein and rhodamine conjugates (Giloh and Sedat, 1982) and that NaN_3 and NaI retard the photobleaching of FITC (Böck *et al.*, 1985), but we have no personal experience with these agents. We also have little experience with the use of chemical antibleaching agents in conjunction with the vital stains discussed in Sections VI and VII; clearly, possible physiological effects of the chemicals would be a concern in such applications.

4. Film

Exposure time is often the limiting factor in getting good immunofluorescence photomicrographs. At the same time, using too fast a film can result in an undesirably grainy micrograph. Thus, it is important to choose an appropriate combination of film and developing conditions.

a. Black-and-White Film. We have taken many satisfactory photomicrographs using standard Kodak Tri-X Pan film (ASA 400), usually "pushing" the development to ASA 1600 using Diafine developer (Acufine Corp.; available in many local photo stores). However, it is clear that the slightly more expensive Kodak T-MAX film (ASA 400) gives a finer-grain negative and can be pushed to ASA 1600 or 3200 by extending developing times with the special T-MAX developer (Kodak). (In our experience to date, we have felt that the best compromise between sensitivity and graininess was achieved at ASA 1600.) Even better in its combination of a fine grain size with high sensitivity is the somewhat more expensive hypersensitized Kodak Technical Pan 2415 film (Lumicon; Schulze and Kirschner, 1986, 1987), developed according to the supplier's instructions.

b. Color Film. Color slides are nice for seminars, especially when double-label experiments are to be shown. We have routinely used Kodak Ektachrome 400 with satisfactory results. It often helps to push the development to higher effective ASA by extending the developing time. On the rare occasions when color prints are needed, they can be prepared from the color slides either by making an internegative or by a direct printing process.

5. The Importance of Taking Many Photographs

Even with the best equipment and the best hands, the outcome of photomicroscopy is somewhat uncertain. Moreover, once a good preparation is available, the cost (both in materials and in investigator time) of taking additional pictures is generally small in relation to the total cost of doing the research. Thus, we strongly recommend that investigators routinely take many more pictures than they expect to need. This allows later documentation of the range of structures observed, as well as providing a wide range of choices from which to select the best micrographs for presentation. Along the same lines, it is convenient to keep two separate cameras loaded with black-and-white and color film, respectively, so that particularly favorable images can be captured in both formats.

III. Staining of Cell Wall Chitin

Shortly before bud emergence, a ring of chitin is formed in the cell wall; the bud then emerges within the confines of this chitin ring, which remains in place as the bud grows (Hayashibe and Katohda, 1973; Seichertová *et al.*, 1973; Cabib and Bowers, 1975; Sloat and Pringle, 1978; Cabib *et al.*, 1982; Roberts *et al.*, 1983). Subsequently, a chitin-rich "primary septum" forms within the confines of the chitin ring as an early step in the separation of mother and daughter cells (Marchant and Smith, 1968; Bowers *et al.*, 1974; Byers and Goetsch, 1976; Cabib *et al.*, 1982; Roberts *et al.*, 1983). After division, the chitin ring and septum remain on the mother cell as the craterlike "bud scar" (Barton, 1950; Bartholomew and Mittwer, 1953; Houwink and Kreger, 1953; Bacon *et al.*, 1966; Beran, 1968; Cabib and Bowers, 1971; Hayashibe and Katohda, 1973; Seichertová *et al.*, 1973; Talens *et al.*, 1973; Sloat and Pringle, 1978; Molano *et al.*, 1980; Holan *et al.*, 1981; Cabib *et al.*, 1982; Roberts *et al.*, 1983).[4] For reasons that remain somewhat unclear (see Section III,B,3), the chitin-rich structures can be stained with high specificity using the fluorescent dyes primulin (Chem. Index 49000; Sigma cat. no. P7522; Polysciences cat. no. 2770) and Calcofluor White M2R New [Chem. Index

[4] A scar (the "birth scar") is also left on the daughter cell and can be visualized by scanning electron microscopy (Talens *et al.*, 1973). However, the birth scar contains little or no chitin and has been difficult (Streiblová and Beran, 1963a,b; Beran, 1968) or impossible to visualize by fluorescence methods. It is not considered further here.

40622; also known as Fluorescent Brightener 28 (Sigma cat. no. F6259) or Cellufluor (Polysciences cat. no. 17353)], as first reported by Streiblová and Beran (1963a,b) and Hayashibe and Katohda (1973), respectively.

Staining with primulin or Calcofluor is thus very useful in distinguishing mother from daughter cells, in distinguishing mother cells of different ages, and in monitoring the success of procedures designed to separate these different classes of cells (Beran, 1968; Hayashibe and Katohda, 1973; Johnston *et al.*, 1979; Lord and Wheals, 1980; Nasmyth, 1983; Brewer *et al.*, 1984; Thomas and Botstein, 1986; Jacobs *et al.*, 1988). In addition, such staining allows determination of the effects of genetic background, mating type, specific genetic lesions, and environmental conditions on the localization of budding sites (Beran, 1968; Streiblová, 1970; Hayashibe, 1975; Thompson and Wheals, 1980; Sloat *et al.*, 1981; Drubin *et al.*, 1988; Johnson and Pringle, 1989), as well as of the effects of mutations, drugs, and physiological effectors on the patterns of chitin deposition (Cabib and Bowers, 1975; Sloat and Pringle, 1978; Schekman and Brawley, 1979; Sloat *et al.*, 1981; Elorza *et al.*, 1983; Kilmartin and Adams, 1984; Novick and Botstein, 1985; Roncero and Durán, 1985; Pringle *et al.*, 1986; Bulawa *et al.*, 1986; Haarer and Pringle, 1987; Roncero *et al.*, 1988a,b; Jacobs *et al.*, 1988; Silverman *et al.*, 1988).

A. Staining Procedures

As we have no personal experience with primulin, the following discussion refers exclusively to Calcofluor. This dye is easily used according to the following protocols.

1. Make a 1 mg/ml stock solution of Calcofluor in water (be sure to get all of the dye into solution) and store at 4°C. This stock usually remains good for several weeks or longer, but occasionally deteriorates more rapidly, as manifested by the appearance of overt precipitate in the stock or of brightly fluorescent specks in stained cell preparations. (Note that some protocols in circulation call for dissolving the Calcofluor in phosphate buffers. However, this approach appears to be less satisfactory than simply using water, perhaps because of reduced solubility of the dye in such buffers.)
2. To simply observe chitin rings and bud scars, collect live cells or cells fixed by the standard protocol (Section II,D,1) by centrifugation. It may sometimes help to wash once with water at this stage, but this does not generally appear necessary. Resuspend the cells in Calcofluor solution (the undiluted stock or a 2- to 10-fold dilution of this) and incubate at room temperature for ~5 minutes. Depending on the strain and the

growth conditions [the relevant variables are not well understood; cf. Section III,B,3(c)], it may be necessary to adjust the Calcofluor concentration, the staining time, or both, in order to get good differential staining of the chitin rings and bud scars. Wash twice with water by centrifugation, mount on a slide, and observe using UV excitatory illumination (Table II).

3. It has sometimes been useful to observe nuclei and bud scars simultaneously (Brewer *et al.*, 1984; Thomas and Botstein, 1986; Haarer and Pringle, 1987; Jacobs *et al.*, 1988). Although several methods have been used, it seems that the best results are obtained with the least effort by simply staining fixed cells with Calcofluor and washing as just described, then mounting them on a slide in DAPI-containing mounting medium (Section X,D,1). Staining of the nuclear and mitochondrial DNA is good after ~30 minutes, and both stains are visible using the same filters (Table II).

4. It has sometimes been useful to observe chitin rings and actin simultaneously (Kilmartin and Adams, 1984; Kim *et al.*, 1989b). This can be accomplished by staining with Calcofluor and washing as described before, then staining with fluorochrome-conjugated phalloidin as described in Section IX.

5. It may sometimes be useful to perform immunofluorescence on cells that have also been stained with Calcofluor (Haarer and Pringle, 1987). This is easily accomplished by staining fixed cells with Calcofluor and washing as described before, then preparing them for immunofluorescence by standard procedures (Section X,D,2). However, it should be noted that the procedures normally used to permeabilize the cells to antibodies destroy enough of the cell wall that most bud scars become detached from their cells of origin (Haarer and Pringle, 1987). Thus, the original spatial relationships are lost, and the double-label images are not very informative. It seems possible that moderation of the normal digestion conditions might define a point at which a loose framework of cell wall still holds the bud scars in place, but is permeable to antibodies, but we are not aware of any successes along these lines.

B. Comments

1. Photobleaching

Photobleaching of Calcofluor is not normally a problem, in part because the staining is so bright to begin with. However, photobleaching is not always negligible, and may be a problem in some applications. We are uncertain of what accounts for the variable photobleaching that we and others have observed, but one factor appears to be the intensity and wavelengths of the excitatory illumination obtained with different microscopes and filter sets. In limited trials, *p*-phenylenediamine did not seem to retard appreciably the moderate rate of photobleaching observed with our Leitz Orthoplan using filter set A.

2. Vital Staining with Calcofluor

Cells are brightly stained after growth for ~2 hours in medium containing 25–500 μg/ml Calcofluor (Elorza *et al.*, 1983; Roncero and Durán, 1985; Roncero *et al.*, 1988a,b). The staining appears still to be specific for chitin (cf. Section III,B,3), but it is distinctly abnormal in pattern, reflecting the fact that Calcofluor produces severe perturbations of cell wall biogenesis, culminating in a complete growth arrest at the higher concentrations. (The severity of these effects is a function of the strain and the pH of the medium, as well as of the Calcofluor concentration.) Thus, any use of Calcofluor as a vital stain should proceed with appropriate caution.

3. Specificity of Calcofluor Staining

The degree to which the Calcofluor and primulin staining of the *S. cerevisiae* cell wall is really specific for chitin per se, as opposed to structural features of the cell wall at the base of the bud and in the bud scar, has long been controversial (Streiblová and Beran, 1963b; Seichertová *et al.*, 1973; Cabib *et al.*, 1974; Holan *et al.*, 1981). For many of the applications of Calcofluor and primulin staining (see earlier), it does not matter how this controversy is resolved. However, it can be an important issue when the dyes are used for studies of cell wall biogenesis. The sources of doubt about the staining specificity of Calcofluor can be summarized briefly as follows.

1. Although both chemical and biological studies indicate that Calcofluor can interact with chitin microfibrils, nascent polymer chains, or both (Maeda and Ishida, 1967; Herth, 1980; and see later), it is clear that the dye interacts well also with a wide range of other β-linked hexapyranose polysaccharides, including especially cellulose (Maeda and Ishida, 1967; Wood, 1980; Benziman *et al.*, 1980; Haigler *et al.*, 1980; Herth and Schnepf, 1980; Brown *et al.*, 1982; Roberts *et al.*, 1982). However, the interaction with β-1,3-linked polymers does appear to be significantly weaker than that with β-1,4-linked polymers (Wood, 1980), which may explain the weak or undetectable [see Section III,B,3(3)] staining of the glucan of the *S. cerevisiae* cell wall.

2. Calcofluor stains *S. pombe* cell walls, including especially the septal regions (B. F. Johnson *et al.,* 1982; Streiblová *et al.,* 1984; Mitchison and Nurse, 1985; Miyata *et al.,* 1986), although *S. pombe* walls appear to contain no chitin (Bulawa *et al.,* 1986).

3. Calcofluor staining of *S. cerevisiae* frequently results in some general staining of the cell wall, the extent of which varies with the Calcofluor concentration, the strain, and apparently with other unidentified factors (as there is often variability between ostensibly identical preparations made on different occasions). However, it remains unclear whether this staining reflects a weak and variable binding of Calcofluor to the glucan layer of the cell wall or a circumcellular distribution of a portion of the chitin that is not localized to the bud sites and bud scars (Bacon *et al.,* 1969; Molano *et al.,* 1980; Holan *et al.,* 1981; Roberts *et al.,*1983; Roncero *et al.,* 1988b).

4. Considerable evidence suggests that the chitin of the bud sites and bud scars, of the general cell wall [if any; cf. Section III,B,3(3)], or both, exists as an intimate complex with glucan (Bacon *et al.,* 1966, 1969; Beran *et al.,* 1970; Cabib and Bowers, 1971; Beran *et al.,* 1972; Molano *et al.,* 1980; Holan *et al.,* 1981; Mol and Wessels, 1987; Roncero *et al.,* 1988b). If this is so, it becomes more difficult to ascribe the binding of the dye to one or the other component of the complex. Moreover, one set of studies indicates that the Calcofluor-stainable ring that forms at the base of the emerging bud contains only glucan at first, and that the chitin is added only later, as the septum begins to form (Holan *et al.,* 1981). Although considerable evidence (see later) argues against the conclusion drawn from these studies, it remains unclear how to explain the data themselves.

5. Silverman *et al.* (1988) show that dying haploid cells containing neither chitin synthase 1 nor chitin synthase 2 nonetheless produce Calcofluor-stainable material. It is not clear whether this reflects the action of residual chitin synthase 2 from the heterozygous mother cell, the action of yet a third chitin synthase, or a staining by Calcofluor that is not chitin-specific.

Although the aforementioned concerns are not trivial, they appear to be outweighed by the large mass of evidence supporting the conclusion that Calcofluor staining, as normally used, reveals essentially only the chitin in the *S. cerevisiae* cell wall. The most telling points are as follows.

6. The ability of bud scars to be stained by Calcofluor is abolished by prior treatment of cell ghosts with chitinase (Hayashibe and Katohda, 1973; Sloat and Pringle, 1978); this chitinase sensitivity is drastically reduced if the scars are stained with Calcofluor prior to exposure to the enzyme (our unpublished observations). In addition, digestion of Calcofluor-stained cells with glusulase removes the bulk of the cell wall

without detectably affecting the brightly stained scars and rings around the necks of budded cells (Haarer and Pringle, 1987). However, from the available data, we cannot be certain that this applies to the very earliest rings [cf. the arguments of Holan *et al.*, 1981; Section III,B,3,(4)].

7. In certain cell cycle mutants, incubation at restrictive temperature results in the deposition of highly disorganized-looking Calcofluor-positive material (Sloat and Pringle, 1978; Sloat *et al.*, 1981; Roberts *et al.*, 1983); this deposition is paralleled by an increase in total chitin as well as by a delocalization of material that can be decorated in thin sections by gold-labeled wheat germ agglutinin [WGA; a lectin (carbohydrate-binding protein) that is specific for certain molecules, including chitin, that contain GlcNAc or *N*-acetylneuraminic acid residues (Bhavanandan and Katlic, 1979; Tartakoff and Vassalli, 1983)]. Moreover, at least in the case of *cdc24*, the staining of this material by Calcofluor is abolished by prior treatment of cell ghosts with chitinase (Sloat and Pringle, 1978). Similarly, both zygotes and shmoos produced by treatment with α factor display disorganized-looking Calcofluor-positive material in their regions of cell wall reorganization; at least in the case of shmoos, this is paralleled by an increase in total chitin (Schekman and Brawley, 1979; Roncero *et al.*, 1988a; Kim *et al.*, 1989a). These observations are difficult to reconcile with the hypothesis (Streiblová and Beran, 1963b; Seichertová *et al.*, 1973; Holan *et al.*, 1981) that the Calcofluor staining of neck rings and bud scars is due to the particular physical organization of polysaccharide microfibrils in these regions.

8. Regenerating spheroplasts that were making glucan but not chitin did not stain with Calcofluor (Elorza *et al.*, 1983).

9. The drug polyoxin D, which inhibits all known chitin synthases *in vitro* and blocks chitin synthesis *in vivo,* also prevents formation of a Calcofluor-stainable ring, even though buds do emerge (Bowers *et al.*, 1974; Cabib and Bowers, 1975; Sburlati and Cabib, 1986; Orlean, 1987).

10. Calcofluor treatment of living cells specifically stimulates chitin synthesis, even though Calcofluor inhibits chitin synthase *in vitro* (Roncero and Durán, 1985; Roncero *et al.*, 1988a); moreover, the treated cells display a delocalization of Calcofluor-positive material. These effects are probably due to interference by Calcofluor with the assembly of nascent chitin chains into normal microfibrils.

11. Under appropriate conditions, treatment with Calcofluor kills *S. cerevisiae* cells but not *S. pombe* cells (whose walls contain β-1,3-glucan but not chitin) (Roncero and Durán, 1985; Roncero *et al.*, 1988a,b). Moreover, *S. cerevisiae* mutants selected as resistant to Calcofluor have markedly reduced chitin contents and display markedly diminished staining with Calcofluor (Roncero *et al.*, 1988b). In addition, mutants

isolated by other means that have drastically reduced chitin contents also fail to stain with Calcofluor and are resistant to killing by this compound (C. E. Bulawa, personal communication).

4. Other Means of Visualizing Chitin-Rich Structures

Although Calcofluor would generally appear to be the stain of choice, other stains may have advantages in particular applications. One possible alternative is the dye Congo red (Chem. Index 22120; Polysciences cat. no. 2736; Sigma cat. no. C6767); its staining specificity and effects on viable cells appear to be similar to those of Calcofluor (Vannini *et al.*, 1983; Roncero and Durán, 1985; Roncero *et al.*, 1988b). In addition, fluorochrome-conjugated WGA (which is commercially available) can yield specific (and apparently very sensitive) staining of the chitin-rich regions of cell ghosts under appropriate conditions (Molano *et al.*, 1980; our unpublished observations), as can fluorochrome-conjugated chitinase (Molano *et al.*, 1980). Yet another possible approach is provided by antibodies, present in the sera of many rabbits, that recognize a component of the yeast cell wall that is probably chitin (Haarer and Pringle, 1987; see Section X,B,2).

IV. Staining of the Cell Surface with Concanavalin A

Concanavalin A is a lectin derived from jack beans that binds avidly to branched, α-linked mannose homopolymers such as those characteristic of the glycoproteins exposed on the surface of the yeast cell wall (Tkacz *et al.*, 1971; Ballou, 1982; Frevert and Ballou, 1985; Kukuruzinska *et al.*, 1987). Thus, conjugation of a fluorochrome to Con A provides a general fluorescent stain for the mannoprotein layer of the cell wall; this stain can be used to determine the patterns of new mannoprotein deposition and to discriminate daughter cells from mother cells (Tkacz and Lampen, 1972; Tkacz and MacKay, 1979; Sloat *et al.*, 1981; Adams and Pringle, 1984; Jacobs *et al.*, 1988). In practice, living cells are stained with the fluorochrome-conjugated Con A, unbound lectin is washed away and/or diluted, and the cells are allowed to continue growth for an appropriate period. Regions of newly formed cell wall are then revealed as non-fluorescent patches or zones on cells that are otherwise uniformly stained. If growth continues long enough after staining, daughter cells are produced that are essentially unstained, while the mother cells remain brightly stained.

A. Staining Protocol

1. Make a 1 mg/ml stock of fluorochrome-conjugated Con A [see comment (1), Section IV,B] in buffer P (10 m*M* sodium phosphate, 150 m*M* NaCl, pH 7.2).
2. Harvest 0.5–2 $\times$ 10^7 cells from growth medium by rapid centrifugation or filtration and wash with buffer P [see comment (2), Section IV,B].
3. Resuspend the cells in 1.5 ml of buffer P at an appropriate temperature (usually that at which the cells had been growing), and add 0.15 ml of the Con A stock solution.
4. Incubate 5–30 minutes [see comment (3), Section IV,B].
5. Wash the cells rapidly with buffer P and resuspend in 5–10 ml of growth medium at the appropriate temperature. Alternatively, simply dilute the stained cells into ~50 ml of growth medium without washing.
6. Immediately after resuspension and ~30 minutes later (to visualize growth zones) or 1.5–3 hours later (to discriminate mother from daughter cells), fix cells by adding formaldehyde directly to portions of the culture (see Section II,D).
7. After ~30 minutes, mount cells in *p*-phenylenediamine-containing mounting medium (Section X,D,1) and observe. Alternatively, the fixed cells can be processed for additional staining such as visualization of actin using fluorochrome-conjugated phalloidin (Section IX; Adams and Pringle, 1984). The fixed cells can also be processed for immunofluorescence (e.g., for the visualization of microtubules). However, the permeabilization step of the immunofluorescence protocol (Section X,D,2,b) ordinarily removes enough of the cell wall that the Con A-staining pattern can no longer be observed. Thus, it is necessary to work with paired samples rather than using a single sample exhibiting true double labeling (see Jacobs *et al.*, 1988).

B. Comments

1. We have successfully used both FITC–Con A and TRITC–Con A (E·Y cat. nos. F-1104 and R-1104, respectively). Other fluorochrome conjugates are also available from E·Y, Molecular Probes, and Vector Labs.
2. Recent observations in other contexts (Figs. 1 and 2; Sections II,D,2,a,VI, and VII) suggest that this protocol would be improved by

including glucose in the buffer used for washing and staining or even by doing the staining directly in growth medium. However, we have not actually tried these variations.

3. In earlier work, we stained the cells for ≥30 minutes. In more recent experiments, 5 minutes of staining were adequate. The exact duration of staining required to give good results may vary with the strain and growth conditions.

4. It might appear that a converse procedure, in which cells are first reacted with nonfluorescent Con A, then allowed to grow, then treated with fluorescent Con A, would also be effective in revealing zones of new growth. However, in limited trials we had no success with this approach; presumably, enough new Con A-binding sites appear in the old wall (through new synthesis, conformational changes, or loss of previously bound Con A) that it is not sufficiently differentiated from newly formed wall in terms of the amounts of fluorescent Con A bound. However, it should be noted that May and Mitchison (1986) apparently had excellent results in using analogous methods with *Bandeiraea simplificolia* lectin to reveal growth zones in *S. pombe* cell walls. Successive coatings either with unlabeled and FITC-labeled lectins or with FITC-labeled and Texas red-labeled lectins were effective.

V. Staining of Nuclei

Several methods are available for the fluorescence visualization of nuclear DNA. Although these methods are convenient and very useful, it should be noted that some workers still feel that staining with the nonfluorescent Giemsa dye (Hartwell *et al.*, 1970; Robinow, 1975; Adams and Pringle, 1984; Huffaker *et al.*, 1988) gives the most reliable light-microscopic visualization of nuclear positions and morphologies. In addition, it is important to realize that the position of the nuclear DNA is not invariably an accurate guide to the position of the nuclear envelope (Thomas and Botstein, 1986). In this regard, a generally available stain (immunofluorescence or other) for the nuclear envelope or nucleoplasm as a whole is much to be desired (see Section V,B).

A. DNA Stains

1. DAPI

The most generally useful of the DNA stains is 4′,6-diamidino-2-phenylindole (DAPI; Sigma cat. no. D-1388; Accurate cat. no. 18860; or other sources), introduced to yeast work by Williamson and Fennell (1975, 1979). The popularity of DAPI is due to its versatility (it can be used on living cells, fixed cells, or chromosome spreads, as described later), its specificity for DNA [which obviates the need to hydrolyze RNA prior to staining; see comment (2), this section], its spectral properties (it is excited and emits at wavelengths short enough that it does not interfere with observation of FITC and/or TRITC fluorescence; see Table II), and its moderate rate of photobleaching. A concentrated stock solution (1–10 mg/ml in water) can be made up and stored for months at 4°C. Three potential problems may be noted.

1. The staining of mitochondria [Section VI and comment (2), this section] is sometimes bright enough to be a problem for observations of the nuclei; this is particularly true of mutants where the number and morphology of the nuclei may be in question. In this situation, we have found it useful to generate petite mutants lacking mitochondrial DNA by growth in medium containing 10μg/ml ethidium bromide (Sherman *et al.*, 1986); the nuclear staining pattern (even if complicated) is then much easier to discern.
2. Although the evidence provided by Williamson and Fennell (1975) and Miyakawa *et al.* (1987) appears to establish the specificity of DAPI for nuclear and mitochondrial DNA in the staining of fixed cells, the DAPI staining of living cells may be more complicated. In particular, the chemical properties of DAPI, the differential staining of mitochondria *vis à vis* nuclei in living cells (see later and Williamson and Fennell, 1975, 1979), and the actual appearance of some stained mitochondria (e.g., Fig. 3, top panels)[5] suggest that DAPI may be able to stain mitochondria in some more general way (perhaps by functioning as a potential-sensitive dye) in addition to staining the mitochondrial DNA. None of the available evidence directly refutes this possibility or the related (if less plausible) one that DAPI might accumulate in the nuclei of living cells in a non-DNA-specific manner. These admittedly speculative concerns illustrate the special difficulties in establishing the specificity of vital staining results (Section II,C,1). In the present case, such concerns might be addressed by comparing the vital staining patterns of wild-type and ρ^--petite strains (both would show staining of mitochondrial DNA, but

[5] In this regard, the diffuse mitochondrial staining observed by Miyakawa *et al.* (1984; see especially their Fig. 3C–J) after treatment of certain glutaraldehyde-fixed cells with DAPI is also of interest. On the one hand, these images might mean that even specific staining of the mitochondrial DNA can result in a very diffuse staining pattern in cells in certain physiological states. In this case, our concerns about the specificity of DAPI as a vital stain might be overblown. On the other hand, it might be the case that glutaraldehyde fixation can preserve the conditions necessary for a more general staining of mitochondria by DAPI.

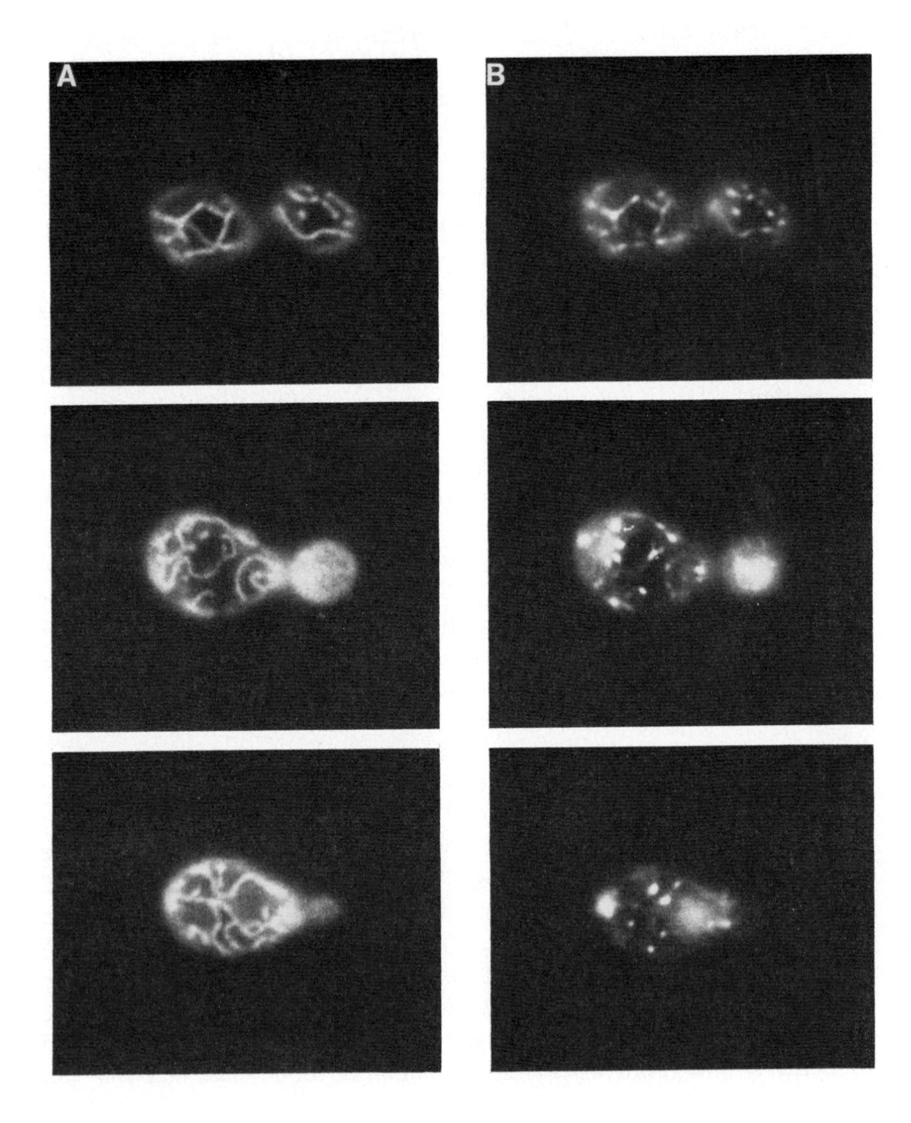

FIG. 3. Double staining of yeast mitochondria with (A) $DiOC_6(3)$ and (B) DAPI. Diploid cells grown in YPG (Sherman *et al.*, 1986; 3% glycerol as carbon source) were stained vitally as described in the text (Section VI). Three cells are pictured, each viewed (A) with a fluorescein filter set to visualize $DiOC_6(3)$ and (B) with a UV (excitation) filter set to visualize DAPI.

the latter might fail to show the hypothetical general mitochondrial staining) and by vital staining of *ndc1* mutant strains (in which the DNA all stays near one pole of the elongating nucleus; Thomas and Botstein, 1986).

3. Allan and Miller (1980) have reported that vacuolar polyphosphate granules can be stained by DAPI in cells grown, fixed, and stained under certain conditions. Although this does not ordinarily seem to be a problem, it may be one in some applications.

a. Vital Staining. Growth in medium containing 0.5–1 μg/ml DAPI is not grossly deleterious to the cells of most strains (although petite mutants are generated in some strains; Williamson and Fennell, 1979) and eventually (after $\geq$1 hours) produces good staining of the nuclei (Williamson and Fennell, 1975, 1979; Moir *et al.*, 1982). In our experience, staining is better in medium at relatively high pH than in medium at relatively low pH, although there may be differences among strains in this regard. It is not necessary to wash the cells before viewing. Mitochondria are also stained under these conditions, rather more rapidly and/or brightly than the nuclei [see Section VI,A and comment (2), this section].

b. Staining of Fixed Cells. For simply observing nuclei, we have found (following Williamson and Fennell, 1975) that the best results are obtained after fixing cells in 70% ethanol. Thus, a culture sample is mixed with two volumes of ethanol, or pelleted cells are resuspended in 70% ethanol. After ~30 minutes, the cells are washed once with water and resuspended in a solution containing 0.1–0.5 μg/ml DAPI in water. Staining of both nuclear and mitochondrial DNA is good within a few minutes and can be viewed without washing the cells. Thomas and Botstein (1986) also obtained good staining using 1 μg/ml DAPI in buffer after fixing cells in 3 : 1 (v/v) methanol–acetic acid. Formaldehyde-fixed cells can also be stained with 0.1–0.5 μg/ml DAPI in water, but the staining requires longer incubations (Williamson and Fennell, 1975, 1979), is more variable from cell to cell (our unpublished observations), or both. In addition, Miyakawa *et al.* (1984) obtained apparently excellent DAPI staining of both nuclear and mitochondrial DNA using a procedure that involved glutaraldehyde fixation.

c. Staining in Conjunction with Immunofluorescence. We normally get excellent staining of both nuclear and mitochondrial DNA in cells that have been prepared for immunofluorescence simply by including ~50 ng/ml DAPI in the mounting medium, as described in Section X,D,1.

d. Staining of Chromosomes in Spread Preparations. Although the fluorescence micrographs published to date have shown chromosomes stained with acridine orange (Dresser and Giroux, 1988), DAPI staining of spread chromosomes is equally satisfactory and more reproducible (C. Giroux, personal communication). The only disadvantage is that DAPI staining does not provide the vivid heterochromatic discrimination between the chromosomes and the nucleolus that is seen with acridine orange. Detailed methods for working with spread chromosome preparations are available from C. Giroux.

2. Other DNA Stains

a. Mithramycin. Slater (1976, 1978) achieved good staining of nuclear DNA with mithramycin using a simple procedure that did not require hydrolysis of RNA. In our limited personal experience with mithramycin, we found photobleaching to be more of a problem than with DAPI; however, this problem may be alleviated by using other filter combinations (Slater, 1976, 1978). In addition, we note that Allan and Miller (1980) found mithramycin superior to DAPI under certain conditions (although it does not appear that these workers had optimized the conditions for DAPI staining).

b. Acridine Orange and Hoechst 33258. Both of these dyes have given good fluorescence staining of chromosomes in spread preparations (Dresser and Giroux, 1988; C. Giroux, personal communication). In addition, M. Snyder has obtained good staining of nuclear DNA in whole cells with Hoechst 33258 using either of two simple procedures (Snyder and Davis, 1988; M. Snyder, personal communication). In one, formaldehyde-fixed cells that have been prepared for immunofluorescence (Section X,D,2) are mounted for viewing in a medium containing 70% glycerol, 2% *n*-propyl gallate, and 0.25 μg/ml Hoechst 33258 in PBS (cf. Sections II,F,3,c, V,A,1,c, and X,D,1). Nuclear staining is good after ~10 minutes. In the other procedure, cells that have simply been fixed by methanol and acetone treatment [cf. Section X,D,2,d(2)] are mounted in the mounting medium just described. Although we have no personal experience with the use of acridine orange on whole cells, it is apparently somewhat cumbersome to use (Slater, 1976; C. Giroux, personal communication), but may offer the advantage of staining the nucleolar region as well as the DNA-rich region of the nucleus (Section V,A,1,d; Royan and Subramaniam, 1960).

c. Propidium Iodide. Yeast nuclei have been stained with propidium iodide for both flow-cytometric (Hutter and Eipel, 1978; Segev and Botstein, 1987; Huffaker *et al.*, 1988; Eilam and Chernichovsky, 1988) and microscopic (Matsumoto *et al.*, 1983a,b, 1985; Uno *et al.*, 1984, 1985; Eilam and Chernichovsky, 1988) applications. The staining is more cumbersome than that with DAPI or mithramycin, as RNA must be hydrolyzed prior to or concomitant with staining [optimal results may require some protein hydrolysis as well (Hutter and Eipel, 1978; Segev and Botstein, 1987; Huffaker *et al.,* 1988)], and the published micrographs are only moderately clear. However, propidium iodide does offer the possibility of quantitative determinations of the amounts of DNA per nucleus. This seems demonstrably effective at the population level, in that either flow cytometry or individual-cell microfluorimetry (Matsumoto *et al.,* 1983a,b, 1985; Uno *et al.,* 1984, 1985) can resolve distinct peaks at the 1C, 2C, and 4C DNA levels within a given population and can document shifts in the relative numbers of cells comprising those peaks. However, it should be noted that the absolute values of the fluorescence intensities associated with the different peaks frequently vary from population to population, and both the flow cytometry profiles and the individual-cell microfluorimetry histograms display considerable variations in fluorescence among cells that presumably have identical DNA contents. Possibly these problems reflect varying efficiencies of the RNase digestion step. In any case, it seems that quantitative DNA determinations (particularly at the individual cell level) should be approached with caution.

B. Other Nuclear Stains

There are as yet no standard methods for fluorescence staining of the nuclear envelope or nucleoplasm as a whole. However, there are several promising approaches that may lead to such standard methods or at least be useful in particular situations. The *ndc1* mutant, in which nuclear segregation and division occur without chromosome segregation (Thomas and Botstein, 1986), appears to provide a useful test for any purported general nuclear stain.

1. Antitubulin immunofluorescence using commercially available antibodies (Section X) allows localization of the spindle-pole bodies (and hence of the associated nuclear envelope), except in cells that have lost their microtubules because of mutation or drug treatment. This approach allowed Thomas and Botstein (1986) to demonstrate the nature of the defect in the *ndc1* mutant.
2. In screening mAbs generated against total nuclear proteins, Aris and Blobel (1988) and K. Armstrong, J. Broach, and M. Rose (personal communication) have identified several that stain nucleoli specifically in whole-cell immunofluorescence. At least some of these antibodies are available to interested investigators. Fluorescence staining of nucleoli

may also be possible using acridine orange (cf. Sections V,A,1,d and V,A,2,b).

3. M. Snyder and C. Copeland (personal communication) have observed that the mAb RL1 (Snow *et al.*, 1987) gives a general staining of the yeast nuclear envelope. The available evidence suggests that this antibody recognizes the O-linked GlcNAc that is found mainly or solely on a group of proteins associated with the nuclear pore complex (at least in mammalian cells).

4. In screening mAbs raised against *Drosophila* heads by S. Benzer and co-workers, B. Weinstein and F. Solomon (personal communication) observed that antibody 8C5 appeared to give a general staining of the yeast nuclear envelope or nucleoplasm. Although the specificity of this antibody is as yet poorly defined, it may still be a useful reagent for morphological studies.

5. In preliminary experiments, P. A. Silver (personal communication) observed that FITC-labeled WGA (see Sections III,B,3–4) appeared to give a general staining of the nuclear envelope or nucleoplasm in cells that had been prepared essentially as for immunofluorescence. The staining was competed with by excess GlcNAc, and might reflect reaction with nuclear pore-associated proteins [cf. item (3), this section], with GlcNAc-containing elements in the endoplasmic reticulum (ER) [which in normal yeast cells may be largely circumnuclear; cf. Wright *et al.* (1988) and item (6), this section], or with GlcNAc-containing or *N*-acetylneuraminic acid-containing elements in the Golgi [cf. the staining of the Golgi in mammalian cells by WGA (Tartakoff and Vassalli, 1983; Segev *et al.*, 1988)]. M. Snyder (personal communication) also had promising results in staining isolated nuclei with WGA, but had difficulties with staining of residual cell wall when he attempted to apply this to whole cells.

6. Immunofluorescence with *KAR2*-specific antibodies yields essentially a circumnuclear staining pattern (M. D. Rose, personal communication). As several lines of evidence suggest that the *KAR2* product is an ER-localized protein, this staining may be viewed more appropriately as ER-specific rather than nuclear envelope-specific. Nonetheless, if used with circumspection, it may be useful in tracking the nuclear envelope in certain applications.

7. In several cases, immunofluorescence with anti-β-galactosidase antibodies has demonstrated the nuclear localization of fusion proteins consisting of part of a yeast nuclear-localized protein plus the bulk of *Escherichia coli* β-galactosidase (Hall *et al.*, 1984; Silver *et al.*, 1984; Moreland *et al.*, 1985). Although not proven, it is likely that in some or all of these cases, the immunofluorescence staining reveals the full extent of the nuclear envelope and/or nucleoplasm.

VI. Staining of Mitochondria[6]

A. Vital Staining

Specific vital staining of mitochondria allows investigation of subjects such as variations in mitochondrial morphology in response to changes in environmental conditions, the role of cytoskeletal elements in mitochondrial movements, etc. Such staining can be accomplished using 3,3′-dihexyloxacarbocyanine iodide [$DiOC_6(3)$; Molecular Probes cat. no. D-273]. This dye is one of a large class of carbocyanine compounds that are sensitive to changes in membrane potential (Waggoner, 1979, 1986; Wang and Taylor, 1989; Molecular Probes Handbook). The mitochondrial DNA can also be visualized in living cells by vital staining with DAPI (Williamson and Fennell, 1975, 1979; Moir *et al.*, 1982; see also Section V,A,1 for further discussion of the specificity of DAPI in vital staining). These two stains—$DiOC_6(3)$ and DAPI—can be used in single-label or double-label experiments (Fig. 3).

1. Staining Protocols

1. Keep $DiOC_6(3)$ as a 0.5 mg/ml stock solution in 100% ethanol at −20°C, and dilute 1 : 100 in H_2O to make a working stock. Keep DAPI as a 1 mg/ml stock solution in water at 4° or −20°C.
2. When desired, add 1 μg/ml DAPI to growing cells in culture medium ~1 hour before viewing or staining with $DiOC_6(3)$.
3. Add $DiOC_6(3)$ to the cells growing in culture medium so that the final concentration is ~50 ng/ml. Typically, 1 μl of the 1 : 100 dilution is added to 100 μl of culture.
4. After incubating the cells for ≥2 minutes, view $DiOC_6(3)$ fluorescence using a standard fluorescein filter set (Table II). When present, view DAPI using UV excitatory illumination (Table II).
5. If cells are to be immobilized for photomicroscopy, mix one volume of culture containing the dye with one volume of the same medium (including carbon source) that has been prepared with 1% low melting point agarose and maintained at 35°C. Place a drop of the mixture quickly on an ordinary slide and cover with a coverslip.

2. Comments

1. Successful staining of yeast and other fungal mitochondria has also been achieved with the dyes 2-(4-dimethylaminostyryl)-*N*-methylpyri-

[6] Questions about this section should be addressed to T. Stearns.

dinium iodide (DASPMI or 2-Di-1-ASP, Molecular Probes cat. no. D-308; Miyakawa *et al.*, 1984; Brakenhoff *et al.*, 1986) and rhodamine 123 (Molecular Probes cat. no. R-302; Oakley and Rinehart, 1985). In our experience, however, $DiOC_6(3)$ is more effective. In addition, we have observed that another carbocyanine dye, 1,1′-dipentyl-3,3,3′,3′-tetramethylindocarbocyanine iodide [$DiIC_5(3)$; Molecular Probes cat. no. D-279], also gives apparently successful vital staining of yeast mitochondria. Although the specificity and reproducibility of this staining have not yet been examined in detail, it is worth noting that $DiIC_5(3)$ uses rhodamine rather than fluorescein filters, and hence may be useful in certain double-label applications.

2. Yeast cells and their mitochondria respond rapidly to changes in their environment; thus, it is best to perform all manipulations of the cells in their growth medium. For example, cells should not in general be washed in water or buffer before staining, as the lack of a carbon source results in rapid morphological changes of the mitochondria. Similarly, as the amount of oxygen available to cells under a coverslip is limited, viable cells soon become anaerobic and stop respiring, and the mitochondrial staining with potential-sensitive dyes becomes weaker. Thus, it is generally best to view and (especially) photograph cells as soon as possible after preparing the slide (see also Section II,C,4).

3. The carbon source in the medium has a large effect both on mitochondrial morphology and on the amount of dye required to visualize the mitochondria. Glucose and (to a lesser extent) other fermentable carbon sources repress mitochondrial function. Cells grown on these carbon sources require ~2- to 5-fold more $DiOC_6(3)$ to get good staining than do cells grown on nonfermentable carbon sources such as glycerol, ethanol, or lactate. The protocol just given was optimized for cells grown on 3% glycerol.

4. The mitochondria of ρ^0, ρ^-, or nuclear petite strains do not maintain a membrane potential, and thus do not stain with $DiOC_6(3)$.

5. Adding too much dye (~3- to 4-fold more than is needed to get good mitochondrial staining in glucose-grown cells) results in nonspecific staining of many or all cell membranes. In mammalian cells, high concentrations of $DiOC_6(3)$ apparently allow effective visualization of the ER (while the mitochondria appear as dark, overstained blobs) (Lee and Chen, 1988).

6. $DiOC_6(3)$ photobleaches at a moderate rate. This is not a serious problem but is somewhat troublesome if photographic exposures >1 minute are necessary. Addition of *p*-phenylenediamine (Section II,F,3,c) did not seem to help much (as well as having unknown effects on viable yeast cells).

7. The mitochondria in living cells do move around, so in photomicroscopy it is best to expose for the shortest possible times in order to get sharp images.

B. Staining of Fixed Cells

The mitochondrial DNA in fixed as well as living cells is stained by DAPI (Williamson and Fennell, 1975, 1979; Miyakawa *et al.*, 1984, 1987; see also Section V,A,1). Methods for visualizing the whole mitochondria in fixed cells would also be useful in certain types of experiments (e.g., colocalization studies using immunofluorescence). It appears that little effort has been expended to date on the development of such methods. Nonetheless, some encouraging results have been obtained. We have observed some weak staining of the mitochondria by immunofluorescence using a polyclonal serum (obtained from T. Mason) against a mixture of F_1-ATPase subunits. C. Kaiser (personal communication) observed that cells carrying a *SUC2* gene with a particular mutant signal sequence showed staining of the mitochondria with antiinvertase antibodies. J. Li and N. Martin (personal communication) observed immunofluorescence staining of mitochondria using anti-β-galactosidase antibodies in cells harboring *TRM1–lacZ* or *ATP2–lacZ* fusion genes (although in the former case, nuclei were also stained, and in the latter case, not all mitochondria were stained). These partial successes suggest that further investigation of antibodies against major components of the mitochondrial membranes, perhaps in conjunction with variations in the standard fixation and permeabilization protocols (cf. Sections II,D,2,b, X,D,2,b and d), would identify generally useful and reliable reagents.

VII. Staining of Vacuoles[7]

A. Vital Staining

Vital staining of vacuoles is proving very useful both in physiological studies (e.g., monitoring vacuole behavior during the cell cycle and in response to changing environmental conditions) and in genetic studies (e.g., isolation and analysis of mutants defective in vacuole biogenesis,

[7] Questions about this section should be addressed to R. A. Preston or E. W. Jones.

endocytosis, or acidification). In presenting the protocols that follow, we emphasize that the points of concern as regards the validity of vital staining results (Section II,C,1–2) have been addressed with varying degrees of confidence for the various procedures. In general, the protocols have been most thoroughly tested in the X2180 genetic background, using cells grown in YP+D rich liquid medium to midexponential phase (1–5 × 10^7 cells/ml) at 30°C. Particular applications may require the use of other strains and of other procedures for cell growth, labeling, and washing. If so, appropriate tests of the staining kinetics, specificity, and stability should be performed.

Evaluation of the vital staining protocols requires attention not only to the usual questions of stain specificity but also to the question of what constitutes normal vacuole morphology in unperturbed cells. In fact, examination of the current and earlier literature reveals considerable disagreement on the latter point. We believe that this confusion reflects the highly dynamic structure of the vacuole; thus, the view obtained can depend on seemingly minor details of the culture conditions and visualization method. When used as will be recommended here, the sulfonate, fluorescein, and *hem* mutant protocols all provide a consistent picture of "the vacuole" as a collection of some 5–30 small vacuoles per cell in the great majority of normal, exponentially growing cells (Fig. 2A). [As reported previously (Wiemken *et al.*, 1970; Hartwell, 1970), large-budded cells tend to have fewer, larger vacuoles; however, in our observations, this tendency is not absolute (e.g., see the cell in the middle of Fig. 2A).] The specificity of the sulfonate and fluorescein dyes for the vacuole is supported by the observation that they show only vestigial labeling of mutants (*pep3, pep5, end1*) that produce only vestigial vacuoles as judged by electron microscopy (Chvatchko *et al.*, 1986; our unpublished observations). The picture of normal vacuole morphology in exponentially growing cells as provided by the fluorescent stains is further supported by phase-contrast and DIC observations on unstained cells (the small vacuoles are indistinct or invisible; see also Wiemken *et al.*, 1970; Hartwell, 1970), by immunofluorescence observations on rapidly fixed cells (Sections II,D,2,a and VII,B), and by electron-microscopic observations on rapidly frozen cells (Wiemken *et al.*, 1970).

However, vacuolar morphology can change rapidly when cells are subjected to stresses, including those arising during handling prior to microscopy. For example, resuspension of cells in glucose-free buffer causes a rapid (1–10 minutes at 24°–30°C) coalescence of the multiple small vacuoles into a single large vacuole that is brightly stained by both the fluorescein (Fig. 2B) and sulfonate (data not shown) dyes and is also conspicuous by phase-contrast or DIC microscopy and by immunofluorescence on fixed cells.[8] A similar coalescence of the vacuole is induced by staining with weakly basic dyes (Section VII,A,4,a), by spheroplasting (Wiemken *et al.*, 1979), and probably by other environmental insults yet to be characterized. This plasticity of vacuolar morphology clearly means that staining methods must be applied with caution if the results obtained are to reflect accurately the behavior of the vacuole in unperturbed cells.

1. Sulfonate Dyes

a. General Remarks. Multiple sulfonate groups cause these dyes to be highly polar and, it is presumed, correspondingly membrane-impermeant. Riezman and co-workers introduced lucifer yellow carbohydrazide (LY-CH; Molecular Probes cat. no. L-453) as a marker for fluid-phase endocytosis in yeast (Riezman, 1985; Chvatchko *et al.*, 1986), but it serves simultaneously as a vacuole marker, as argued earlier. Pyranine (8-hydroxypyrene-1,3,6-trisulfonic acid, trisodium salt; Wolfbeis *et al.*, 1983; Molecular Probes cat. no. H-438) has labeling properties similar to those of LY-CH and is much less expensive. The same staining protocol is suitable for either dye.

b. Staining Protocol. Harvest 1–2 × 10^7 cells by rapid filtration. Resuspend in 1 ml of growth medium containing 5 mg/ml LY-CH or Pyranine. Shake 60–90 minutes at 30°–37°C or for ~2 hours at 24°C. Filter and wash (on the filter) with 2–5 ml PBS + 2% glucose or 0.1 *M* sodium phosphate + 2% glucose (pH 7) at 0°C. Resuspend in 0.1 ml buffer + 2% glucose at 0°C. Place a drop on an ordinary microscope slide (if desired, pretreat the slide with Con A solution to immobilize the cells—see Section II,F,2,b) and examine immediately by fluorescence microscopy using a filter set that provides violet excitatory illumination (see Table II).

c. Comments. The following points should be kept in mind when using the sulfonate protocol.

1. Note that LY-CH is different from lucifer yellow vinylsulphone (LY-VS). We have no experience with LY-VS. Both dyes are covalent labeling reagents, but LY-CH appears to be unreactive and nontoxic under physiological conditions (Stewart, 1981). Labeling of yeast vacuoles by LY-CH is reversible at least to the extent that washing reduces the amount of label to below the limit of detectability by fluorescence microscopy.

2. The protocol requires high concentrations of LY-CH or Pyranine,

[8] After the coalescence of vacuoles induced by glucose deprivation, a few very small, noncoalesced vesicles remain detectable by fluorescence. The nature of these vesicles is an open question.

so it is particularly important, in some applications, to control for labeling by trace impurities (Section II,C,1,b).

3. The cell wall region is stained to some extent with LY-CH, particularly if cells are grown in YPD rather than YP+D medium (Section II,B). Vacuoles stain more intensely in YP+D than in YPD, in our experience, but the more conspicuous cell wall staining observed in YPD-grown cells might be useful for some purposes.

4. For fluorimetric, rather than microscopic assays, background fluorescence should be minimized by resuspending cells (washed on a filter) in 5 ml additional wash buffer (0° or 24°C) in a test tube, then refiltering.

5. Pyranine may be useful as an *in vivo* pH indicator for yeast vacuoles. The fluorescence excitation spectrum of this dye changes over a pH range (6–8) appropriate for ratio imaging or dual-excitation flow cytometry. Leakage of the dye renders it less suitable for pH measurements by conventional fluorimetry. Also, Pyranine fluorescence is particularly sensitive to conditions unrelated to pH (e.g., ionic strength), so that use of the dye as a pH indicator *in vivo* would require careful calibration.

2. Fluorescein Dyes via Nonfluorescent, Membrane-Permeant Precursors

a. General Remarks. Fluorescein-based dyes have a long history of biological applications, particularly in relation to intracellular and intraorganellar pH measurements. We have observed that 5(6)-carboxyfluorescein diacetate (CFDA; Bruning *et al.*, 1980; Molecular Probes cat. no. C-195) and 5(6)-carboxy-2′,7′-dichlorofluorescein diacetate (CDCFDA; Molecular Probes cat. no. C-369) label yeast vacuoles more-or-less specifically, as argued before. The mechanism of labeling presumably depends on removal of the acetate ester groups by nonspecific esterases, followed by trapping or accumulation of polyanionic products, possibly with the assistance of the vacuolar membrane potential. The relative contributions of cytoplasmic and vacuolar esterases to this process are not known, although pleiotropic mutations that eliminate many vacuolar hydrolases do not eliminate staining by CDCFDA.

b. Staining Protocol. Harvest 1–2 × 10^7 cells by rapid filtration. Resuspend in 1 ml of YP+D that has been made 50 m*M* in citric acid and 5 μ*M* in CFDA (final pH 3.0) or 1 ml of YP+D that has been made 50 m*M* in citric acid and 1–10 μ*M* in CDCFDA (final pH 3.0–6.5). Shake at 30°C for 10–30 minutes (CDCFDA) or 30 minutes (CFDA). Examine directly by fluorescence microscopy using a standard fluorescein filter set (Table II), or, for increased vacuole specificity, filter, wash, and resuspend as in the sulfonate protocol. This latter step should remove essentially all of the cytoplasmic label, and will concentrate the cells for efficient photomicrography. In addition, the washed cells can be immobilized on the slide using the Con A method (Section II,F,2,b) if desired.

c. Comments. The following points should be kept in mind when using the fluorescein protocol.

1. The CDCFDA procedure yields bright vacuolar staining under a wide range of conditions, as long as the pH of the staining mixture is kept below 7. For example, cells scraped from a colony into a mildly acidic staining mixture (~10 μl) on a microscope slide should give easily visible vacuoles a few minutes later. In a pH 3 staining medium, 2 μM CDCFDA is an appropriate concentration; higher concentrations may be required in media that have a pH >3. In medium with pH approaching 7 (e.g., in fresh, unbuffered YP+D), CDCFDA is unstable, giving high background fluorescence unless the wash step (see earlier) is included. With CFDA, pH and timing are more critical to obtaining good staining; its advantage over CDCFDA is that its pK_a is more appropriate for *in vivo* vacuolar pH measurements [see comment (4), this section].

2. A mixture of the 5-carboxy and 6-carboxy isomers of these dyes [indicated as 5(6)-carboxy-] has been used in most applications to date. However, we found marked differences in the rates of staining of yeast vacuoles with the two isomers of CFDA (we have not tested the separate isomers of CDCFDA). Vacuoles label more slowly with 6-CFDA than with 5-CFDA, but 6-CFDA also appears to be more uniformly retained after washing. The 30-minute staining period in the protocol is appropriate for 6-CFDA.

3. Medium YP+D plus citrate forms a precipitate when adjusted to pH 3. Decant or filter after autoclaving to remove it.

4. The CFDA procedure permits determination of vacuolar pH either by flow cytometry (Wilson and Murphy, Chap. 16, this volume) or by fluorescence ratio microscopy (Taylor and Wang, 1989). Although flow cytometry provides more accurate pH measurements, fluorescence microscopy is more useful for screening large numbers of mutants and provides morphological information not obtainable by flow cytometry. Thomas *et al.* (1979) first described the use of 6-CFDA as an intracellular pH probe, and Bright *et al.* (1987) described pH assays by fluorescence ratio microscopy of mammalian cells. We are using the following simplified variant of those methods to screen for yeast mutants that have abnormally acidic or alkaline vacuoles. A standard Zeiss epifluorescence microscope is equipped for both 450-nm and 495-nm excitation, using 10-nm bandpass filters (Omega). (We use the standard Zeiss dual-position

filter cube with two fluorescein filter sets no. 487709, but with the excitation filters replaced with the aforementioned bandpass filters, and with an extra LP520 emission filter to reduce the leakage of the 495-nm excitatory illumination.) Using wild-type cells that have been labeled with 6-CFDA and washed as described before, the intensity of the 495-nm excitation is adjusted so that the emission from the vacuoles is the same intensity at both excitatory wavelengths. This is accomplished by adding one or more gelatin neutral-density filters (Kodak), cut to 18 mm diameter, to the objective side of the 495-nm interference filter. Very slight differences in emission intensity can be detected by rapidly switching between the 450- and 495-nm excitations (the filter cube should be well greased). Given the spectral characteristics of carboxyfluorescein (CF, the product of CFDA hydrolysis; see Thomas *et al.*, 1979, and Table II), mutants labeled in parallel cultures will now be differentially dimmer at 450 nm (relatively alkaline vacuoles) or brighter at 450 nm (relatively acidic vacuoles) than they are at 495 nm.

Under the labeling conditions just described, our wild-type strain accumulates 6 pmol of CF per 10^6 cells, based on fluorimetric assays of cell extracts. In some cases, mutants labeled in the same concentration of 6-CFDA as used for wild type will accumulate greater or lesser absolute amounts of CF, and hence be more or less intensely fluorescent than the wild type at both wavelengths. This need not interfere with the ratio pH assay, unless the dye concentration is so high that quenching occurs, or so low that differences in emission are difficult to see. With mutants that seem much brighter than wild type, we relabel using a lower 6-CFDA concentration to avoid possible quenching. The cells used for these assays must be sufficiently well washed (and resuspended in a non-fluorescent medium) to avoid changes in background fluorescence (450 versus 495 nm) that could obscure or be confused for changes of vacuole fluorescence. This is less of a problem with well-labeled cells, because these allow the use of low-level bright-field or phase-contrast illumination simultaneously with fluorescence, thereby making small differences in fluorescence background negligible.

The assay just described is sufficiently sensitive to detect relative differences of 0.3–0.5 pH unit, depending on the observer's visual acuity. The sensitivity can be tested and the observed differences in emission ratio calibrated to known pH values by observing labeled cells suspended in buffered ionophore cocktails at various pH values. We use cocktails containing 50 m*M* HEPES, 50 m*M* MES, 50 m*M* KCl, 50 m*M* NaCl, 200 m*M* ammonium acetate, 10 m*M* NaN_3, 10 m*M* 2-deoxyglucose, and 50 μM carbonylcyanide chlorophenylhydrazone, at pH 5–7.5 in steps of 0.5 pH unit. After determining the density of the neutral-density filters required to equalize emission intensities at the two excitatory wavelengths with cells in ionophore cocktails at a series of pH values, the pH in any vacuole (in the absence of ionophores) is assayed by determining the filter density required to equalize emission and comparing that density to the calibration curve obtained from cells in the ionophore buffers.

We have identified several mutants that have abnormal vacuole acidity using the aforementioned assay. In two of these mutants, the abnormality becomes apparent only after a slide has "incubated" several minutes on the microscope stage at room temperature. This effect may or may not be related to the anaerobiosis that eventually occurs with such slides; in any case, it provides yet another example of an unanticipated complication attending the use of a vital stain.

3. Endogenous Fluorophores

In some applications, it would be convenient to be able to identify vacuoles without using synthetic stains. One possible approach is provided by various mutations that cause the accumulation of fluorescent metabolic intermediates or by-products in the vacuole. This lets the yeast do the work, in the sense that labeling is obtained under specific physiological conditions without further manipulation. However, because of the mutant background, the physiological conditions that must be employed, or both, the vacuolar labeling observed will not necessarily reflect the behavior of the vacuoles of normal wild-type cells, even under the same growth conditions.

a. ade1 and ade2 Mutations. These mutations have been observed to result in fluorescent vacuoles in cells grown on SC medium containing limiting amounts of adenine (10–12 μg/ml) (Weisman *et al.*, 1987; Bruschi and Chuba, 1988). The fluorescence was reported to be present in vacuoles in "late logarithmic" or "early stationary" cultures. The state of the vacuole under these conditions of starvation is likely to be different from that in healthy cells, especially as the vacuole appears to be involved in coping with starvation (Jones, 1984). Indeed, the morphology of the vacuoles in the cited reports differs markedly from that observed in midexponential-phase, unstarved cells (Fig. 1A). The specific causes of this difference are not known. The red pigmentation of adenine auxotrophs has been a very convenient marker in genetic analyses; possibly their fluorescence could also be useful in this way.

b. Porphyrin Mutations. Mutant alleles of several genes in the heme-biosynthetic pathway cause colonies to fluoresce strongly under near-UV illumination (Kurlandzka and Rytka, 1985). We isolated a mutant strain with that phenotype and found that the fluorescence was localized to the vacuole, as identified by simultaneous labeling with CDCFDA. The mutation in this strain is allelic to *hem12* mutations (J. Rytka, personal communication). In some applications, vacuolar labeling caused by *hem12* alleles may be preferable to that due to *ade* mutations, as overt starvation is not required for labeling and vacuole morphology is not grossly altered by (at least some) *hem12* mutations. Nonetheless, the physiological effects of these *hem12* mutations remain to be described in detail, so they should be used with appropriate caution.

4. Dyes That Have Known or Suspected Problems

Because it is impossible to imagine all of the potential applications of fluorescent vacuole markers, we mention the following dyes that may be useful in specific situations, but that have known or suspected disadvantages in other situations.

a. Cationic ("Lysosomotropic") Dyes. Chloroquine, quinacrine, neutral red, and a host of other weak bases accumulate in vacuoles and can be detected by fluorescence. However, these dyes have a variety of physiological effects in yeast (Pena *et al.*, 1979; Lenz and Holzer, 1984; Kalisz *et al.*, 1987) and other systems (de Duve *et al.*, 1974; Ohkuma and Poole, 1978) and thus are far from being inert markers. In particular, they promote coalescence of vacuoles in exponential-phase cells, giving an artifactual view of vacuolar morphology. While potentially useful as probes of vacuolar physiology, these dyes certainly do not qualify as nonperturbing vacuole markers.

b. FITC-Conjugated Dextran. Vacuolar labeling by impure samples of 70-kDa FITC–dextran is due to fluorescent impurities (including FITC) in the dextran (Preston *et al.*, 1987). There is no evidence that pure macromolecular FITC–dextran can label the vacuole.

c. FITC. This dye, usually used as a covalent protein-labeling agent, labels vacuoles at low concentrations (Preston *et al.*, 1987). However, it has been reported to inhibit certain ATPases (Pick and Bassilian, 1981) and vesicle fusion at the plasma membranes of growing pollen tubes (Picton and Steer, 1985) at micromolar concentrations. These findings (and seemingly common sense) argue against attempts to use this reactive substance as a nonperturbing vital stain.

B. Staining of Fixed Cells

For the reasons already discussed (Section II,C,1–2 and introduction to Section VII,A), it is important to corroborate by independent methods the results obtained using vital vacuole stains. In addition, certain applications (e.g., double-label experiments involving immunofluorescence) require unambiguous identification of the vacuole in fixed cells. Antibodies to vacuole-specific markers should provide the means to both of these ends. The use of rapid fixation methods (Section II,D,2,a) during the preparation of cells for immunolabeling should presumably circumvent artifacts due to handling or the vital-staining procedures themselves. The feasibility of using antibodies to the vacuolar protease carboxypeptidase Y (CPY) for immunoelectron-microscopic studies of the vacuole has been demonstrated (van Tuinen and Riezman, 1987). We have also used affinity-purified anti-CPY (Stevens *et al.*, 1982) to label the vacuole successfully by indirect immunofluorescence (Section X,D) in both unstressed and glucose-starved cells. As indicated earlier (introduction to Section VII,A), the results were fully consistent with those of vital staining using the sulfonate, fluorescein, or *hem* mutant protocols.

Of the synthetic vital stains, only LY-CH has been reported to be fixable, and, in our hands, fixation caused redistribution of LY-CH to the cytoplasm. Thus, double labeling of fixed cells using the vital stains and immunofluorescence seems unlikely to be effective, although it remains possible that the porphyrin derivatives accumulated in *hem12* mutants may be fixable.

VIII. Staining of Other Membrane-Bounded Organelles

Reliable, generally available methods for fluorescence visualization of the ER, Golgi, secretory vesicles, and peroxisomes in yeast would clearly be of value in a wide variety of studies. Although such standard methods do not exist as yet, there are several promising approaches that may lead to their development or at least be useful in particular applications. The *sec* mutants that exhibit hypertrophy of ER, Golgi, or secretory vesicles (Schekman, 1985) should be useful in evaluating the specificity of possible probes for these organelles.

1. Apparently specific visualization of the ER in living mammalian cells has been achieved with the dye $DiOC_6(3)$ (Lee and Chen, 1988), using concentrations at which the mitochondria are grossly overstained (cf. Section VI,A). It is not clear whether this would work with yeast.

Visualization of the ER in fixed cells should be possible using immunofluorescence with antibodies specific for ER-localized proteins. However, interpretation of the results obtained to date is complicated by uncertainty about the localization of the ER in normal cells. For example, it is not clear whether the circumnuclear localization of the *KAR2* product [Section V,B(6)] reflects the actual distribution of the ER or a restriction of this protein to a particular portion of the nuclear envelope–ER system. A similar uncertainty afflicts interpretation of the circumnuclear staining observed with WGA [Section V,B(5)] and with antibodies to HMG–CoA reductase in cells that are overproducing this protein (Wright *et al.*, 1988; Wright and Rine, Chap. 23, this volume; see also Section X,E,2). It seems likely that further studies employing these and similar reagents will soon provide satisfactory methods for visualizing the ER.

2. Apparently specific visualization of the Golgi in living mammalian cells has been achieved using a fluorescent sphingolipid (Lipsky and Pagano, 1985). It is not clear whether this would work with yeast; administration of the probe to whole cells seems likely to be a problem (it was administered to the mammalian cells in liposomes or as a complex with BSA), but spheroplasts might be more tractable. Visualization of the Golgi in fixed cells should be possible using antibodies to Golgi-localized proteins. For example, antibodies to the *YPT1* product appear to stain the Golgi specifically both in yeast and in mammalian cells (Segev *et al.*, 1988). The staining observed with WGA may also be relevant to the problem of Golgi localization [cf. Section V,B(5)].

3. Specific visualization of secretory vesicles has apparently been achieved using antibodies to the *SEC4* product (Goud *et al.*, 1988), and should also be possible with antibodies to other vesicle-localized proteins. We are not aware of any promising approaches to the specific visualization of secretory vesicles in living cells.

4. The peroxisomes of the yeast *Candida tropicalis* have been visualized effectively by immunofluorescence using an antiserum raised against total *C. tropicalis* peroxisomal proteins (Small *et al.*, 1987; P. Lazarow, personal communication). As this worked even with glucose-grown *C. tropicalis,* whose peroxisomes are small and inconspicuous, it might also work with *S. cerevisiae,* whose peroxisomes are also modest in number and size under most growth conditions (Veenhuis *et al.*, 1987). The targeting of firefly luciferase to peroxisomes (Gould *et al.*, 1987) might conceivably provide an approach to the visualization of peroxisomes in appropriately engineered living cells.

IX. Staining of Actin with Fluorochrome-Conjugated Phalloidin

Phallacidin and phalloidin are bicyclic heptapeptide (MW 800–850) mushroom toxins that interact specifically with yeast (Greer and Schekman, 1982) and other (Wieland and Faulstich, 1978; Vandekerckhove *et al.*, 1985) actins; the interaction appears to be specific for polymerized (F) rather than unpolymerized (G) actin. The coupling of fluorochromes to these phallotoxins provides a quick and convenient alternative to antiactin immunofluorescence for visualization of the actin cytoskeleton in various types of cells; in some cases, staining of living as well as fixed cells has been achieved (Wulf *et al.*, 1979; Barak *et al.*, 1980, 1981; Faulstich *et al.*, 1983). In yeast, phallotoxin staining has been accomplished to date only with fixed cells; this staining reveals patterns of cytoplasmic actin fibers and cortical actin patches very similar to those seen by immunofluorescence (Adams and Pringle, 1984; Kilmartin and Adams, 1984). A particular advantage of phallotoxin staining in work with yeast is that the low molecular weight stain (in contrast to antibodies) can enter cells whose walls have not been digested. This allows double-label experiments in which the distribution of actin and the sites of incorporation of new cell wall material are visualized simultaneously (Adams and Pringle, 1984; Kilmartin and Adams, 1984). We have had good results with both the fluorescein and tetramethylrhodamine derivatives of phalloidin (Wieland *et al.*, 1983; Molecular Probes cat. nos. F-432 and R-415, respectively). 7-Nitrobenz-2-oxa-1,3-diazole (NBD)-phallacidin (Barak and Yocum, 1981; Molecular Probes cat. no. N-354) also stains yeast actin (Adams and Pringle, 1984), but is less satisfactory because its very rapid photobleaching is not retarded by *p*-phenylenediamine.

We have obtained good results with both of the following protocols. For reasons that were not fully clear (but seemed to involve difficulties in getting effective washes), we had less success with a third protocol in which formaldehyde-fixed cells were fixed to a multiwell slide with polylysine and then stained without removing their cell walls.

a. First Protocol. This is quick, easy, and readily combined with staining of cell wall components using Calcofluor (Section III) or fluorochrome-conjugated Con A (Section IV). All steps are conducted at room temperature.

1. Fix cells with formaldehyde according to the standard procedure

(Section II,D,1). If desired, the cells can be stained prior to fixation with fluorochrome-conjugated Con A or after fixation with Calcofluor.

2. Collect ~10^7 cells by centrifugation and wash two or three times in water or PBS, then resuspend in 100 μl PBS.

3. Stain cells with 0.2–1.5 μM fluorochrome-conjugated phalloidin for 30–90 minutes in the dark. [Note that the fluorochrome-conjugated phalloidins are presently supplied by Molecular Probes as 3.3 μM stock solutions in methanol. We have achieved good staining both (a) by simply adding 6–80 μl of this stock solution to the 100 μl of cell suspension and (b) by first drying down the stock solution in a Speed-Vac Concentrator (Savant), then redissolving at a higher concentration in PBS, then diluting appropriately into the cell suspension.]

4. Wash the cells five times with PBS by centrifugation, resuspend in a drop of *p*-phenylenediamine-containing mounting medium (Section X,D,1), and observe using the standard fluorescein or rhodamine filter sets (Table II).

b. Second Protocol. This is designed specifically for combining phalloidin staining of actin with immunofluorescence localization of some other component of interest.

1. Follow the immunofluorescence protocol (Section X,D,2) until the washes following treatment with secondary antibody have been completed. [One caution should be noted: the methanol and acetone treatments necessary to visualize actin (and presumably some other proteins) by immunofluorescence (Section X,D,2,d) appear to destroy the ability of actin to interact with phalloidin.]

2. Immediately following removal of the last wash, add to each well 6–10 μl of a solution of fluorochrome-conjugated phalloidin (the 3.3 μM stock supplied by Molecular Probes diluted 5- to 15-fold with PBS or with 40 mM potassium phosphate, 0.5 mM $MgCl_2$, pH 6.5).

3. Stain for 5–30 minutes (depending on the phalloidin concentration) at room temperature in the dark.

4. Remove the phalloidin solution by aspiration, wash several times with the buffer used in step (2), and mount the cells in *p*-phenylenediamine-containing mounting medium (Section X,D,1). (In working with the lower concentrations of fluorochrome-conjugated phalloidin, the washes are not essential.)

X. Immunofluorescence

A. General Remarks

1. The Broad Utility of Immunofluorescence in Yeast

Before immunofluorescence methods had been applied to yeast, there was considerable pessimism as to how well these methods (which were developed for large, flat, wall-less animal cells) would work on yeast (which are small, round, and heavily walled). Happily, it is now clear from the steadily expanding list of successful applications (Table III) that this pessimism was unwarranted, and that essentially the full panoply of intracellular structures can be visualized in yeast, as in other cell types. Moreover, although some of the successful applications have involved staining of spheroplasts, isolated organelles, or cell sections, many others have involved staining of whole cells whose shape and spatial organization have been preserved by fixation prior to the removal of the cell wall. In practice, of course, the detectability of particular antigens will depend on their abundance, degree of localization, and retention of antigenicity after fixation, as well as on the properties of the particular antisera used (see also Section X,E). Moreover, when spatial resolution rather than simply detection is at issue (e.g., in double-label experiments), the inherent limit of resolution of the light microscope may also be a factor. The protocols to be provided have worked well in many specific applications, and should work well in many more. However, it seems likely that other applications will require idiosyncratic alterations of these protocols to yield optimal results; as in other areas, a flexible, empirical approach is strongly recommended.

2. Direct and Indirect Immunofluorescence

Most immunofluorescence experiments done with yeast and other cell types have utilized *indirect* immunofluorescence. That is, the fluorochrome ultimately visualized is present not on the *primary* antibodies (i.e., the antibodies that recognize the antigen of interest) but on *secondary* antibodies that recognize and bind to the primary antibodies. (For example, to visualize actin, one might use rabbit antiactin primary antibodies and fluorochrome-conjugated goat anti-rabbit IgG secondary

TABLE III

SOME SUCCESSFUL APPLICATIONS OF IMMUNOFLUORESCENCE IN YEAST[a]

Structure visualized	Antibody used	References
Microtubules (intranuclear and cytoplasmic) in normal whole vegetative cells, sporulating cells, spheroplasts, detergent-extracted spheroplasts, and isolated nuclei; residual tubulin-containing structures in mutant and drug-treated cells	Rat monoclonal (YOL1/34) and rabbit polyclonal versus purified yeast tubulin; mouse monoclonal versus purified pig brain tubulin; rabbit polyclonals versus synthetic yeast tubulin peptides coupled to keyhole limpet hemocyanin	Kilmartin *et al.* (1982); Kilmartin and Adams (1984); Adams and Pringle (1984); Pillus and Solomon (1986); Hašek *et al.* (1986, 1987); Rose and Fink (1987); Huffaker *et al.* (1988); Jacobs *et al.* (1988); L. Hogan and R. Easton Esposito (personal communication); P. Schatz (personal communication)
KAR1–lacZ fusion protein localized to the spindle-pole body	Mouse monoclonal anti-β-galactosidase	M. Rose (personal communication)
Actin-containing structures in normal and mutant cells	Rat and rabbit polyclonals versus purified yeast actin	Kilmartin and Adams (1984); Novick and Botstein (1985); Drubin *et al.* (1988)
67-kDa and 85-kDa actin-binding proteins localized to actin-containing structures	Rabbit polyclonals versus purified actin-binding proteins	Drubin *et al.* (1988)
CDC12 and *CDC3* products localized to vicinity of the 10-nm filaments in the mother-bud neck	Rabbit polyclonals versus *CDC12–lacZ*, *trpE–CDC12*, *lacZ–CDC3*, and *trpE–CDC3* fusion proteins purified from *E. coli*	Haarer and Pringle (1987); Kim *et al.* (1989a)
CDC28 product localized to cytoplasm and perhaps cytoplasmic matrix	Rabbit polyclonal versus *lacZ–CDC28* fusion protein purified from *E. coli*; rabbit polyclonal versus synthetic *CDC28* C-terminal peptide	Wittenberg *et al.* (1987); S. Richardson and S. Reed (personal communication)
SNF1 product localized to cytoplasm or cytoplasmic membrane	Rabbit polyclonal versus *trpE–SNF1* fusion protein purified from *E. coli*	Celenza and Carlson (1986)
SNF3–lacZ fusion protein localized to the plasma membrane	Mouse monoclonal anti-β-galactosidase	Celenza *et al.* (1988)

FUS1–lacZ fusion protein localized to shmoo tips and conjugation bridges	Mouse monoclonal anti-β-galactosidase	Trueheart *et al.* (1987)
MATα2–lacZ, GAL4–lacZ, and ribosomal protein L3-*lacZ* fusion proteins, localized (or not) to the nucleus, in spheroplasts or whole cells	Rabbit polyclonal anti-β-galactosidase	Hall *et al.* (1984); Silver *et al.* (1984); Moreland *et al.* (1985)
tRNA ligase localized to nucleus, perhaps to nuclear pores, in whole cells, spheroplasts, and isolated nuclei	Rabbit polyclonals versus yeast tRNA ligase purified from yeast or *E. coli*	Clark and Abelson (1987)
RNA2 and *RNA3* products localized to nucleus	Rabbit polyclonals versus *ompf–RNA2–lacZ* and *ompf–RNA3–lacZ* fusion proteins purified from *E. coli*	Last and Woolford (1986)
HMG–CoA reductase localized to circumnuclear membranes in overproducing cells using whole cells or sections of LR-White-embedded cells	Rabbit polyclonal versus *lacZ–HMG1* fusion protein purified from *E. coli*	Wright *et al.* (1988); R. Wright (personal communication)
Topoisomerase I localized to the chromosomes and nucleoli in spread preparations of meiotic prophase nuclei	Rabbit polyclonal versus purified topoisomerase I	Dresser and Giroux (1988); C. Giroux (personal communication)
YPT1 product localized to Golgi apparatus	Rabbit polyclonal versus *trpE–YPT1* fusion protein purified from *E. coli*	Segev *et al.* (1988)
SEC4 product localized to plasma membrane and secretory vesicles	Rabbit polyclonal versus *trpE–SEC4* fusion protein purified from *E. coli*	Goud *et al.* (1988)
Vacuoles	Rabbit polyclonal versus carboxypeptidase Y purified from yeast and deglycosylated	Stevens *et al.* (1982); R. Preston (unpublished results)
Peroxisomes of repressed (glucose-grown) or derepressed (oleate-grown) *Candida tropicalis*	Rabbit polyclonal versus total *C. tropicalis* peroxisomal proteins	Small *et al.* (1987); P. Lazarow (personal communication)

[a] This listing is not intended to be exhaustive but merely to illustrate the range of successful applications.

antibodies.) This approach offers some amplification of signal (several secondary antibody molecules, each with its complement of fluorochromes, can bind to one primary antibody molecule) as well as convenience (the fluorochrome-conjugated secondary antibodies, each of which may be used with a variety of different primary antibodies, can be purchased readymade from commercial suppliers). However, despite these advantages, *direct* immunofluorescence, in which the fluorochrome is conjugated directly to the primary antibodies, may sometimes be preferable (Talian *et al.*, 1983; Kilmartin and Adams, 1984). The coupling reactions with conventional fluorochromes such as FITC and TRITC are not difficult (Haaijman, 1983; Talian *et al.*, 1983). Although the signal is likely to be weaker than in indirect immunofluorescence, the signal–noise ratio may actually be better (thus facilitating video enhancement of the weak signal—Section X,E,1), as nonspecific binding of secondary antibodies appears to be a frequent source of troublesome "background" fluorescence. Moreover, direct immunofluorescence can greatly facilitate double-label experiments, which are otherwise difficult if the primary antibodies available are all derived from the same type of animal (see also Section X,F).

B. Primary Antibodies and Affinity Purification

1. Sources of Primary Antibodies

Sometimes useful primary antibodies can be obtained from colleagues or commercial sources. More often, however, the desired primary antibodies must be prepared by the investigator. This requires tactical decisions as to what type of antibodies (polyclonal, monoclonal, or oligopeptide-directed) is to be prepared, as well as to the nature of the immunogen (protein purified from yeast, fusion protein purified from *E. coli,* etc.) to be used. Full consideration of these issues and of the details of raising antibodies is beyond the scope of this review; novices are referred to the relevant volumes of *Methods in Enzymology* (Van Vunakis and Langone, 1980; Langone and Van Vunakis, 1981, 1983, 1986), as well as to their local immunologists. We offer only the following possibly useful comments.

a. Type of Antibodies. It is clear that monoclonal and oligopeptide-directed antibodies can offer unexcelled specificity, which is essential for some purposes [e.g., distinguishing between the intracellular locations of two closely related antigens (Phillips *et al.*, 1985; Bond *et al.*, 1986; Piperno *et al.*, 1987; Deng and Storrie, 1988)]. However, it should not be forgotten that such antibodies can sometimes recognize an epitope shared by two or more related or unrelated antigens (Pruss *et al.*, 1981; Kilmartin *et al.*, 1982; Elledge and Davis, 1987), so that uncritical acceptance of their specificity is not warranted. Moreover, for most immunofluorescence purposes, it appears that properly purified (monospecific) polyclonal antibodies are equally good or even superior reagents, as well as being (usually) easier and cheaper to prepare. The potential superiority of polyclonal antibodies derives from the fact that in most cases, the serum will contain multiple different antibody species that recognize different determinants on the antigen of interest. Thus, although any one antigenic determinant may be fixation-sensitive, or inaccessible to antibody in the native structure, it is unlikely that all will be. Moreover, a single molecule of a macromolecular antigen should be able to bind several different antibody molecules, thus increasing the ultimate strength of the immunofluorescence signal.

b. Type of Immunogen. The choice will in most cases be conditioned by whether the original identification of an antigen of interest was biochemical or genetic. In the latter case, there are usually several ways to approach the acquisition of an immunogen that can be used to raise antibodies specific for that gene product. Although every case should be evaluated on its own merits, it is worth noting that the use of fusion proteins frequently offers advantages (see the examples of successful applications in Table III). Given modern vector systems (notably for constructing fusions of the gene of interest to the *E. coli lacZ* and *trpE* genes), the necessary constructions at the DNA level are usually straightforward, especially if the sequence of the gene of interest is known. The fusion proteins are then produced in large amounts by *E. coli* and can usually be purified sufficiently to serve as immunogens by simple one- or two-step procedures. As any contaminating proteins derive from *E. coli* rather than yeast, they are less likely to elicit antibodies that will subsequently give confusing results. Moreover, the fusion proteins themselves then provide convenient bases for affinity purification (Haarer and Pringle, 1987; Segev *et al.*, 1988; Wright *et al.*, 1988; Kim *et al.*, 1989a). Especially valuable is the fact that two quite different fusion proteins can usually be prepared (nearly) as cheaply as one. This increases the probability of eliciting a strong immune response and allows a particularly effective affinity purification, in which antibodies raised against one fusion protein are purified on a matrix composed of the other (Haarer and Pringle, 1987; Kim *et al.*, 1989a; Section X,B,2).

2. The Importance of Affinity Purification

The serum of a mammal or bird is an extremely complex reagent. Numerous examples make clear that the serum of an immunized animal may contain, in addition to the antibodies specifically elicited by the immunization procedure, other antibodies that react with antigens in the organism of interest. For example, many rabbits' preimmune sera contain antibodies that react with some component of mammalian spindle poles (Connally and Kalnins, 1978; Neighbors *et al.*, 1988). In yeast work, Payne and Schekman (1985) observed that the sera of rabbits immunized with purified clathrin heavy chain also contained antibodies (apparently recognizing carbohydrate structures) that were a problem in immunoblotting experiments unless they were first removed by preadsorption of the sera with whole cells. Similarly, attempts to immunolocalize the yeast *CDC12* gene product were at first thoroughly obfuscated by antibodies, present in the preimmune sera of all rabbits tested (~20, including both males and females), that reacted with a component (apparently chitin) of the cell wall (Haarer and Pringle, 1987). In one rabbit, the titer of this cell wall-reactive antibody was boosted ~50-fold during the course of immunization with fusion protein purified from *E. coli,* so that the comparison of preimmune to immune serum was initially quite misleading. Lillie and Brown (1987) also observed that the comparison of preimmune to immune sera is not always an adequate test of the specificity of an immunofluorescence result. One of their immune sera produced a striking staining pattern of spots and blotches that was not apparent with the preimmune serum used at comparable concentrations; however, affinity purification showed clearly that this staining pattern was not due to the antibodies recognizing the protein that had been used as immunogen. (Indeed, such staining patterns have proved to be a rather common artifact in attempts to do immunofluorescence with insufficiently purified sera.)

Lillie and Brown (1987) also observed that the preimmune sera of various rabbits frequently recognized a variety of yeast proteins in immunoblotting experiments, and that the pattern of proteins recognized could change on a time scale of weeks even in the absence of any intentional immunization of the rabbits.

These results make clear that it is very dangerous to take seriously an immunofluorescence result until there is good evidence that the antibodies used are in fact monospecific for the antigen of interest. Immunoblotting experiments can provide important evidence for such monospecificity, but it should be noted that this test is not foolproof; for example, antibodies reactive with cell wall carbohydrates or nucleic acids presumably would not be detected by testing antisera on blots of cellular proteins. Thus, it seems to us that careful affinity purification should *always* precede attempts to determine the intracellular localization of an antigen by immunofluorescence. When possible, the affinity purification should use a matrix other than the original immunogen (e.g., antibodies raised against one fusion protein can be purified using a different fusion protein), to minimize the chances that antibodies recognizing a contaminating antigen in the original immunogen preparation will be affinity-purified right along with the antibodies recognizing the antigen of interest. In any case, the success of the affinity purification should be monitored by performing both immunoblotting and immunofluorescence experiments using both the purified and "depleted" fractions (see later).

3. Affinity Purification on Nitrocellulose Blots

The use of blots of electrophoretically separated proteins as affinity-purification matrices was pioneered by Olmsted (1981). In comparison with column purification methods (Section X,B,4), this approach offers simplicity and facilitates effective work with low-abundance antigens and their corresponding antibodies. Although the yields from blot purification are lower, they are sufficient for numerous immunofluorescence experiments, the screening of λgt11 libraries, etc. Olmsted's original procedure and some subsequent successful applications (e.g., Green *et al.*, 1988) have utilized diazotized paper; this may possibly offer some advantage in terms of the number of times a given blot can be reused. However, for most purposes, it appears simpler and at least as satisfactory to use nitrocellulose blots (Talian *et al.*, 1983; Smith and Fisher, 1984; Lillie and Brown, 1987; Haarer and Pringle, 1987). These blots also can be reused many times; in one case, a blot used >100 times is still giving satisfactory affinity purification (S. H. Lillie, personal communication).

a. Purification Protocol. We have consistently had satisfactory results with the following protocol. After the initial preparation of the blots, all steps are performed at room temperature except as noted.

1. Preparation of blots. Separate proteins by SDS–PAGE and transfer electrophoretically to nitrocellulose paper using standard procedures (e.g., Towbin *et al.*, 1979; Burnette, 1981). Allow blot to dry without rinsing out the residual transfer buffer.
2. Identification of protein bands. Immediately after preparation of the blot or whenever desired (the dried blots are stable for months if not years), visualize protein bands of interest on the blot by staining with Ponceau S (Goldstein *et al.*, 1986; Moreau *et al.*, 1986). Immerse the blot in 0.2% Ponceau S (Sigma cat. no. P-3504) in 0.3% trichloroacetic acid (TCA) for ~10 minutes, then destain with several changes of distilled water until bands are visible (typically 5–10 minutes), using gentle agitation at each step. At this point, the band(s) of interest can be cut out using a clean razor blade and the remaining stain removed by washing with several changes of PBS (to minimize the chance of interference with the antigen–antibody reaction). The strip of blot can then be used immediately for affinity purification or dried and stored indefinitely.

Alternatively, the Ponceau-stained blot can be dried and stored until needed; the stain is then removed by washing with PBS as just described just before proceeding with the affinity purification.

3. "Blocking" (to minimize nonspecific binding of antibodies). Incubate the strip in 3–10 ml blocking solution (5% nonfat dry milk in PBS) for 50 minutes with gentle agitation in a plastic Petri dish or other suitable container.

4. Washing. Wash the strip three times, using 3–10 ml PBS and 5 minutes of gentle agitation per wash.

5. Incubation with antiserum. Place a piece of parafilm on the bottom of a Petri dish and lay the nitrocellulose strip on top (drained, but wet; protein side up). Carefully layer crude serum (undiluted or diluted several-fold with PBS) or IgG fraction [prepared by chromatography on protein A–Sepharose (Pharmacia cat. no. 17-0963-03), protein G–Sepharose (Pharmacia cat. no. 17-0618), or Affi-Gel Protein A (Bio Rad cat. no. 153-6153), following the manufacturer's instructions] on top of the nitrocellulose, using ~200 μl for a 1 × 3 cm strip. Place on shaker for 2–3 hours, shaking fast enough to see the antibody solution move back and forth. It may help to tape a wet Kimwipe to the top of the Petri dish to prevent excessive evaporation.

6. Removal of "depleted fraction." Lift the strip slowly from one corner while removing liquid with pipetman; save this "depleted fraction" at ≤4°C as a control on the effectiveness of the affinity purification.

7. Washing. Wash the strip three times, using 3–10 ml PBS and 10 minutes of gentle agitation per wash.

8. Elution of purified antibodies. Place the drained, but wet, nitrocellulose strip on a fresh piece of parafilm in a Petri dish. Layer on 200 μl of low-pH buffer (0.2 *M* glycine, 1 m*M* EGTA, pH 2.3–2.7) and shake as in step (5) for 10–20 minutes. (The optimal pH and time are probably compromises between elution and inactivation of the antibodies of interest, and may vary from case to case.) Remove the liquid as in step (6), quickly neutralize either by adding an equal volume of cold 100 m*M* Tris base or by adding 3 *N* NaOH (~4–4.5 μl) to bring to ~pH 7 (check by spotting 1-μl aliquots onto pH paper), and store at ≤4°C. In some cases, a significant amount of additional purified antibody may be recovered by repeating the elution procedure.

9. Washing and storing filter. Wash strip three times with PBS, as before, and store at ≤4°C in PBS or after drying for reuse.

b. Comments. As with other protocols, the conditions necessary to give optimal results will almost certainly vary from case to case. The following considerations may be helpful.

1. Although our own experience has all been with blots of proteins separated by SDS–PAGE, it seems likely that blots of proteins separated by other means (e.g., urea–gel electrophoresis or two-dimensional gel electrophoresis) would also form satisfactory matrices for affinity purification. In addition, spotting purified native proteins onto nitrocellulose might provide satisfactory matrices in some cases; this approach could be especially useful in purifying antibodies that do not react well with denatured proteins.

2. Although we have had good success to date with the Ponceau S method of visualizing the protein bands of interest prior to excision, it seems likely that in some cases this stain or the associated exposure to TCA may be inimical to subsequent antigen–antibody reactions. In such a case, the bands of interest can be located by staining outside lanes and/or a central lane. Staining for 1 minute in an aqueous solution containing 0.1% (w/v) amido black, 25% (v/v) isopropanol, and 10% (v/v) acetic acid, followed by destaining for 20 minutes in the same solution without the amido black, has given us satisfactory results. Also, in some cases involving complex protein mixtures, it may be necessary to use antibody staining (e.g., with the relevant primary antibody plus ^{125}I-labeled protein A or peroxidase-coupled secondary antibody) to locate the protein of interest prior to excision of the appropriate region of the blot. In such a case, however, special caution would be needed to ensure that the antibodies purified really recognized only the protein of interest, and not also one or more comigrating proteins.

3. Although nonfat dry milk seems particularly effective as a blocking agent, other proteins or protein mixtures (e.g., BSA or gelatin solutions) may also give satisfactory results.

4. Although to date we have had excellent success in eluting the affinity-purified antibodies with the low-pH buffer described earlier, several lines of evidence suggest that this protocol will not always give an acceptable yield of active antibodies. First, in attempts to purify antibodies specific for the mammalian microtubule-associated protein tau using affinity columns, elution with low-pH buffers consistently failed to release active antibodies (D. G. Drubin, unpublished results). However, elution with 4.5 *M* $MgCl_2$ did release active antibodies (Pfeffer *et al.*, 1983). Second, in using affinity columns to purify antibodies specific for the yeast *CDC46* gene product, K. Hennessy (personal communication) observed that with one rabbit's serum, the desired antibodies were eluted effectively with 4.5 *M* $MgCl_2$, whereas with a second rabbit's serum, 4.5 *M* $MgCl_2$ was not effective but the antibodies were eluted by a subsequent wash with low-pH buffer. Third, in attempting to use blots of purified yeast profilin to purify profilin-specific antibodies, S. Brown

(personal communication) recovered little active antibody after eluting with low-pH buffer; apparently the antibodies remained bound to the profilin (as opposed to being eluted but inactivated). Surprisingly, however, good yields of profilin-specific antibodies were obtained when the same protocol was used with a blot of *trpE*–profilin fusion protein as the affinity-purification matrix. Given these precedents, we suggest that if low-pH buffer does not seem to be eluting the desired antibodies satisfactorily in a given case, other elution regimens (such as one using 4.5 *M* $MgCl_2$; see also Section X,B,4,b) should be tried before the procedure is abandoned.

5. In some cases, better results seem to be obtained if the desired antibodies are carried through two cycles of affinity purification. For example, we significantly improved the signal–noise ratio observed both in immunoblots and in immunofluorescence with *CDC10*-specific antibodies by carrying the antibodies through two cycles of affinity purification, one on a blot of *trpE–CDC10* fusion protein, the other on a blot of *lacZ–CDC10* fusion protein (H. B. Kim and J. R. Pringle, unpublished results).

6. At least in some cases, blots of the appropriate protein can be used to concentrate as well as purify the antibodies of interest. To do this, simply incubate the nitrocellulose strip (or strips) in a relatively large volume (e.g., 1–2 ml) of serum, and then elute with a smaller volume of low-pH buffer. It may help in such applications to increase the incubation times. In this case, be sure to keep the environment humid; it may also help to conduct the longer incubations at 4°C rather than room temperature.

4. Affinity Purification on Columns

Affinity purification on columns is somewhat more laborious and requires more antigen than does affinity purification on blots, but offers higher yields of purified antibodies. (If large amounts of antibodies are required, it may be less work in the long run to do a single large preparation by affinity column than to do repeated small preparations using blots.)

a. Purification Protocol. Protocols like the following have been effective in a variety of specific applications (e.g., Pfeffer *et al.*, 1983; Wittenberg *et al.*, 1987; Segev *et al.*, 1988; Drubin *et al.*, 1988); except as noted, all steps are conducted at room temperature.

1. Prepare buffer T (50 m*M* Tris-HCl, pH 7.4) and buffer M (4.5 *M*

$MgCl_2$, 0.1% BSA in buffer T; prepare by adding 91.5 g of $MgCl_2 \cdot 6H_2O$ to 40 ml buffer T containing 2.5 mg/ml BSA).

2. Prepare a 0.5–1 ml column containing 0.2–2 mg of antigen coupled to CNBr-activated Sepharose 4B (Pharmacia cat. no. 17-0430-01) according to the protocol supplied by Pharmacia.
3. Wash the column with 15 ml of 6 *M* guanidine-HCl, then equilibrate with buffer T by passing 25 ml through the column.
4. Wash the column with 20 ml buffer M, then wash again with 50 ml buffer T.
5. Run 5–30 ml crude serum or purified IgG fraction (see Section X,B,3,a) over the column in ≥2 hours. Save the flowthrough (depleted fraction) as a control to determine how efficiently the antibodies of interest were bound.
6. Wash the column successively with 20 ml buffer T, 40 ml 1.0 *M* guanidine-HCl, and 20 ml buffer T.
7. Elute purified antibodies with buffer M. Collect six or more 1-ml fractions and dialyze immediately (separately or after pooling) against 1 liter PBS for ≥3 hours. To assay antibody in the fractions, dilute 1 μl of each fraction (before dialysis) with 10 μl PBS and spot on nitrocellulose. When the filter is dry, probe it with ^{125}I-labeled protein A or with enzyme-conjugated secondary antibody.
8. After dialyzing against PBS (step 7) dialyze for an additional ~12 hours against PBS containing 35% (v/v) glycerol, then store at −20°C. The solution should not freeze because of the glycerol, and the antibodies are stabilized by the BSA.
9. Flush the column with buffer T containing 0.02% NaN_3 and store at ≤4°C.

b. Comments. (i) Although we have no personal experience with them, the Affi-Gel supports (Bio Rad cat. nos. 153-6046 and 153-6052) should provide a satisfactory alternative to CNBr-activated Sepharose. (ii) Although the elution of purified antibodies from affinity columns with 4.5 *M* $MgCl_2$ is generally effective, it is not always so [see comments in Section X,B,3,b(4)]. Other elution regimens that are reportedly effective in certain applications are as follows: high-pH buffer (50 m*M* diethylamine-HCl, pH 11.5); high-pH buffer containing 10% dioxane; low-pH buffer [as in Section X,B,3,a(8)]; and low-pH buffer containing 10% dioxane. See the very helpful Bio Rad Bulletin 1099 for additional discussion. (iii) As in the case of purification on blots [Section X,B,3,b(5)], better results may sometimes be obtained by using more than one cycle of affinity purification. For example, Segev *et al.* (1988)

used two cycles of preadsorption on a column of *trpE* protein plus affinity purification on a column of *trpE–YPT1* fusion protein to prepare *YPT1*-specific antibodies.

C. Secondary Antibodies

In these modern times, high-quality, fluorochrome-conjugated secondary antibodies are available from a variety of sources. We have worked mainly with antibodies obtained from Cappel (now Organon Teknika–Cappel), Accurate, Bio Rad, and Sigma, but other suppliers' products may be equally good. Although the non-affinity-purified products are often adequate, we recommend in general that affinity-purified secondary antibodies be used, as the modest additional expense is likely to be repaid by a diminution of annoying background fluorescence. (Recall that the ability to localize a product of interest, and to produce convincing photomicrographs documenting this localization, depends not just on the signal but on the signal–noise ratio.) Even with affinity-purified secondary antibodies, absolute specificity for the primary IgG of interest should not be assumed without trial; for example, some goat anti-rabbit IgG sera cross-react also with rat IgG, causing potential problems in double-label immunofluorescence experiments (Drubin *et al.*, 1988; Section X,F).

Secondary antibodies are also available with a choice of fluorochromes. Most work has been done using FITC and TRITC conjugates, which are equally satisfactory. (FITC photobleaches more rapidly, but inclusion of *p*-phenylenediamine in the mounting medium solves this problem for most purposes.) However, it should be noted that other fluorochromes (including other fluorescein and rhodamine derivatives) are also available or being developed (Section II,E) and may offer advantages in certain applications. Worthy of special note is Texas red (Table II), currently available in a variety of commercial secondary antibody conjugates, which offers a comparable quantum yield (hence brightness) to TRITC and better spectral separation from FITC (thus reducing potentially troublesome "crossover" fluorescence in double-label experiments; see Section X,F). The commercial availability of biotin-conjugated secondary antibodies should also be noted; these can be used in conjunction with fluorochrome-conjugated avidin or streptavidin in schemes to amplify immunofluorescence signals (Section X,E).

D. Immunofluorescence Procedures

1. Solutions

a. Solution A. Mix 1 *M* K_2HPO_4 with 1 *M* KH_2PO_4 to obtain a solution at pH 6.5. Dilute with H_2O to 40 m*M*. Add $MgCl_2$ to 0.5 m*M* and sorbitol to 1.2 *M*.

b. Solution B. Mix 1 *M* K_2HPO_4 with 1 *M* KH_2PO_4 to obtain a solution at pH 7.5. Dilute with H_2O to 100 m*M*. Add sorbitol to 1.2 *M*.

c. Solution C. Mix together 4 ml 1 *M* Tris-HCl (pH 9.0), 4 ml 0.1 *M* disodium EDTA (pH 8.0), 10 ml 2*M* NaCl, and 2 ml H_2O. Add 120 μl β-mercaptoethanol just before use.

d. Solution D. Make phosphate–citrate buffer, pH 5.8 (contains 22.32 g KH_2PO_4 and 9.41 g sodium citrate per liter), then mix 1 : 1 with 2 *M* NaCl.

e. Solution E. Dissolve 180 g sorbitol in 250 ml phosphate–citrate buffer, pH 5.8 (as in solution D). Dilute to 1 liter with H_2O.

f. Solution F. Add 10 mg KH_2PO_4 to 90 ml H_2O and titrate to pH 7.4 with 0.1 *N* KOH. Dilute to 100 ml with H_2O. Add 0.85 g NaCl, 0.1 g BSA, and 0.1 g NaN_3.

g. Polylysine Stock. Dissolve 10 mg polylysine (MW >300,000; Sigma cat. no. P-1524 or comparable) in 10 ml H_2O. This solution can be stored for several months at −20°C, with multiple freeze–thaw cycles, without obvious detriment.

h. Phosphate-Buffered Saline (PBS). Dissolve 160 g NaCl, 4 g KCl, 22.8 g Na_2HPO_4, and 4 g KH_2PO_4 in H_2O to a final volume of 1 liter. Adjust to pH 7.3 with 10 *N* NaOH. Dilute 20× with H_2O prior to use.

i. Mounting Medium. Our standard mounting medium contains *p*-phenylenediamine to retard photobleaching (Section II,F,3). As this is a somewhat nasty chemical (reportedly carcinogenic), we prefer to weigh it out infrequently; thus, we make up mounting medium in large batches as follows: (i) Dissolve 100 mg *p*-phenylenediamine (Sigma cat. no. P-6001 or comparable) in 10 ml PBS and adjust to pH 9 if needed. (Stir vigorously at room temperature to facilitate dissolving the *p*-phenylenediamine.) (ii) To this solution, add 90 ml glycerol and stir until homogeneous. (iii) If desired for DNA staining (Sections V and VI), add 2.25 μl of fresh DAPI stock solution (1 mg/ml DAPI in H_2O). Note that this is frequently useful and rarely harmful, so that we almost always include this step. (iv) Store at low temperature in the dark. We have stored the full stock in a bottle at −20°C (it stays liquid) and removed samples as needed without warming.

This is reasonably satisfactory, but the solution does gradually deteriorate (time scale of a few months), developing a dark color and producing apparent fluorescence artifacts. Thus, we recommend storing multiple small aliquots of the mounting medium in capped microcentrifuge tubes at −70°C, and retrieving these one at a time as needed. Under these conditions the mounting medium appears to stay good for many months if not years.

2. Immunofluorescence Protocols

The following protocols are written for standard immunofluorescence on whole cells. Appropriate modifications can adapt them for use with spheroplasts, isolated organelles, or cell sections (for examples of these latter approaches, see Table III).

a. Harvesting and Fixation. Unless otherwise indicated (Section II,D,2,b), fix cells according to the standard protocol (Section II,D,1). Note that the cells can generally be stored for some time at this stage before proceeding, without obvious adverse effects (Section II,D,2,e).

b. Permeabilization. When ready to proceed, spin down $\sim 2 \times 10^8$ cells and wash twice with the solution appropriate to one of the following cell wall-removal protocols. For ordinary growing cells, we have generally found either protocol (1) or (2) to be satisfactory. For cells with walls that are more resistant to digestion (e.g., stationary-phase cells or arrested cell-cycle mutants), protocol (2) or (3) has generally given better results.

1. Wash cells with solution A (see Section X,D,1) and resuspend in 1 ml solution A containing 10 μl β-mercaptoethanol and 55 μl Glusulase (DuPont NEN cat. no. NEE-154). Incubate for 2 hours at 36°C with gentle agitation (e.g., in a roller drum).
2. Wash cells with solution B and resuspend in 1 ml solution B containing 2 μl β-mercaptoethanol and 20 μl of a Zymolyase stock [1 mg/ml Zymolyase 100T (ICN Immunobiologicals cat. no. 32093-1)[9] in H_2O]. Incubate 30 minutes at 37°C with gentle agitation.
3. Wash cells with H_2O, then incubate in 1 ml solution C for 10 minutes at room temperature. Spin down and wash once with solution D, then twice with solution E. Resuspend cells in 1 ml solution A containing 10 μl β-mercaptoethanol, 110 μl Glusulase, and 22 μl of a Zymolyase stock

[9] In these and other protocols calling for Zymolyase, it may be possible to substitute Sigma Lyticase (cat. nos. L8012, L8137, and L5263) with similar efficacy and less expense; however, we have not checked this systematically.

[27 mg/ml Zymolyase 20T (ICN Immunobiologicals cat. no. 32092-1)[9] in H_2O]. Incubate 30 minutes at 37°C with gentle agitation.

After digestion by any of these procedures, spin cells down at low speed and wash once with solution A or B, then resuspend gently in 1 ml solution A or B. Note: (1) this final wash may not be necessary, unless the cells are going to be stored for a time before proceeding; (2) once washed and resuspended, the cells can be stored for some time before proceeding (at least for some antigens; see Section II,D,2,e), so long as they are kept cold; (3) even though the cells are fixed, once their walls are removed they lose their shapes with rough handling (hard centrifugation, vortexing), which should therefore be avoided; (4) although additional treatments (e.g., with acetone and methanol or detergent) do not appear to be necessary for permeabilization per se, such treatments do facilitate the visualization of some antigens, presumably by contributing to fixation and/or denaturation of the antigen [Sections II,D,2,b and X,D,2,d(2)].

c. Preparation of Slides. Put ~10 μl of polylysine stock solution (Section X,D,1,g) in each well of a multiwell slide (Section II,F,2,a). After 5–10 seconds, aspirate the solution off and air-dry. Wash each well three times with drops of water that are removed by aspiration. Air-dry completely. Normally, the slide is now ready for use. However, if background fluorescence proves to be a problem (this may vary with different batches of slides and antibodies), it often helps to wash the slides more extensively. To do this, take the slide after polylysine treatment and drying, and place it in distilled water in a capped plastic tube. Shake this or place it in a roller drum for ~10 minutes, then air-dry completely. Slides can be prepared at least several hours before use.

d. Staining of Cells with Antibodies. The following steps describe standard indirect immunofluorescence. The obvious modifications would be necessary to use direct immunofluorescence (Section X,A,2) or "sandwiching" methods (Section X,E).

1. Place 10 μl of cell suspension (see earlier) in each well. After ~10 seconds, aspirate off the fluid and allow the slide to air-dry. Check the slide microscopically to ensure that the cells have retained their shapes and are at a suitable density and not clumped.
2. For most antigens, the cells can now be reacted directly with the primary antibodies, step (3). However, for some antigens (or at least for some antigen–antibody combinations, for example, actin and the antiactin antibodies that have been used to date), additional treatment for fixation and/or denaturation is essential before treatment with antibodies (Drubin *et al.*, 1988). Use fresh methanol and acetone that have been

chilled to −20°C. Immerse the wells of the slide in methanol at −20°C for 6 minutes, then in acetone at −20°C for 30 seconds, then air-dry completely.

3. Place 5–10 μl of primary antiserum [diluted as appropriate in solution F (see Section X,D,1) or with PBS containing 1 mg/ml BSA (PBS–BSA)] in each well. In some cases, eventual background fluorescence may be reduced by incubating briefly with solution F or PBS–BSA alone before the incubation with primary antiserum. Use a control antiserum in one well per slide (or per batch of slides) as needed (Section X,D,3).

4. Incubate the slide at room temperature in a moist environment (e.g., in a Petri dish containing a wet Kimwipe) for 0.5–1.5 hour. (The exact time required may vary with different antigen–antibody combinations.)

5. Aspirate off the primary antiserum and wash cells ~10 times with solution F or PBS–BSA by placing a drop of solution in each well and aspirating it off after a few seconds. Do not let the wells dry out completely during these washes; that is, put in a new drop of solution as soon as the old drop is aspirated off. We generally wash the first well ~7 times, leave it under a drop of wash solution while we wash the remaining wells ~7 times apiece, then return to the beginning and wash each well 3 more times, again leaving each well under wash solution. Then these final washes are removed and replaced immediately by the secondary antibody solution.

6. Place 5–10 μl of fluorochrome-conjugated secondary antiserum (diluted as appropriate in solution F or PBS–BSA) in each well and incubate at room temperature in a moist environment for 0.5–1.5 hour [see note at step (4)]. Note that it is important to conduct these manipulations in low light and to incubate in the dark, to avoid photobleaching of the fluorochrome (Section II,F,3).

7. Aspirate off the secondary antiserum and wash ~10 times with solution F or PBS–BSA [as described in step (5); remove the last drop of wash solution immediately before adding mounting medium].

Note that in some applications (e.g., the screening of large numbers of antisera), it is feasible and convenient to conduct staining with a method using larger puddles of solution that cover several or all wells at one time. The slides can then be washed either by using similar puddles of wash buffer or by simply immersing the slides in wash buffer. However, it should be noted that once wells have been "fused" in this way they cannot subsequently be treated separately without danger of cross-contamination of solutions.

e. Mounting of Slides. Place a drop of mounting medium (Section X,D,1,i) on the slide (one small drop per four wells is adequate, and too much causes problems such as floating coverslips or mixing of immersion oil with mounting medium), and cover with coverslips. View immediately or after storage at −20°C in the dark. The images deteriorate little or not at all (opinions differ) during storage under these conditions for days or even weeks. If the coverslips are sealed around the edges with clear nail polish, slides can be stored at −20°C in the dark for months or years without gross deterioration of immunofluorescence images.

3. The Value of a Positive Control

Even in experienced hands, the results of immunofluorescence are not always equally successful, especially when applying the method to a new strain or to cells grown under nonstandard conditions. (The vagaries of permeabilization appear to be one major source of variable results.) Thus, we recommend that a positive control be included routinely each time that immunofluorescence is done. One convenient and effective control is provided by the Kilmartin monoclonal antitubulin antibody YOL1/34; this antibody is commercially available (Accurate cat. no. MAS078; we purchase the supernatant form and use it at 1 : 100 to 1 : 500 dilution), is derived from a rat cell line (making it convenient for double-label experiments with rabbit antibodies), and gives a well-characterized pattern of staining of both cytoplasmic and intraorganellar structures (Kilmartin *et al.*, 1982; Kilmartin and Adams, 1984; Huffaker *et al.*, 1988; Jacobs *et al.*, 1988). However, it should be noted that this particular control is not a panacea; for example, conditions that routinely give excellent antitubulin immunofluorescence have required modification to give acceptable antiactin immunofluorescence (Drubin *et al.*, 1988; Section X,D,2,d). Thus, as in other areas, a flexible, empirical approach is required for the selection and evaluation of appropriate positive controls.

E. Detection of Nonabundant Antigens

It is clear from Table III that immunofluorescence is capable of localizing a wide variety of antigens of unequal abundance in the cell. However, it is also clear that sufficiently nonabundant antigens will be difficult or impossible to localize. We mention briefly here some approaches that may be helpful in the borderline cases.

1. Video Enhancement of Signals

Recent spectacular results with animal cells and subcellular systems *in vitro* show clearly the potential power of electronic enhancement of the optical signals (Allen *et al.*, 1985; Vale *et al.*, 1985; Horio and Hotani, 1986; Inoué, 1986; Taylor *et al.*, 1986; Sammak and Borisy, 1988; Wang and Taylor, 1989; Taylor and Wang, 1989). However, it should be noted that in immunofluorescence of whole yeast cells, immunodetection is likely to be limited more often by the ratio of signal (from the object of interest) to noise (contributed mainly by nonspecific background fluorescence) than by the strength of the signal per se. Thus, simple amplification of the optical signals (which will amplify the nonspecific background as well as the signal of interest) will probably not be useful in many cases. However, the video systems can also improve signal–noise ratios in some situations. Although much of the technology is directed toward reducing noise from within the optical and video systems themselves, the systems can also enhance image contrast electronically, seek discrete structures using algorithms such as those for thresholding and edge detection, and play tricks such as electronically subtracting an out-of-focus image from an in-focus image. It remains unclear how generally useful such approaches will be in practical work with yeast. The only successful applications of which we are aware are the studies of *Schizosaccharomyces pombe* nuclei and chromosomes by Yanagida and his co-workers (Yanagida *et al.*, 1986), and even these workers have relied primarily on conventional fluorescence micrographs. It should be noted that use of direct rather than indirect immunofluorescence (Section X,A,2) should sometimes facilitate the application of video enhancement methods by improving the starting ratio of signal to nonspecific background fluorescence. Confocal microscopy also has great promise in this regard (see Section II,F,1,b).

2. Overexpression of Gene Products and of Fusion-Gene Products

It seems clear that overexpression of a gene product (by introducing the gene on a high-copy-number plasmid or linking it to a strong promoter) should in some cases push a weak immunofluorescence signal above the threshold of detectability. However, it also seems clear that overexpression will sometimes lead to mislocalization of the gene product, perhaps accompanied by a general cellular pathology, so that the immunofluorescence results obtained may be misleading. For example, Wright *et al.* (1988) were unable to immunolocalize HMG–CoA reductase in normal cells, but were successful with cells that overproduced the enzyme.

However, the overproduced enzyme localized to a distinctly abnormal structure, namely a set of stacked circumnuclear membranes not present in normal cells. Thus, although the results were interesting in a variety of ways, they did not answer the question of the normal localization of HMG–CoA reductase. Similarly, Clark and Abelson (1987) observed that cells overproducing tRNA ligase under *GAL10* control yielded a detectable nuclear staining with antiligase antibodies under fixation and permeabilization conditions with which normal or uninduced cells yielded no detectable signal. Although this observation was useful in helping to verify the results obtained with wild-type cells using other fixation and permeabilization conditions, detailed examination of the immunofluorescence and immunoelectron microscopy results suggested that tRNA ligase was partially mislocalized in the overproducing cells.

In another case, overexpression of *KAR1* (Rose and Fink, 1987) yielded an immunofluorescence signal (a small dot) with *KAR1*-specific antibodies that was not detectable when normal cells were examined (M. D. Rose, personal communication). However, the overexpressing cells were arrested in the cell cycle and dying, and the dots of fluorescence were not consistently localized with respect to the spindle poles, so that the significance of the results was not clear. More satisfying results were obtained using an alternative approach that may also be useful with other nonabundant antigens (M. D. Rose, personal communication). Cells overexpressing a *KAR1–lacZ* fusion protein (which was *not* lethal to the cells) were examined using anti-β-galactosidase antibody. Under these conditions, immunofluorescence revealed a dot of staining that associated consistently with the spindle poles in such a way as to suggest strongly that the bona fide localization of the *KAR1* product had been revealed.

In summary, if immunofluorescence on normal cells yields no detectable signal, overexpression of the gene product of interest or of an appropriate fusion protein is probably worth a try. However, any results obtained must be interpreted with considerable caution.

3. "Sandwiching" Methods

Schulze and Kirschner (1987) described a method in which up to four successive layers of secondary antibodies were built up on a single primary antibody; in particular, microtubules containing biotinylated tubulin were reacted with rabbit antibiotin antibody, then successively with goat anti-rabbit IgG, rabbit anti-goat IgG, goat anti-rabbit IgG, and rabbit anti-goat IgG. Two, three, or all four of the layers of secondary antibodies utilized fluorescein-conjugated antibodies. This method served both to amplify the immunofluorescence signal over what was obtained with a single layer of fluorochrome-conjugated secondary antibodies and to block the microtubules containing biotinylated tubulin from reaction with antitubulin antibodies during a subsequent incubation (used to reveal microtubules that did not contain biotinylated tubulin). In yeast work, this approach has been used successfully by K. Hennessy (personal communication) to visualize the intracellular localization of the CDC46 gene product; each layer of antibodies was applied using standard procedures (Section X,D). Similar methods have been used by K. Redding and R. Fuller (personal communication) to localize the KEX2 product in normal cells and by L. Pillus and J. Rine (personal communication) to localize the SIR2 product in cells overproducing this protein.

The major potential problem with this powerful approach is that nonspecific "background" binding of antibodies at any stage will be amplified in the subsequent stages, with deleterious consequences to the signal–noise ratio. Thus, it may be necessary to invest some effort in optimizing the dilutions at which the various layers of secondary antibodies are applied. Another variable is the number of layers of fluorochrome-conjugated secondary antibodies to be used. Schulze and Kirschner (1987) reported that beyond two such layers their signal–noise ratio actually decreased, but K. Hennessy (personal communication) has had good success with three such layers, whereas K. Redding and R. Fuller (personal communication) used successfully four layers of antibodies in which only the antibodies of the final layer were fluorochrome-conjugated.

A related approach to signal amplification involves the use of a biotin-conjugated primary or secondary antibody and fluorochrome-conjugated avidin or streptavidin (Fuccillo, 1985; Greenwalt and Mather, 1985; Phillips *et al.*, 1985; Kobayashi *et al.*, 1986; reagents available from Accurate, BRL, Molecular Probes, Vector Labs, Zymed, and other sources). This approach has been used successfully to localize the *SPA1* and *SPA2* gene products in mitotic and meiotic yeast cells (M. Snyder, C. Copeland, and B. Page, personal communication). Further amplification can be attempted by adding a biotinylated carrier such as BSA, or biotinylated antiavidin or antistreptavidin, after the first avidin or streptavidin treatment, then adding another layer of fluorochrome-conjugated avidin or streptavidin (Phillips *et al.*, 1985). A potential problem with this general approach is background due to endogenous biotin-containing macromolecules, other avidin- or streptavidin-binding substances, or biotin-binding proteins in the cells. It is not clear whether these problems will be significant in yeast work. If they are, they can probably be alleviated using the methods that have worked in other systems, such as preliminary incubation of the cells with avidin or streptavidin, followed by incubation with free biotin (Wood and Warnke, 1981; Duhamel and Johnson, 1985; K. Hennessy, personal communication).

4. Other Methods

Clark and Abelson (1987) reported that removal of cell walls prior to fixation allowed them to detect the nuclear localization of tRNA ligase in normal cells (cf. Section X,E,2), which was not detectable if the cells were fixed first as in the usual protocol. Although this approach may be useful in other cases, it should be used with caution because of the danger of rearrangement of cell constituents during the prolonged incubations prior to fixation (cf. Section II,D,2,a). The use of fluorochromes with superior quantum yields (Section II,E) may also be of some value in detecting nonabundant antigens.

F. Double-Label Immunofluorescence

In attempting to compare the intracellular localizations of two cellular constituents, it is frequently valuable to perform double-label experiments so that these localizations can be compared in the same individual cells. This is relatively straightforward if primary antibodies from different types of animals are available or if the available antibodies are of different immunoglobulin classes (e.g., an IgG and an IgM; see Edwards *et al.*, 1984); in either case, appropriate secondary antibodies can effect the necessary discrimination. Thus, for example, one could incubate the cells (sequentially or simultaneously) with rabbit antiactin and rat antitubulin, then apply FITC-conjugated goat anti-rabbit IgG and TRITC-conjugated goat anti-rat IgG. Two potential problems should be noted; both are more troublesome if the antigens of interest appear to colocalize than if they show obviously different localizations. First, commercially available secondary antibodies are not always entirely specific for their target IgG; for example, Drubin *et al.* (1988) observed that their goat anti-rabbit IgG antibodies cross-reacted with rat IgG. Fortunately, the rat-reactive component could be effectively removed by preadsorption with immobilized rat IgG. Second, illumination intended for the fluorochrome that absorbs and emits at shorter wavelengths often results also in some detectable fluorescence from the fluorochrome that absorbs and emits at longer wavelengths. For example, illumination for FITC generally results also in some visible TRITC fluorescence from a double-labeled sample. If this is a problem, it can usually be solved either by choosing different fluorochromes (e.g., Texas red gives better spectral separation from FITC than does TRITC) or by using additional (or different) filters to circumscribe the exciting wavelengths, the emitted wavelengths that are allowed to pass through to the eyepieces, or both. (See the lists of filters available from the microscope manufacturers or from Omega Optical.)

Double-label immunofluorescence is more difficult if both of the available primary antibodies are of the same immunoglobulin class and are derived from the same type of animal. One solution is to resort to direct immunofluorescence, conjugating an appropriate fluorochrome directly to one or both primary antibodies (Section X,A,2). Thus, for example, J. Kilmartin achieved excellent double labeling with rat anti-yeast actin and anti-yeast tubulin antibodies using an approach adapted from that of Hynes and Destree (1978); cells were incubated successively with the rat antitubulin, TRITC-conjugated goat anti-rat IgG, excess unlabeled rat IgG (to block residual binding sites on the goat anti-rat IgG), and FITC-conjugated rat antiactin (Kilmartin and Adams, 1984). In related approaches that may allow stronger signals to be obtained (see Section X,E,3), one or both primary antibodies can be biotinylated or one can be biotinylated and the other dinitrophenylated (Edwards *et al.*, 1984; Miller *et al.*, 1985; Phillips *et al.*, 1985). Appropriate sequences of incubations with fluorochrome-conjugated avidin, streptavadin, or anti-dinitrophenyl group antibodies, plus appropriate blocking solutions (free biotin or unlabeled IgGs), then allow the double-label fluorescence to be visualized.

Another possible approach to double labeling (we do not know if this has actually been used successfully) is based on the "sandwiching" method of Schulze and Kirschner (1987; see also Section X,E,3). Thus, for example, one could react the cells with rabbit antiactin, then FITC-conjugated goat anti-rabbit IgG, then successive layers of nonfluorescent chicken anti-goat IgG and goat anti-chicken IgG. One would then react the cells with rabbit antitubulin followed by TRITC-labeled goat anti-rabbit IgG, hoping that the previously applied rabbit antiactin antibodies would be hidden from the new secondary antibodies by the layers of goat and chicken IgGs.

Finally, it should be noted that multiple-label immunofluorescence (localizing three or more antigens simultaneously) is possible at least in some situations using presently available fluorochromes (Hardy *et al.*, 1983; Lanier and Loken, 1984; Parks *et al.*, 1984) and may be further facilitated by new fluorochromes presently under development (Section II,E).

ACKNOWLEDGMENTS

We thank the numerous colleagues who have participated in the development of the ideas and procedures described herein, provided us with unpublished information, or commented on this manuscript. These include, but are not limited to, J. Allen, K. Armstrong, D. Botstein, J. Broach, S. Brown, C. Bulawa, C. Copeland, K. Corrado, M. Douglas, A. Farewell, R. Fuller, C. Giroux, K. Hennessy, T. Huffaker, D. Johnson, J. Kilmartin, H. Kim, P. Lazarow, S. Lillie, N. Martin, R. Murphy, P. Novick, B. Page, L. Pillus, K. Redding, S. Reed, J. Rine, M. Rose, P. Schatz, P. Silver, M. Snyder, F. Solomon, T. Stevens, P. Takasawa, A. Waggoner, B. Weinstein, M. Welsh, and R. Wright. Unpublished work from our laboratories was supported by NIH grants GM31006 (to J. R. P), AM18090 and GM29713 (to E. W. J.), and GM21253 and GM18973 (to D. Botstein), American Cancer Society grant MV 90 (to D. Botstein), and postdoctoral fellowships from the NIH (GM11329 to R. A. P.), Burroughs Wellcome Fund of the Life Sciences Research Foundation (to A. E. M. A.), and Helen Hay Whitney Foundation (to D. G. D.).

REFERENCES

Adams, A. E. M., and Pringle, J. R. (1984). *J. Cell Biol.* **98,** 934–945.

Allan, R. A., and Miller, J. J. (1980). *Can. J. Microbiol.* **26,** 912–920.

Allen, R. D., Weiss, D. G., Hayden, J. H., Brown, D. T., Fujiwake, H., and Simpson, M. (1985). *J. Cell Biol.* **100,** 1736–1752.

Anderson, J. M., and Soll, D. R. (1986). *J. Gen. Microbiol.* **132,** 2035–2047.

Aris, J. P., and Blobel, G. (1988). *J. Cell Biol.* **107,** 17–31.

Bacon, J. S. D., Davidson, E. D., Jones, D., and Taylor, I. F. (1966). *Biochem. J.* **101,** 36c–38c.

Bacon, J. S. D., Farmer, V. C., Jones, D., and Taylor, I. F. (1969). *Biochem. J.* **114,** 557–567.

Ballou, C. E. (1982). *In* "The Molecular Biology of the Yeast *Saccharomyces*: Metabolism and Gene Expression" (J. N. Strathern, E. W. Jones, and J. R. Broach, eds.), pp. 335–360. Cold Spring Harbor Lab., Cold Spring Harbor, New York.

Barak, L. S., and Yocum, R. R. (1981). *Anal. Biochem.* **110,** 31–38.

Barak, L. S., Yocum, R. R., Nothnagel, E. A., and Webb, W. W. (1980). *Proc. Natl. Acad. Sci. U.S.A.* **77,** 980–984.

Barak, L. S., Yocum, R. R., and Webb, W. W. (1981). *J. Cell Biol.* **89,** 368–372.

Bartholomew, J. W., and Mittwer, T. (1953). *J. Bacteriol.* **65,** 272–275.

Barton, A. A. (1950). *J. Gen. Microbiol.* **4,** 84–86.

Benziman, M., Haigler, C. H., Brown, R. M., Jr., White, A. R., and Cooper, K. M. (1980). *Proc. Natl. Acad. Sci. U.S.A.* **77,** 6678–6682.

Beran, K. (1968). *Adv. Microb. Physiol.* **2,** 143–171.

Beran, K., Řeháček, J., and Seichertová, O. (1970). *Acta Fac. Med. Univ. Brun.* **37,** 171–182.

Beran, K., Holan, Z., and Baldrián, J. (1972). *Folia Microbiol. (Prague)* **17,** 322–330.

Bhavanandan, V. P., and Katlic, A. W. (1979). *J. Biol. Chem.* **254,** 4000–4008.

Böck, G., Hilchenbach, M., Schauenstein, K., and Wick. G. (1985). *J. Histochem. Cytochem.* **33,** 699–705.

Bond, J. F., Fridovich-Keil, J. L., Pillus, L., Mulligan, R. C., and Solomon, F. (1986). *Cell (Cambridge, Mass.)* **44,** 461–468.
Botstein, D., and Fink, G. R. (1988). *Science* **240,** 1439–1443.
Bourne, H. R. (1988). *Cell (Cambridge, Mass.)* **53,** 699–671.
Bowers, B., Levin, G., and Cabib, E. (1974). *J. Bacteriol.* **119,** 564–575.
Boyde, A. (1985). *Science* **230,** 1270–1272.
Brakenhoff, G. J., van der Voort, H. T. M., van Spronsen, E. A., Linnemans, W. A. M., and Nanninga, N. (1985). *Nature (London)* **317,** 748–749.
Brakenhoff, G. J., van der Voort, H. T. M., van Spronsen, E. A., and Nanninga, N. (1986). *Ann. N.Y. Acad. Sci.* **483,** 405–415.
Brewer, B. J., Chlebowicz-Sledziewska, E., and Fangman, W. L. (1984). *Mol. Cell. Biol.* **4,** 2529–2531.
Bright, G. R., Fisher, G. W., Rogowska, J., and Taylor, D. L. (1987). *J. Cell Biol.* **104,** 1019–1033.
Brown, R. M., Jr., Haigler, C., and Cooper, K. (1982). *Science* **218,** 1141–1142.
Bruning, J. W., Kardol, M. J., and Arentzen, R. (1980). *J. Immunol. Methods* **33,** 33–44.
Bruschi, C. V., and Chuba, P. J. (1988). *Cytometry* **9,** 60–67.
Bulawa, C. E., Slater, M., Cabib, E., Au-Young, J., Sburlati, A., Adair, W. L., Jr., and Robbins, P. W. (1986). *Cell (Cambridge, Mass.)* **46,** 213–225.
Burnette, W. N. (1981). *Anal. Biochem.* **112,** 195–203.
Byers, B., and Goetsch, L. (1976). *J. Cell Biol.* **69,** 717–721.
Cabib, E., and Bowers, B. (1971). *J. Biol. Chem.* **246,** 152–159.
Cabib, E., and Bowers, B. (1975). *J. Bacteriol.* **124,** 1586–1593.
Cabib, E., Ulane, R., and Bowers, B. (1974). *Curr. Top. Cell. Regul.* **8,** 1–32.
Cabib, E., Roberts, R., and Bowers, B. (1982). *Annu. Rev. Biochem.* **51,** 763–793.
Celenza, J. L., and Carlson, M. (1986). *Science* **233,** 1175–1180.
Celenza, J. L., Marshall-Carlson, L., and Carlson, M. (1988). *Proc. Natl. Acad. Sci. U.S.A.* **85,** 2130–2134.
Chaffin, W. L. (1984). *J. Gen. Microbiol.* **130,** 431–440.
Chen, L. B., Rosenberg, S., Nadakavukaren, K. K., Walker, E. S., Shepherd, E. L., and Steele, G. D., Jr. (1985). *In* "Hybridoma Technology in the Biosciences and Medicine" (T. A. Springer, ed.), pp. 251–268. Plenum, New York.
Chvatchko, Y., Howald, I., and Riezman, H. (1986). *Cell (Cambridge, Mass.)* **46,** 355–364.
Clark, M. W., and Abelson, J. (1987). *J. Cell Biol.* **105,** 1515–1526.
Connolly, J. A., and Kalnins, V. I. (1978). *J. Cell Biol.* **79,** 526–532.
de Duve, C., de Barsy, T., Poole, B., Trouet, A., Tulkens, P., and Van Hoof, F. (1974). *Biochem. Pharmacol.* **23,** 2495–2531.
Deng, Y., and Storrie, B. (1988). *Proc. Natl. Acad. Sci. U.S.A.* **85,** 3860–3864.
Dresser, M. E., and Giroux, C. N. (1988). *J. Cell Biol.* **106,** 567–573.
Drubin, D. G., Miller, K. G., and Botstein, D. (1988). *J. Cell Biol.* **107,** 2551–2561.
Duhamel, R. C., and Johnson, D. A. (1985). *J. Histochem. Cytochem.* **33,** 711–714.
Edwards, P. A. W., Brooks, I. M., and Monaghan, P. (1984). *Differentiation* **25,** 247–258.
Eilam, Y., and Chernichovsky, D. (1988). *J. Gen. Microbiol.* **134,** 1063–1069.
Elledge, S. J., and Davis, R. W. (1987). *Mol. Cell. Biol.* **7,** 2783–2793.
Elorza, M. V., Rico, H., and Sentandreu, R. (1983). *J. Gen. Microbiol.* **129,** 1577–1582.
Faulstich, H., Trischmann, H., and Mayer, D. (1983). *Exp. Cell Res.* **144,** 73–82.
Frevert, J., and Ballou, C. E. (1985). *Biochemistry* **24,** 753–759.
Fuccillo, D. A. (1985). *BioTechniques* **3,** 494–501.
Giloh, H., and Sedat, J. W. (1982). *Science* **217,** 1252–1255.
Glazer, A. N., and Stryer, L. (1984). *Trends Biochem. Sci.* **9,** 423–427.

Goldstein, L. S. B., Laymon, R. A., and McIntosh, J. R. (1986). *J. Cell Biol.* **102,** 2076–2087.
Goud, B., Salminen, A., Walworth, N. C., and Novick, P. J. (1988). *Cell (Cambridge, Mass.)* **53,** 753–768.
Gould, S. J., Keller, G.-A., and Subramani, S. (1987). *J. Cell Biol.* **105,** 2923–2931.
Gown, A. M., and Vogel, A. M. (1982). *J. Cell Biol.* **95,** 414–424.
Green, K. J., Goldman, R. D., and Chisholm, R. L. (1988). *Proc. Natl. Acad. Sci. U.S.A.* **85,** 2613–2617.
Greenwalt, D. E., and Mather, I. H. (1985). *J. Cell Biol.* **100,** 397–408.
Greer, C., and Schekman, R. (1982). *Mol. Cell. Biol.* **2,** 1270–1278.
Haaijman, J. J. (1983). *In* "Immunohistochemistry" (A. C. Cuello, ed.), pp. 47–85. Wiley, Chichester, England.
Haarer, B. K., and Pringle, J. R. (1987). *Mol. Cell. Biol.* **7,** 3678–3687.
Hagan, I. M., and Hyams, J. S. (1988). *J. Cell Sci.* **89,** 343–357.
Haigler, C. H., Brown, R. M., Jr., and Benziman, M. (1980). *Science* **210,** 903–906.
Hall, M. N., Hereford, L., and Herskowitz, I. (1984). *Cell (Cambridge, Mass.)* **36,** 1057–1065.
Hardy, R. R., Hayakawa, K., Parks, D. R., and Herzenberg, L. A. (1983). *Nature (London)* **306,** 270–272.
Hartwell, L. H. (1970). *J. Bacteriol.* **104,** 1280–1285.
Hartwell, L. H., Culotti, J., and Reid, B. (1970). *Proc. Natl. Acad. Sci. U.S.A.* **66,** 352–359.
Hašek, J., Svobodová, J., and Streiblová, E. (1986). *Eur. J. Cell Biol.* **41,** 150–156.
Hašek, J., Rupeš, I., Svobodová, J., and Streiblová, E. (1987). *J. Gen. Microbiol.* **133,** 3355–3363.
Haugland, R. P. (1983). *In* "Excited States of Biopolymers" (R. F. Steiner, ed.), pp. 29–58. Plenum, New York.
Hayashibe, M. (1975). *In* "Growth and Differentiation in Microorganisms" (T. Ishikawa, Y. Maruyama, and H. Matsumiya, eds.), pp. 165–191. University Park Press, Baltimore, Maryland.
Hayashibe, M., and Katohda, S. (1973). *J. Gen. Appl. Microbiol.* **19,** 23–39.
Herth, W. (1980). *J. Cell Biol.* **87,** 442–450.
Herth, W., and Schnepf, E. (1980). *Protoplasma* **105,** 129–133.
Hirano, T., Funahashi, S., Uemura, T., and Yanagida, M. (1986). *EMBO J.* **5,** 2973–2979.
Hirano, T., Hiraoka, Y., and Yanagida, M. (1988). *J. Cell Biol.* **106,** 1171–1183.
Holan, Z., Pokorný, V., Beran, K., Gemperle, A., Tuzar, Z., and Baldrián, J. (1981). *Arch. Microbiol.* **130,** 312–318.
Horio, T., and Hotani, H. (1986). *Nature (London)* **321,** 605–607.
Houwink, A. L., and Kreger, D. R. (1953). *Antonie van Leeuwenhoek* **19,** 1–24.
Huffaker, T. C., Hoyt, M. A., and Botstein, D. (1987). *Annu. Rev. Genet.* **21,** 259–284.
Huffaker, T. C., Thomas, J. H., and Botstein, D. (1988). *J. Cell Biol.* **106,** 1997–2010.
Hutter, K.-J., and Eipel, H. E. (1978). *Antonie van Leeuwenhoek* **44,** 269–282.
Hynes, R. O., and Destree, A. T. (1978). *Cell (Cambridge, Mass.)* **15,** 875–886.
Inoué, S. (1986). "Video Microscopy." Plenum, New York.
Jacobs, C. W., Adams, A. E. M., Szaniszlo, P. J., and Pringle, J. R. (1988). *J. Cell Biol.* **107,** 1409–1426.
Johnson, B. F., Calleja, G. B., Yoo, B. Y., Zuker, M., and McDonald, I. J. (1982). *Int. Rev. Cytol.* **75,** 167–208.
Johnson, D. I., and Pringle, J. R. (1989). Submitted for publication.
Johnson, G. D., and Nogueira Araujo, G. M. de C. (1981). *J. Immunol. Methods* **43,** 349–350.

Johnson, G. D., Davidson, R. S., McNamee, K. C., Russell, G., Goodwin, D., and Holborow, E. J. (1982). *J. Immunol. Methods* **55,** 231–242.
Johnston, G. C., Ehrhardt, C. W., Lorincz, A., and Carter, B. L. A. (1979). *J. Bacteriol.* **137,** 1–5.
Jones, E. W. (1984). *Annu. Rev. Genet.* **18,** 233–270.
Kalisz, H., Pohlig, G., and Holzer, H. (1987). *Arch. Microbiol.* **147,** 235–239.
Kilmartin, J. V., and Adams, A. E. M. (1984). *J. Cell Biol.* **98,** 922–933.
Kilmartin, J. V., Wright, B., and Milstein, C. (1982). *J. Cell Biol.* **93,** 576–582.
Kim, H. B., Haarer, B. K., and Pringle, J. R. (1989a). *J. Cell Biol.* (in press).
Kim, H. B., Haarer, B. K., and Pringle, J. R. (1989b). In preparation.
Kobayashi, T., Sugimoto, T., Itoh, T., Kosaka, K., Tanaka, T., Suwa, S., Sato, K., and Tsuji, K. (1986). *Diabetes* **35,** 335–340.
Kukuruzinska, M. A., Bergh, M. L. E., and Jackson, B. J. (1987). *Annu. Rev. Biochem.* **56,** 915–944.
Kurlandzka, A., and Rytka, J. (1985). *J. Gen. Microbiol.* **131,** 2909–2918.
Langone, J. J., and Van Vunakis, H., eds. (1981). "Methods in Enzymology," Vol. 73. Academic Press, New York.
Langone, J. J., and Van Vunakis, H., eds. (1983). "Methods in Enzymology," Vol. 92. Academic Press, New York.
Langone, J. J., and Van Vunakis, H., eds. (1986). "Methods in Enzymology," Vol. 121. Academic Press, Orlando, Florida.
Lanier, L. L., and Loken, M. R. (1984). *J. Immunol.* **132,** 151–156.
Last, R. L., and Woolford, J. L., Jr. (1986). *J. Cell Biol.* **103,** 2103–2112.
Lee, C., and Chen, L. B. (1988). *Cell (Cambridge, Mass.)* **54,** 37–46.
Lemmon, S. K., and Jones, E. W. (1987). *Science* **238,** 504–509.
Lenz, A., and Holzer, H. (1984). *Arch. Microbiol.* **137,** 104–108.
Lillie, S. H., and Brown, S. S. (1987). *Yeast* **3,** 63–70.
Lipsky, N. G., and Pagano, R. E. (1985). *Science* **228,** 745–747.
Lord, P. G., and Wheals, A. E. (1980). *J. Bacteriol.* **142,** 808–818.
Maeda, H., and Ishida, N. (1967). *J. Biochem. (Tokyo)* **62,** 276–278.
Marchant, R., and Smith, D. G. (1968). *J. Gen. Microbiol.* **53,** 163–169.
Marks, J., and Hyams, J. S. (1985). *Eur. J. Cell Biol.* **39,** 27–32.
Matsumoto, K., Uno, I., and Ishikawa, T. (1983a). *Cell (Cambridge, Mass.)* **32,** 417–423.
Matsumoto, K., Uno, I., and Ishikawa, T. (1983b). *Exp. Cell Res.* **146,** 151–161.
Matsumoto, K., Uno, I., Kato, K., and Ishikawa, T. (1985). *Yeast* **1,** 25–38.
May, J. W., and Mitchison, J. M. (1986). *Nature (London)* **322,** 752–754.
Miller, J. B., Crow, M. T., and Stockdale, F. E. (1985). *J. Cell Biol.* **101,** 1643–1650.
Mitchison, J. M., and Nurse, P. (1985). *J. Cell Sci.* **75,** 357–376.
Miyakawa, I., Aoi, H., Sando, N., and Kuroiwa, T. (1984). *J. Cell Sci.* **66,** 21–38.
Miyakawa, I., Sando, N., Kawano, S., Nakamura, S., and Kuroiwa, T. (1987). *J. Cell Sci.* **88,** 431–439.
Miyata, M., Miyata, H., and Johnson, B. F. (1986). *J. Gen. Microbiol.* **132,** 883–891.
Moir, D., Stewart, S. E., Osmond, B. C., and Botstein, D. (1982). *Genetics* **100,** 547–563.
Mol, P. C., and Wessels, J. G. H. (1987). *FEMS Microbiol. Lett.* **41,** 95–99.
Molano, J., Bowers, B., and Cabib, E. (1980). *J. Cell Biol.* **85,** 199–212.
Moreau, N., Angelier, N., Bonnanfant-Jais, M.-L., Gounon, P., and Kubisz, P. (1986). *J. Cell Biol.* **103,** 683–690.
Moreland, R. B., Nam, H. G., Hereford, L. M., and Fried, H. M. (1985). *Proc. Natl. Acad. Sci. U.S.A.* **82,** 6561–6565.
Nasmyth, K. (1983). *Nature (London)* **302,** 670–676.

Neighbors, B. W., Williams, R. C., Jr., and McIntosh, J. R. (1988). *J. Cell Biol.* **106,** 1193–1204.
Novick, P., and Botstein, D. (1985). *Cell (Cambridge, Mass.)* **40,** 405–416.
Novick, P., Osmond, B. C., and Botstein, D. (1989). *Genetics* **121,** 659–674.
Oakley, B. R., and Rinehart, J. E. (1985). *J. Cell Biol.* **101,** 2392–2397.
Ohkuma, S., and Poole, B. (1978). *Proc. Natl. Acad. Sci. U.S.A.* **75,** 3327–3331.
Oi, V. T., Glazer, A. N., and Stryer, L. (1982). *J. Cell Biol.* **93,** 981–986.
Olmsted, J. B. (1981). *J. Biol. Chem.* **256,** 11955–11957.
Orlean, P. (1987). *J. Biol. Chem.* **262,** 5732–5739.
Paradiso, A. M., Tsien, R. Y., and Machen, T. E. (1987). *Nature (London)* **325,** 447–450.
Parks, D. R., Hardy, R. R., and Herzenberg, L. A. (1984). *Cytometry* **5,** 159–168.
Payne, G. S., and Schekman, R. (1985). *Science* **230,** 1009–1014.
Payne, G. S., Baker, D., van Tuinen, E., and Schekman, R. (1988). *J. Cell Biol.* **106,** 1453–1461.
Pena, A., Mora, M. A., and Carrasco, N. (1979). *J. Membr. Biol.* **47,** 261–284.
Petersen, N. O., Felde, S., and Elson, E. L. (1986). *In* "Handbook of Experimental Immunology" (D. M. Weir, ed.), 4th ed., Vol. 1, pp. 24.1–24.22. Blackwell, Oxford.
Pfeffer, S. R., Drubin, D. G., and Kelly, R. B. (1983). *J. Cell Biol.* **97,** 40–47.
Phillips, H. S., Nikolics, K., Branton, D., and Seeburg, P. H. (1985). *Nature (London)* **316,** 542–545.
Pick, U., and Bassilian, S. (1981). *FEBS Lett.* **123,** 127–130.
Picton, J. M., and Steer, M. W. (1985). *Planta* **163,** 20–26.
Pillus, L., and Solomon, F. (1986). *Proc. Natl. Acad. Sci. U.S.A.* **83,** 2468–2472.
Piperno, G., LeDizet, M., and Chang, X.-J. (1987). *J. Cell Biol.* **104,** 289–302.
Platt, J. L., and Michael, A. F. (1983). *J. Histochem. Cytochem.* **31,** 840–842.
Preston, R. A., Murphy, R. F., and Jones, E. W. (1987). *J. Cell Biol.* **105,** 1981–1987.
Pringle, J. R., and Mor, J.-R. (1975). *In* "Methods in Cell Biology" (D. M. Prescott, ed.), Vol. 11, pp. 131–168. Academic Press, New York.
Pringle, J. R., Lillie, S. H., Adams, A. E. M., Jacobs, C. W., Haarer, B. K., Coleman, K. G., Robinson, J. S., Bloom, L., and Preston, R. A. (1986). *In* "Yeast Cell Biology" (J. Hicks, ed.), pp. 47–80. Liss, New York.
Pruss, R. M., Mirsky, R., Raff, M. C., Thorpe, R., Dowding, A. J., and Anderton, B. H. (1981). *Cell (Cambridge, Mass.)* **27,** 419–428.
Riezman, H. (1985). *Cell (Cambridge, Mass.)* **40,** 1001–1009.
Roberts, E., Seagull, R. W., Haigler, C. H., and Brown, R. M., Jr. (1982). *Protoplasma* **113,** 1–9.
Roberts, R. L., Bowers, B., Slater, M. L., and Cabib, E. (1983). *Mol. Cell. Biol.* **3,** 922–930.
Robinow, C. F. (1975). *In* "Methods in Cell Biology" (D. M. Prescott, ed.), Vol. 11, pp. 1–22. Academic Press, New York.
Roncero, C., and Durán, A. (1985). *J. Bacteriol.* **163,** 1180–1185.
Roncero, C., Valdivieso, M. H., Ribas, J. C., and Durán, A. (1988a). *J. Bacteriol.* **170,** 1945–1949.
Roncero, C., Valdivieso, M. H., Ribas, J. C., and Durán, A. (1988b). *J. Bacteriol.* **170,** 1950–1954.
Rose, M. D., and Fink, G. R. (1987). *Cell (Cambridge, Mass.)* **48,** 1047–1060.
Royan, S., and Subramaniam, M. K. (1960). *Proc.—Indian Acad. Sci., Sect. B* **51B,** 205–210.
Sammak, P. J., and Borisy, G. G. (1988). *Nature (London)* **332,** 724–726.
Sburlati, A., and Cabib, E. (1986). *J. Biol. Chem.* **261,** 15147–15152.
Schekman, R. (1985). *Annu. Rev. Cell Biol.* **1,** 115–143.

Schekman, R., and Brawley, V. (1979). *Proc. Natl. Acad. Sci. U.S.A.* **76,** 645–649.
Schulze, E., and Kirschner, M. (1986). *J. Cell Biol.* **102,** 1020–1031.
Schulze, E., and Kirschner, M. (1987). *J. Cell Biol.* **104,** 277–288.
Segev, N., and Botstein, D. (1987). *Mol. Cell. Biol.* **7,** 2367–2377.
Segev, N., Mulholland, J., and Botstein, D. (1988). *Cell (Cambridge, Mass.)* **52,** 915–924.
Seichertová, O., Beran, K., Holan, Z., and Pokorný, V. (1973). *Folia Microbiol. (Prague)* **18,** 207–211.
Sherman, F., Fink, G. R., and Hicks, J. B. (1986). "Methods in Yeast Genetics: A Laboratory Manual." Cold Spring Harbor Lab., Cold Spring Harbor, New York.
Silver, P. A., Keegan, L. P., and Ptashne, M. (1984). *Proc. Natl. Acad. Sci. U.S.A.* **81,** 5951–5955.
Silverman, S. J., Sburlati, A., Slater, M. L., and Cabib, E. (1988). *Proc. Natl. Acad. Sci. U.S.A.* **85,** 4735–4739.
Slater, M. L. (1976). *J. Bacteriol.* **126,** 1339–1341.
Slater, M. L. (1978). *In* "Methods in Cell Biology" (G. Stein, J. Stein, and L. J. Kleinsmith, eds.), Vol. 20, pp. 135–140. Academic Press, New York.
Sloat, B. F., and Pringle, J. R. (1978). *Science* **200,** 1171–1173.
Sloat, B. F., Adams, A. E. M., and Pringle, J. R. (1981). *J. Cell Biol.* **89,** 395–405.
Small, G. M., Imanaka, T., Shio, H., and Lazarow, P. B. (1987). *Mol. Cell. Biol.* **7,** 1848–1855.
Smith, D. E., and Fisher, P. A. (1984). *J. Cell Biol.* **99,** 20–28.
Snow, C. M., Senior, A., and Gerace. L. (1987). *J. Cell Biol.* **104,** 1143–1156.
Snyder, M., and Davis, R. W. (1988). *Cell (Cambridge, Mass.)* **54,** 743–754.
Stevens, T., Esmon, B., and Schekman, R. (1982). *Cell (Cambridge, Mass.)* **30,** 439–448.
Stewart, W. W. (1981). *Nature (London)* **292,** 17–21.
Streiblová, E. (1970). *Can. J. Microbiol.* **16,** 827–831.
Streiblová, E., and Beran, K. (1963a). *Exp. Cell Res.* **30,** 603–605.
Streiblová, E., and Beran, K. (1963b). *Folia Microbiol. (Prague)* **8,** 221–227.
Streiblová, E., Hašek, J., and Jelke, E. (1984). *J. Cell Sci.* **69,** 47–65.
Tague, B. W., and Chrispeels, M. J. (1987). *J. Cell Biol.* **105,** 1971–1979.
Talens, L. T., Miranda, M., and Miller, M. W. (1973). *J. Bacteriol.* **114,** 413–423.
Talian, J. C., Olmsted, J. B., and Goldman, R. D. *J. Cell Biol.* **97,** 1277–1282.
Tartakoff, A. M., and Vassalli, P. (1983). *J. Cell Biol.* **97,** 1243–1248.
Taylor, D. L., and Wang, Y.-L., eds. (1989). "Methods in Cell Biology," Vol. 30. Academic Press, San Diego, California.
Taylor, D. L., Waggoner, A. S., Murphy, R. F., Lanni, F., and Birge, R. R., eds. (1986). "Applications of Fluorescence in the Biomedical Sciences." Liss, New York.
Thomas, J. A., Buchsbaum, R. N., Zimniak, A., and Racker, E. (1979). *Biochemistry* **18,** 2210–2218.
Thomas J. H., and Botstein, D. (1986). *Cell (Cambridge, Mass.)* **44,** 65–76.
Thompson, P. W., and Wheals, A. E. (1980). *J. Gen. Microbiol.* **121,** 401–409.
Tkacz, J. S., and Lampen, J. O. (1972). *J. Gen. Microbiol.* **72,** 243–247.
Tkacz, J. S., and MacKay, V. L. (1979). *J. Cell Biol.* **80,** 326–333.
Tkacz, J. S., Cybulska, E. B., and Lampen, J. O. (1971). *J. Bacteriol.* **105,** 1–5.
Towbin, H., Staehelin, T., and Gordon, J. (1979). *Proc. Natl. Acad. Sci. U.S.A.* **76,** 4350–4354.
Trueheart, J., Boeke, J. D., and Fink, G. R. (1987). *Mol. Cell. Biol.* **7,** 2316–2328.
Tsien, R. Y., Rink, T. J., and Poenie, M. (1985). *Cell Calcium* **6,** 145–157.
Uemura, T., Ohkura, H., Adachi, Y., Morino, K., Shiozaki, K., and Yanagida, M. (1987). *Cell (Cambridge, Mass.)* **50,** 917–925.

Uno, I., Matsumoto, K., Adachi, K., and Ishikawa, T. (1984). *J. Biol. Chem.* **259,** 12508–12513.

Uno, I., Matsumoto, K., Hirata, A., and Ishikawa, T. (1985). *J. Cell Biol.* **100,** 1854–1862.

Vale, R. D., Schnapp, B. J., Mitchison, T., Steuer, E., Reese, T. S., and Sheetz, M. P. (1985). *Cell (Cambridge, Mass.)* **43,** 623–632.

Vandekerckhove, J., Deboben, A., Nassal, M., and Wieland, T. (1985). *EMBO J.* **4,** 2815–2818.

Vannini, G. L., Poli, F., Donini, A., and Pancaldi, S. (1983). *Plant Sci. Lett.* **31,** 9–17.

van Tuinen, E., and Riezman, H. (1987). *J. Histochem. Cytochem.* **35,** 327–333.

Van Vunakis, H., and Langone, J. J., eds. (1980). "Methods in Enzymology," Vol. 70. Academic Press, New York.

Veenhuis, M., Mateblowski, M., Kunau, W. H., and Harder, W. (1987). *Yeast* **3,** 77–84.

Waggoner, A. S. (1979). *Annu. Rev. Biophys. Bioeng.* **8,** 47–68.

Waggoner, A. S. (1986). *In* "Applications of Fluorescence in the Biomedical Sciences" (D. L. Taylor, A. S. Waggoner, R. F. Murphy, F. Lanni, and R. R. Birge, eds.), pp. 3–28. Liss, New York.

Wang, Y.-L., and Taylor, D. L., eds. (1989). "Methods in Cell Biology," Vol. 29. Academic Press, San Diego, California.

Weisman, L. S., Bacallao, R., and Wickner, W. (1987). *J. Cell Biol.* **105,** 1539–1547.

White, J. G., Amos, W. B., and Fordham, M. (1987). *J. Cell Biol.* **105,** 41–48.

Wieland, T., and Faulstich, H. (1978). *CRC Crit. Rev. Biochem.* **5,** 185–260.

Wieland, T., Miura, T., and Seeliger, A. (1983). *Int. J. Pept. Protein Res.* **21,** 3–10.

Wiemken, A., Matile, P., and Moor, H. (1970). *Arch. Mikrobiol.* **70,** 89–103.

Wiemken, A., Schellenberg, M., and Urech, K. (1979). *Arch. Microbiol.* **123,** 23–35.

Williamson, D. H., and Fennell, D. J. (1975). *In* "Methods in Cell Biology" (D. M. Prescott, ed.), Vol. 12, pp. 335–351. Academic Press, New York.

Williamson, D. H., and Fennell, D. J. (1979). *In* "Methods in Enzymology" (S. Fleischer and L. Packer, eds.), Vol. 56, pp. 728–733. Academic Press, New York.

Wittenberg, C., Richardson, S. L., and Reed, S. I. (1987). *J. Cell Biol.* **105,** 1527–1538.

Wolfbeis, O. S., Fürlinger, E., Kroneis, H., and Marsoner, H. (1983). *Fresenius' Z. Anal. Chem.* **314,** 119–124.

Wood, G. S., and Warnke, R. (1981). *J. Histochem. Cytochem.* **29,** 1196–1204.

Wood, P. J. (1980). *Carbohydr. Res.* **85,** 271–287.

Wright, R., Basson, M., D'Ari, L., and Rine, J. (1988). *J. Cell Biol.* **107,** 101–114.

Wulf, E., Deboben, A., Bautz, F. A., Faulstich, H., and Wieland, T. (1979). *Proc. Natl. Acad. Sci. U.S.A.* **76,** 4498–4502.

Yanagida, M., Morikawa, K., Hiraoka, Y., Matsumoto, S., Uemura, T., and Okada, S. (1986). *In* "Applications of Fluorescence in the Biomedical Sciences" (D. L. Taylor, A. S. Waggoner, R. F. Murphy, F. Lanni, and R. R. Birge, eds.), pp. 321–345. Liss, New York.

Chapter 20

Preservation of Biological Specimens for Observation in a Confocal Fluorescence Microscope and Operational Principles of Confocal Fluorescence Microscopy

ROBERT BACALLAO

Division of Nephrology
UCLA Medical Center
Los Angeles, California 90024

ERNST H. K. STELZER

Confocal Light Microscopy Group
European Molecular Biology Laboratory
D-6900 Heidelberg, Federal Republic of Germany

METHODS IN CELL BIOLOGY, VOL. 31

I. Introduction

Recent advances in the field of microscopy have caused a revival of interest in the correlation of cellular structure with genetic and biochemical data. Work performed on microtubules using video-enhanced microscopy has allowed for the unique combination of biochemical assays with structural and functional information (Allen and Kreis, 1986; Matteoni and Kreis, 1987; Dabora and Sheetz, 1988; Lee and Chen, 1988). Confocal scanning-laser microscopy combined with sophisticated image-processing techniques allow the cell biologist to explore the three-dimensional infrastructure of the cell (Wijnaendts-van Resandt *et al.*, 1985; White *et al.*, 1987; van Meer *et al.*, 1987), thus conferring a high degree of precision in localization experiments.

These new microscope technologies can only fulfill their promise if a sufficient effort is made to ensure that the information obtained from biological specimens is accurate. This chapter will address some of the inherent problems in preservation of biological specimens so that reliable data can be acquired with a confocal fluorescence microscope. The methodology that will be described is the result of an effort to study the genesis of epithelial polarity in the Madin–Darby canine kidney (MDCK) cell line. While specific methods have been developed to study these particular cells, the lessons learned about their preservation should be widely applicable. A description of the confocal principle introduces basic concepts of confocal fluorescence microscopy. One can then understand how images are recorded and how the information may be interpreted.

II. The Confocal Principle

The principal configuration of a confocal fluorescence microscope is shown in Fig. 1c. A laser beam is focused into a pinhole P1. The light penetrating the pinhole is reflected by a dichroic mirror and focused into the sample with the microscope objective. Fluorescent light emitted in the sample is collected with the same microscope objective. Passing through the dichroic mirror, the light is focused into the detector pinhole P2 (Fig. 1c). A photomultiplier tube, behind the pinhole P2, is used to measure the intensity of the light. As indicated in Fig. 1b, the light penetrates other areas above and below the focal volume. The images of

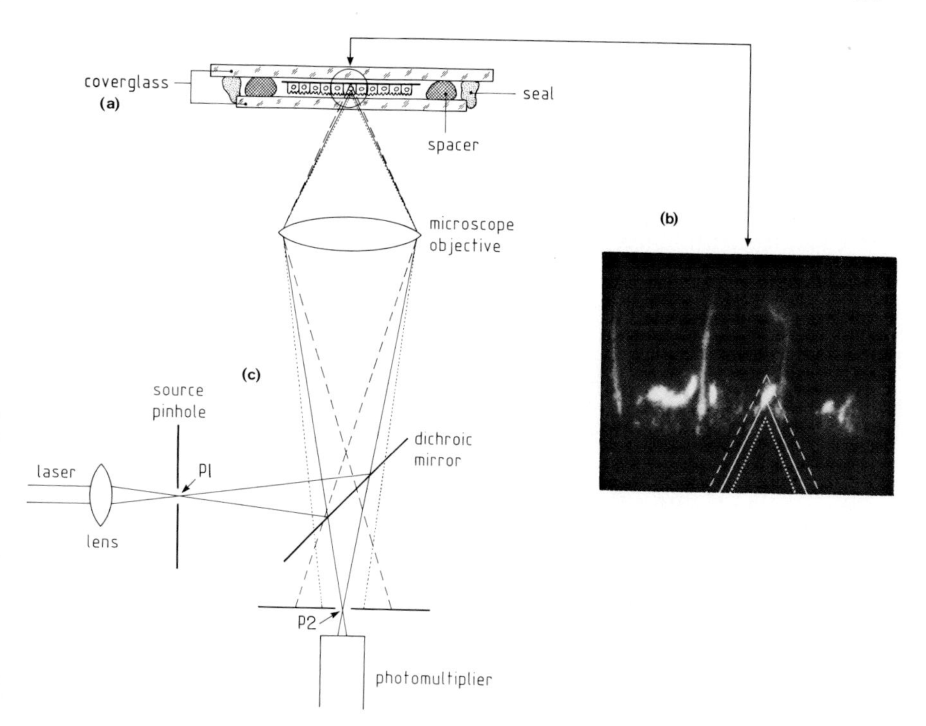

FIG. 1. Configuration and depth discrimination in a confocal microscope. (a) A sample in which MDCK cells grown on filters are mounted. The spacers make sure that the cells will not be squashed, and the seal fixes the position of the cover glass. The apical membrane of MDCK cells is oriented toward the microscope objective. (b) An X/Z image of MDCK cells labeled with C6-NBD-ceramide for 1 hour at 20°C and then incubated for an additional hour at 37°C. This labels both the Golgi apparatus and the plasma membrane. The laser beam is focused into the Golgi apparatus. (c) The principal configuration of a confocal fluorescence microscope.

these areas are either in front of or behind the detector pinhole P2; therefore, their contribution is discrminated (Sheppard and Wilson, 1978). An alternative way to understand the confocal principle is to regard the microscope objective as the lens that forms an image of the source pinhole and the detector pinhole as the sample. If the two images overlap and are diffraction-limited (i.e., the size of the pinholes is adjusted to the power of the lens), the arrangement is confocal.

III. Criteria for Cell Preservation

Cell Culture Conditions

Cell morphology is profoundly affected by the conditions of the growth media and supporting surfaces. Every attempt must be made to define precisely the growth conditions so that the cellular morphology closely matches the morphology seen in the organism. For the purposes of our studies, MDCK cells are grown on Costar polycarbonate 0.4 μm pore diameter, tissue culture-treated filters. The cells are grown with MEM + 10% FCS supplemented with 2 m*M* glutamine (Fuller *et al.*, 1984). Penicillin and streptomycin are added as antibiotic agents. The filters with their holders are placed in mini-Marbrook chambers, which ensure that the cells are supplied with nutrients from both the apical and basolateral cell surface (Fuller *et al.*, 1984; Misek *et al.*, 1984). Under these conditions, the cells will grow to a height of 15–20 μm when they reach their final columnar, polarized state. This typically occurs 1 day after the formation of a transepithelial resistance (R. Bacallao, unpublished observations). Supplying the cells with an artificial basement membrane such as Matrigel, which promotes terminal differentiation of neuronal cells (Kleinman *et al.*, 1987), does not change the kinetics of polarization or the final cell height in the columnar polarized state.

Because the data obtained on a confocal microscope can be reconstructed to provide a three-dimensional image of the cell, the method of cell fixation is absolutely crucial to the success of the preservation. There is a trade-off between the fixation of the cell and the preservation of antigenic epitopes within the cell. Specimens that have been overfixed cannot be stained using immunofluorescence labeling methods. Underfixation causes diffuse localization of the target proteins one might wish to study. Once the sample is stained it is important to determine whether the three-dimensional integrity of the cell has been maintained during the fixation and staining. The ultimate goal is to match the *in vivo* state of the cell as closely as possible. Figure 2 shows a stereo picture of MDCK cells studied on the confocal microscope after *in vivo* labeling of the Golgi apparatus and plasma membrane using the fluorescent lipid analog C6-NBD-ceramide (Lipsky and Pagano, 1985; van Meer *et al.*, 1987). One can clearly see the relative height of the cells and the perinuclear Golgi apparatus. The two cells in the middle of the picture with a diffuse punctate staining pattern are cells undergoing cell division. It is notable that the plasma membrane of these cells is not labeled.

We have compared various fixation methods with a view to the previously mentioned criteria. The methods were evaluated by examining

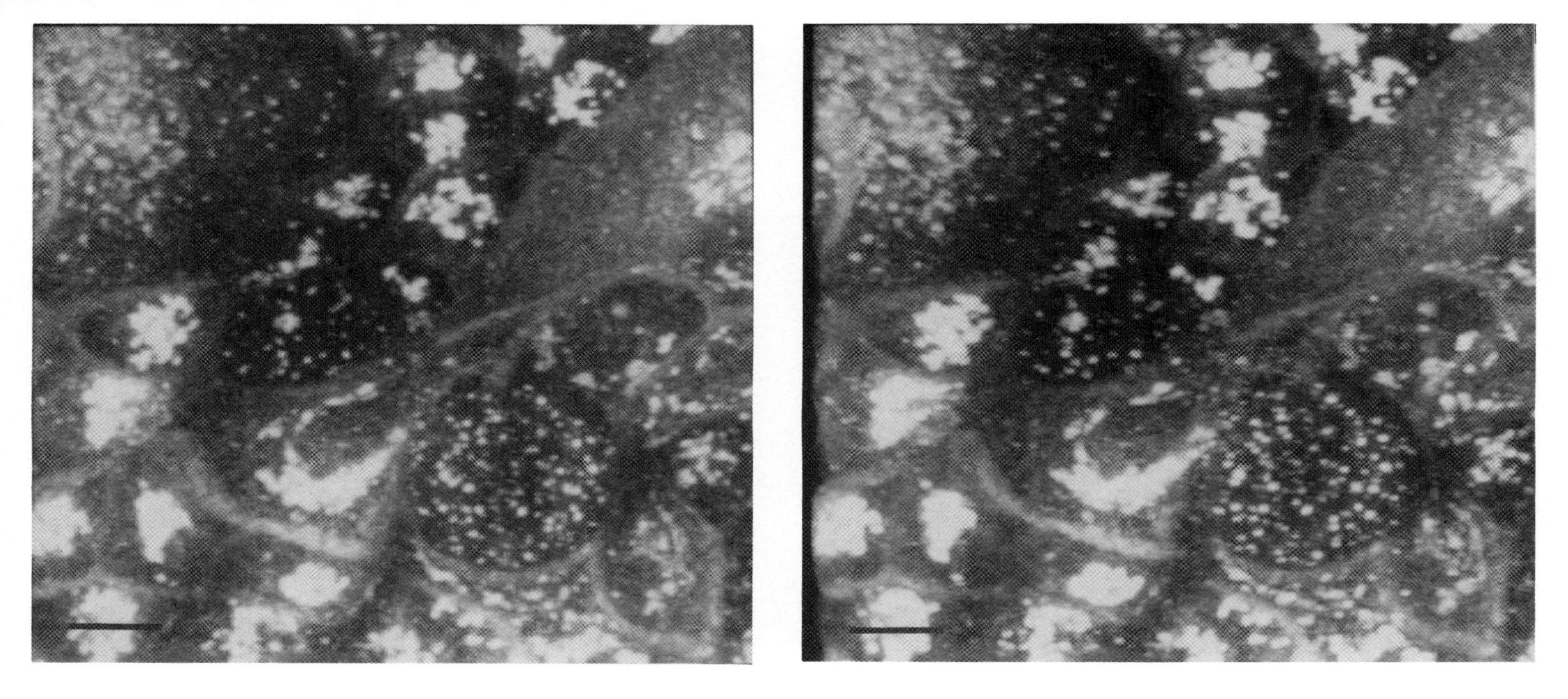

FIG. 2. Stereo images of MDCK cells labeled with C6-NBD-ceramide. MDCK I cells were grown on Costar polycarbonate filters and incubated with liposomes containing C6-NBD-ceramide. The cells were labeled with liposomes containing C6-NBD-ceramide as previously described (van Meer *et al.*, 1987). The cells were incubated for 1 hour at 20°C and shifted to 37°C for 15 minutes. This allows both the plasma membrane and the Golgi apparatus to be labeled. The image is slightly overexposed to bring out the plasma membrane label (the gain is too high). The black areas in the cells show that there is little background contribution from the other focal planes. Gray areas are therefore the result of plasma membrane staining with NBD-ceramide, revealing important and accurate information. The stereo images are computer reconstructions of a series of x–y images taken from consecutive "z" positions. No further image processing was performed. The change in z position between each x–y image is 0.4 μm. The distance between the apical and basal membranes is 12 μm. The field is made up of 512×512 pixels, the image is the average of eight scans per line. Bar = 10 μm.

cell height and shape of mitotic spindles. The fixation methods described below preserve cell height well, when compared to *in vivo*-labeled cells (R. Bacallao, unpublished observation). In addition, when the cells are stained to examine the microtubules in the mitotic spindles, the microtubules are well preserved. The microtubule staining is a good indication of the preservation efficacy of a fixation methodology.

An inherent problem with immunofluorescence staining of cells grown on filter supports is an increase in background fluorescence. This seems to be caused by two separate parameters. First, the filter supports tend to trap the antibodies. The filters must be extensively washed during the immunostaining procedure to minimize this problem. Second, MDCK cells grow to a greater height and a higher density on the polycarbonate filter supports. This means that antibody solutions must be incubated with the sample for longer time periods. Third, the filters are opaque and tend to reflect the background fluorescence. Therefore, the fixation method must have as low a level of background fluorescence as possible. Sometimes the growth conditions may influence the quality of immunofluorescence staining. For example, Matrigel increases the background staining seen with second antibodies specific for mouse IgG (R. Bacallao, unpublished observation).

IV. Methods

A. Glutaraldehyde Fixation

1. Stock Solutions

8% Glutaraldehyde E.M. grade (Polysciences)

80 m*M* K-PIPES (pH 6.8), 5 m*M* EGTA, 2 m*M* $MgCl_2$, 0.1% Triton X-100

Phosphate-buffered saline without Ca^{2+}/Mg^{2+} [PBS(−)]

Phosphate-buffered saline without Ca^{2+}/Mg^{2+}, pH 8.0 [PBS(−), pH 8.0]

2. Preparation of Stock Solutions

E.M.-Grade glutaraldehyde was obtained from Polysciences. It is supplied as an 8% aqueous solution. When a new vial is opened, the glutaraldehyde is diluted to 0.3% in a solution of 80 m*M* K-PIPES, pH 6.8, 5 m*M* EGTA, 2 m*M* $MgCl_2$, 0.1% Triton X-100. The aliquots are stored at

−20°C. Prior to each experiment a fresh aliquot is used soon after thawing. These aliquots are never frozen again for reuse, because this causes the fixation to suffer. Adding a few drops of 6 *N* NaOH to normal PBS produces PBS, pH 8.0.

3. Fixation Protocl

1. Warm 100 ml of 80 m*M* K-PIPES (pH 6.8), 5 m*M* EGTA, 2 m*M* $MgCl_2$ to 37°C in a beaker.
2. Pour off the media in the apical well of the Costar filter. Dip the entire filter plus filter holder in the prewarmed 80 m*M* K-PIPES (pH 6.8), 5 m*M* EGTA, 2 m*M* $MgCl_2$ buffer for 5 seconds.
3. Transfer the filter to the six-well plate supplied with the polycarbonate filters to permit convenient fixation and washing steps.
4. Fix the cells for 10 minutes with 0.3% glutaraldehyde + 0.1% Triton X-100 at room temperature. The glutaraldehyde fixative is added to the apical (2 ml) and basal (3 ml) portion of the filter. During all the incubation steps and washes the six-well plate is agitated on a rotary shaker.
5. During the fixation period, weigh out fresh $NaBH^4$. Weigh out 3–10 mg aliquots per filter. The $NaBH_4$ should be kept in an anhydrous state and should be weighed out immediately prior to use. The $NaBH_4$ is poured into 50-ml sterilized conical tubes with screwcaps.
6. Aspirate the fixative and dip the entire filter successively in three 100-ml beakers containing PBS(−).
7. Add PBS(−), pH 8.0 to the $NaBH_4$. (Final concentration 1 mg/ml). Add 3 ml to the apical portion of the cell and 4 ml to the basal chamber. Incubate 15 minutes at room temperature. Repeat this step two more times using freshly dissolved $NaBH_4$. The adjustment of the pH to 8.0 increases the half-life of $NaBH_4$ in the solution. This step is essential to decrease the autofluorescence of the glutaraldehyde-fixed cells.
8. Wash the cells with PBS(−) by dipping in three beakers containing PBS(−). Return the filters to the six-well plate with PBS(−) bathing the apical and basal side. The filter is now ready for immunofluorescence staining.

B. The pH-Shift Paraformaldehyde Fixation

This method is a variation of the method described by Berod (Berod *et al.*, 1981).

1. Stock Solutions

40% Paraformaldehyde (Merck) in H_2O
80 m*M* K-PIPES (pH 6.5), 5 m*M* EGTA, 2 m*M* $MgCl_2$
100 m*M* NaB_4O_7 (pH 11.0) + 0.02% saponin

2. Preparation of the Stock Solutions

The preparation of the paraformaldehyde stock solution is based on the description by Robertson *et al.* (1963). Forty grams of paraformaldehyde are added to 100 ml of H_2O. While continuously stirring, the mixture is heated to >67°C. A few drops of 6 *N* NaOH are added to dissolve the paraformaldehyde. The stock solution is divided into aliquots and stored at −20°C. Prior to use, aliquots are thawed by warming in a water bath. Above 75°C the paraformaldehyde will go into solution. The paraformaldehyde solution should not be allowed to boil, however. The paraformaldehyde is diluted to 2–4% in both the K-PIPES and sodium borate buffers. For our purposes, a 3% solution of paraformaldehyde was adequate for preserving both the structures and antigenic determinants of a wide variety of cell organelles (see Section V). The pH of the K-PIPES buffer is brought to 6.5 with 1 *N* HCl after the paraformaldehyde has been added.

Adding a few drops of 6 *N* NaOH to normal PBS produces PBS(−), pH 8.0. Sodium borate (100 m*M*) is titrated to pH 11.0 by adding 6 *N* NaOH to the buffer.

3. Fixation Protocol

1. Pour off the medium in the apical well of the filter.
2. Dip the filters in 80 m*M* K-PIPES (pH 6.8), 5 m*M* EGTA, 2 m*M* $MgCl_2$ prewarmed to 37°C.
3. Add 3 ml of 3.0% paraformaldehyde in 80 m*M* K-PIPES (pH 6.5), 5 m*M* EGTA, 2 m*M* $MgCl_2$ to the basal chamber of the filter in a six-well dish, and add 2 ml of the fixative on the apical surface of the cells. Incubate the cells with agitation on a rotary table for 5 minutes at room temperature.
4. Aspirate the paraformaldehyde–K-PIPES solution. Add 3 ml of 3% paraformaldehyde in 100 m*M* NaB_4O_7 (pH 11.0), containing 0.025% saponin to the basal side and 2 ml to the apical side of each filter. Incubate with agitation on a rotary table for 10 minutes at room temperature.

5. Weigh out two 10-mg aliquots of $NaBH_4$ for each filter and store in a conical tube with a screwcap.
6. Aspirate the fixation solution. Wash the filters by successively dipping the filters in three beakers containing 100 ml of PBS(−).
7. Dissolve each aliquot in 10 ml PBS(−), pH 8.0. (Final concentration of $NaBH_4$ should be 1 mg/ml.) Vortex the solution briefly and add the solution to the apical (2 ml) and basal (3 ml) portions of the filters. Incubate 15 minutes while shaking the filters on a rotary table. Repeat this step one more time using a fresh solution of $NaBH_4$–PBS(−), pH 8.0.
8. Wash the filters by successively dipping the filters in three beakers containing 100 ml of PBS(−). The filters can be stored overnight at 4°C with PBS(−)+0.1% NaN_3.

C. Immunofluorescence Staining

1. Cut the filter from its plastic holder. Be sure to note which side of the filter has cells! Cut the filter into four squares using a sharp scalpel, keeping the filter wetted with PBS while cutting. To ensure that the cell side of the filter can be readily identified, we routinely cut a slit in the right upper corner of the filter. The filter is cut into squares because this tends to give a flat field of cells after the filter has been mounted. Dividing a filter into wedges with one rounded edge causes the filter to ripple during mounting of the specimen and results in multiple overlapping focal planes.
2. Wash the filter squares in PBS(−) containing 0.2% fish skin gelatin (Sigma) (FSG), and the appropriate percentage of detergent. Wash the filter squares in a six-well plate. All washes are done in 4 ml of solution, at room temperature, with agitation. Unless otherwise stated, the filters are washed for 15 minutes after every change of washing buffer.
3. Place a 50-μl drop of your first antibody dissolved in PBS(−) containing 0.2% FSG on a piece of parafilm on the bottom of a Petri dish.
4. Place a filter square, cell-side down, on the antibody solution. Place a piece of wet Whatman filter paper in the Petri dish well away from the antibody solution. Cover the Petri dish, to form a small humidified chamber.
5. Incubate at 37°C for 1 hour.
6. Wash the filter twice with PBS(−) containing 0.2% FSG. Follow this with three successive washes with PBS(−).
7. Wash the filter once more with PBS(−) containing 0.2% FSG.

8. Add the second antibody as described for the first antibody.
9. Incubate at 37°C for 1 hour.
10. Wash the filters in PBS(−) containing 0.2% FSG twice.
11. Wash three times in PBS(−).
12. Incubate once in PBS containing 0.1% Triton X-100 for 5 minutes.
13. Wash twice in PBS(−) for 5 minutes each time.

The filter squares are ready for mounting.

D. Mounting the Specimen

Place four drops of clear acrylic nail polish on a microscope slide to make support mounts for a coverslip. Each drop should be at a point corresponding to the corner of a coverslip. We use 22 × 22 mm coverslips. The acrylic when dry should be no thicker than 40 μm. Place the filter square in the center of the area demarcated by the nail polish. Take care to ensure that the cells are facing up. Put a drop of 50% glycerol–2× PBS–0.1% NaN_3 on the filter. Carefully place the coverslip over the filter. Make sure that the corners lie on the drops of acrylic polish. Aspirate the excess glycerol–PBS–NaN_3. Put four drops of nail polish on the four corners of the coverslp to stabilize the mount. Once the nail polish has dried, the entire mount can be sealed with nail polish. The specimen should be viewed within 24 hours, as these are not permanent mounts. Semipermanent samples can be made by post-fixing the filter in 4% paraformaldehyde dissolved in 100 mM NaCacodylate pH 7.5 for 30 minutes at room temperature followed by quenching with 50 mM NH_4Cl in PBS for 15 minutes.

V. General Notes and Caveats

The glutaraldehyde fixation gives excellent preservation of the microtubules and other cytoskeletal elements (Fig. 3). The paraformaldehyde pH-shift method also preserves microtubules quite well when a polyclonal antitubulin antibody is used (the antibody was generously supplied by Dr. Jan De Mey from EMBL). In addition, excellent staining has been achieved with a wide variety of antibodies using this fixation method. The paraformaldehyde fixation method has allowed us successfully to stain, with antibodies, marker proteins in the tight junction (Stevenson *et al.*, 1986), microtubules, RER (antidocking protein antibody was generously supplied by Dr. Peter Walter from UCSF), Golgi apparatus (generously

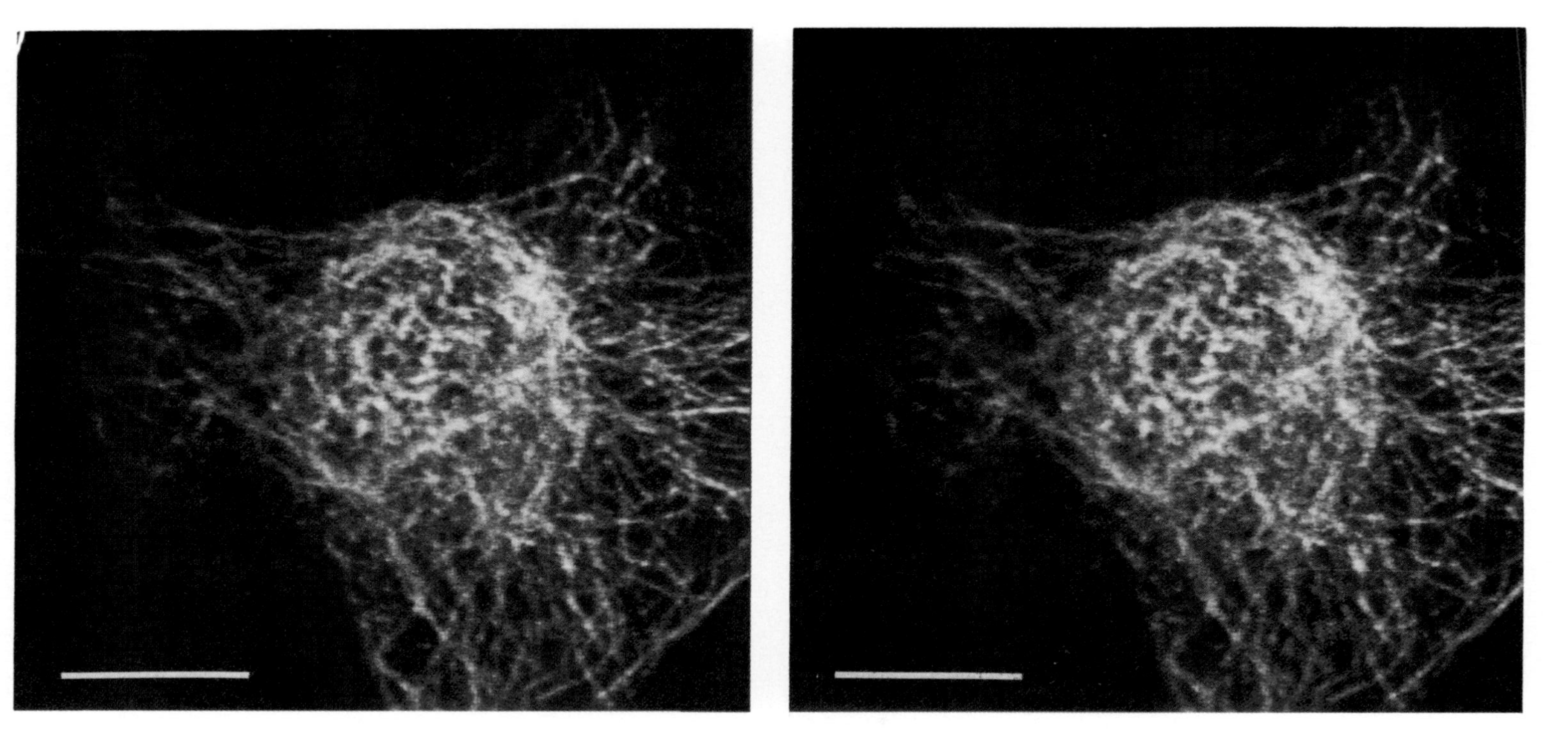

FIG. 3. Stereo images of immunofluorescence staining of microtubules in MDCK cells fixed with glutaraldehyde. MDCK II cells were grown on Costar filters and fixed as described using the glutaraldehyde fixation method. The stereo images are computer reconstructions of a series of x–y images taken from consecutive "z" positions. The change in z position between each x–y image is 0.4 μm. No further image processing was performed. The distance between the apical and basal sides of the cell is 7μm. Bar = 10 μm. The field is made up of 512 × 512 pixels. The image is the average of six scans per line.

supplied by Dr. M. Bornens from the CNRS), and the nucleus (generously supplied by Dr. E. Hurt from EMBL). One may have to vary the amount of paraformaldehyde and saponin for a particular cell system, however. Methanol causes severe shrinkage of the cells when it was tested as a fixative or as a permeabilization agent (R. Bacallao, unpublished observation).

The repeated use of $NaBH_4$ in the fixation protocols is critical to diminish background fluorescence. For similar reasons the extensive washes are absolutely required when immunofluorescence is performed on cells plated on filter supports. Prolonged incubation of the antibodies is required, especially with thick cells. It may be necessary to add some of the antibody solution to the basal portion of the filter as well.

We have tested commercially available mounting solution, such as Gelvatol and Mowiol. We have found that these agents cause an unacceptable shrinkage of the specimens. We have not tested any antibleaching agents as yet, because bleaching has not been a problem with the confocal microscope at EMBL. This is true even for samples labeled with FITC. A description of the EMBL confocal microscope is in preparation (Stelzer *et al.*, 1989).

The choice of which fluorophore to use for staining purposes is constrained by the particular confocal microscope available for use. Because a laser is used as a light source, wavelengths different from those found in a mercury lamp are used to excite common dyes such as FITC, NBC, rhodamine, and Texas red. If for example an argon-ion laser is installed in the microscope, the 514.5-nm line will be used to excite rhodamine. The same line will also excite FITC, thus complicating the use of the two dyes for double-labeling experiments. Replacing rhodamine with Texas red allows the use of filters with a higher cutoff (~580 nm in contrast to 530 nm). The FITC is still subjected to illumination, however, and photobleaching cannot be avoided in principle when observing the other dye with 514.5 nm. There are several possible solutions to these problems:

a. Use different lasers. High-powered (usually water-cooled) argon-ion lasers provide a line at 528.9 nm.
b. Change the dyes. Use dyes that are well matched to the lines provided by the laser.
c. Customize the filters. The usual filter packs are dedicated units optimized for a specific set of specifications (such as light source, dye, reflection of the signal, etc.). These filter packs can be replaced, leading to marked improvements in the light intensity entering the detector.

Under any circumstances it will be necessary to check the use of the dyes under biological conditions.

VI. Presentation of the Data and Image Processing

The optical transfer functions for conventional and confocal fluorescence microscopes differ dramatically (Wilson and Sheppard, 1984, p. 51). This affects the resolution in the x–y plane and the depth discrimination properties. There are several theoretical resolution criteria that lead to different conclusions when a comparison is made between the theoretical resolution obtainable in a conventional versus confocal microscope. Examination of the first-order zero-intensity crossing in the image of a single point shows no difference in resolution between the two microscopes (Brakenhoff *et al.*, 1979). The resolution in a confocal microscope is 27% better than a conventional microscope when the full-width half-maximum of the zero-order maximum is studied (Wilson and Sheppard, 1984, p. 48). The confocal microscope's resolution is twice that of a conventional microscope when the cutoff frequency of the optical transfer function is used to analyze the two systems (Wilson and Sheppard, 1984, p. 51). In practice the thereotical advantages translate into a negligible improvement in resolution, especially when observing flat samples with a confocal fluorescence microscope. The main difference between a confocal and a conventional fluorescence microscope is the improved depth discrimination. The depth of field in a conventional microscope can be observed easily: as soon as the image is out of focus the field becomes dark. The lens can be focused to observe certain areas (spaced along the optical axis). These become brighter while the other areas (below and above) become dimmer. The bright area is usually surrounded by a "foglike background." A common technique to reduce the loss in contrast is the use of field apertures. The aperture decreases the size of the illuminated area and the size of the detected area. This process is inherent in the design of a confocal microscope. Because "field pinholes" are in the source and in the detector path, the confocal fluorescence microscope achieves the best possible discrimination against out-of-focus light. The confocal fluorescence microscope is always in focus. This causes a reduction of the "foglike background" observed in conventional microscopy and hence an increase in contrast (Stelzer and Wijnaendts-van Resandt, 1986).

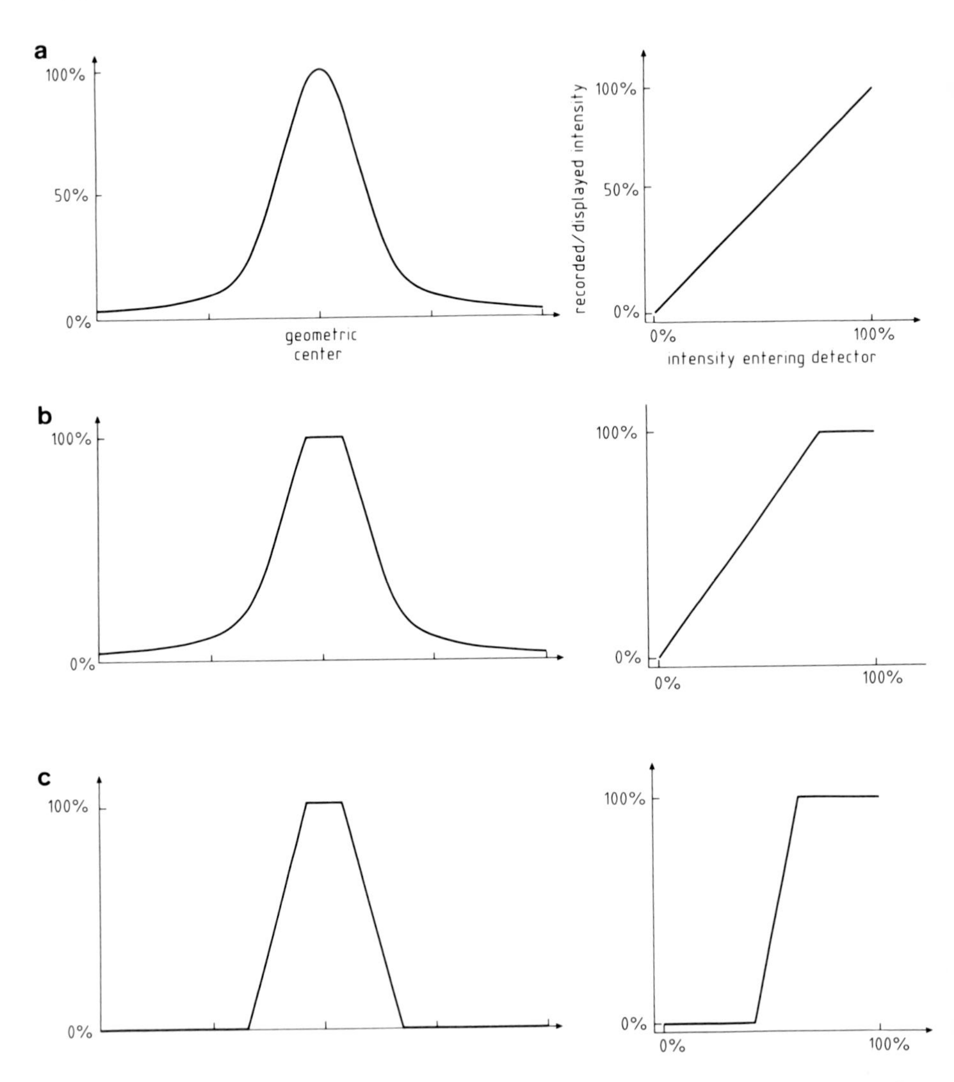

FIG. 4. Influence of gain and offset on depth discrimination. Gain and offset can be used to influence the apparent resolution along the optical axis. (a) The intensity of an "infinitely" small spot as a function of the distance of the focus from the geometric center of the spot along the optical axis. The offset is zero and the gain does not saturate the amplifier. The relation between the number of emitted photons and the displayed intensity is linear. (b) The gain saturates the amplifier. A center region has identical contrast, independent of the distance. The relation is linear only below the saturation level of the amplifier. (c) Applying a negative offset discriminates against the lower 90%, apparently "improving the resolution" along the optical axis. The relation is not linear any longer. The electroncis operate in "thresholding mode." The same argument applies to an apparent "improvement of the resolution" in the x–y plane.

A well-known technique from video-enhanced microscopy is the use of gain and negative offset to improve the contrast (Allen and Allen, 1981). Gain increases the signal strength from the photomultiplier tube. The negative offset applies a negative voltage to the signal. The combination of the two increases the signal strength and removes the background noise from the signal, thus increasing the contrast. The same technique can be used when recording images on the confocal microscope and is in fact used whenever a 400-ASA film is pushed to 1600 ASA. Exposing a 400-ASA film like this for tens of seconds has exactly this effect. There is a tendency for biologists with a background in immunofluorescence microscopy to expect confocal images to display contrasted images equivalent to the contrast seen in conventional fluorescence microscope images. This can be achieved by setting the gain very high (saturating the amplifier) and adding a negative voltage (negative offset). The principle is described in Fig. 4. Since such operations are usually not documented in the final data presentation, the effect on the final image will vary and could lead to the conclusion that *z*-discrimination and resolution have improved. However such images will provide no quantitative information.

Although it is usually desirable to set the gain and offset such that the dynamic range is fully utilized, the dynamic range of an 8-bit memory may not be sufficient to display all the details in the picture. The use of a 16-bit memory to acquire the image will allow a more complete visualization. Alternatively, one can set the gain high (overexposing the bright areas) or color-code the signal displayed by modifying the display look-up table.

The maximum displayable field size is determined by the resolution [i.e., primarily by the numerical aperture (NA) of the microscope objective] and the number of recordable-displayable picture elements. Therefore, maximum field size = (diameter of spot) × (number of picture elements). If every picture element is to be resolved with a high-NA lens (NA ~1.3) on a display of 500 lines by 500 picture elements per line, the field size should not exceed 100 × 100 μm. This means that there is no need to work with lenses that have magnifications <100× to increase the field size (field size >200–300 μm).

ACKNOWLEDGMENTS

The authors would like to thank Jan DeMey, Eric Karsenti, and Kai Simons for their stimulating discussions and suggestions. We thank Morgane Bomsel for independently verifying the fixation methods and for her helpful discussions. We would like to thank Clemens Storz, Reiner Stricker, and Reinhard Pick for their superb technical assistance. We also thank Angela Wandinger-Ness for critically reviewing the manuscript. R. Bacallao is supported by a Physician Scientist Award, NIH grant DK01777-02.

References

Allen, R. D., and Allen, N. S. (1981). *J. Microsc. (Oxford)* **129,** 3–17.

Allen, V. J., and Kreis, T. E. (1986). *J. Cell Biol.* **103,** 2229–2239.

Berod, A., Hartman, B. K., and Pujol, J. F. (1981). *J. Histochem. Cytochem.* **29,** 844–850.

Brakenhoff, G. J., Blom, P., and Barends, P. (1979). *J. Microsc. (Oxford)* **177,** 219–232.

Dabora, S. L., and Sheetz, M. P. (1988). *Cell (Cambridge, Mass.)* **54,** 27–35.

Fuller, S. D., von Bonsdorff, C.-H., and Simons, K. (1984). *Cell (Cambridge, Mass.)* **38,** 65–77.

Kleinman, H. K., Luckenbill-Edds, L., Cannon, F. W., and Shephel, G. C. (1987). *Anal. Biochem.* **166,** 1–13.

Lee, C., and Chen, L. B. (1988). *Cell (Cambridge, Mass.)* **54,** 37–46.

Lipsky, N., and Pagano, R. E. (1985). *J. Cell Biol.* **100,** 27–34.

Matteoni, R., and Kreis, T. E. (1987). *J. Cell Biol.* **105,** 1253–1265.

Misek, D. E., Bard, E., and Rodriquez-Boulan, E. (1984). *Cell (Cambridge, Mass.)* **39,** 537–400.

Robertson, J. D., Bodenheimer, T. S., and Stage, D. E. (1963). *J. Cell Biol.* **19,** 159–199.

Sheppard, C. J. R., and Wilson, T. (1978). *Opt. Lett.* **6,** 625.

Stelzer, E. H. K., and Wijnaendts-van Resandt, R. W. (1986). *Proc. SPIE—Int. Soc. Opt. Eng.* **602,** 63–70.

Stelzer, E. H. K. *et al.* (1989). Submitted.

Stevenson, B. R., Siliciano, J. D., Mooseker, M. S., and Goodenough, D. A. (1986). *J. Cell Biol.* **103,** 755–766.

van Meer, G., Stelzer, E. H. K., Wijnaendts-van Resandt, R. W., and Simons, K. (1987). *J. Cell Biol.* **105,** 1623–1635.

White, J. G., Amos, W. B., and Fordham, M. (1987). *J. Cell Biol.* **105,** 41–48.

Wijnaendts-van Resandt, R. W., Marsman, H. J. B., Kaplan, R., Davoust, J., Stelzer, E. H. K., and Stricker, R. (1985). *J. Microsc. (Oxford)* **138,** 29–34.

Wilson, T., and Sheppard, C. J. R. (1984). "Theory and Practice of Scanning Optical Microscopy." Academic Press, London.

Chapter 21

Organic-Anion Transport Inhibitors to Facilitate Measurement of Cytosolic Free Ca^{2+} with Fura-2

FRANCESCO DI VIRGILIO

C.N.R. Center for the Study of the Physiology of Mitochondria and
Institute of General Pathology
University of Padua
Padua, Italy

THOMAS H. STEINBERG

Department of Medicine
Columbia University College of Physicians and Surgeons
New York, New York 10032

SAMUEL C. SILVERSTEIN

Rover Physiology Laboratories
Department of Physiology and Cellular Biophysics
Columbia University College of Physicians and Surgeons
New York, New York 10032

METHODS IN CELL BIOLOGY, VOL. 31

I. Introduction

The fluorescent reporter dyes quin2 and fura-2 are used by many investigators to quantitate the cytosolic free-calcium concentration ($[Ca^{2+}]_i$) in different types of cells. These dyes are introduced into the cytoplasmic matrix of cells as membrane-permeant acetoxymethyl esters (quin2/AM and fura-2/AM) that are hydrolyzed by cytoplasmic esterases to yield membrane-impermeant free acids. For the dyes to measure accurately $[Ca^{2+}]_i$, they must be selectively and uniformly distributed within the cytoplasmic matrix of the cells. However, in a number of cell types fura-2 does not remain within the cytoplasmic matrix; it accumulates within intracellular compartments, is secreted from the cells entirely, or both. We have discovered that mouse macrophages possess organic-anion transporters that remove fluorescent dyes, including fura-2, from the cytoplasmic matrix of these cells. The dyes are sequestered within cytoplasmic vacuoles and secreted into the extracellular medium. The transporters that promote dye sequestration and secretion are inhibited by the drugs probenecid and sulfinpyrazone, which block transport of organic anions in polarized epithelial cells. In our hands, probenecid and sulfinpyrazone prevent fura-2 from leaving the cytoplasmic matrix of mouse macrophages and other cells, and facilitate the use of fura-2 in measuring $[Ca^{2+}]_i$ in these cells. In this chapter, we review the evidence for fura-2 sequestration and secretion by cells and describe our use of organic-anion transport inhibitors to ameliorate problems caused by these processes.

II. The Problem: Sequestration and Secretion of Fura-2

Early studies using quin2 to measure $[Ca^{2+}]_i$ addressed the problem of the intracellular localization of this dye. Cells loaded with quin2 were permeabilized using digitonin or high-voltage electric discharge, and the quin2 released was compared to the release of contents of mitochondria and other organelles. Nearly all quin2 was released by digitonin treatment of lymphocytes and of Yoshida and Ehrlich tumor cells, suggesting that the dye was contained exclusively within the cytosol (Tsien *et al.*, 1982; Arslan *et al.*, 1985).

Studies using fura-2 have yielded different results. Investigators have reported inhomogeneous distribution of this dye in several different types of cells, including smooth muscle cells (Williams *et al.*, 1985), rat mast cells (Almers and Neher, 1985), PTK_1 cells (Poenie *et al.*, 1986), fibroblasts (Malgaroli *et al.*, 1987), inflammatory mouse macrophages (Di Virgilio *et al.*, 1988a), J774 macrophages (Di Virgilio *et al.*, 1988b), cultured renal proximal tubular cells (Goligorsky *et al.*, 1986), rat astrocytoma cells (D. Milani, personal communication), bovine endothelial cells (S. F. Steinberg *et al.*, 1987), and lymphokine-activated killer cells (T. Pozzan, personal communication). In most of these instances fura-2 accumulates within membrane-bound cytoplasmic organelles.

When $[Ca^{2+}]_i$ measurements in single cells are made using a fluorescence microscope, dye inhomogeneity does not necessarily present a problem to the investigator, because suitable areas of the cytoplasm can be selected for fluorescence measurement. However, when average measurements are taken from a large number of cells using a fluorescence spectrophotometer, fura-2 trapped within cytoplasmic organelles will affect the readings. Under the latter conditions, the fluorescence measurements will not reflect the true $[Ca^{2+}]_i$ of the cytoplasmic matrix.

Different types of cells appear to accumulate fura-2 in different organelles. Fura-2 accumulates within secretory graunles in rat peritoneal mast cells, and degranulation is accompanied by loss of dye from the cells (Almers and Neher, 1985). In endothelial cells, fura-2 was reported to colocalize with rhodamine 123, which stains mitochondria (S. F. Steinberg *et al.*, 1987). In fibroblasts, fura-2 accumulates in vesicles that take up acridine orange, implying that they are lysosomes (Malgaroli *et al.*, 1987). The vacuoles that sequester fura-2 and other fluorescent dyes in mouse macrophages and J774 cells are early endocytic organelles that deliver their contents to lysosomes, as described later.

In some cells fura-2 not only is sequestered within intracellular compartments, it is also released from the cells entirely. Extracellular dye will report the Ca^{2+} concentration in the medium rather than in the cells and can be misinterpreted as a steady rise in the $[Ca^{2+}]_i$ if the problem is not recognized. In most instances where accumulation of dye within intracellular compartments has been reported, release of dye into the extracellular medium has also been a problem. This finding suggests that accumulation of dye within intracellular compartments and release of dye from cells may be caused by the same mechanism. Fura-2 release has also been a problem in some cells where intracellular dye segregation has not been noted. These cells include cytotoxic T lymphocytes (Treves *et al.*, 1987), PC12 cells (Di Virgilio *et al.*, 1988c; Fasolato *et al.*, 1988), and N2A neuroblastoma cells (Di Virgilio *et al.*, 1988c).

We have quantitated the relase of fura-2 from several cell types. The amount of dye that is excreted from the cells during a 10-minute incubation at 37°C ranges from 15% in J774 macrophages (Di Virgilio *et al.*, 1988b) to 40% in N2A neuroblastoma cells (Di Virgilio *et al.*, 1988c). Dye efflux is reduced at temperatures <37°C. It is the membrane-impermeant fura-2 and not fura-2/AM that is excreted by these cells. Two pieces of evidence support this statement. First, the excitation spectrum and the 340/380 ratio of dye excreted from the cells at saturating $[Ca^{2+}]$ is that of fura-2 and not fura-2/AM. Second, when fura-2 is introduced into cells by reversible ATP-permeabilization of the plasma membrane, the dye is rapidly excreted.

III. Mechanisms of Fura-2 Sequestration and Secretion

How is fura-2 compartmentalized within cells? Malgaroli *et al.* (1987) proposed three possible mechanisms by which this could occur: diffusion of the membrane-permeant fura-2/AM across intracellular membranes before it is hydrolyzed, movement of fura-2 across these membranes by a transporter, and pinocytosis of the dye. Each of these mechanisms has been invoked as a cause of fura-2 compartmentalization in different cells. Accumulation within mast cell granules takes place when the cells are loaded with fura-2/AM, but does not occur when the free acid fura-2 is introduced into mast cells through a patch pipet. This observation suggested to Almers and Neher (1985) that accumulation of fura-2 within granules is caused by diffusion of fura-2/AM into the granules followed by hydrolysis of the dye by esterases contained within the granules. Malgaroli *et al.* (1987) found that fura-2 accumulates within vesicles when fibroblasts are incubated with fura-2/AM at 37°C, but remains within the cytoplasmic matrix if the cells are incubated with fura-2/AM at 15°C. The dye does not accumulate within vesicles when the cells are subsequently rewarmed to 37°C. Therefore, they inferred that accumulation of dye within vesicles is not due to diffusion of fura-2/AM into vesicles, but rather occurs by pinocytic uptake of the dye.

We have investigated the sequestration of fura-2 in elicited mouse peritoneal macrophages and cells of the J774 mouse macrophagelike cell line. When these cells are loaded with fura-2/AM, some of the dye is subsequently sequestered within phase-lucent cytoplasmic vacuoles. Sequestration is not caused by diffusion of the membrane-permeant AM form of the dye into vacuoles, because when membrane-impermeant fura-2 is introduced into the cytoplasmic matrix of these cells by reversible ATP permeabilization as described elsewhere in this volume, the dye is also sequestered. Sequestration of fura-2 is not caused by endocytosis of the dye. Immediately after the dye is introduced into the cytoplasmic matrix by ATP permeabilization, phase-lucent vacuoles inside the cells are devoid of dye; accumulation within these vacuoles occurs after the dyes have been removed from the extracellular medium.

IV. Organic-Anion Transport in Macrophages

We have found that the dyes fura-2, lucifer yellow, and carboxyfluorescein are cleared from the cytoplasmic matrix of J774 cells and mouse macrophages in the same manner (T. H. Steinberg *et al.*, 1987). In studying this process, we have relied mostly on the dye lucifer yellow, a highly fluorescent anionic dye. When J774 cells are incubated in medium containing lucifer yellow, washed, and viewed by fluorescence microscopy, the dye appears inside the cells only within cytoplasmic vesicles characteristic of pinosomes. Lucifer yellow can be introduced into the cytoplasmic matrix of J774 cells by ATP permeabilization, and trapped within this compartment by removing extracellular ATP or by adding excess Mg^{2+}. As with fura-2, lucifer yellow does not remain within the cytoplasmic matrix, but rather is sequestered within phase-lucent cytoplasmic vacuoles and secreted from the cells. The vacuoles that sequester lucifer yellow are distinct from lysosomes: when the lysosomes of J774 cells are labeled with a Texas red–ovalbumin conjugate before lucifer yellow is loaded into the cells, the vacuoles that sequester lucifer yellow are distinct from the Texas red-labeled lysosomes (Steinberg *et al.*, 1988). Nevertheless, these vacuoles subsequently deliver lucifer yellow into the lyosomal compartment. The lucifer yellow-sequestering vacuoles are at least partly composed of plasma membrane constituents, receive bulk-phase markers by pinocytosis, and are of light buoyant density. This evidence indicates that these sequestering vacuoles are endosomes (Steinberg *et al.*, 1988).

Lucifer yellow introduced into the cytoplasmic matrix of J774 cells is rapidly excreted by the cells as is fura-2 (T. H. Steinberg *et al.*, 1987). When lucifer yellow is introduced into the cytoplasmic matrix of J774 cells and the cells are incubated in fresh medium for 30 minutes at 37°C, 90% of the dye is excreted. Excretion of lucifer yellow occurs by an active transport process; it is temperature-sensitive and occurs against a concentration gradient. Because the processes of dye sequestration and dye secretion take place simultaneously and are both inhibited by the same conditions (i.e., decrease in temperature) and agents, we consider them to be caused by the same transport process.

Lucifer yellow, fura-2, and carboxyfluorescein are all organic anions. This fact suggested to us that the transporters that clear these dyes from the macrophage cytoplasm might be similar to those that transport organic anions, including uric acid, *p*-aminohippurate, and penicillin, across epithelia such as renal proximal tubule and hepatocytes. We confirmed this hypothesis by blocking clearance of the fluorescent dyes from mouse macrophages with probenecid or sulfinpyrazone, drugs that are known to inhibit organic-anion transport in epithelia (T. H. Steinberg *et al.*, 1987). When cells loaded with lucifer yellow by ATP permeabilization are incubated in medium containing 5 m*M* probenecid for 60 minutes at 37°C, both vacuolar sequestration and secretion of dye from the cells are inhibited. Both of these processes resume when probenecid is removed from the extracellular medium.

Sequestration of fura-2 within cytoplasmic vacuoles and secretion of fura-2 into the medium are also inhibited by probenecid and sulfinpyrazone as shown in Fig. 1 (Di Virgilio *et al.*, 1988b). In the absence of probenecid, 90% of intracellular fura-2 is released into the medium within 60 minutes. When probenecid is added to the medium, only 20% of intracellular fura-2 is secreted by the cells during this time. In addition, incubation of cells in medium containing 2.5 m*M* probenecid during the loading of fura-2/AM increases the percentage of fura-2 that is taken up by the cells from 6% to 15% (Di Virgilio *et al.*, 1988b). This finding implies that dye secretion limits the amount of dye that can be loaded into the cells.

We have found probenecid and sulfinpyrazone to be useful in facilitating the measurement of $[Ca^{2+}]_i$ with fura-2 (Fig. 2). When cells secrete fura-2, the resting $[Ca^{2+}]_i$ appears to increase constantly, because the fura-2 is steadily released into the medium where it encounters millimolar concentrations of Ca^{2+}. During the course of a typical experiment, a significant fraction of the dye can be excreted from the cell. When this efflux of fura-2 is inhibited with probenecid or sulfinpyrazone, the resting $[Ca^{2+}]_i$ determination remains constant, and any changes in $[Ca^{2+}]_i$ induced by Ca^{2+}-mobilizing agonists can be accurately measured.

V. Method for Inhibiting Fura-2 Efflux with Probenecid and Sulfinpyrazone

1. Preparation of stock solution: Probenecid and sulfinpyrazone are relatively insoluble in aqueous solutions at neutral pH. Make 100 m*M*

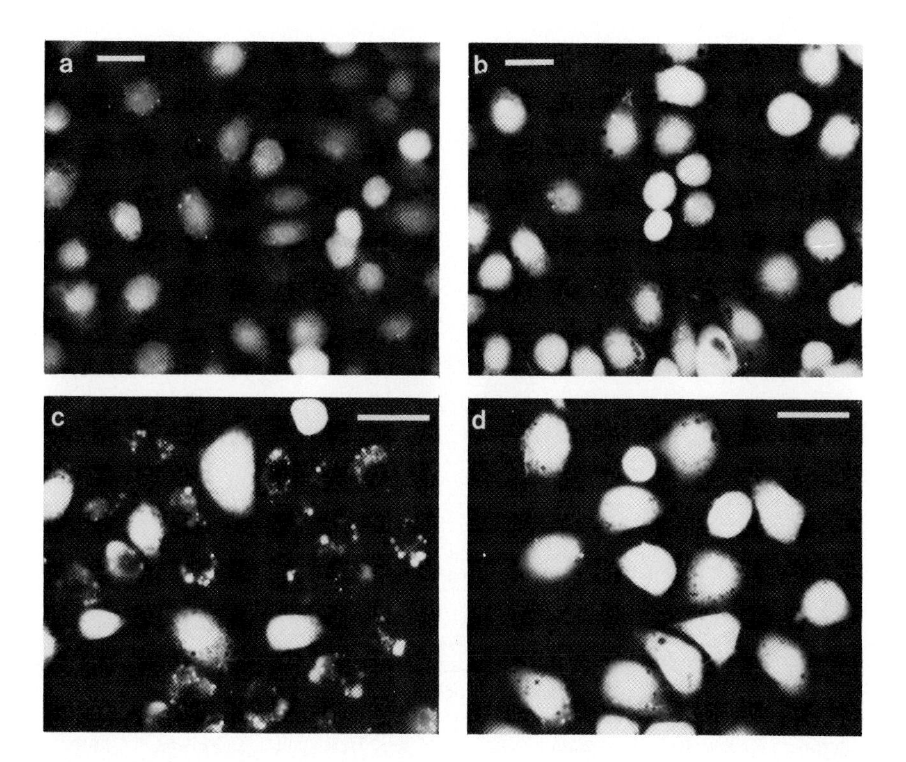

FIG. 1. Probenecid blocks sequestration of fura-2 in J774 cells. Adherent J774 cells were incubated in HEPES-buffered saline containing 10 μM fura-2/AM for 20 minutes at 37°C, washed, and incubated in HEPES-buffered saline without fura-2/AM. (a, b) Thirty minutes after the initiation of the loading procedure; (c, d) 90 minutes after the initiation of loading. (a, c) Cells maintained in the absence of probenecid; (b, d) cells incubated in 2.5 mM probenecid throughout the experiment. Bar = 20 μm. From Di Virgilio *et al.* (1988b), by permission.

(sulfinpyrazone, 10 mM) stock solutions by adding the appropriate quantity of solid compound to water, and then adding NaOH to alkalinize the solution until the drug dissolves. A 100 mM solution of probenecid prepared this way has a pH of ~9.5–10.0.

2. Because the stock solution is alkaline, it should be added to the medium before the pH is adjusted to the desired value. This precaution is not necessary if micromolar quantities of sulfinpyrazone are sufficient to inhibit fura-2 efflux in a particular cell.

3. We have used one of the following experimental routines. For macrophages we add 2.5 mM probenecid both during the loading period with fura-2/AM and during the experiment. In PC12 and N2A cells, we

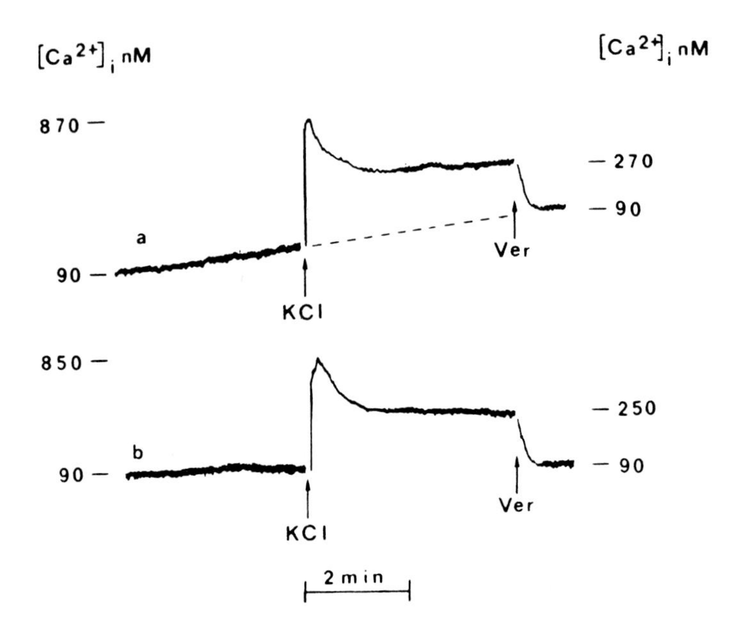

FIG. 2. The effect of sulfinpyrazone on changes in $[Ca^{2+}]_i$ induced by depolarization of the plasma membrane in PC12 cells. PC12 cells were loaded with 6 μM fura-2/AM and resuspended in a fluorimeter cuvet at a concentration of 7.5×10^5 cells/ml. Cells were incubated in the presence or absence of sulfinpyrazone for 10 minutes at 37°C before the addition of 60 m*M* KCl. Verapamil (Ver) was 20 μM. In tracing a, there was an apparent increase in the resting $[Ca^{2+}]_i$ caused by secretion of fura-2 into the extracellular medium. In tracing b, the addition of 250 μM sulfinpyrazone eliminated the secretion of fura-2 and stabilized the resting $[Ca^{2+}]_i$ measurements. From Di Virgilio *et al.* (1988c), by permission.

add 250 μM sulfinpyrazone after fura-2 has been loaded into the cells, and immediately before we begin monitoring $[Ca^{2+}]_i$.

VI. The Use of Organic-Anion Transport Inhibitors in Different Cell Types

The concentrations of probenecid or sulfinpyrazone required to inhibit organic-anion transport vary among different types of cells. In J774 cells, sulfinpyrazone is about twice as potent as probenecid, but is more toxic. In N2A and PC12 cells, sulfinpyrazone is considerably more potent than probenecid, and effectively blocks fura-2 efflux at a concentration of 250 μM (Di Virgilio *et al.*, 1988c). Sulfinpyrazone also inhibits fura-2 efflux in Hit cells, an insulin-secreting cell line (C. B. Wollheim, personal communication) and in lymphocytes (T. Pozzan, personal communication).

VII. Effects of Organic-Anion Transport Inhibitors on Cells

The physiological significance of organic anion transporters in nonpolarized cells such as macrophages is unclear. These cells secrete metabolites such as lactate and bilirubin, and inflammatory mediators such as leukotrienes and prostaglandins. It is possible that secretion of these compounds is mediated by probenecid-inhibitable organic-anion transporters. Whether or not this suggestion is correct, inhibition of organic-anion transport may alter cell function. Therefore, the effects of probenecid and sulfinpyrazone on each cell type should be assessed before using these drugs to facilitate measurements of $[Ca^{2+}]_i$.

In J774 cells and mouse macrophages, 5 m*M* probenecid has no effect on cell viability for as long as 3 hours. In mouse macrophages, 2.5 m*M* probenecid induces a 20% increase in resting $[Ca^{2+}]_i$, but the increases in $[Ca^{2+}]_i$ elicited by Ca^{2+}-mobilizing agonists such as PAF (Platelet Activating Factor) or heat-aggregated IgG are not altered. Similarly, the rise in $[Ca^{2+}]_i$ caused by Fc receptor-mediated phagocytosis is not affected by probenecid. J774 cells incubated in medium containing 2.5 m*M* probenecid ingest IgG-coated erythrocytes as efficiently as do cells incubated without probenecid (Di Virgilio *et al.*, 1988b).

In other cells, organic-anion transport blockers have more pronounced side effects. In PC12 cells and N2A neuroblastoma cells, probenecid reduces the rise in $[Ca^{2+}]_i$ caused by plasma membrane depolarization or the neuropeptide bradykinin. Sulfinpyrazone, however, does not have these unwanted effects on $[Ca^{2+}]_i$ for incubation times up to 15 minutes. As shown in Table I, resting $[Ca^{2+}]_i$ and elicited rises in $[Ca^{2+}]_i$ are not altered by 250 μM sulfinpyrazone. At this concentration of sulfinpyrazone, fura-2 excretion is almost completely blocked.

VIII. Other Potential Uses for Probenecid and Sulfinpyrazone

In addition to fura-2, other fluorescent anionic dyes are used to report cytosolic conditions. A number of fluorescein derivatives, notably BCECF, can be used to assess cytosolic pH. We have found that many fluorescein derivatives are excreted by mouse macrophages and J774 cells, and that this process is blocked by probenecid. It is possible that organic-anion transport blockers will find a role in facilitating intracellular pH measurements, as they have in measuring $[Ca^{2+}]_i$.

TABLE I

EFFECT OF SULFINPYRAZONE ON $[Ca^{2+}]_i$ HOMEOSTASIS[a]

Cell Type	Stimulus	Sulfinpyrazone (250 μM)	$[Ca^{2+}]_i$ (nM)	
			Basal	Stimulated
PC12	Depolarization	−	100 ± 24 (6)	740 ± 220 (6)
PC12	Depolarization	+	90 ± 10 (9)	750 ± 140 (9)
N2A	Bradykinin	−	80 ± 10 (5)	380 ± 80 (5)
N2A	Bradykinin	+	80 ± 10 (4)	430 ± 90 (4)

[a] PC12 and N2A cells were loaded with fura-2/AM and $[Ca^{2+}]_i$ was measured as described in Di Virgilio *et al.* (1988c). PC12 cells were depolarized by adding 60 mM KCl to the medium. Bradykinin was added at a concentration of 400 nM. Data are mean ± SD. Number of experiments is given in parentheses.

ACKNOWLEDGMENTS

Supported by grants AI20516, and HL07018 from the NIH, and by the Cystic Fibrosis Research Development Program at Columbia Unviersity. Dr. Di Virgilio was the recipient of an EMBO fellowship. Dr. Steinberg is the recipient of a Clinical Investigator Award (AI00893).

REFERENCES

Almers, W., and Neher, E. (1985). *FEBS Lett.* **192,** 13–18.
Arslan, P., Di Virgilio, F., Beltrame, M., Tsien, R. Y., and Pozzan, T. (1985). *J. Biol. Chem.* **260,** 2719–2727.
Di Virgilio, F., Meyer, B. C., Greenberg, S., and Silverstein, S. C. (1988a). *J. Cell Biol.* **106,** 657–666.
Di Virgilio, F., Steinberg, T. H., Swanson, J. A., and Silverstein, S. C. (1988b). *J. Immunol.* **140,** 915–920.
Di Virgilio, F., Fasolato, C., and Steinberg, T. H. (1988c). *Biochem. J.* **256,** 959–963.
Fasolato, C., Pandiella, A., Meldolesi, J., and Pozzan, T. (1988). *J. Biol. Chem.* **263,** 17350–17359.
Goligorsky, M. S., Hruska, K. A., Loftus, D. J., and Elson, E. L. (1986). *J. Cell. Physiol.* **128,** 466–474.
Malgaroli, A., Milani, D., Meldolesi, J., and Pozzan, T. (1987). *J. Cell Biol.* **105,** 2145–2155.
Poenie, M., Alderton, J., Steinahrdt, R., and Tsien, R. (1986). *Science* **233,** 886–889.
Steinberg, S. F., Bilezikian, J. P., and Al-Awqati, Q. (1987). *Am. J. Physiol.* **253,** C744–C7547.
Steinberg, T. H., Newman, A. S., Swanson, J. A., and Silverstein, S. C. (1987). *J. Cell Biol.* **105,** 2695–2702.
Steinberg, T. H., Swanson, J. A., and Silverstein, S. C. (1988). *J. Cell Biol.* **107,** 887–896.
Treves, S., Di Virgilio, F., Cerundolo, V., Zanovello, P., Collavo, D., and Pozzan, T. (1987). *J. Exp. Med.* **166,** 33–42.
Tsien, R. Y., Pozzan, T., and Rink, T. J. (1982). *J. Cell Biol.* **94,** 325–334.
Williams, D. A., Fogarty, K. E., Tsien, R. Y., and Fay, F. S. (1985). *Nature (London)* **318,** 558–561.

Chapter 22

Postembedding Detection of Acidic Compartments

RICHARD G. W. ANDERSON

Department of Cell Biology and Anatomy
The University of Texas Southwestern Medical Center
Dallas, Texas 75235

I. Introduction

Acidic membrane-bound compartments are stable elements of all eukaryotic cells. Endocytic vesicles (Anderson *et al.*, 1984; Galloway *et al.*, 1983; Maxfield, 1982; Tycko and Maxfield, 1982; Yamashiro *et al.*, 1983), lysosomes (Ohkuma and Poole, 1978; Poole and Ohkuma, 1981), portions of the trans-Golgi apparatus (Anderson and Pathak, 1985; Orci *et al.*, 1985, 1986), certain secretory vesicles (Fishkes and Rudnick, 1982;

Hutton, 1982; Johnson and Scarpa, 1984; Orci *et al.,* 1986; Russell, 1984), and plant tonoplasts (Boller and Wiemken, 1986) are listed among the known acidic compartments. The function of low pH in each compartment is not entirely understood. In some compartments the low pH helps to maintain ionic gradients across the vacuole membrane, whereas in others the high H^+ concentration unfolds proteins, permitting the release of bound ions such as iron or exposing a hydrophobic site that facilitates the passage of the molecule across the membrane. The low-pH environment can activate the proteolytic enzymes of both lyosomes and trans-Golgi vesicles or control the proper sorting of molecules that travel either the endocytic or the exocytic pathway. (These and other functions are reviewed in Mellman *et al.,* 1986, and Anderson and Orci, 1988.)

The identification of acidic compartments in living cells has long depended on vital staining techniques (Metchnikoff, 1968). Vital dyes that accumulate in acidic compartments are weak bases. At neutral pH they are hydrophobic and readily cross membranes; however, in an acidic environment they acquire a charge (an absorbed H^+) and leave the compartment slowly. As an example, acridine orange is a fluorescent weak base that has been used extensively to study, by light microscopy, the dynamic distribution of acidic compartments in living cells.

The proton gradient across the membrane of each acidic vacuole is generated by an ATP-dependent proton pump (Mellman *et al.,* 1986; Al-Awqati, 1985; Rudnick, 1986). The pump must be active for vital dyes to accumulate. When the pump is inactivated, the H^+ gradient dissipates and any vital indicator that had accumulated leaves the compartment. Thus, vital indicators have been severely limited in their use because of the requirement that cells remain alive during the experimental procedure.

To overcome these limitations, weak bases have been found that accumulate in living cells but are retained in acidic compartments following fixation with aldehyde fixatives (Anderson *et al.,* 1984; Anderson and Pathak, 1985; Orci *et al.,* 1986, 1987a,b; Schwartz *et al.,* 1985). The reagent can then be localized by immunocytochemical techniques after the cells or tissues have been embedded in plastic. In this way, the H^+ gradient can be captured as it existed in the living cell and made visible with either the light or electron microscope. The postembedding detection of acidic compartments (PEDAC) has proved to be a valuable adjunct to other techniques for studying this interesting aspect of organelle physiology.

II. Materials and Methods

A. Materials

3-(2,4-Dinitroanilino)-3′-amino-*N*-methyldipropylamine (DAMP) and monoclonal antidinitrophenol (anti-DNP) IgG (Oxford Biomedical Research, Oxford, MI)

Affinity-purified anti-mouse IgG or anti-mouse IgG coupled to tetramethylrhodamine isothiocyanate (Zymed, San Francisco, CA)

Saponin, ovalbumin, and 3′-diaminobenzidine tetrahydrochloride (Sigma)

Anti-mouse IgG coupled to horseradish peroxidase (HRP) (Cappell, Westchester, PA)

Protein A–gold or goat anti-rabbit IgG–gold conjugates (Janssen, Olen, Belgium)

Lowicryl K4M (Polysciences, Warringtin, PA)

Epon (Fluka, Hauppauge, NY)

Monesin (Calbiochem, San Diego, CA)

B. Indirect Immunofluorescence Microscopy

Cultured cells are easily labeled with the weak base, DAMP, by first growing the monolayers on coverslips and incubating the coverslips with 30–50 μM DAMP in normal medium for 30 minutes at 37°C. Following the incubation, cells are washed with culture medium and then fixed for 15 minutes at room temperature with 3% (w/v) paraformaldehyde in buffer A (10 mM sodium phosphate, 150 mM NaCl, 2 mM $MgCl_2$, pH 7.4). To control for specificity, a set of cells is subsequently incubated with 25 μM monensin for 5 minutes at 37°C before fixation, which dissipates the proton gradient and causes DAMP to leave the compartment. Cells are then fixed and the coverslips are washed with 2 ml of 15 mM NH_4Cl and twice with buffer A. Each monolayer is then permeabilized by overlaying the coverslip with 2 ml of 0.1% (v/v) Triton X-100 in buffer A for 5 minutes at −10°C. Each coverslip is then placed in a Petri dish, cell-side up, covered with 60 μl of monoclonal mouse anti-DNP IgG (15 μg/ml) and incubated for 60 minutes at 37°C in a moist chamber. Following four washes with buffer A (15 minutes each) the cells are incubated with 50 μl of tetramethylrhodamine isothiocyanate-labeled rabbit anti-mouse IgG (40 μg/ml) for 60 minutes at 37°C. Coverslips are then washed once more, mounted on glass slides, and viewed with a fluorescence microscope.

Tissue samples can also be labeled with DAMP. There are two general methods of labeling: tissue perfusion with DAMP or incubation of tissue slices with the DAMP. Although 30–50 μM DAMP is still a workable concentration, penetration is a problem and *thin* tissue slices are recommended. Regardless of the method of administering the DAMP, to localize sites of DAMP accumulation, tissues are fixed for 2 hours at room temperature with 1% glutaraldehyde in 0.1 *M* sodium phosphate buffer (pH 7.3), dehydrated, and embedded in Epon. Thick sections (0.5–1.0 μm) are prepared of the Epon-embedded material, collected on glass slides, and processed to remove Epon (Orci *et al.*, 1986). Sections are then incubated with monoclonal anti-DNP IgG (5 μg/ml) for 1 hour at 37°C, washed with phosphate buffer (2×, 5 minutes each), and then incubated with rhodamine-labeled anti-mouse IgG for 1 hour at 37°C. Sections can then be mounted and viewed by either epifluorescence or transmitted-fluorescence microscopy (Orci *et al.*, 1986).

C. Indirect Immunoelectron Microscopy

1. Immunoperoxidase

For cultured cells, cells are incubated with DAMP as just described, and then fixed with either 2% paraformaldehyde in buffer B (10 m*M* sodium periodate, 0.75 *M* lysine, 37.5 m*M* sodium phosphate, pH 6.2) or with a 3% paraformaldehyde in buffer C [100 m*M* sodium phosphate (pH 7.8), 3 m*M* KCl, 3 m*M* $MgCl_2$, and 3 m*M* 2,4,6-trinitrophenol]. The cells are then processed while attached to the dish for indirect immunoperoxidase localization of DAMP (Anderson *et al.*, 1984), using 50 μg/ml of anti-DNP IgG, 0.5 mg/ml of HRP-conjugated goat anti-mouse IgG. Saponin concentration for permeabilization must be adjusted for the cell type. To reveal the HRP sites, cells are incubated at room temperature for 10 minutes with 0.2% (w/v) diaminobenzidine and 0.01% (v/v) H_2O_2. The cells are then fixed in 2% (w/v) osmium tetroxide, and 1% (w/v) potassium ferrocyanide in 0.1 *M* sodium cacodylate (pH 7.3), dehydrated, released from the dish in propylene oxide, pelleted, and embedded in Epon (Anderson *et al.*, 1981).

The same method can be used to localize DAMP in tissues.

2. Immunogold Localization

Incubation of either cultured cells or tissue samples with DAMP is carried out as described earlier. Cells or tissues are then fixed with 1% glutaraldehyde in 0.1 *M* sodium phosphate buffer (pH 7.3) for 1 hour at room temperature. Cultured cells are pelleted in the fixative. Pellets or tissue samples are washed with 0.1 *M* sodium phosphate buffer and incubated in 0.5 *M* NH_4Cl in phosphate buffer for 30 minutes at room temperature. Samples are washed once again with phosphate buffer and embedded in Lowicryl K4M at −20°C as described (Anderson and Pathak, 1985; Orci *et al.*, 1986).

Thin sections are prepared and mounted on formvar-carbon-coated nickel grids. The grids are floated face-down on buffer C [0.5 *M* NaCl, 0.1% NaN_3, 1% (w/v) ovalbumin, 0.01 *M* Tris-HCl buffer, pH 7.2] for 30 minutes at room temperature. Grids are then edge-dried on filter paper (Whatman, no. 50) and transferred to a drop (70 μl) of buffer C containing 5 μg/ml of anti-DNP IgG for 16 hours at 4°C. Grids are then rinsed thoroughly in buffer D (0.1 *M* Tris-HCl, 0.15 *M* NaCl, pH 7.2) for 10 minutes, quickly dried on filter paper, and immediately floated face-down on buffer E (buffer D + 0.02% PEG-20, 0.1% NaN_3) in a porcelain spot plate. Grids are then incubated for 2 hours at 37°C with 5 μg/ml of rat anti-mouse IgG in buffer E followed by 1 hour incubation at room temperature with 1 : 70 dilution of protein A–gold (10 ± 2 nm diameter) in buffer E. All antibody solutions are centrifuged at 100,000 *g* for 30 minutes prior to use. After the protein A–gold incubation, grids are washed with a stream of distilled water for 20 seconds and air-dried. Labeled sections are double-stained at room temperature with 5% aqueous uranyl acetate (10 minutes) and lead citrate (3 minutes).

Tissue samples or cells incubated in the presence of DAMP can also be embedded in Epon. Epon-embedded samples are then labeled by the immunogold procedure according to the protocol just outlined, with the following modifications. Before any labeling, the sections are etched by floating the grids, section-side down on saturated solution of sodium metaperiodate for 1 hour at room temperature, followed by thorough rinsing with distilled water, before processing for immunolabeling.

3. Quantification of Gold Label

The number of gold particles per square micron of compartment can be evaluated directly in electron micrographs by standard procedures (Orci *et al.*, 1986).

The density of gold particles due to anti-DNP IgG binding can be used to calculate pH if the number of gold particles is proportional to the proton concentration (Orci *et al.*, 1986). Apparently DAMP accumulates in proportion to the proton concentrations in a compartment; however, the quantitative retention of DAMP is related to the number of available crosslinking sites in the compartment. For protein-rich secretory vacuoles, most likely all of the accumulated DAMP becomes crossolinked during fixation; however, for a protein-poor compartment such as the tonoplast, most likely all of the DAMP would be released after fixation. Compartments such as transitional vesicles of the Golgi apparatus, endosomes, and lysosomes seem to have ample crosslinking sites.

The pH of a compartment can be estimated using the formula: pH = 7.0 − log D_1/D_2, where D_1 = density of DAMP-specific gold particles in the compartment of interest, and D_2 = density of gold particles in a pH 7.0 compartment such as the nucleus.

III. Results and Discussion

A. Light Microscopy

Although the acidic compartments of tissue culture cells can be easily identified with reagents such as acridine orange, PEDAC offers a broader range of experimental options because the cells have been fixed. Figure 1 shows a typical micrograph of cultured fibroblasts that have been incubated with DAMP. Acidic compartments appear as brightly fluo-

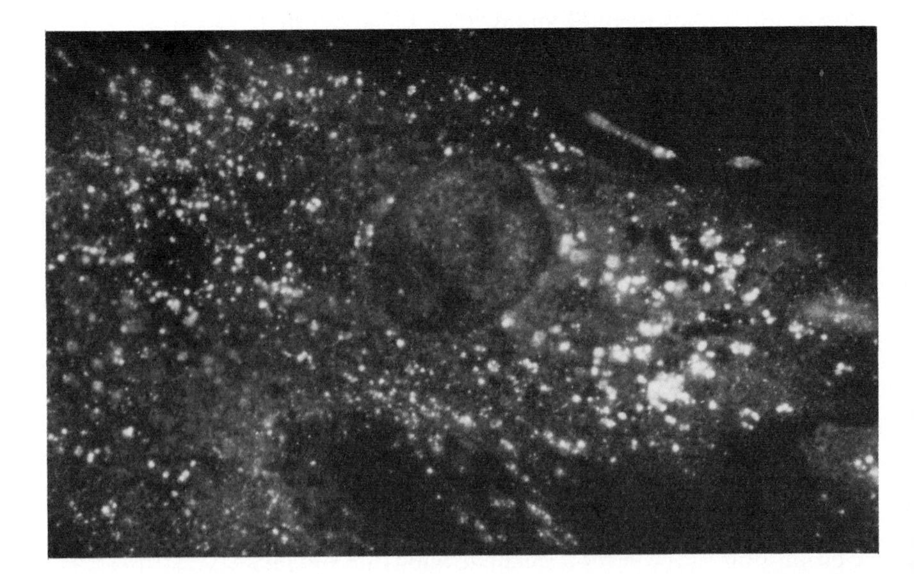

FIG. 1. Light-microscopic visualization of acidic compartments in cultured human fibroblasts. The bright dots represent sites of DAMP accumulation. 50 μM DAMP, 30 minutes at 37°C. ×2500.

rescent vacuoles that are often clustered around the nucleus of the cell. A similarly prepared sample can be used to colocalize antigens that are suspected of residing, either permanently or transiently, in acidic compartments. Moreover, with a suitably labeled ligand, the dynamics of acidification during endocytosis can be studied.

There is very little known about the distribution of acidic compartments in the various tissues. The PEDAC technique provides a convenient way to survey by light microscopy various tissues for the presence of acidic compartments. The antigenicity of the DNP group on DAMP survives fixation and embedding in plastic; therefore, immunolocalization can be performed directly on thick (0.5–1.0 μm) sections. The option is available to do colocalization studies if there is an interest in identifying molecules that might reside in the acidic compartments.

The major technical difficulty with applying PEDAC to tissue samples is obtaining thorough penetration of DAMP into cells. Perfusion of tissues with DAMP is recommended. Tissue slices can be incubated with DAMP, although we have found that DAMP may not reach the inner cells of the slice. In this case, caution is recommended: only the outer cell layer should be used for analysis.

B. Electron Microscopy

The reason for developing the PEDAC procedure was to be able to visualize acidic compartments with the resolution of the electron microscope. Two basic formats are available: immunoperoxidase, which gives qualitative information about the distribution of these compartments, and immunogold, which can be used to quantify the distribution of acidic compartments. The choice of techniques depends on the type of study being performed. Generally, immunogold gives better resolution than immunoperoxidase.

The PEDAC procedure can provide high-resolution morphological information about the distribution of acidic compartments in tissues. We already know that lysosomes, endosomes, certain secretory vacuoles, and portions of the Golgi apparatus are acidic. However, little information is available as to whether certain tissues utilize special acidic compartments for specific functional needs. There is also the possibility that certain disease processes cause abnormal regulation of intracellular vacuolar pH.

Electron-microscopic PEDAC has a major application in mapping the function of low-pH compartments. Double-labeling experiments allow, for example, the colocalization of acidic compartments and the movement of endocytic markers (Fig. 2). Likewise, double-labeling with antibodies to a specific antigen can give valuable information about the traffic pattern

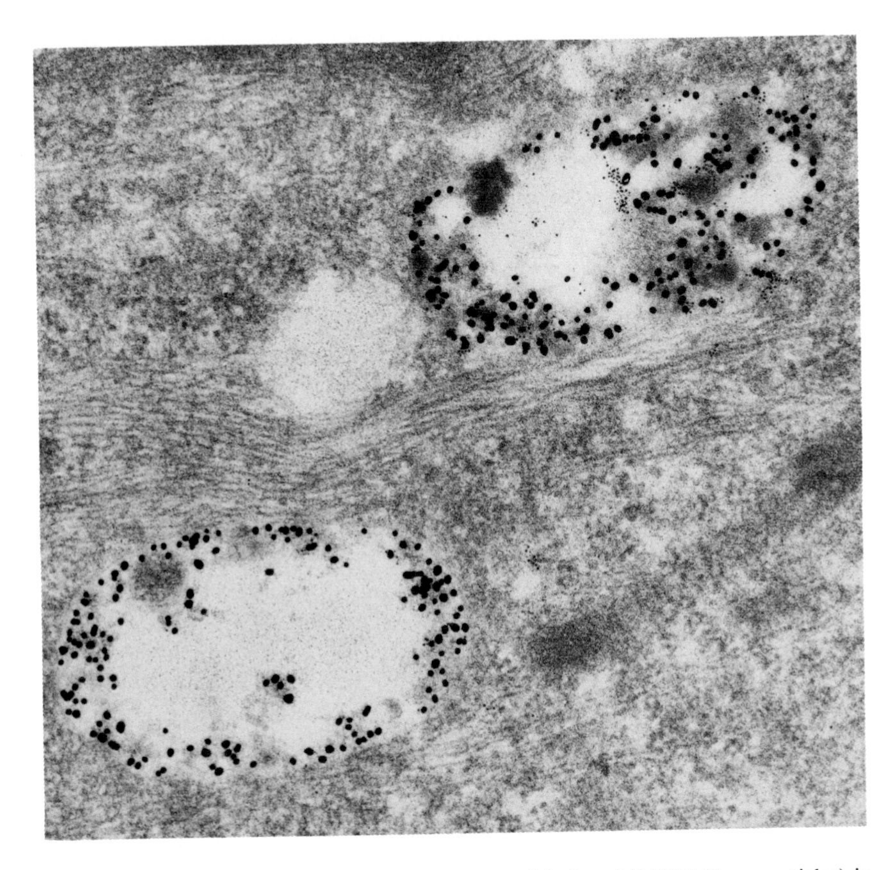

FIG. 2. Colocalization of LDL–gold (15-nm particles) and DAMP (5-nm particles) in endosomes of human fibroblasts. Cultured human fibroblasts grown according to standard conditions (Anderson *et al.*, 1984), were incubated in the presence of LDL–gold (20 μg/ml) plus 50 μM DAMP for 30 minutes at 37°C. Cells were washed, fixed, and embedded in Lowicryl K4M. Thin sections were then processed to localize DAMP. Notice that the endosome on the right contains both LDL–gold and DAMP, whereas the endosome on the left contains little DAMP-specific labeling, indicating that the pH of the former is quite a bit lower than the latter. ×55,000.

of antigens relative to the regulation of acidification. An extreme refinement of this application is to utilize conformation-specific monoclonal antibodies (mAb) to detect the relationship between the shape of a molecule and the pH of the compartment in which it resides (Orci *et al.*, 1986). For example, there is good biochemical evidence that transferrin loses its iron in an acidic endosomal compartment (Mellman *et al.*, 1986). With mAb that can distinguish iron-loaded transferrin from apotransferrin, one could obtain important information about the relationship between the delivery of iron and the acidification of the endosome.

Aside from being able to identify compartments that are acidic, the PEDAC procedure also offers the opportunity to estimate the pH of intracellular compartments that are inaccessible to measurement with fluoresceinated molecules (Tycko and Maxfield, 1982). This method relies on a simple numerical relationship between the pH and the density of immunogold labeling due to DAMP accumulation in a particular compartment. As with any technique, these calculations are dependent on several assumptions. Critical among these is that DAMP accumulates proportionally to the proton concentration in the compartment and that fixatives quantitatively retain DAMP at sites of accumulation. For some tissues these conditions appear to hold true (Orci *et al.*, 1986); however, caution should be applied in approaching any new system. A highly acidic compartment that does not have sufficient crosslinking sites for DAMP to bind to during fixation will not retain the marker after fixation even though it may have accumulated to a high concentration in the living cell.

Another area of investigation where PEDAC could be valuable is studying the regulation of pH in these compartments. There is considerable evidence that the magnitude of the H^+ gradient within vacuoles of a continuous exocytic or endocytic pathway are quite different (see Fig. 2). For example, whereas trans-Golgi vesicles of certain exocrine cells are slightly acidic (pH 5.5), the mature secretory vesicles derived from these vesicles have a neutral pH. There are several ways that the pH might be regulated. The pH may be regulated by controlling the number of proton pumps in the membrane (Al-Awqati, 1985). Alternatively, the leakiness of the membrane to protons may be adjusted at various stages in the life cycle of the vesicle. Because proton pumps are electrogenic (Al-Awqati, 1985), the modulation of permeant anion fluxes (e.g., Cl^-) could also be an important way to control the movement of protons in response to the activity of the pump. Whereas there will be an important application of biochemical techniques to determining the mechanism of pH regulation, the PEDAC technique can play a central role in distinguishing among various hypotheses.

IV. Conclusions

Postembedding detection of acidic compartments (PEDAC) is a technique that has the potential for unraveling some of the mysteries surrounding the function of low-pH compartments in health and disease.

There are many applications that have not been touched on in this brief review. When combined with cell fractionation, biochemical, and physiological techniques, PEDAC offers the interested investigator a method for understanding the dynamic function of low-pH compartments in living cells.

Acknowledgments

I would like to thank Dr. Ravindra Pathak for preparing the light and electron micrographs. Some of these techniques were developed in collaboration with Drs. Lelio Orci, Michael Brown, and Joseph Goldstein. I greatly appreciate the help of Ms. Mary Surovik in preparing this manuscript.

References

Al-Awqati, Q. (1985). *Annu. Rev. Cell Biol.* **2,** 179.
Anderson, R. G. W., and Orci, L. (1988). *J. Cell Biol.* **106,** 539.
Anderson, R. G. W., and Pathak, R. K. (1985). *Cell (Cambridge, Mass.)* **40,** 635.
Anderson, R. G. W., Brown, M. S., and Goldstein, J. L. (1981). *J. Cell Biol.* **88,** 441.
Anderson,R. G. W., Falck, J. R., Goldstein, J. L., and Brown, M. S. (1984). *Proc. Natl. Acad. Sci. U.S.A.* **81,** 4838.
Boller, T., and Wiemken, A. (1986). *Annu. Rev. Plant Physiol.* **37,** 137.
Fishkes, H., and Rudnick, G. (1982). *J. Biol. Chem.* **257,** 5671.
Balloway, C. J., Dean, G. E., Marsh, M., Rudnick, G., and Mellman, I. (1983). *Proc. Natl. Acad. Sci. U.S.A.* **80,** 3334.
Hutton, J. C. (1982). *Biochem. J.* **204,** 171.
Johnson, R. G., and Scarpa, A. (1984). *In* "Electrogenic Transport: Fundamental Principles and Physiological Implications" (M. P. Blaustein and M. Lieberman, eds.), pp. 71–91. Raven Press, New York.
Maxfield, F. R. (1982). *J. Cell Biol.* **95,** 676.
Mellman, I., Fuchs, R., and Helenius, A. (1986). *Annu. Rev. Biochem.* **55,** 663.
Metchnikoff, E. (1968). "Lectures on the Comparative Pathology of Inflammation." Dover, New York.
Ohkuma, S., and Poole, B. (1978). *Proc. Natl. Acad. Sci. U.S.A.* **75,** 3327.
Orci, L., Ravazzolla, M., Amherdt, M., Madsen, O., Vassalli, J.-D., and Perrelet, A. (1985). *Cell (Cambridge, Mass.)* **42,** 671.
Orci, L., Ravazzola, M., Amherdt, M., Madsen, O., Perrelet, A., Vassalli, J.-D., and Anderson, R. G. W. (1986). *J. Cell Biol.* **103,** 2273.
Orci, L., Ravazzola, M., and Anderson, R. G. W. (1987a). *Nature (London)* **326,** 77.
Orci, L., Ravazzola, M., Storch, M.-G., Anderson, R. G. W., Vassalli, J.-D., and Perrelet, A. (1987b). *Cell (Cambridge, Mass.)* **49,** 865.
Poole, B., and Ohkuma, S. (1981). *J. Cell Biol.* **90,** 665.
Rudnick, G. (1986). *Annu. Rev. Physiol.* **48,** 403.
Russell, J. T. (1984). *J. Biol. Chem.* **259,** 9496.
Schwartz, A. L., Strous, G. J. A. M., Slot, J. W., and Geuze, H. J. (1985). *EMBO J.* **4,** 899.
Tycko, B., and Maxfield, F. R. (1982). *Cell (Cambridge, Mass.)* **26,** 643.
Yamashiro, D. J., Fluss, S. R., and Maxfield, F. R. (1983). *J. Cell Biol.* **97,** 929.

Chapter 23

Transmission Electron Microscopy and Immunocytochemical Studies of Yeast: Analysis of HMG–CoA Reductase Overproduction by Electron Microscopy

ROBIN WRIGHT AND JASPER RINE

Department of Biochemistry
University of California
Berkeley, California 94720

I. Introduction

Immunocytochemistry provides a powerful technique for analysis of molecular organization within cells. Using specific antibody probes to-

METHODS IN CELL BIOLOGY, VOL. 31

gether with electron microscopy (EM), the localization of a protein within individual cells can be analyzed. For proteins of known function, this positional information reveals where the function takes place within the cytoplasm. In the case of proteins with unknown functions, knowledge of the protein's subcellular localization can often provide clues concerning the role of the protein. Such quantal information complements that obtained from more classic techniques, such as cell fractionation, which furnish an average view of the protein within a population of cells.

Such a wealth of potential information does not come without a price. Even under the best conditions, immunocytochemical analyses require compromises between two generally opposing goals: optimal preservation of ultrastructure and maximal retention of antigenicity. A variety of techniques have been developed to reduce the conflict between these two aims. Many of these, such as the use of frozen sections of chemically unfixed tissue, require special equipment and special skills. Immunolabeling of thin sections of cells following chemical fixation can be accomplished by anyone with both fair proficiency in general electron-microscopic techniques and a fair level of tolerance and patience.

The purpose of this chapter is to encourage the faint-of-heart that, sometimes, the easiest way proves to be sufficient. During our electron-microscopic analysis of yeast cells, we found that samples prepared for routine EM could be labeled quite successfully with affinity-purified antiserum against the membrane protein, 3-hydroxy-3-methylglutaryl coenzyme A (HMG–CoA reductase), and colloidal gold reagents (Wright *et al.*, 1988). Detailed protocols concerning these methods are provided here. A number of excellent reviews concerning the history, preparation, utilization, and application of colloidal gold probes are available, and this information will not be reiterated (De Mey, 1983b; Polak and Van Noorden, 1987; Smit and Todd, 1986; Romano and Romano, 1984; Roth, 1982, 1983, 1986). For general electron-microscopic techniques, books by Robinson *et al.* (1987), Glauert (1975), and Hayat (1981a) are highly recommended.

II. Sample Preparation

A. Culture Conditions

1. General Considerations

It is not unexpected that cells grown in different media might have different ultrastructural features, and this is the case for yeast. The strain used in the examples presented here is JRY1239, a derivative of YM256 (obtained from M. Johnston: Johnston and Davis, 1984) that is closely related to the common yeast strain S288C. This strain has the genotype: *a ura3-52 his3Δ200 ade2-101 lys2-901*. It also carries plasmid pJR59, a YEp24-based plasmid containing the coding sequences for one of the yeast HMG–CoA reductase isozymes (i.e., the *HMG1* gene) and, consequently, is generally grown in yeast minimal medium (0.17% yeast nitrogen base without amino acids or ammonium sulfate, Difco Laboratories; 5% ammonium sulfate; 2% glucose; 30 mg/liter adenine, histidine, lysine, and methionine) under selective conditions (without uracil supplementation). Figure 1 is a example of JRY1239 fixed directly following growth in minimal medium. The cytoplasm appears very electron dense and it is often difficult to resolve the nuclear envelope and other membranes within the cell. In addition, the cell wall is much less susceptible to enzyme digestion (if this step is performed). To enhance the visibility of internal membranes and improve overall ultrastructural characteristics, strain containing plasmids are grown overnight in supplemented minimal medium under selection and then diluted into rich medium (YPD: 1% yeast extract, 2% Bacto peptone, 1% glucose). After 2–4 hours' growth, the cells are processed for EM. The other micrographs presented here are of JRY1239 fixed after dilution into rich medium and 2 hours' growth. A number of plasmid-containing strains yield similar results following these culture conditions. Plasmid loss during the period of nonselective growth should be minimal, but can be monitored by plating an aliquot of the culture onto YPD plates just prior to fixation and subsequently replica-plating onto selective minimal plates. This control experiment may be especially important if the plasmid confers a growth disadvantage to the cell.

Generally, cell density should be limited to $\sim 5 \times 10^6$/ml ($OD_{600} = 0.5$). Such cells are in early log phase and typically display uniform ultrastructure. Stationary-phase cells possess large, electron-dense vacuoles and numerous electron-dense inclusion bodies that obscure ultrastructural details. A 50- to 100-ml culture at this density provides a pellet of sufficient size for easy processing.

2. Experimental Procedures

a. A single colony is inoculated into 2 ml selective minimal medium and grown overnight at 30°C with vigorous shaking.
b. A 50-ml volume of selective minimal medium is inoculated with a sufficient amount of the overnight culture to provide a culture at OD 0.5 after 12–24 hours' growth. This culture is incubated at 30°C with vigorous shaking for the appropriate length of time.

c. YPD broth (100 ml) is inoculated with a sufficient volume of the second overnight culture to yield an OD_{600} of 0.5 in 2–4 hours. This culture is then grown at 30°C with vigorous shaking, until time of fixation.

B. Aldehyde Fixation

1. Goals and Selection of Fixation Conditions

This is probably the most critical step of the many stages of EM sample preparation. Although obvious, it should be pointed out that "fixing" a cell is actually killing it—hopefully, in a special way so that its physical structure is preserved. To accomplish this *coup de grace,* cellular components must be rapidly crosslinked to prevent autolysis, changes in morphology, and extraction or diffusion of molecules. Furthermore, the degree of crosslinking must be adequate to maintain this state throughout the harsh treatments that follow, so that the structures observed later in the electron beam accurately reflect the condition of the living cell at the time of fixation. When the investigator further requires retention of sufficient tertiary structure to allow maximal immunoreactivity, initial fixation becomes, arguably, the major hurdle to surmount.

Currently, the monofunctional and bifunctional aldehydes, formaldehyde and glutaraldehyde, are the most widely and successfully used reagents for this purpose. These compounds react with amino groups of proteins and stabilize them by forming interchain and intrachain crosslinks, but demonstrate little, if any, reactivity with lipids or carbohdyrates. The molecular mechanisms of fixation are not yet well understood, but discussions of the current theory and practice of chemical fixation can be found in Hayat (1981b).

An enormous number of conditions (pH, temperature, concentration and composition of fixative, duration of fixation, presence of additives such as sucrose, Ca^{2+}, picric acid, etc.) are known to affect fixation quality and can be explored to improve or to restore immunoreactivity in specific cases. However, initial immunocytochemical attempts should employ the most widely used, routine fixation conditions. If these prove successful, no further experimentation is necessary. However, if the initial attempts are disappointing, the wealth of the immunocytochemist's experience can be brought to bear on the particular problem. Brandtzaeg (1983) is recommended as a reference for studies of the effects of various fixation regimes on antigenicity. Unfortunately, until the actual chemistry of fixation is much more fully understood, it is likely that determining optimal fixation conditions will remain empirical and may require fine-tuning for each particular antigen, antiserum, or strain.

2. Safe Laboratory Practices

Electron microscopy requires careful use of many chemicals that are hazardous or extremely allergenic. Fixatives are toxic, often volatile, and should be handled and disposed of with proper care (Smithwick, 1985). Cacodylate is an arsenate; resins are toxic and known or suspected carcinogens (Ringo *et al.*, 1982). Lead and uranyl salts used for staining are toxic. Use of these chemicals in hoods and with adequate protection (gloves, eye protection if necessary) is essential. Staff at EM facilities are excellent sources of information concerning use and disposal of EM reagents. Carolyn Schooley and Douglas Davis at the EM Facility, Life Sciences [University of California, Berkeley (UCB)] have developed brief handouts on safety that can be obtained by mail. In addition, the Electron Microscopy Society of America and commercial suppliers of reagents for EM are useful resources for this important information.

3. General Comments Concerning Initial Fixation of Yeast

The speed of initial dispersion of the cells into the fixative solution is important for optimal fixation of yeast. When the fixative is added directly to a cell pellet that is subsequently resuspended, the quality of fixation varies greatly from cell to cell. In the worst cases severe autolysis occurs (Fig. 4A). To avoid this artifact, the fixative solution should be added to a cell suspension (or vice versa) with sufficient rapidity to ensure rapid equilibration.

The temperature at which fixation of yeast occurs is another parameter that merits attention. There are two opposing viewpoints on the matter. A temperature of 0°–4°C is advocated to inhibit undesired enzyme activity (such as protease activity) during the fixation process. On the other hand, lower temperatures slow diffusion of the fixative into the cell, prolonging the time between cell death and optimal crosslinking and, thus potentially creating an environment ripe for undesired ultrastructural alterations. We generally favor the low-temperature viewpoint, because the yeast cell's small size makes the decrease in diffusion rates at lower temperatures less important than for many other organisms. Using mixtures of the rapidly penetrating (but less effective) crosslinker, formaldehyde, combined with the more slowly penetrating (but more efficient) crosslinker, glutaraldehyde, tend to offset the disadvantages encountered with low-temperature fixation. However, in practice, no difference in cells fixed at 4°C versus at room temperature are detectable in our hands. The temperature of fixation may then be a matter of convenience for the investigator.

4. Reagents

For consistency, EM-grade glutaraldehyde (Ted Pella, Redding, CA) is used for immunocytochemistry. When only ultrastructural studies are required, biological-grade glutaraldehyde (Ted Pella) can be substituted with no obvious alteration in results and is considerably cheaper. Formaldehyde is made by depolmerizing paraformaldehyde with heat and should be used within 2 days (Hayat, 1981b). We have also used methanol-free formaldehyde (Ted Pella) with good results. Both solutions should be colorless and should be discarded if a yellow cast is present.

Glass-distilled water of highest possible purity is used for all solutions or washes used in EM. A 4× stock of cacodylate (0.4 *M*) and 25× stock of potassium phosphate (1 *M*) are prepared and adjusted to pH 6.8. If used, divalent cations ($CaCl_2$ and/or $MgCl_2$) are added to a final concentration of 1 m*M* just prior to use of the buffer. The same buffer is used throughout the protocol (for prefixation, buffer washes, and postfixation).

5. Experimental Procedures

The favored protocol for fixation has undergone a considerable evolution during the course of our studies. Several fixation protocols that produce good results are presented here, with the current method of choice outlined in Section II,B,5,c. Centrifugations are performed in a Sorvall (5 minutes, 3000 rpm, SS34) or a clinical (3–5 minutes, highest speed) centrifuge.

a. Washed Cells–Cacodylate Buffer

1. Cells are harvested by centrifugation and resuspended in one-half volume minimal medium. The cells are washed twice by centrifugation–resuspension in minimal medium.
2. The washed cell pellet is resuspended completely in minimal medium, (1/20 original volume) and transferred to a heavy-walled, conical glass test tube (Corex). An equal volume of ice-cold, 2× fixative (4% glutaraldehyde, 4% freshly depolymerized paraformaldehyde, 2 m*M* $CaCl_2$, 2 m*M* $MgCl_2$, 0.2 *M* cacodylate, pH 6.8) is added, and the suspension rapidly mixed. Fixation is allowed to proceed for 5 minutes on ice.
3. The cells are pelleted and the first fixative solution removed. The cells are then uniformly resuspended in 1× fixative. Fixation continues for 30 minutes on ice.
4. Excess fixative is removed by three washes in buffer. The cells should remain in each wash for at least 5 minutes before pelleting.

b. Washed Cells–Phosphate Buffer. Alison Adams (Massachusetts Institute of Technology, Cambridge, MA, personal communication) has recommended the use of phosphate buffer rather than the widely used arsenate, cacodylate. We have tested this buffer and find that it gives identical results to cacodylate. Adams uses 4 m*M* potassium phosphate (pH 6.8) and adds 0.1 g $MgCl_2$ per liter. The protocol outlined in Section II,B,5,a is followed, substituting this buffer for cacodylate.

c. Cells in Culture–Phosphate Buffer. Gross morphological alterations and changes in the presence and/or location of proteins can be induced by subjecting yeast to centrifugal forces (R. Preston, Carnegie-Mellon University, Pittsburgh, PA, personal communication; L. Pillus, UCB, personal communication). To avoid such changes during fixation for immunofluorescence, the fixative can be added directly to the growing yeast culture (L. Pillus, personal communication; Stearns and Botstein, 1988). Encouraged by this observation, Chris Kaiser (UCB) and I have tested its feasibility for EM with excellent results. Because the technique is considerably more rapid than others that require multiple washes prior to fixation, this direct fixation method has become our favored protocol.

1. A one-tenth volume of 10× fixative solution at room temperature is placed into a centrifuge bottle. (It is good practice to dedicate reusable plastic and glassware for EM, to avoid any possibility of contamination of cultures, etc., with traces of fixatives).
2. The culture is poured rapidly into the fixative and allowed to sit at room temperature for 5 minutes. During this time, the medium will turn quite dark (YPD) or yellow (YM).
3. The cells are pelleted, resuspended in one-tenth volume of ice-cold, 1× prefixative, and allowed to complete fixation on ice for 30 minutes.
4. Excess fixative is removed by three buffer washes, leaving the cell suspension in each change of buffer for at least 5 minutes.
5. Note: Either cacodylate or phosphate buffer gives comparable results, and the presence or absence of Ca^{2+} and/or Mg^{2+} does not make any obvious differences.

C. Cell Wall Removal

1. Cell Considerations

After prefixation and removal of excessive fixative with buffer, it is general practice to digest partially the cell wall. This treatment facilitates resin infiltration and can be accomplished by zymolyase, glusulase, or lyticase (Scott and Schekman, 1980). Unfortunately, the degree of digestion is inconsistent, varying from cell to cell in the same preparation as well as from sample to sample or day to day. This variability can be extreme, resulting in samples that contain both cells with apparently intact walls and cells with no cell wall at all. In the latter case, damage to the plasma membrane and internal ultrastructure is often apparent. Even in samples displaying minimal cell wall digestion, subtle damage is observed in those cells that have been more affected by the enzyme treatment. These cells are invariably less well stained and appear quite extracted. Furthermore, the bud cell wall appears more sensitive to the enzyme than the mother cell wall. This produces a mother cell with well-defined structures and normal staining characteristics, and a bud of very different appearance and staining sensitivity. Apart from the serious questions these conditions raise concerning the uniformity of fixation or extraction, it is extremely difficult to obtain electron micrographs of publication quality from cells with such variable contrast.

A method for lyticase digestion of fixed cells, derived from that of Daniela Brada (Biozentrum, Basel, Switzerland), is presented here. In view of the problems described in the previous paragraph, we prefer to eliminate cell wall digestion completely if at all possible. Infiltration of intact yeast cells is more difficult, however, and special care must be taken to avoid a sample that produces sections with an unacceptable number of holes (Section II,E,1).

2. Experimental Procedures

1. Following fixation and buffer washes, the cells are resuspended in 10 ml TMS [50 m*M* Tris-HCl (pH 7.5), 5 m*M* $MgCl_2$, 1.4 *M* sorbitol, 50 m*M* β-mercaptoethanol], spun down, and resuspended in 0.5 ml TMS.
2. Lyticase (a gift from the Schekman lab) is added to a final concentration of 1 k unit/50 A_{600}. The mixture is gently shaken at room temperature for 10–15 minutes.
3. The cell suspension is then diluted by addition of 9.5 ml buffer (i.e., the same as used for fixation). The cells are pelleted, resuspended in 10 ml buffer, and repelleted.
4. Glusulase (diluted 1 : 20 in 0.1 *M* phosphate–citrate buffer, pH 5.8; Alison Adams, personal communication) and zymolyase (0.5 mg/ml in 0.1 *M* phosphate–citrate buffer, pH 5.8; Lois Banta and Scott Emr, California Institute of Technology, Pasadena, CA, personal communication) are also effective.

D. Postfixation and en Bloc Staining

1. General Considerations

The quality of preservation obtained by initial fixation with glutaraldehyde followed by postfixation with osmium tetroxide is currently unrivaled by other fixation schemes. Osmium reacts with many classes of molecules, including lipids, proteins, and nucleic acids (Hayat, 1981b). Because it is a heavy metal, it also imparts electron density to the components with which it reacts, thus acting as a stain as well as a fixative. It is widely assumed that immunoreactivity is destroyed by osmication, although actual experimental data are scant. Thus, for immunocytochemistry, use of osmium solutions is generally proscribed (see Hayat, 1981b; Smit and Todd, 1986).

The poor preservation of membranous structures with aldehyde fixation alone, however, can make interpreting the localization of membrane proteins difficult. In the case of HMG–CoA reductase, a membrane protein of the endoplasmic reticulum, postfixation did not appear to inhibit immunoreactivity severely and the increase in resolution of the novel membrane structure (karmellae) induced by this protein made up for any quantitative decrease in labeling density. A reasonable and cautious approach would be to postfix a portion of the aldehyde-fixed sample, and process both aliquots in parallel through the subsequent steps. The osmium-treated sample will provide the best ultrastructural information and, possibly, satisfactory immunolabeling results. If immunolabeling of the postfixed sample fails, the aldehyde-fixed sample will be immediately available for additional localization attempts.

Exposure of the sample to heavy-metal salts prior to infiltration and embedding with resin (that is, en bloc staining) is a common procedure in general EM. The use of uranyl acetate for en bloc staining provides electron contast and acts as a fixative for nucleic acid (Kellenberger *et al.*, 1958; Todd, 1986). Similar arguments to those just discussed for osmium treatment also apply when considering whether or not to use en bloc staining for immunocytochemistry. Again, in the case of HMG–CoA reductase, en bloc staining produced an observable improvement in ultrastructure with little or no reduction in immunolabeling. Therefore, we generally include this step in our sample preparation.

2. Reagents

The source of osmium is of critical importance. Premixed aqueous stocks should be avoided, because they can contain "stabilizers" of unknown (i.e., proprietary) identity that adversely affect fixation quality. In addition, osmium solutions made from solid osmium can deteriorate and produce very disappointing results. The osmium stock solution should be very pale, straw-colored, and clear. Darker, yellowed solutions should be discarded. Once again, safety should be of utmost consideration in both use and disposal of this volatile and toxic chemical.

Enhanced preservation of lipids has been noted with use of mixtures of osmium tetroxide and potassium ferricyanide (Karnovsky, 1971). This mixture is also found to improve microfilament preservation (McDonald, 1984). In our hands, this "OsFeCN" formula is superior to osmium alone for yeast postfixation. The solution should be mixed just prior to use. It contains 0.5% osmium tetroxide, 0.8% potassium ferricyanide, in 0.04 *M* phosphate buffer, pH 6.8. (The buffer should be the same as used for prefixation and washes.).

Uranyl acetate can be used at concentrations ranging from 0.5 to 2%. The solution is freshly mixed in distilled water and filtered through a 0.2-μm filter (Acrodisc) before use. If either cacodylate or phosphate buffers are used, care should be taken to remove the buffers with water washes prior to uranyl acetate treatment, because they can cause precipitation of the uranium salt onto or in the sample.

3. Experimental Procedures

a. OsFeCN Postfixation

1. The prefixed, washed (and if desired, digested) cell pellet is resuspended in postfixative (1/25 of the original culture volume) and allowed to sit on ice for 5 minutes.
2. The cells are pelleted and the postfixative removed. A fresh aliquot of postfixative is gently pipeted into the tube, taking care not to disturb the pellet. Fixation continues on ice for 15 minutes.
3. The postfixative is removed and the cells are washed thoroughly in distilled water, gently pelleted as required to maintain the pellet (1 minute in a clinical centrifuge at top speed).
4. The cell pellet should be deep gray to black in color after OsFeCN postfixation.

b. Permanganate Postfixation. For samples that will be used for ultrastructural analysis alone, postfixation in potassium permanganate is recommended (Stevens, 1981; Hayat, 1981b). This procedure gives an excellent view of cytoplasmic organization, especially of membranes. In addition, material that has been postfixed in potassium permanganate infiltrates perfectly to produce uniform sections with no holes, probably due to extraction of or reaction with cell wall components.

Cells treated with permanganate are incompatible with colloidal gold reagents. A dense, uniform precipitate forms over the entire permanganate-fixed cell when exposed to these chemicals. In addition, permanganate treatment following osmication produces very poor fixation. Mixed postfixatives containing both permanganate and osmium have not been tried but could conceivably produce cells with the good qualities of each fixation alone.

1. An aqueous 0.5% solution of potassium permanganate is prepared about 30 minutes prior to use and filtered through a 0.22-μm filter as it is added to the cells.
2. The prefixed, washed cell pellet is resuspended in permanganate (one-fifth original culture volume) in a conical glass test tube and allowed to react for 5 minutes at room temperature. The cells are pelleted and fixation continued for 20 minutes at room temperature.
3. Extensive washing in distilled water is performed prior to en bloc staining overnight at 4°C (see next section). The cell pellet will be deep, golden-brown. A brown scum may form on the wall of the test tube, but this does not affect fixation.

c. *En Bloc Staining.* Either a 1-hour incubation at room temperature in 2% aqueous uranyl acetate or an overnight incubation at 4°C in 0.5% aqueous uranyl acetate produce similar results, so convenience or impatience can make the decision.

1. After postfixation, special care must be taken to remove cacodylate or phosphate buffers by distilled-water washes. Usually three to five resuspension–centrifugation cycles are sufficient.
2. The cell pellet should be gently resuspended in the uranyl acetate solution. At this point, the pellet will break into fine chunks, and no effort should be directed to uniform resuspension because the clumping characteristic is desirable for later infiltration steps. The cells are repelleted and left as a pellet for the remainder of the incubation period.
3. Excess uranyl salts must be removed by three to five distilled-water washes following en bloc staining to eliminate precipitation of uranyl acetate in or on the cells during dehydration.

E. Dehydration, Infiltration, and Embedding

1. General Considerations

The goal of these sample preparation steps is to remove all water from the cell interior and to replace it with a plastic resin. This resin is subsequently polymerized to support the cell's ultrastructure during thin-sectioning and electron bombardment. The yeast cell, especially with an intact cell wall, is often refractory to resin infiltration, evidenced by the fenestrated sections often produced. The steps presented later minimize infiltration-related problems.

Discussions of the chemistry of resins and choice of resin for immunocytochemistry are available in papers by Luft (1973), Aldrich and Mollenhauer (1986), and Causton (1984). The two most common classes of resins are epoxies and acrylates. Epoxy resins (such as Spurr's and Pelco Ultra-low viscosity) infiltrate the yeast cell relatively well to produce blocks that are easy to section and sections that stain well and have good stability in the electron beam. Their chief drawback, as far as immunocytochemistry is concerned, is their considerable hydrophobicity, a characteristic that can inhibit availability of the antigen to the antibody. In addition, the temperatures necessary to cure these resins (50°–70°C) are thought to affect the immunoreactivity of particular antigens.

Acrylic resins (such as LR White and Lowicryl K4M) are much more hydrophilic and will tolerate residual water in the sample. For immunocytochemical applications, the ability of these resins to be polymerized at low temperatures (by UV light) is important for minimizing extraction and maintaining maximal immunoreactivity, because the high temperatures required for epoxy polymerization can cause considerable denaturation. These resins infiltrate the interior of the intact yeast cell quite readily. However, shrinkage of these resins during polymerization is often considerable, and this produces breaches between the plasma membrane and cell wall as the cytoplasm contracts away from the rigid cell wall. The problem is exacerbated by the pronounced movement or melting of the section by the electron beam, often resulting in gaping holes. In the worst cases, the entire cytoplasm pops out, leaving only a cell wall shell and a void where the cell once was. Consequently, cell wall removal is absolutely essential if these resins are used. Even when this precaution is taken, section quality of acrylic-embedded yeast is quite poor in our hands. The quality of infiltration might benefit from dehydration with acetone or from "clearing" with propylene oxide, but I have no direct information concerning this possibility nor the effect of such chemicals on immunocytochemistry.

We initially compared immunolabeling of cells embedded in Spurr's resin (Spurr, 1969: 10 g vinyl cyclohexene dioxide; 6 g DER 736 resin; 26 g nonenyl succinic anhydride, 0.4 g dimethylaminoethanol) to Lowicryl K4M (Ted Pella; used according to Daniela Brada's protocol, personal communication) with essentially identical results. For our particular antigen, HMG–CoA reductase, the temperatures of polymerization and hydrophobicity of the epoxy resin did not appear to be critical factors for maintaining immunoreactivity. Subsequently we also examined LR White resin (Polysciences, Warrington, PA) and the Spurr derivative, Pelco Ultra-low Viscosity resin (Pelco: 5 g vinyl cyclohexene dioxide; 10.5 g *n*-hexenyl succinic anhydride; 1.5 g DER 736 resin; 0.2 g dimethylaminoethanol). The best labeling quality was produced from LR White-embedded samples. Unfortunately, the sections obtained from these samples were nearly unusable, with so many large holes that they appeared like Swiss cheese (see Section VI for recent modifications solving this problem). Best section quality (absence of pericellular or internal holes) was obtained with the Pelco Ultra-low Viscosity mixture. Best ultrastructure resolution was obtained with Spurr's formula. Thus, choice of resin must be dictated by the objective of the particular experiment. The wide variation in cellular appearance observed with different resin formulas was unexpected. It may be prudent to include more than one resin in sample preparation.

During dehydration, fixed yeast cells will form small clumps that can be manipulated directly for infiltration and embedding steps. This feature is convenient because it avoids the necessity of embedding the cell suspension in agar. In addition, the sample is composed entirely of yeast cells, minimizing "search and discover" time at the microscope: a single grid space (300-mesh) will contain profiles of ~200 cells.

2. Experimental Procedures

The protocol presented here can be used with Spurr, Pelco Ultra-low, or LR White resins. Section II,E,2,d contains information on the variations required for LR White. Lowicryl K4M embedding procedures are described by Roth in Chapter 24 of this volume.

a. Dehydration

1. Following the fixation and en bloc-staining steps, the cells are resuspended in 10 ml 50% ethanol and then spun in a conical, *glass*

test tube, in a clinical centrifuge for 5 minutes. (The use of glass is important for maximizing clumping of the pellet into aggregates of convenient size.)

2. The supernatant is replaced with 1 ml of 70% ethanol. At this point, the cells will form a firm pellet that is gently dislodged with a Pasteur pipet and chipped into small, pepper-size fragments. An aliquot containing 10–20 small chunks is dispensed into glass, snap-cap vials containing 70% ethanol. The remaining sample is stored at −20°C in Eppendorf tubes and is available for subsequent processing, if necessary. We have removed aliquots from samples stored in this manner after several months and noted no alteration in ultrastructure or immunoreactivity.
3. The glass vial is capped and placed on a rotating drum for 5 minutes. Dehydration is completed in subsequent incubations (5-minute rotations) in 95% and three changes of 100% ethanol. For the absolute ethanol incubations, a new bottle of ethanol is opened to ensure absence of water.
4. A convenient way to change solutions is to use a Pasteur pipet to remove most of the liquid from the vial while it is tipped to allow the sample fragments to gather at one side of the vial. Then place the end of the pipet flush with the bottom of the vial on the side opposite the sample. The vial is tipped to allow fluid to reach the pipet, and the fluid is slowly aspirated. Sample fragments of appropriate size for processing will gather around the bore of the pipet, but will not enter, *if* the pipet is properly oriented. This procedure allows solution changes to be relatively rapid and complete.

b. Infiltration

1. Patience is a virtue. The last 100% ethanol is replaced with a solution of two parts ethanol to one part resin. The vial is capped and returned to the rotator for 1 hour.
2. The resin mixture is replaced with 1 : 1 resin–ethanol mix and the capped vial rotated for at least 1 hour.
3. The resin is exchanged with a fresh 1 : 1 mix and the samples are rotated, *uncapped*, overnight in a hood. The ethanol slowly evaporates, increasing resin concentration gradually.
4. The next day, the residual resin is replaced with fresh 100% resin and the samples are rotated for an hour. The resin is changed and the samples placed under vacuum (20 psi) for 15 minutes to degas, then returned to the rotator for 1 hour.
5. Small pellet fragments are removed from the vial, using a sharpened applicator stick. The fragment is gently "rolled" across a Kimwipe

to remove residual resin and placed into 5-ml plastic embedding cups (Ted Pella) containing resin. These cups are placed under vacuum for 1–2 hours.

c. Embedding

1. The sample fragments are then removed, blotted on a Kimwipe, and transferred to labeled BEEM capsules containing resin (one fragment per capsule).
2. The sample is allowed to sink to the bottom, positioned in the center of the capsule, and then placed under vacuum for 15 mintues.
3. Polymerization is accomplished in a vacuum oven at 15 psi, 60°C (2 days).

d. Special Techniques for LR White

1. Gelatin capsules are recommended as embedding molds because they exclude oxygen better than BEEM capsules and are not penetrated by LR White resin. These capsules should be dried in an oven (60°C) overnight before use. The longer, narrower portion of the capsule is filled with resin and the wider, shorter portion used as a cap after the sample is introduced.
2. It is important to maintain the temperature ±2°C during polymerization of LR White, and most ovens do not have this degree of control. However, the complete capsules will fit into the small holes of a temperature block normally used to hold 0.5-ml Eppendorf tubes for molecular genetic manipulations. (Note that the gelatin capsule must be capped; the narrow portion of the capsule will sink completely to the bottom of the temp block.) A temperature of 60°C is recommended by the manufacturer, but this produces a considerable amount of shrinkage. A temperature of 45°C overnight followed by 50°C for 24 hours may minimize this undesired effect. During curing, the gelatin capsules are covered with several layers of foil to maintain temperature at the surface.
3. Lois Banta and Scott Emr recommend a 3-day polymerization of LR White at 4°C using UV irradiation.

F. Grid Preparation and Sectioning

1. General Considerations

Our first immunocytochemistry attempt was a resounding disaster, because the sections floated off all 20 grids during the first wash. That experience underscored the necessity of using a technique that ensures the sections will stay in place during the entire procedure. A formvar-coating technique of Bonnie Chojnacki (Carnegie-Mellon University, Pittsburgh, PA) has proved wonderfully effective for this purpose (see later). Nickel grids (200–300 mesh) are used, because they are less "reactive" than the standard copper ones. However, nickel grids readily become magnetized and it will save a great deal of frustration to have nonmagnetic forceps on hand for handling them. The "tennis racket"-style grids with handles (Ted Pella) are highly recommended for the ease of handling they afford.

Both Spurr and Pelco Ultra-low resins produce blocks that are very easy to section by standard techniques. The hydrophilic nature of LR White requires that the water level in the boat be kept very low to prevent water from leaping onto the block face as it passes the knife edge. Other special handling is not necessary. For serial sections, Fahrenbach's method (trimming the block to have wide leading and trailing edges and using diluted rubber cement to coat these edges) works amazingly well (Fahrenbach, 1984).

2. Experimental Procedures

a. *Grid Preparation*

1. Nickel grids are placed into a glass vial containing 100% ethanol and sonicated for 1 minute. They are then rinsed several times in ethanol, dumped onto filter paper in a glass Petri dish, and dried in a 50°–60°C oven.
2. The washed grids are placed in a glass Petri dish containing a dilute formvar solution (1 ml 2% formvar in ethylene dichloride into 25 ml 24 : 1 ethylene dichloride–chloroform). A Pasteur pipet can be used to wash the grids into one edge of the dish.
3. The grids are individually removed with forceps and placed onto clean filter paper to dry. Only the bars are coated with formvar, leaving the entire grid space open for view of the section. "Sticky-bar" grids prepared in this manner are usable as soon as they have dried and are effective indefinitely.

b. *Sectioning and Section Mounting*

1. It is wise to prepare a sufficient number of grids for several immunolabeling experiments at one time, taking into consideration all the controls and parameters to be tested. Each grid should contain as many sections as possible. Silicon mats with divided

sections (available from any EM supplier) are very convenient for storing grids securely.

2. Sections with gold interference color are cut and then exposed to chloroform vapors wafted over the surface of the water in the knife-boat, using a cotton-tipped applicator stick. This will "spread" the sections, eliminating compression that occurs during sectioning. It also reduces section thickness about one interference color (i.e., they will be silver).
3. Sections are manipulated into a group using an eyelash glued to a sharpened applicator stick (clear nail polish works well). The dull side of a coated grid is carefully lowered over a group of sections floating in the boat and pressed into the water surface, without breaking surface tension. The grid is removed and inverted onto a Kimwipe to blot excess moisture. After air-drying, the grid is placed onto a silicon mat for storage.
4. In view of the many grids required to do a complete immunolabeling experiment, We have incorporated Alice Taylor's (UCB) "assembly-line" method. Sections are allowed to accumulate in the boat until a sufficient number have been cut. During the later stages of cutting, grids are loaded into 10 forceps, each of which has been fitted with a narrow piece of tygon tubing or an o-ring. The tubing is pushed down toward the forcep points, holding the forcep closed and keeping the grid in place until needed. The loaded forceps are propped up, ready for use. By having 10 forceps available with grids, the time for loading the sections onto the grids is reduced considerably.

c. Securing the Sections

1. If the sections are to be used immediately, it is *essential* to secure the sections onto the grid. The dish containing the grids loaded with sections is placed into a 50°–60°C oven for 1–2 minutes. This treatment does not affect immunolabeling or ultrastructure and ensures that the sections do not leave the grid, even under harsh conditions.
2. Air-drying the grids for 1–2 days is also usually effective, but we have lost sections when the heating step is omitted.

d. Storing Sections. Polymerized resins are surprisingly fluid (Aldrich and Mollenhauer, 1986). Movement of embedded material occurs both in blocks and on sections. For optimal resolution, the sections should probably be used within a few days. In practice, we have detected no noticeable changes in immunolabeling or resolution after 3 months, but the possibility of alterations should be kept in mind.

III. Immunolabeling

A. General Considerations

Generation of reagent antibodies of high specificity is of utmost importance for accurate immunolocalization at the EM level. While this requirement cannot be overstated, it is beyond the scope of this article to review the techniques for preparation of the antibody probes. De Mey (1983a) is an appropriate initial source for this information.

The antiserum should be affinity-purified. While theoretically feasible, immunoadsorption to remove undesired antibodies does not produce a serum with the required specificity for immunocytochemistry. Attempts to utilize preadsorbed sera in collaboration with Johanna Reneke and Jeremy Thorner (UCB) and with Alex Fransusoff and Randy Schekman (UCB) have convinced us of the necessity of affinity purification. Even a miniscule percentage of "contaminating" antibodies left after immunoadsorption can produce severe problems in interpretation. Background staining of the cell wall, vacuole, and nucleus are especially problematic. The time involved in preparation of cells for immunolabeling merits use of the best possible antibody.

The antiserum used in these examples was prepared by Linda D'Ari and Michael Basson (Wright *et al.*, 1988) from a fusion protein containing the amino terminus of β-galactosidase and the carboxyl terminus of HMG–CoA reductase (encoded by the *HMG1* gene). Polyclonal antiserum was produced in rabbits and subsequently affinity-purified on a Sepharose 4B column containing the antigen. The serum is used at a dilution of 1 : 200 ($\sim$2 μg/ml) for immunoblots. For most of the immunolabeling experiments, it was used 10 times more concentrated (i.e., 1 : 20, 20 μg/ml).

A prudent investigator would attempt immunofluorescent localization before immunocytochemistry, because it is easier, quicker, and might provide sufficient data without resorting to the more challenging, time-consuming steps of immunocytochemical studies. In addition, immunofluorescence will provide a standard against which immunocytochemical data can be interpreted. In fact, we initially suspected that visibility of an antigen–antibody complex by immunofluorescence would prove a good indicator of the probable success of immunocytochemistry. Collaboration with Kevin Redding and Robert Fuller (Stanford Unviersity, Stanford, CA) to localize the *KEX2* protein by immunocytochemistry demonstrated that predictions based on successful immunofluorescence are not always correct. Redding had obtained dazzling immunofluorescence results, but we failed to localize the protein on thin section of his cells. Suspecting that the antigen might be sensitive to glutaraldehyde fixation, Redding examined antigenicity of the KEX2p on immunoblots fixed in 2% glutaraldehyde. He found that his antigen–antibody interaction was not affected by this treatment, suggesting that the antigen might be sensitive to *subsequent* steps in processing. Thus, immunofluorescence and immunoblotting results may not assure immunocytochemical success, and the cause of failure may be difficult to determine.

Preparation of colloidal gold reagents conjugated to protein A is quite simple (see Roth, 1982, 1983; Smit and Todd, 1986). Immunoglobulin-congugated gold reagents purchased from commercial sources (Janssen, Piscataway, NJ) give less background in our hands, however. Whether this difference is due to better conjugation techniques or to inherent increases in specificity of immunoglobulin as compared to protein A is not clear. The results presented here utilize Goat anti-rabbit immunoglobulin-conjugated gold (GARG), purchased from Janssen. We have noticed some problems with clumping and non-specific staining with 15-nm GARG particles from this supplier (see results), but reagents with smaller gold particle size were satisfactory.

B. Controls: Convincing Yourself and Colleagues that the Results Are Real

For control experiments, the same rules apply to all immunochemical techniques. Labeling of sections in the absence of primary antibody will control for immunoreactivity of the secondary antibody alone. Using preimmune serum in the primary incubations will provide evidence that the labeling observed is dependent on the immune response induced after the antigen is introduced into the animal. In the case of affinity purification, the passthrough from the affinity column contains those antibodies in the animal that do not react with the antigen and provides a similar control to the preimmune serum and is analogous to controls based on immunoadsorption. In addition to controls based on varying the antibody probe, yeast offers a wealth of possibilities to test whether or not the observations are valid. For example, strains that either overproduce or that lack the particular antigen can be immunolabeled and the patterns compared to that of the wild-type strain. Controls based on reproducibility merit mentioning: conclusions should be based on multiple experiments, including observation of duplicates from a single experiment, labeling of sections from different blocks of a single fixation, and labeling of sections from blocks of cells fixed on another day and/or with a different fixation procedure.

C. Experimental Procedures

1. Setup of Incubation Chamber

1. It is very helpful to set up the incubation chamber with labels at the start of the experiment. The chamber consists of a box with a tightfitting lid, of sufficient size to contain all the grids and all the solution droplets. A padding of paper towels or sponge-cloths (very thin sponges ~6 × 6 in., from the grocery store) is placed into the bottom of the dish and thoroughly wetted to maintain humidity throughout the labeling steps. The surface should be fairly flat. Paper towels are folded, pressed onto the sides of the dish, and moistened.

2. Onto the wet pad, a length of parafilm is positioned. Small colored adhesive dots ("sticky dots") labeled with the identity of the grid (i.e., strain, fixation variation, etc.) are placed on the left side, at ~1-in. intervals down the length of the parafilm strip. Across the top of the parafilm, four sticky dots are also positioned. The first and third represent the position where droplets of blocker will be placed. The second is the position of the primary antibody (1°) and the fourth is the position of the gold-conjugated secondary antiserum (2°). It is good practice to do duplicates of each incubation, so that a backup is available if technical problems occur.

2. Reagents

Throughout the immunolabeling protocol, PBST (140 m*M* NaCl, 3 m*M* KCl, 8 m*M* Na_2HPO_4, 1.5 m*M* KH_2PO_4, and 0.05% TWEEN-20) is used for all washes and as the vehicle for the blocker. Blocker is PBST containing 2% ovalbumin. Glass-distilled water of high purity (as for EM) is used, and solutions are filtered through a 0.22-μm filter (Acrodisc) before use. Blocker solution can be prepared, filtered, aliquoted, and stored frozen at −20°C. If the blocking solution has been frozen, it is refiltered before use. Both 1° and 2° are diluted into blocker. The dilution factor for the 1° must be empirically determined, using a dilution series. The gold-conjugated 2° should be adjusted to A_{525} = 0.3 for 15-nm particles and to A_{525} = 0.13 for 5- to 10-nm particles.

3. Blocking

Twenty-microliter droplets of 2% ovalbumin in PBST are positioned in the appropriate position on the parafilm sheet. The appropriate grid is submerged into the solution and allowed to incubate for 15 minutes at room temperature. Submerging is preferable to floating, because it will allow labeling of exposed antigen on both sides of the section.

4. Incubation in Primary Antiserum

1. After blocking, the grid is removed from the blocker and touched to a Kimwipe to removed excess fluid. For this and subsequent blottings, the forceps tip and grid should be held sideways on the tissue surface, so that fluid between the froceps tips is also removed. This step should be done rapidly, not allowing the sections to dry.
2. The grid is then submerged in a 20-μl droplet of diluted primary antiserum. Length of incubation is probably a matter of convenience. The results presented here were from 2-hour incubations at room temperature. Overnight incubations at room temperature or 4°C have also been successful and may allow a lower concentration of primary to be used.

5. Washes

1. Washes are performed in the wells of porcelain or glass spot plates. The wells are marked and filled with PBST, and the spot plate is placed on an orbital shaker.
2. After removal and blotting of excess fluid from the grid as just described, the grid is submerged in the appropriate well. The shaker is adjusted so that the solution is moving as rapidly as possible without spilling out of the well.
3. After 5 minutes, the grid is removed, blotted, and transferred to the next well. A total of three (5-minute) washes are performed, and the grid is blotted and transferred to a second droplet of blocker.

6. Incubation in Secondary (Gold-Conjugated) Antiserum

1. After a 15-minute incubation in blocker, the grid is blotted and transferred to the diluted secondary antiserum. The grid is incubated for 1 hour at room temperature and then washed as before (Section III,B,5).

2. After the final PBST wash, the grid is washed in distilled water by dipping 10 times with rapid up-and-down motion in a 5-ml beaker of water. This wash removes salts that would crystallize on the section, obscuring the view in the electron microscope. The grid is blotted on a Kimwipe, transferred to a labeled silicon mat, and allowed to air-dry.

D. Observation

1. General Considerations

The immunolabeled grids should be observed prior to exposure of the section to the typical heavy-metal stains necessary for proper resolution of ultrastructure. This order of events allows the overall disposition of the gold label to be readily observed, because the gold particles might be masked if the cell is darkly stained (compare Figs. 5 and 6).

2. Experimental Procedures

Following immunolabeling, standard staining methods are used to increase contrast of the sections.

1. The grids are immersed in a droplet of 1% uranyl acetate (filtered through a 0.22-μm filter before use) and incubated for 5–15 minutes at room temperature or at 50°–60°C.
2. The excess fluid is removed by blotting with a Kimwipe and the grids are washed by dipping (10 rapid up-and-down movements) in two changes of distilled water.
3. The grid is rapidly blotted and immersed in Reynolds' lead citrate (Reynolds, 1963) for 30 seconds–5 minutes. This step is carried out in small plastic Petri dishes, on islands of dental wax sitting on a bed of NaOH prill.
4. The grids are washed as just described, omitting the blotting step.
5. Immersion of the grid in the staining solution is very helpful for eliminating problems of lead precipitation that occur frequently. Use of freshly made lead citrate is also helpful. Should lead precipitates form, the sections can sometimes be rescued by dipping in 0.1 *M* NaOH followed by thorough water rinses. The grid can then be restained. On occasion, it is difficult to achieve adequate contrast of Spurr-embedded material using Reynolds' lead citrate. In this case, Millonig's lead tartrate (see Hayat, 1981a) works very well.

3. Presentation and Interpretation of Data

For the studies presented here, no quantitative analysis was required. Our object was merely to determine the ultrastructural localization of HMG–CoA reductase. However, all microscopic studies, especially such qualitative ones, require care to ensure unbiased observations. One way to accomplish this is to take a predetermined number of pictures at relatively low magnifications, such that the labeling patterns cannot be detected by the observer but will be evident when the negative is enlarged. Cells are selected for photography based on a gross morphological characteristic (so that the level of the section is approximately the same from cell to cell). For example, the condition might be that the section passes through the nucleus. The section is scanned in a systematic way, and each cell that meets the section-level requirement is photographed. After printing, the pattern of gold particles (or other feature) is analyzed. In the case of a specific structure, such as karmellae, specific labeling is readily apparent. Chris Kaiser (UCB) has used this method to analyze epistasis relationships among *sec* mutations. When the labeling pattern is more complex, a statistical analysis such as that used by Clark and Abelson (1987) may be useful.

Absolute quantitation (i.e., number of antigen molecules in an given cell or organelle) by immunocytochemistry is rarely possible. Relative quantitation, however, can be readily performed. For example, the average number of gold grains in a strain that overproduces a particular antigen can be compared to that in a wild-type strain to determine the level of overproduction.

The data obtained from immunocytochemistry, or from any microscopic technique, are inherently visual. In spite of a picture being worth a thousand words, visual data representing conditions within individual cells are subject to higher degrees of skepticism than are those that represent population averages. Space available in journals is generally insufficient to provide micrographs of a substantial number of examples together with control experiments. Including such quasi-quantitative methods as outlined earlier can be helpful, but the scientific community is left to rely on the investigator's integrity and observational accuity.

IV. Results

The electron micrographs presented here compare several fixation protocols and resins, and the effect of these variables on immunoreactivity of HMG–CoA reductase with affinity-purified antiserum.

A. Ultrastructural Studies

1. Cells Grown in Minimal Medium

The cell in Fig. 1 was processed directly after growth in a minimal medium. When this cell is compared to those allowed to grow in a rich medium for 2 hours before fixation (all other micrographs in this communication), it is clear that the type of medium in which cells are cultured can exert a pronounced effect on ultrastructure. Type of medium also affects cell size: cells grown in rich medium are larger than those in minimal medium. The investigator may wish to consider growth conditions when choosing fixation conditions.

2. Effect of Various Fixations

Figure 2 compares the ultrastructure of Spurr-embedded yeast (JRY1239) fixed in 1% formaldehyde–0.1% glutaraldehyde (1%/0.1%) or 1% formaldehyde–1% glutaraldehyde (1%/1%) alone, and those prefixed in 1% formaldehyde–1% glutaraldehyde and postfixed in osmium–

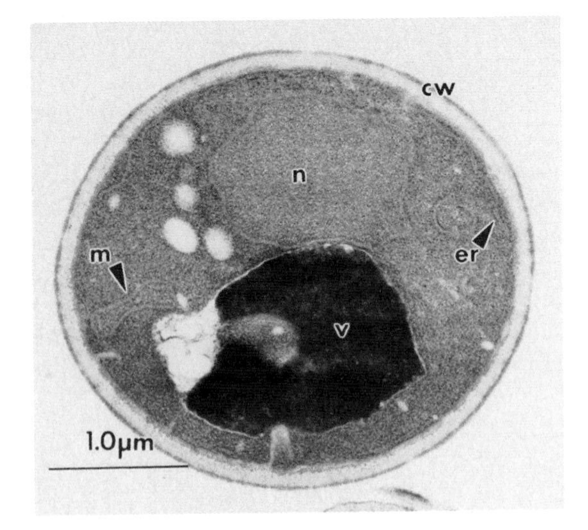

Fig. 1. Ultrastructure of yeast fixed directly after growth in minimal medium. Note the dense cytoplasm, very large vacuole, and little contrast between nucleoplasm and cytoplasm. Compare this micrograph with Figs. 2–8, which show cells following a 2-hour shift to rich medium. Growth was in minimal medium; cells were treated with lyticase and fixed with OsFeCN. They were then embedded with Spurr's resin. cw, cell wall; er, endoplasmic reticulum; m, mitochondria; n, nucleus; v, vacuole.

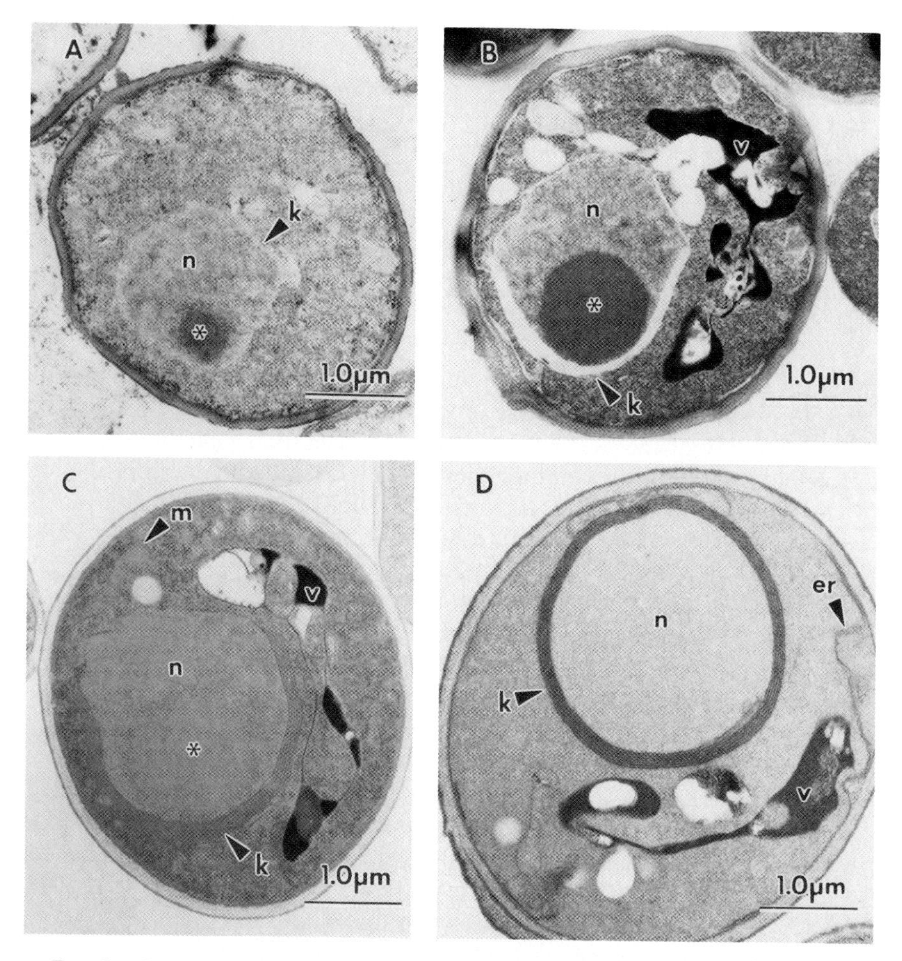

FIG. 2. Comparison of ultrastructure following various fixation protocols. All cells shown were embedded in Spurr's resin and differ only in the manner in which they were fixed. (A) Fixation in only 1% formaldehyde–0.1% glutaraldehyde does not adequately preserve ultrastructure. The cell shown here represents one of the best preserved cells from this fixation protocol. (B) Fixation in 1% formaldehyde–1% glutaraldehyde preserves general ultrastructure very well. The membranes are not retained, however, as illustrated by the extracted karmellar membranes around the nucleus. (C) Excellent ultrastructural preservation is observed in cells after fixation in 1% formaldehyde–1% glutaraldehyde followed by postfixation in OsFeCN and en bloc staining with uranyl acetate. (D) Although extraction of ribosomes and nucleoplasm is pronounced, fixation in 1% formaldehyde–1% glutaraldehyde followed by postfixation with KMnO provides a good view of the general organization of cytoplasm, especially of cytoplasmic membranes. The asterisk marks the location of a zone of unusual staining density present in the nucleoplasm of karmellae-containing cells. This structure, though of unknown composition, serves as a convenient marker to illustrate fixation effects on ultrastructure in this and subsequent figures. Note the variations in appearance, including its total absence the in KMnO-fixed cell (D). For normal nucleolar structure, see Fig. 3A. er, Endoplasmic reticulum; k, karmellae; m, mitochondria; n, nucleus; v, vacuole.

potassium (OsFeCN) ferricyanide or potassium permanganate (KMNO). Karmellae-containing nuclei have an unusual structure of unknown identity within the nucleoplasm. This structure is positioned in a specific location within the nucleus, in relation to the karmellar membranes that encircle the nucleoplasm. It is marked by an asterisk in the electron micrographs presented here. The varied appearance of this structure after different fixations illustrates the profound effect the fixative and/or resin can have on ultrastructure.

a. 1% Formaldehyde–0.1% Glutaraldehyde Alone. This fixation did not sufficiently preserve ultrastructure (Fig. 2A). In fact, most cells were unable to survive the remainder of the protocol intact; the cell shown in Fig. 2A is one of the better preserved. The cells were heavily extracted, probably during the dehydration steps. They were also quite fragmented. Cells fixed in this manner, surprisingly, required cell wall removal in order to achieve adequate infiltration. Infiltration of intact yeast cells after 1%/0.1% fixation was so poor that it was impossible to obtain sections at all.

b. 1% Formaldehyde–1% Glutaraldehyde Alone. Ultrastructure was well preserved in the 1%/1% samples (Fig. 2B). In the absence of postfixation, membranes are not preserved and therefore appear as clear areas. Mitochondria are not optimally preserved. Infiltration is adequate, although small holes are often present in the vacuoles of intact cells.

c. 1% Formaldehyde–1% Glutaraldehyde Prefixation with OsFeCN Postfixation. Postfixation using the OsFeCN procedure produced very well-defined membranes as Fig. 2C illustrates. Other subcellular structure is generally well preserved. Infiltration results were analogous to that of the 1%/1% sample—generally adequate, but with some holes.

d. 1% Formaldehyde/1% Glutaraldehyde Prefixation with Permanganate Postfixation. Permanganate postfixation produced perfectly infiltrated samples of intact yeast cells. These cells generally demonstrate cellular structure quite well (Fig. 2D). The nucleoplasm and ribosomes are quite extracted, however, and thus may not be the method of choice for certain studies.

e. Additional Results. A combined glutaraldehyde–osmium prefixation followed by postfixation in osmium alone was attempted with poor results. In addition, cells were fixed by the standard OsFeCN procedure and subsequently treated with permanganate. It was hoped that this protocol would combine the preservative qualities of OsFeCN treatment with the unknown properties of permanganate treatment that allow ease of infiltration. The ultrastructure of such cells proved disappointing, however, appearing almost as if not subjected to any postfixation at all.

3. Comparison of Resin Effects on Ultrastructure

The type of resin used to infiltrate yeast produces a surprising degree of alteration on the appearance of the cell's ultrastructure. Figure 3 compares cells fixed by the 1%/1% protocol and embedded in Spurr's (Fig. 3A), Pelco Ultra-low Viscosity (Fig. 3B), and LR White resins (Fig. 3C). For ease of penetration, the Pelco Ultra-low is recommended. Sections of cells embedded in this resin are very easy to stain and have high contrast, so they are easy to photograph. Spurr's resin produced similar results to those of the Pelco resin, but stained less readily. LR White resin was plagued with hole problems, and the resin melted significantly when in the electron beam. Once stabilized, however, it was possible to photograph the cells, even when only partially supported by the resin. In addition, cells embedded in LR White were generally smaller than those embedded in Spurr's resin. In turn, Spurr-embedded cells were smaller than those in Pelco. The size variations probably represent differing degrees of shrinkage associated with these resins.

4. Fixation–Infiltration Artifacts

As mentioned in Section II,B, speed of equilibration of the fixative with the cells is important for optimal fixation. Figure 4A shows an example of a cell that has undergone severe autolysis induced by poor equilibration. There isn't a lot there to point out, which *is* the point.

Inadequate infiltration can manifest itself in a variety of ways. Sometimes, holes form between the cell wall and the plasma membrane. In other instances, breaches can develop in association with internal membranes. This latter case is illustrated in Fig. 4B, which shows a large fissure between vacuolar contents and the vacuolar membrane (arrowheads).

B. Immunocytochemistry

1. Observation of Sections Prior to Staining

The cells in Fig. 5 were photographed directly following immunolabeling. Every gold particle is easily observed against a background of very low contrast. The cytoplasmic staining, so obvious in these micrographs, is nearly undetectable following lead citrate staining (compare to Figs. 6–8). Figure 5 also illustrates the labeling pattern in a dilution series of the 1° antibody (Fig. 5A, 1 : 20; Fig. 5B, 1 : 40; Fig. 5C, 1 : 100). The

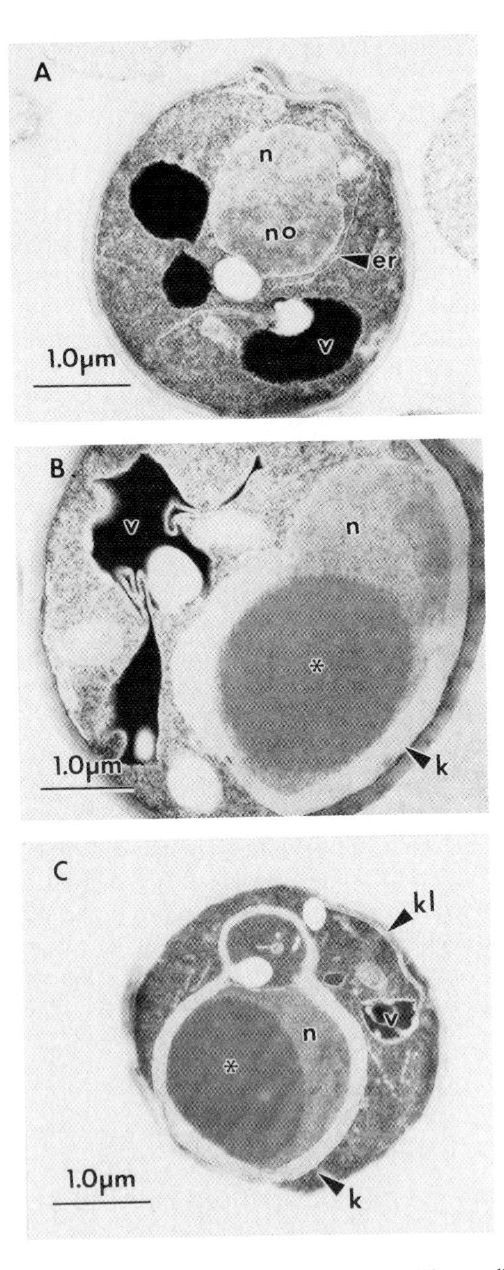

FIG. 3. Comparison of the effect of resin on ultrastructure. These cells were fixed in 1% formaldehyde–1% glutaraldehyde, dehydrated, and then divided into three aliquots. (A) Embedded in Spurr's resin. This resin infiltrates well, producing blocks that are easy to section. A high degree of ultrastructural detail is retained. Shrinkage is generally minimal. (B) Embedded in Pelco Ultra-low Viscosity formula. Intact yeast cells were infiltrated easily with this resin, and ultrastructure was well preserved. Extraction of lipids in the absence of

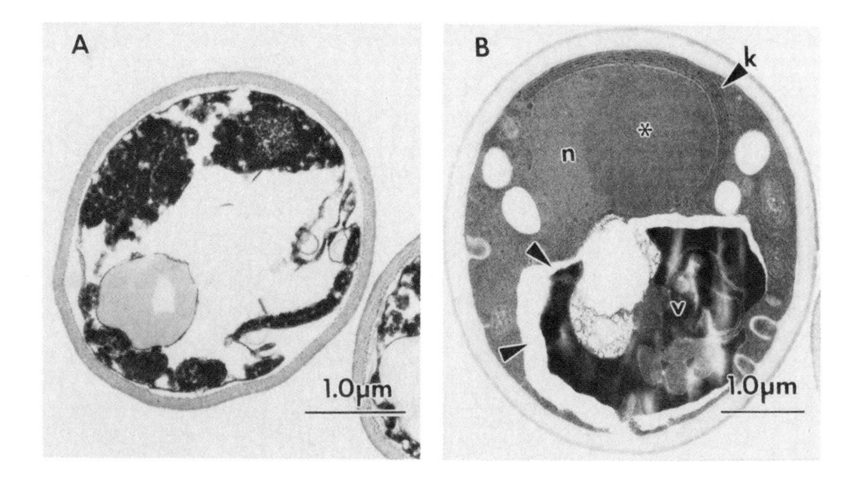

FIG. 4. Fixation and infiltration artifacts. (A) An extreme example of autolysis that occurs when initial mixing of fixative with cells is slow. (B) An example of poor infiltration, resulting in a rift between the vacuolar contents and the vacuolar membrane (arrowheads). Both were fixed by the OsFeCN procedure and embedded in Spurr's resin. Abbreviations as in Fig. 3.

decrease in gold particles in the cytoplasm with relatively little decrease in karmellae labeling may indicate that the cytoplasmic labeling is nonspecific.

2. Effects of Postfixation on Immunolabeling

Figure 6 compares the effect of postfixation with OsFeCN on immunolocalization of HMG–CoA reductase in karmellae. A higher labeling density is observed in cells fixed only in 1% formaldehyde–1% glutaraldehyde (Fig. 6A), although fair labeling density was also present in OsFeCN-fixed karmellae (Fig. 6B). It is clear that for this particular antigen, treatment with osmium is not particularly detrimental.

---

postfixation was more pronounced with this resin than with LR White, but was similar to that observed with Spurr's resin. This resin produced the least cell shrinkage of those tested. (C) Embedded in LR White resin. This resin infiltrates cells poorly and undergoes considerable shrinkage, producing fenestrated sections that are very difficult to examine. Extraction of lipids in the absence of postfixation appears minimal with this resin, however. n, Nucleus; no, nucleolus; k, karmellae; kl, karmellae loop (a portion of karmellae extending away from the nucleus into the cytoplasm); v, vacuole.

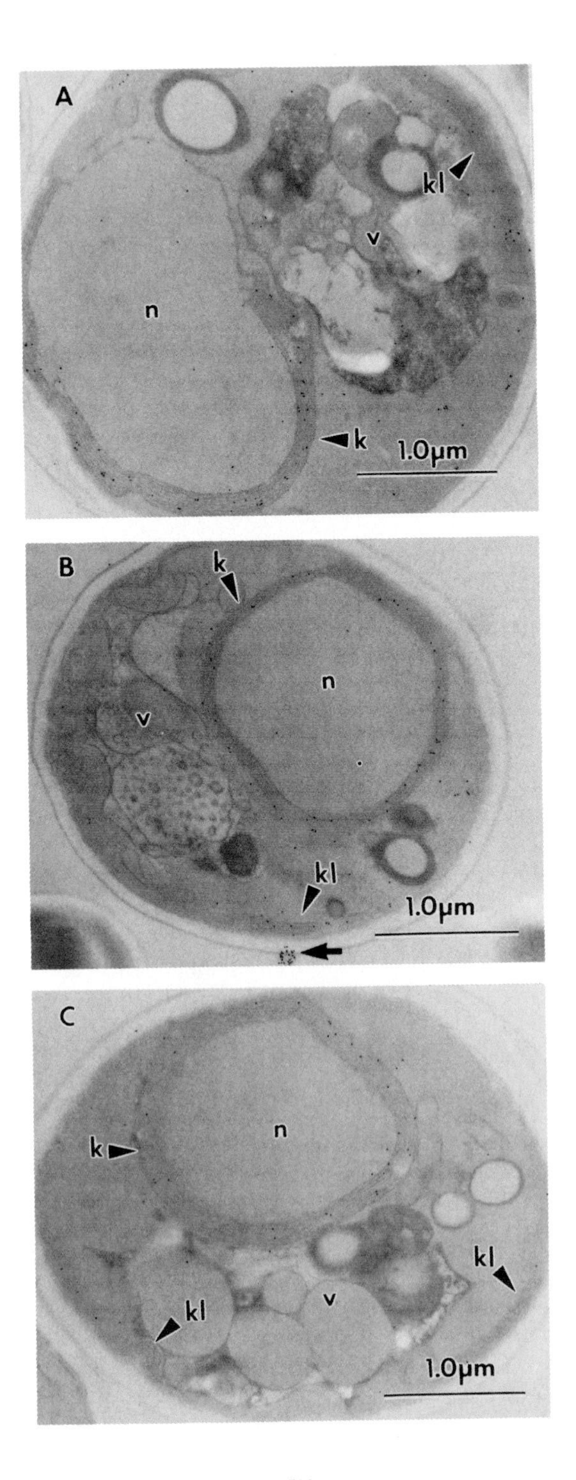
A
kl
v
n
k
1.0μm
B
k
n
v
kl
1.0μm
C
k
n
kl
kl
v
1.0μm

Several techniques to "etch" sections and restore antigenicity following osmium fixation have been reported (Bendayan and Zollinger, 1983). We tried both hydrogen peroxide (10% for 15 minutes) and sodium metaperiodate treatment (saturated aqueous solution, 15- to 60-minute incubations). Neither treatment resulted in a significant increase in labeling density. In addition, the hydrogen peroxide treatment was deleterious to ultrastructural preservation.

Use of sodium borohydride has been recommended to block free aldehyde following glutaraldehyde fixation and, thus, increase retention of antigenicity (Weber *et al.*, 1978). This treatment did not adversely affect fixation quality, nor did it increase labeling of karmellae with antiserum against HMG–CoA reductase.

Essentially no labeling was observed in cells fixed with 1% formaldehyde–0.1% glutaraldehyde, and ultrastructural preservation was so poor that identification of karmellae was difficult.

3. Effect of Resin on Immunolabeling

The effect of resin on immunolabeling is illustrated in Fig. 7, which shows cells fixed in 1% formaldehyde–1% glutaraldehyde and embedded in different resins. The highest density of labeling, with little or no nonspecific background, was observed in cells embedded in LR White resin and this resin is recommended when the maximal labeling is essential (Fig. 7B). Though demonstrating a reasonable level of specific staining, Pelco Ultra-low-embedded cells also had high, nonspecific background labeling of the cytoplasm and, especially, the nucleus (Fig. 7A). These background problems preclude its use for immunocytochemistry. Cells embedded in Spurr's resin (Fig. 7C) were labeled with high specificity, but the density was less than those embedded in LR White.

Fig. 5. Observation of immunogold localization of HMG–CoA reductase in karmellae. This series of micrographs shows immunolabeling using a dilution series of the primary antibody. A decrease in cytoplasmic gold particles with little decrease in particle density in karmellae is observed with increasing dilution. (A) 1 : 20 dilution; (B) 1 : 40 dilution; (C) 1 : 100 dilution. These micrographs also illustrate the value of observing immunolabeled sections before staining with lead citrate. Each gold particle is clearly resolved (compare with the difficulty of resolving cytoplasmic gold particles in Figs. 6–8). These cells were fixed by the OsFeCN protocol, en bloc stained with uranyl acetate, and embedded in Spurr's resin. Immunolabeling: 1 : 20 primary antiserum and 10-nm GARG. k, Karmellae; kl, karmellae loop; n, nucleus; v, vacuole.

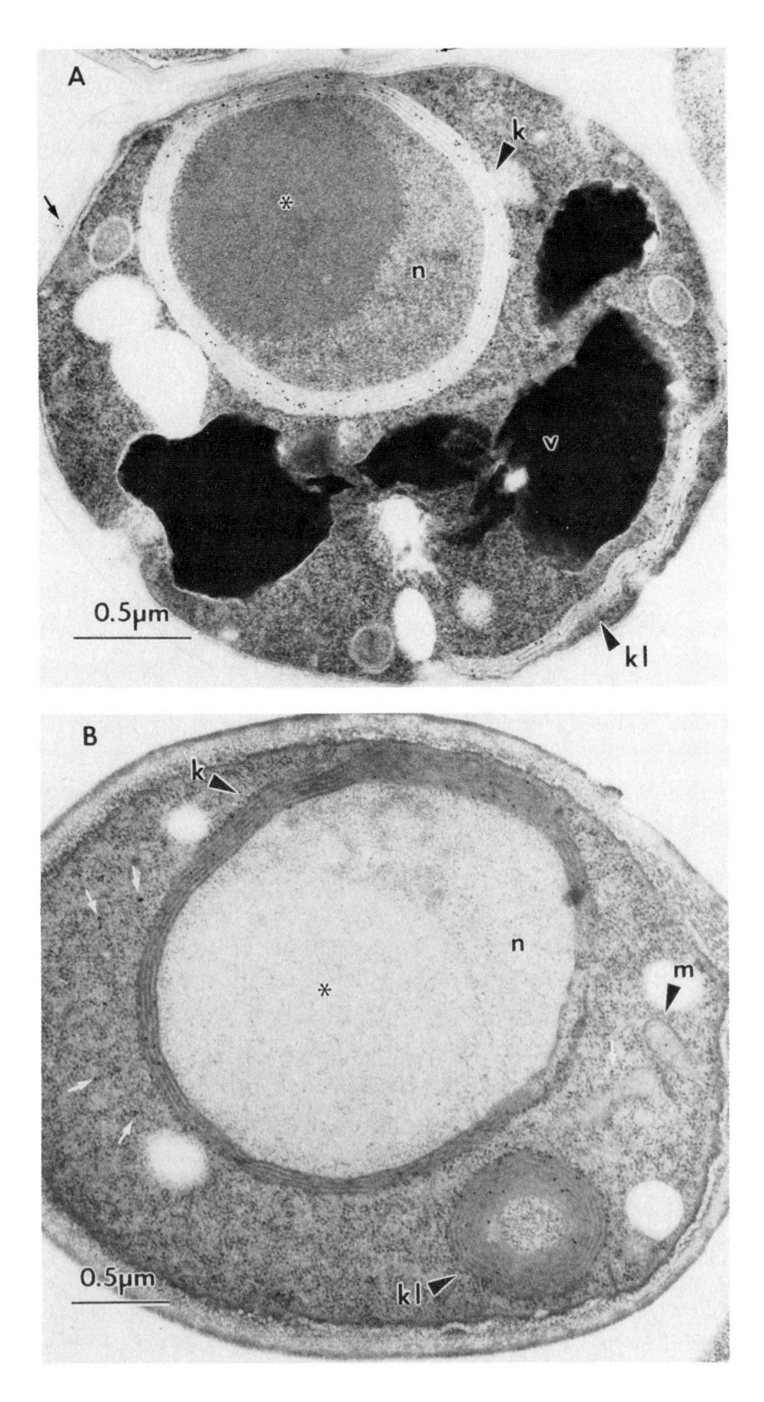
A
k
n
v
0.5μm
kl
B
k
n
m
0.5μm
kl

4. Some Artifacts Encountered in Immunocytochemistry

A troubling, artifactual labeling pattern consisting of patches of dense labeling throughout the cytoplasm was observed on occasion. No labeling was present over mitochondria, nucleus, or vacuole, producing a quite specific appearance of label (Fig. 8A and B). This pattern was traced to a purchased gold reagent (Janssen GARG, 15 nm) nearing the end of its recommended shelf-life and probably reflects aggregation of the particles. Why it should produce such a pattern is not understood, but that it *can* underscores the importance of appropriate controls.

The opposite problem, complete absence of labeling, can occur by blocking with solutions containing high protein concentration (Fig. 8C). This cell is from the same block as those in Figs. 6A and 7C, but was blocked with 10% calf serum and 2.5% BSA in PBST. Complete absence of labeling (that is, *no* gold particles whatsoever), is observed, in spite of overnight incubation in the primary serum.

5. Additional Comments

The same variations employed to reduce background in immunoblotting or immunoprecipitation can be utilized for immunocytochemistry. For example, including 2 *M* urea in PBST in the initial wash after incubation in the primary antibody can reduce nonspecific labeling. Daniela Brada (personal communication) recommends increasing the NaCl concentration to 0.5 *M* during primary incubation to achieve a similar result.

FIG. 6. Comparison of the effects of osmium treatment on immunolocalization of HMG–CoA reductase in karmellae. Both cells were prefixed in 1% formaldehyde–1% glutaraldehyde and embedded in Spurr's resin. The cell in (B) was postfixed in OsFeCN prior to dehydration and embedding, whereas the cell in (A) was processed immediately after initial fixation. Osmium treatment does not severely affect the immunoreactivity of HMG–CoA reductase, although labeling density is somewhat decreased. The tiny arrows point out gold particles in the cytoplasm. Note how difficult it is to distinguish the particles from the surrounding ribosomes, which are of similar dimensions. Immunolabeling: 1 : 20 primary antiserum and 10-nm GARG. k, Karmellae; kl, karmellae loop; m, mitochondria; n, nucleus.

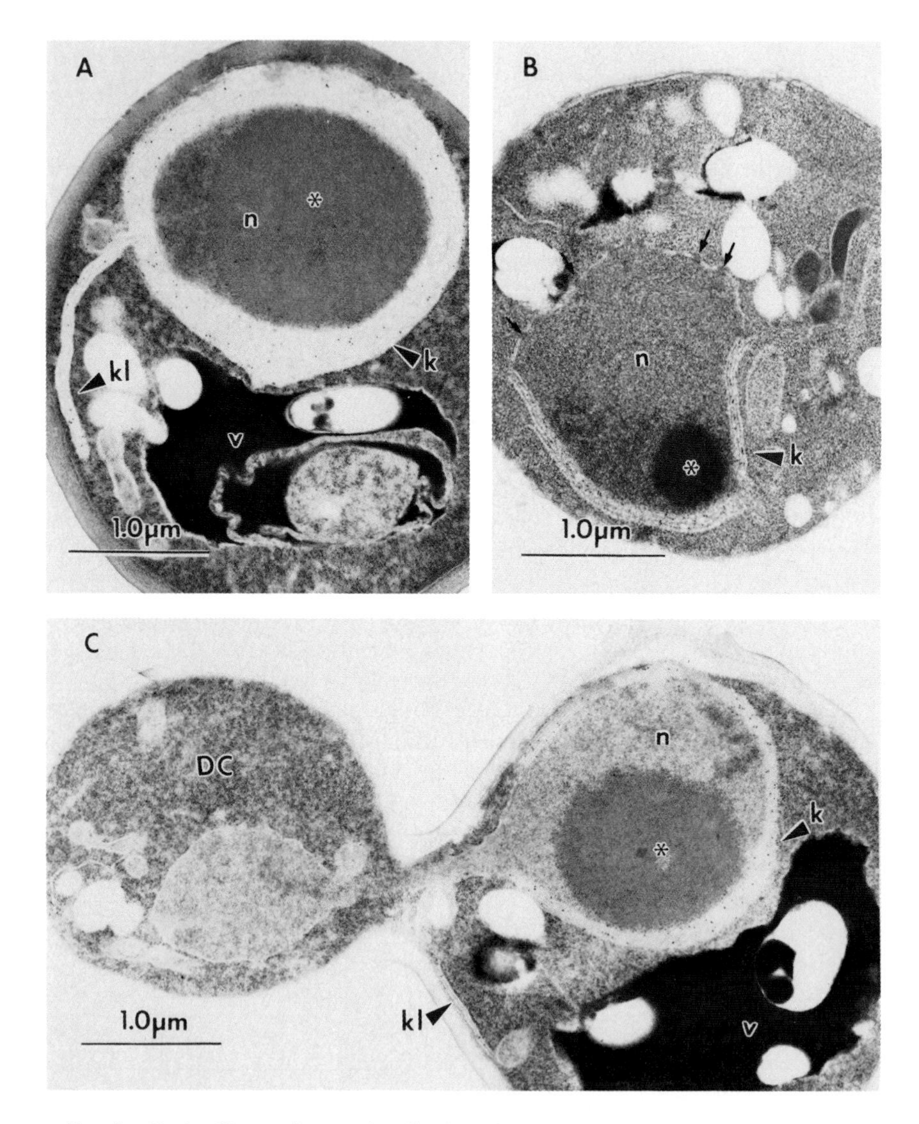

FIG. 7. Resin effect on immunolocalization of HMG–CoA reductase in karmellae. The cells in this figure were fixed in 1% formaldehyde–1% glutaraldehyde, dehydrated, and then embedded in Pelco Ultra-low Viscosity (A), LR White (B), or Spurr's (C) resins. Highest labeling density was consistently observed in cells embedded in LR White. Extraction of cytoplasmic components appears minimal in this resin (note the nuclear pores marked by the tiny arrows in B). Although specific labeling of karmellae was observed in cells embedded in Pelco Ultra-low resin, the density of label is markedly lower, and nonspecific background staining is increased (note the gold particles over the nucleoplasm). Specificity of label was very good in Spurr-embedded cells (C), although the label density was slightly decreased in comparison with LR White resin. Labeling: 1 : 20 primary antibody and 10-nm GARG. k, Karmellae; kl, karmellae loop; n, nucleus; v, vacuole.

V. Summary

The results and anecdotes presented here are intended only as a general guide to other would-be immunocytochemists, because other proteins will undoubtedly respond at least somewhat differently than does HMG–CoA reductase. Nevertheless, based on these experiences, we offer the following suggestions:

1. Antiserum of high specificity should be raised and affinity-purified. Using this antiserum, immunofluorescence microscopy should be attempted before resorting to electron microscopic localization. In the absence of immunolocalization at the light-microscope level, it may be a waste of time to pursue the problem to higher levels of resolution.
2. Cells should be prefixed in 1% formaldehyde–1% glutaraldehyde. Direct fixation of the growing culture and use of phosphate buffer are recommended. The prefixed sample can then be divided into two or three aliquots. One aliquot should receive no postfixation (for optimal immunoreactivity), while the others can be postfixed in osmium–potassium ferricyanide (for possible immunolocalization) or permanganate (for ultrastructural analysis). Because of its ease of use, Spurr's resin should be tried initially. If immunocytochemistry is successful, no further preparations are necessary. If unsuccessful, LR White resin is recommended, but the sample must be treated to remove the cell wall.

Electron microscopy and immunocytochemistry offer views into the molecular arrangement of individual cells, a view not easily obtained by other means. It is satisfying and often enlightening to be able to see the extremes as well as the average. In studies of the organization of karmellae, for example, ultrastructural analysis easily revealed the asymmetric segregation pattern, while immunoblots and cell fractionation could not even demonstrate the existence of this membrane organization. The richness of the information available to those who can avert reductionist tendencies, even for a short time, is remarkable.

VI. Recent Developments: Use of LR White Resin on Whole Cells

We have recently developed a method using LR White resin that consistantly produces excellent infiltration of whole cells. Consequently, we use this method for routine electron microscopy as well as for

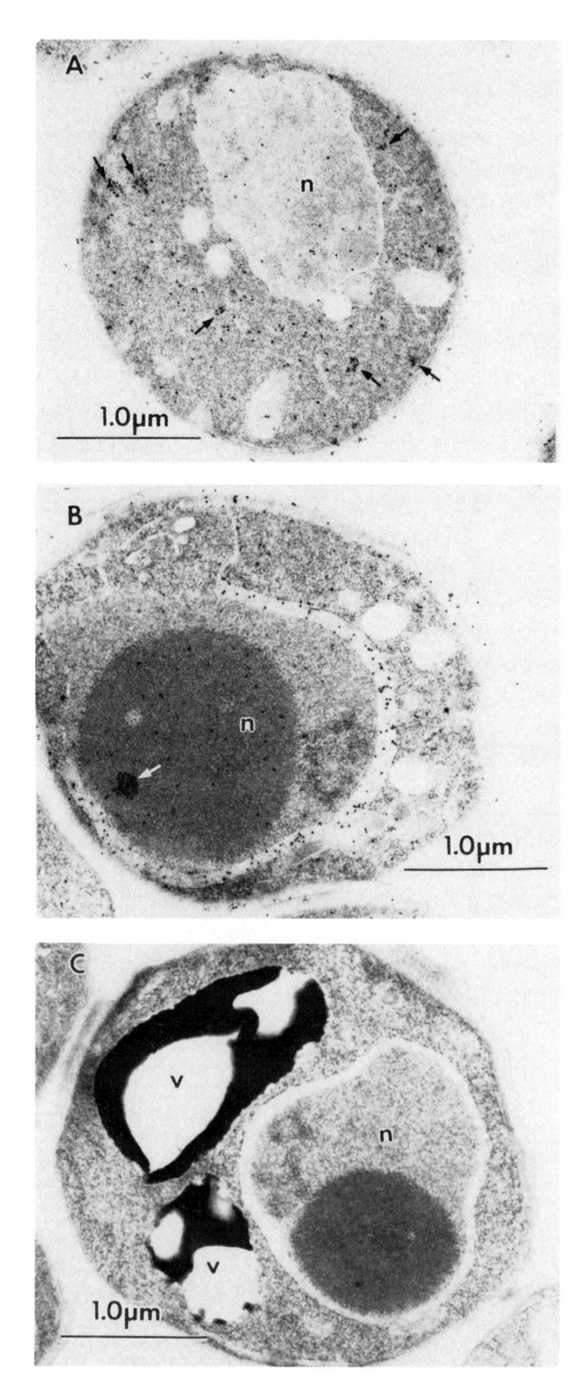
A
n
1.0μm
B
n
1.0μm
C
v
n
v
1.0μm

immunolocalization. The critical factor involves treatment of fixed cells with sodium periodate, a method pioneered by van Tuinen and Riezman (1987). To date, we have successfully localized three different antigens, using five different polyclonal antisera, in cells processed in this manner. Background staining is essentially nonexistent and labeling density quite improved over Spurr-embedded material. This procedure is outlined below and an electron micrograph showing an example of this technique is shown in Fig. 9.

Periodate/LR White Method

1. Cells are fixed as described in section II,B,5,c. Following fixation, the cells are washed three times in phosphate buffer and then pelleted an additional time.
2. The washed cell pellet is resuspended in 5 ml of 1% sodium metaperiodate (aqueous solution, freshly mixed). After a 15 min incubation at room temperature, the cells are pelleted, washed once in phosphate buffer, and repelleted.
3. The cell pellet is resuspended in 50 mM ammonium chloride to quench free aldehyde groups. The suspension is incubated for 15 min at room temperature after which the cells are then pelleted and washed once in distilled water.
4. Dehydration, infiltration, and embedding are completed as described in section II,E,2. It is essential to allow the resin to warm to room temperature before use and to use gelatin capsules that have been dried in an oven (60°C) overnight.

FIG. 8. Immunocytochemical artifacts. A gold-labeled secondary antibody can produce primary antibody-dependent patterns with obvious, but spurious, specificity. (A, B) Cells were immunolabeled in exactly the same manner as those in Figs. 5–7, using a 1 : 20 dilution of primary antiserum and 2 hours incubation. Although karmellae are specifically labeled (B), the overall pattern of staining seen here is very different from that observed in the other cells. Note the appearance of patches of gold particles (tiny arrows) in the cytoplasm. This pattern is insidious because it looks so specific: the labeling generally does not occur over the nucleus (although see B), the mitochondria, or the vacuole, but rather in patches throughout the cytoplasm. The pattern illustrated in (A) and (B) was observed only with a gold-conjugated antibody (15-nm GARG) nearing the end of its recommended shelf-life. Thus, the quality of the secondary is of utmost importance and must be demonstrated by careful control experiments. At the other extreme (C), using high concentrations of protein to block nonspecific sites can completely eliminate all labeling reactivity. (C) A cell that was incubated in 10% calf serum, 2.5% BSA in PBST for 10 minutes prior to incubation overnight in primary antiserum (1 : 20 dilution in 2% ovalbumin in PBST). Overzealous blocking should be avoided.

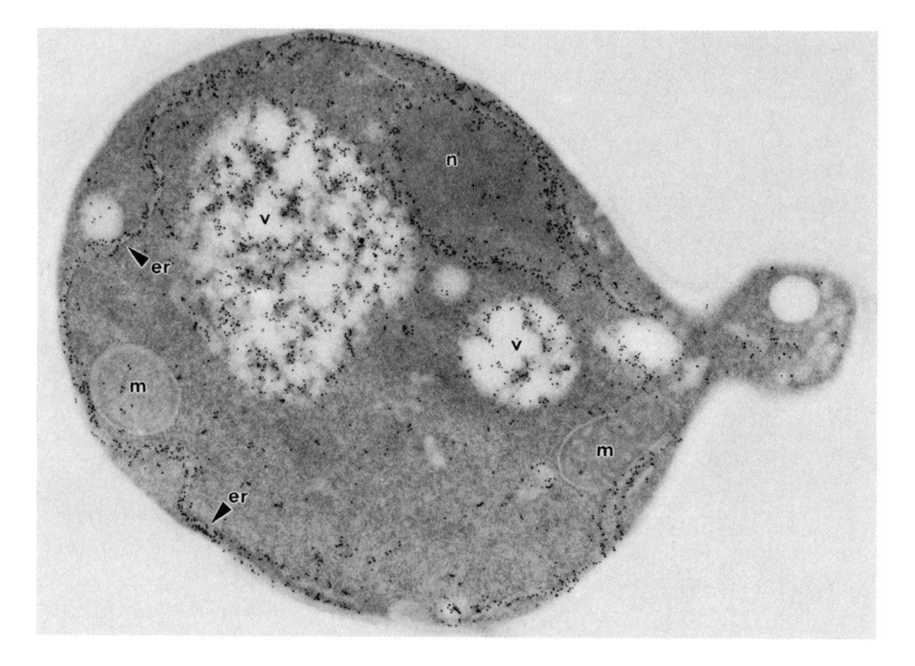

FIG. 9. Immunolabeling using the Periodate/LR White procedure. Polyclonal antisera against carboxypeptiase Y and 10 nm colloidal gold-conjugated goat anti-rabbit secondary sera were used to demonstrate the location of a fusion protein consisting of the first 6 membrane-spanning domains of the HMG–CoA reductase isozyme encoded by *HMG1* and a carboxyl terminus containing a truncated version of pro-carboxypeptidase Y that lacks vacuolar targetting information. This protein is present in the nuclear envelope, the ER, and inside the vacuole. Note the quality of ultrastructural preservation and the excellent immunocytochemical labeling.

5. A tempblock that has holes drilled specifically to accommodate gelatin capsules should be used for polymerization, since standard tempblocks have cavities that are too deep for this purpose and produce blocks that are not optimally hardened. Polymerization is allowed to progress for two days at 45°C, resulting in blocks that are uniformly polymerized and easy to section.

6. The water level in the trough should be kept quite low during sectioning to minimize the possibility of wetting the blockface. If the blockface is inadvertantly wetted, it is useful to allow it to dry for 30 min before attempting to continue sectioning. Procedures for mounting sections and immunolabeling are as previously described (Section II,E).

Acknowledgments

Daniela Brada introduced us to the prospect of immunocytochemistry in yeast, and we gratefully acknowledge her contribution to this work. Bonnie Chojnacki and Alice Taylor provided training, advice, and support concerning EM techniques for which we are indebted. The authors also thank Lorraine Pillus for critical reading of the manuscript.

This work was supported by grants from the California Biotechnology and Research Education Program and from the American Cancer Society to R. W. and by grants from the National Institutes of Health to J. R.

References

Aldrich, H. C., and Mollenhauer, H. H. (1986). *In* "Ultrastructure Techniques of Microorganisms" (H. C. Aldrich and W. J. Todd, eds.), pp. 101–132. Plenum, New York.

Bendayan, M., and Zollinger, M. (1983). *J. Histochem. Cytochem.* **31,** 101–109.

Brandtzaeg, P. (1983). *In* "Techniques for Immunocytochemistry" (G. R. Bullock and P. Petrusz, eds.), Vol. 2, pp. 1–75. Academic Press, New York.

Causton, B. E. (1984). *In* "Immunolabeling for Electron Microscopy" (J. M. Polak and I. M. Varndell, eds.), pp. 29–36. Elsevier/North Holland, Amsterdam.

Clark, M. W., and Abelson, J. (1987). *J. Cell Biol.* **105,** 1515–1526.

De Mey, J. (1983a). *In* "Immunocytochemistry" (J. M. Polak and S. Van Noorden, eds.), pp. 43–52. Wright, Bristol, England.

De Mey, J. (1983b). *In* "Immunocytochemistry" (J. M. Polak and S. Van Noorden, eds.), pp. 82–112. Wright, Bristol, England.

Fahrenbach, W. H. (1984). *J. Electron Microsc. Tech.* **1,** 387–398.

Glauert, A. M. (1975). *In* "Practical Methods in Electron Microscopy" (A. M. Glauert, ed.), pp. 1–207. Elsevier/North-Holland, Amsterdam.

Hayat, M. A. (1981a). "Principles and Techniques of Electron Microscopy: Biological Application," 2nd ed., Vol. 1. University Park Press, Baltimore, Maryland.

Hayat, M. A. (1981b). "Fixation for Electron Microscopy." Academic Press, New York.

Johnston, M., and Davis, R. W. (1984). *Mol. Cell. Biol.* **4,** 1440–1448.

Karnovsky, M. J. (1971). *In* "Proceedings of the 14th Annual Meeting of Cell Biologists," p.146. Rockefeller Univ. Press, New York.

Kellenberger, E., Ryter, A., and Sechaud, J. (1958). *J. Biophys. Biochem. Cytol.* **4,** 671–680.

Luft, J. H. (1973). *In* "Advanced Techniques in Biological Electron Microscopy" (J. K. Koehler, ed.), Springer–Verlag, New York.

McDonald, K. (1984). *J. Ultrastruct. Res.* **86,** 107–118.

Polak, J. M., and Van Noorden, S., eds. (1987). "An Introduction to Immunocytochemistry: Current Techniques and Problems." Oxford Univ. Press, London and New York.

Reynolds, E. S. (1963). *J. Cell Biol.* **17,** 208–212.

Ringo, D. L., Brennan, E. F., and Cota-Robles, E. H. (1982). *J. Ultrastruct. Res.* **80,** 280–287.

Robinson, D. G., Ehlers, U., Herken, R., Herman, B., Mayer, F., and Schurmann, F.-W. (1987). "Methods of Preparation for Electron Microscopy." Springer-Verlag, Berlin.

Romano, E. L., and Romano, M. (1984). *In* "Immunolabeling for Electron Microscopy" (J. M. Polak and I. M. Varndell, eds.), pp. 3–15. Elsevier/North-Holland, Amsterdam.

Roth, J. (1982). *In* "Techniques in Immunocytochemistry" (G. R. Bullock and P. Petrusz, eds.), Vol. 1, pp. 107–133. Academic Press, New York.

Roth, J. (1983). *In* "Techniques for Immunocytochemistry" (G. R. Bullock and P. Petrusz, eds.), Vol. 2, pp. 215–284. Academic Press, New York.

Roth, J. (1986). *J. Microsc. (Oxford)* **143,** 125–137.

Scott, J. H., and Schekman, R. (1980). *J. Bacteriol.* **142,** 414–423.

Smit, J., and Todd, W. J. (1986). *In* "Ultrastructure Techniques for Microorganisms" (H. C. Aldrich and W. J. Todd, eds.), pp. 469–516. Plenum, New York.

Smithwick, E. B. (1985). *J. Electron Microsc. Tech.* **2,** 193–200.

Spurr, A. R. (1969). *J. Ultrastruct. Res.* **26,** 31–43.

Stearns, T., and Botstein, D. (1988). *Genetics* **119,** 249–260.

Stevens, B. (1981). *In* "The Molecular Biology of the Yeast *Saccharomyces cerevisiae*: Life Cycle and Inheritance" (J. N. Strathern, E. W. Jones, and J. R. Broach, eds.), pp. 471–504. Cold Spring Harbor Lab., Cold Spring Harbor, New York.

Todd, W. J. (1986). *In* "Ultrastructure Techniques for Microorganisms" (H. C. Aldrich and W. J. Todd, eds.), pp. 87–100. Plenum, New York.

van Tuinen, E., and Riezman, H. (1987). *J. Histochem. Cytochem.* **35,** 327–333.

Weber, K., Rathke, P. C., and Osborn, M. (1978). *Proc. Natl. Acad. Sci. U.S.A.* **75,** 1820–1824.

Wright, R., Basson, M., D'Ari, L., and Rine, J. (1988). *J. Cell Biol.* **107,** 101–114.

Chapter 24

Postembedding Labeling on Lowicryl K4M Tissue Sections: Detection and Modification of Cellular Components

J. ROTH

Interdepartmental Electron Microscopy and Department of Cell Biology
Biocenter
University of Basel
CH-4056 Basel, Switzerland

I. Introduction

During the last two decades, major achievements in the localization of cellular components by light and electron microscopy became possible through the introduction of the colloidal gold marker system and the development of hydrophilic resins such as Lowicryl K4M for low-temperature embedding of biological matter. The use of particles of colloidal gold as an electron-dense marker by Faulk and Taylor (1971) can be considered a milestone in immunoelectron microscopy that has revolutionized the entire discipline. This marker has been proven most reliable and versatile: it can be easily prepared in sizes between 2 and 40 nm and larger, provides high contrast under the electron beam, permits excellent resolution, allows for multiple labeling procedures, is amenable to quantification, can be used to form complexes with almost every type of macromolecule, and last but not least, is applicable for both light and electron microscopy (for review, see Roth, 1983a). A major breakthrough for postembedding immunolabeling was the application of the protein A–gold (pAg) complex for the detection of intracellular antigens on ultrathin sections of resin-embedded tissue by Roth *et al.* (1978). The pAg technique allowed the localization of practically all classes of cellular proteins (for review, see Roth, 1983a, 1984, 1986; Bendayan, 1984a) and the detection of nucleic acids by *in situ* hybridization (Binder *et al.*, 1986). Though originally worked out with Epon-embedded tissue, this type of tissue processing soon severely hindered certain applications of the pAg technique because of inadequate preservation of cellular fine-structural details and drastic loss in reactivity of cellular components with antibodies and other reagents. To this end the development of the Lowicryl resins, in particular the hydrophilic Lowicryl K4M, for low-temperature embedding by Carlemalm *et al.* (1982) and their successful introduction for postembedding immunolabeling (Roth *et al.*, 1981a) can be considerd another major achievement. This low-temperature embedding technique has not only provided superior preservation of antigens, but has also resulted in improved preservation of fine-structural details of mildly aldehyde-fixed tissues and cells, and a drastically reduced background staining. In addition, it dramatically improved the detectability of glycoconjugates with lectins, monoclonal antibodies, and glycosyltransferases (for review, see Roth, 1987a). In the following, the detailed procedures for low-temperature embedding in Lowicryl K4M will be given, together with various protocols for the light and electron microscopic localization of cellular constituents as they have been used in our laboratory.

II. Some Physicochemical Characteristics of Lowicryl K4M

The commercially available Lowicryl resin family currently consists of two polar (hydrophilic) and two apolar (hydrophobic) formulations: (i) Lowicryl K4M for work to −40°C and Lowicryl K11M usable to −60°C, and (ii) Lowicryl HM 20 usable to −50°C and Lowicryl HM 23 to −80°C (Carlemalm *et al.*, 1982, 1985a; Acetarin *et al.*, 1986). These resins, exclusively produced by Chemische Werke Lowi GmbH (Waldkraiburg, FRG), can be purchased from all major suppliers of material for electron microscopy. The polar Lowicryl K4M is most often used in postembedding labeling studies.

Lowicryl K4M (and the other Lowicryl resins) is an acrylate–methacrylate mixture of low viscosity that consists of four components (for details, see Carlemalm *et al.*, 1982). Of critical importance for the many-fold applications of Lowicryl K4M is its constant behavior with respect to tissue infiltration, polymerization, and sectioning under a broad range of environmental conditions. Polymerization is usually achieved by long-wave (~360 nm) UV irradiation in conjunction with a photochemical initiator. Large amounts of pigments in tissues, especially the yellowish ones, are suspected to interfere with the polymerization by adsorbing UV photons, thereby resulting in incomplete polymerization and blocks of less than optimal quality (Acetarin and Carlemalm, 1982). However, as shown for the retinal pigment epithelium, intensely black-stained tissues pose no problems for polymerization (Rungger-Brändle *et al.*, 1987), which may also be the case with osmicated tissue. However, excess osmium tetroxide in the specimen will attack the unsaturated bonds in the resin. In case of difficulties, polymerization can be initiated chemically or by high temperature (see Appendix to Carlemalm *et al.*, 1982).

The UV irradiation-initiated polymerization is a free-radical addition reaction that forms a saturated-vinyl type of carbon–carbon backbone. The resulting highly crosslinked Lowicryl K4M is quite stable in the electron beam in contrast to classical methacrylate-based resins (Kellenberger *et al.*, 1956). Specimen shrinkage during polymerization is negligible. An important aspect of the free-radical polymerization is the very weak bonding of the resin with the biological material. Unsaturated carbon–carbon bonds exist rarely in biological matter for copolymerization with the resin. This lacking or weak cohesion between resin and tissue results during sectioning in a cleavage most often occurring at the interface of Lowicryl K4M–tissue (Carlemalm *et al.*, 1985b; Kellenberger *et al.*, 1986; Acetarin *et al.*, 1987). The resulting surface relief seems to be an important factor in postembedding labeling and section contrasting (Carlemalm *et al.*, 1985b; Kellenberger *et al.*, 1986, 1987).

Finally, one among other advantages provided by the hydrophilic properties of Lowicryl K4M is worth mentioning. During dehydration and resin infiltration the specimens may be kept in a partially hydrated state, because Lowicryl K4M may be polymerized with up to 5% (by weight) water in the block. That means in practical terms that the water has to be added to the resin–solvent mixtures and the pure resin. Otherwise the specimen will be effectively dehydrated by the hydrophilic resin.

III. Low-Temperature Embedding in Lowicryl K4M

The primary rationale for developing low-temperature embedding techniques is that low temperature during dehydration with organic solvents, infiltration with the resin, and resin polymerization are known to improve preservation of molecular and supramolecular structures in tissues and cells (for review, see Kellenberger *et al.*, 1985). Improved conformation at the molecular level provides superior preservation of cellular structures and reactivity of cellular constituents with antibodies, lectins, etc., both of which are parameters of critical importance for successful postembedding immunolabeling.

A. Fixation Protocols

The various cellular constituents exhibit differential sensitivity against chemical fixatives. A relatively simple and reliable test to estimate the effect of different fixatives and fixation times on the reactivity of a given antigen with antibodies is as follows:

1. Antigens are exposed to varying types and concentrations of fixatives (paraformaldehyde or glutaraldehyde) for varying times either in solution or after spot-blotting onto nitrocellulose.
2. The reaction is stopped with 50 m*M* NH_4Cl for 30 minutes. Antigens in solution are quantitatively spot-blotted onto nitrocellulose.
3. Nitrocellulose is blocked with either 5% BSA or 4% defatted milk powder for 1 hour, and incubated with primary antibody followed by ^{125}I-, enzyme-, or gold-labeled protein A, or secondary antibody using standard Western blotting procedures.

4. Densitometric scanning or visual inspection of the protein blots as compared to spot-blots from native antigens is performed.

1. Chemical Fixation

We perform fixation by vascular perfusion as whole-body perfusion via the left cardiac ventricle or organ perfusion via the main artery as follows:

1. Perfusion with Millipore-filtered (0.5 μm) oxygenated Hanks balanced salt solution (HBSS, pH 7.2–7.4) warmed to 37°C containing 4% polyvinyl pyrrolidone (MW 30,000; Fluka, Buchs, Switzerland) and 70 m*M* $NaNO_2$ (Merck, Darmstadt, FRG) at a hydrostatic pressure of ~130 cm water column for 2–3 minutes (whole-body perfusion) or until the perfused organ is blanched.

2. Perfusion with the just-mentioned Millipore-filtered solution that contains the fixative for 10 minutes at 37°C. As fixative we use routinely a mixture of 3% formaldehyde freshly prepared from paraformaldehyde (Merck) and 0.1% glutaraldehyde (purissimum grade; Fluka). However, the fixative may vary depending on the antigen. We have applied the following other formulations: (i) 4% (para-)formaldehyde, (ii) 3% (para-)formaldehyde with 0.05%, 0.2%, or 0.5% glutaraldehyde, and (iii) 0.5% or 1% glutaraldehyde. Then, tissue slices were either further fixed by immersion for a total time of 1 or 2 hours, or immediately processed to step 3. All different fixatives are used during the day they were made up.

3. Thin slices or small pieces of the tissue are rinsed quickly two times with PBS and placed in 50 m*M* NH_4Cl in PBS for 30–60 minutes at room temperature to amidinate free-aldehyde groups.

4. Tissue is stored in PBS containing 0.02% NaN_3 at 4°C overnight or immediately processed.

Cell cultures (after two quick rinses with HBSS at 37°C to remove the culture medium) and surface epithelia like the mucosal linings of internal organs are fixed by immersion in one of the aforementioned fixatives initially at 37°C and afterward processed to steps 3 and 4.

2. Cryofixation and Freeze Substitution

Because tissue pieces or cells can be infiltrated and polymerized in Lowicryl K4M or HM 20 between −35°C and −40°C and at −50°C, respectively, it is possible to omit chemical fixation and dehydration by stepwise lowering of the temperature and, instead, to process the material initially by cryofixation and freeze substitution. The introduction of these cryomethods followed by low-temperature embedding has provided some promising results regarding fine cellular structure and immunolabeling (Carlemalm *et al.*, 1985a; Humbel *et al.*, 1983; Humbel and Müller, 1985; Hunziker and Schenk, 1984; Hunziker and Herrmann, 1987; Hunziker *et al.*, 1984; Hobot *et al.*, 1984, 1985; Wroblewski and Wroblewski, 1984, 1985), and the interested reader is referred to these publications for adequate equipment and details of the procedures. Because these techniques are far from being fully exploited and cannot be regarded at present as a routine method, they will not be further considered here.

B. Dehydration at Low Temperatures

Individual tissue pieces should be $\leq 1\ mm^3$ to ensure good infiltration with the resin, adequate penetration of the UV light, and an even polymerization. Monolayer of cultured cells can be embedded *in situ* provided that the plastic is resistant to Lowicryl K4M, such as tissue culture dishes with film liner (Falcon 3006 Optical, Becton Dickinson), Petriper hydrophilic or hydrophobic (Heraeus AG, Zürich, Switzerland), or Teflon-based filters. Otherwise, the cells may be gently mechanically removed and pelleted by low-speed centrifugation. The cell pellet is then resuspended in a minimal volume of buffer and enclosed in 2% agar. To the agar, a colored marker such as immobilized Cibacron Blue F3GA (Pierce Chemical Corp., Rockford, IL) should be added to allow easy localization of the cells in the polymerized blocks.

1. Choice of the Dehydrating Liquid

Most polar and nonpolar dehydrating agents are miscible with Lowicryl K4M and may be used. From theoretical considerations, nonpolar organic liquids may be superior to polar ones. The latter may compete for the hydration shell of the biological material and because of its removal induce conformational changes.

We have used methanol, ethanol, ethylene glycol, and dimethyl formamide to dehydrate various animal tissues by progressive lowering of temperature, and have been unable to detect significant differences in structural preservation or intensity of immunolabeling with the pAg technique. In certain instance, however, it may be advantageous to use a particular solvent. The polysaccharide capsule of *Escherichia coli* K29 consists of >95% water and conventional dehydration for electron microscopy causes material collapse. The capsule could be preserved in uncollapsed form when glutaraldehyde-fixed (2% glutaraldehyde for 1 hour), gelatin-enrobed cells were dehydrated in dimethyl formamide and embedded in Lowicryl K4M (Bayer *et al.*, 1985). Furthermore, methanol dehydration was demonstrated to be highly suitable for the preservation of DNA structures in herpesviruses (Puvion-Dutilleul *et al.*, 1987).

2. Progressive Lowering of Temperature (PLT) Technique

The PLT technique[1] involves the stepwise reduction in temperature as the concentration of the dehydrating agent is increased. The temperature at each step is above the freezing point of the dehydrating agent concentration used in the step before. It is important to agitate the samples during dehydration. (See Tables I–III.)

Dehydration (as well as resin infiltration) can be performed with Balzer's low-temperature embedding (LTE 020) apparatus, which provides four sample-holding blocks, each of which may be preset to any temperature from 0°C to −50°C. It comes with stirring heads for continuous sample agitation. Other methods to achieve low temperatures are as follows:

1. For −20°C, use ice–NaCl, 3 : 1 (w/w), or appropriate freezer.
2. For temperatures of −35° to −40°C, use household chest-type freezer. Other means are mixtures of *o*- and *m*-xylene in combination with crushed dry ice. To achieve ≈ −35°C, crushed dry ice is added to a mixture of eight volume parts of *o*-xylene and two volume parts of *m*-xylene to form a thick slurry. When kept in a Dewar flask, temperature will remain constant for ~8–10 hours.

TABLE I

DEHYDRATION SCHEDULE FOR ETHANOL OR METHANOL

Alcohol in Water (vol %)	Temperature (°C)	Time (minutes)
30	0	30
50	−20	60
80	−35	60
100	−35 (−40)	60
100	−35 (−40)	60

[1] Altman *et al.* (1984) have published a rapid procedure for Lowicryl K4M embedding. They performed dehydration and resin infiltration at room temperature and UV polymerization at 4°C over very short distance.

TABLE II

DEHYDRATION SCHEDULE FOR DIMETHYL FORMAMIDE

Dimethyl Formamide (%)	Temperature (°C)	Time (minutes)
30	0	30
50	−20	60
70	−35	60
100	−35 (−40)	60
100	−35 (−40)	60

In any instance, to minimize temperature gradients, it is desirable to place the sample vials in an aluminum block with drilled holes, which is then inserted in the cooling bath. The samples are periodically agitated by stirring with a toothpick.

C. Preparation of Resin and Low-Temperature Infiltration

A note of caution: Lowicryl resins may cause eczema on sensitive individuals. It is highly recommended to use Lowicryl K4M-resistant gloves during all manipulations and to avoid inhaling of the resin vapors by performing mixing in a well-ventilated fume hood.

Because of its very low viscosity, Lowicryl K4M does not require vigorous stirring to mix the resin components. *Note:* Mixing too vigorously or for prolonged periods will result in the incorporation of oxygen into the resin, thereby preventing complete polymerization. We recommend the following procedure: Weight out, into a dark vial, the appropriate amount of crosslinker, monomer, and initiator. Mix gently by

TABLE III

DEHYDRATION SCHEDULE FOR ETHYLENE GLYCOL

Ethylene Glycol (%)	Temperature (°C)	Time (minutes)
50	0	15
80	−20 or −30	60

bubbling a continuous stream of dry nitrogen gas into the mixture by a Pasteur pipet for ~3 minutes or until the initiator is dissolved.

We routinely use the following mixture to prepare the resin, which will produce blocks of average hardness:

Crosslinker A, 2.5 g
Monomer B, 17.5 g
Initiator C, 0.1 g

The hardness of the blocks may be varied by using more (to produce harder blocks) or less crosslinker to the mixture. Its concentration may be varied from 0.8 to 3.6 g for 20 g resin.

Yeast cells are notoriously difficult to embed because their cell wall is an effective barrier for infiltration with resins. A simple procedure to overcome this problem has been published in conjunction with Lowicryl HM 20 embedding (van Tuinen and Riezman, 1987), which works as well when Lowicryl K4M embedding is performed (M. Binder, personal communication). After aldehyde fixation, the cells are incubated with 1% sodium metaperiodate for 30–60 minutes at room temperature, which, after PLT ethanol dehydration (Table I), permits excellent infiltration with the resin under conditions described later (Section III,C, 1).

1. Infiltration Schedule with Ethanol or Methanol

One part alcohol and one part resin for 60 minutes at −35°C (−40°C)
One part alcohol and two parts of resin for 60 minutes at −35°C (−40°C)
100% Resin two times for 60 minutes at −35°C (−40°C)
100% Resin overnight at −35°C (−40°C)
100% Resin for 6–8 hours at −35°C (−40°C)

2. Infiltration Schedule with Dimethyl Formamide

Two parts of dimethyl formamide and one part of resin for 60 minutes at −35°C

One part of dimethyl formamide and two parts of resin for 60 minutes at −35°C

100% Resin for 60 minutes at −35°C
100% resin overnight at −35°C

D. Low-Temperature Polymerization

The resin-infiltrated samples are now transferred into gelatin capsules at −35°C (−40°C). BEEM capsules or any other UV light-transparent type may be used as well. Transfer and subsequent polymerization can be done in Balzer's low-temperature polymerization apparatus or in a low-temperature chest freezer. In order to minimize the condensation of water and the crystallization of ice in the sample vials and capsules, all apparatus should be precooled. To minimize temperature fluctuations, a flat block of aluminum with holes to fit the gelatin capsules should be used and placed on top of a small illumination box for better view during sample transfer. Pieces of dry ice should be placed in the cold chamber if *in situ* embedding of cell cultures in film line dishes is performed in order to prevent impairment of UV polymerization by oxygen.

1. Fill capsules with freshly prepared, precooled resin to the top, to minimize dead air space (oxygen) over the resin.
2. Transfer samples to the capsules with a toothpick. It is recommended to place only one piece per capsule. Close the capsules and let them stand for ~1 hour. The capsules should be placed in a holder that guarantees that the capsules receive UV irradiation from all sides. We use stands made from heavy-gauge wire onto which finger-gauge twisted wire loops are soldered.
3. Polymerize for 24 hours by indirect diffuse UV irradiation at −35° to −45°C. The light source consists of two 360-nm long-wavelength UV, 15-W fluorescent tubes. The capsule holder is placed 30–40 cm below the fluorescent lamps. To provide diffuse illumination, a right-angle reflector is suspended below the UV lamps and all inner surfaces are constructed of a UV-reflective material, or lined with aluminum foil.
4. Transfer the capsules with the holder to room temperature and continue UV irradiation for 2–3 days.
5. Remove blocks from the capsule holder. Blocks should be stored under dry conditions. Air-conditioning provides sufficient conditions. Under humid conditions, it is often necessary to store the blocks under vacuum in an exsiccator with a desiccant.

IV. Sectioning and Section Storage

The basic principles of sectioning resin-embedded materials apply to Lowicryl K4M. Lowicryl K4M blocks can be sectioned with glass or diamond knives. The angle of the pyramids should be in the range of 55°–60°. It is recommended to trim the final pyramids with glass knives or on a trimming machine. For sectioning, precautions have to be taken to prevent wetting of the pyramid because Lowicryl K4M is hydrophilic. This is best accomplished by sectioning with a level of fluid in the trough that is slightly below normal (instead of silver gray it should be dark gray level) but not so low that the knife edge becomes dry. Sectioning speed should be 5–10 mm/second at the beginning, and as soon as the sections have reached the size of the pyramid surface it should be reduced to 2 mm/second. If a block is too soft to be useful for sectioning, it should be further cured under UV light at room temperature for 2–3 days. If a block became wet during sectioning, it should be immediately removed and placed in an exsiccator with a desiccant for 1 day. Before continuing sectioning, the pyramid needs to be retrimmed.

We have stored Lowicryl K4M blocks for as long as 9 years in certain cases. They could be sectioned as easily as recently prepared blocks.

A. Semithin Sections for Light Microscopy

We routinely prepare 0.5- to 1.5-μm-thick sections. For best results with phase contrast, 1- to 1.5-μm-thick sections are preferable. Sections are mounted on poly-L-lysine-activated glass slides and dried overnight at 40°C. Polylysine-activated slides are prepared as follows: Clean glass slides are marked on one side with a diamond, and this region on the other side is covered with poly-L-lysine solution (1 mg/ml; MW 300,000–500,000) for 5 minutes at room temperature. After a quick rinse with distilled water, the slides are air-dried and stored until use.

Mounted sections can be stored for prolonged periods of time at room temperature in a slide box without special precautions and show no apparent loss in reactivity with antibodies or lectins.

B. Ultrathin Sections for Electron Microscopy

Thin sections are cut at a nominal thickness of 60–70 nm on Reichert or LKB ultramicrotomes. Because Lowicryl K4M is a hydrophilic resin, the sections should be soon collected after they have been cut. Sections are placed on nickel grids (75–150 square mesh) covered with Parlodion and coated with carbon to give them additional support during observation in the electron microscope. The presence of a supporting film also prevents the grids from sinking in the incubation fluids during labeling. Sections on the grids can be stored without special precautions at room temperature in a grid box. We have stored unlabeled thin sections in grid boxes for several years without change in the labeling for several protein or carbohydrate antigens.

V. Protocols for Labeling on Sections

Lowicryl K4M sections are suitable for postembedding labeling with the pAg technique, various lectin–gold techniques, gold-labeled primary or secondary antibodies, toxins, enzymes, streptavidin, etc. (for reviews, see Roth, 1982a, 1983a, 1986, 1987a; Bendayan, 1984a,b; Bendayan *et al.*, 1987). The reactivity of cellular components is generally sufficient for their detection with monoclonal and polyclonal antibodies. The detectability of glycoconjugates with lectins, monoclonal antibodies, or certain toxins such as cholera toxin is drastically improved as compared to thin sections from Epon, Vestopal, or glycol methacrylate-embedded materials. There are some data to indicate that the low-temperature condition is of importance for the degree of labeling intensity (Armbruster *et al.*, 1983; Carlemalm *et al.*, 1985a; Hobot *et al.*, 1985). However, depending on the antigen to be localized, this may be highly variable. For example, on sections from rapidly room temperature-embedded *Bufo marinus* kidney, high labeling intensity with anti-Na^+,K^+-ATPase α-chain catalytic subunit antibodies could be observed (Altman *et al.*, 1984). Semithin or thin Lowicryl K4M need *not* to be etched prior to immunolabeling or lectin labeling. As mentioned in Section II, the surface of the Lowicryl K4M sections shows a specimen-related relief that is in the range of 2–6 nm and a consequence of sectioning. Because during the sectioning of the blocks cleavage occurs that follows the interface between the biological matter and the Lowicryl K4M, epitopes are supposed to be exposed for subsequent interaction with the labeling reagents (Kellenberger *et al.*, 1987). On the basis of theoretical estimations, a significant labeling for randomly dispersed globular proteins can be expected if the concentration of the antigen is $\sim 10\ \mu M$ or more (for a detailed discussion, see Kellenberger *et al.*, 1987). Labeling on sections provides access to the reagents for interaction with components present in the various intracellular organelles and the cytoplasmic matrix. But it should be always remembered that only those components can be detected that are sufficiently exposed on the section surface and that no reagent penetration occurs in Lowicryl K4M sections. This may be a limiting factor for labeling of fibrillar proteins. Staining with gold-labeled reagents at the light microscope level may give unsatisfactory results, because small amounts of particles of colloidal gold will produce only a faint incomplete pink staining or nonvisible staining. The photochemical silver reaction (Danscher, 1981) has been shown to render such staining visible in most cases (Danscher and Rytter Nörgaard, 1983; Lucocq and Roth, 1985; Taatjes *et al.*, 1987a). Furthermore, enhanced sensitivity as compared to immunoperoxidase could be observed (Holgate *et al.*, 1983; Springall *et al.*, 1984).

Detailed protocols for the preparation of various types of colloidal gold and their complex formation with different macromolecules have been published (Roth, 1982a, 1983a; Slot and Geuze, 1985; de Mey, 1986). Colloidal gold and various protein–gold complexes are commercially available form Janssen Pharmaceutica, E·Y Laboratories, Amersham, Sigma, Polyscience, Cambridge Research Biochemicals, and other companies.

A. The Protein A–Gold (pAg) Technique

Protein A from the cell wall of *Staphylococcus aureus* interacts with high affinity with the Fc portion of immunoglobulins, notably IgG (for review, see Goding, 1978). This interaction is a "pseudoimmune" (non-antibody-type) reaction that is rapid and reaches saturating levels in ~30 minutes at temperatures between 4° and 37°C (Langone, 1980). Protein A is highly reactive with IgG from rabbit, guinea pig, and human, whereas it is generally believed that it is at the most weakly reactive with mouse, rat, sheep, horse, and goat IgG's (for review, see Langone, 1982). However, it needs to be stressed that protein A reacts very well with certain IgG subclasses such as mouse IgG_{2a} and that enormous variability in reactivity with sheep and goat IgG's have been found (Richman *et al.*, 1982; Guss *et al.*, 1986). In agreement with these data, highly intense pAg labeling could be observed in conjunction with monoclonal mouse IgG_{2a} antibodies and certain antisera from sheep and goat (Roth, 1984; Taatjes *et al.*, 1987b). Protein G from *Streptococcus* strain G148 is claimed to possess an avidity for a broader spectrum of animal immunoglobulins if compared in immunochemical test systems with protein A (Åckerström and Björck, 1986; Åckerström *et al.*, 1985; Björck *et al.*, 1987). However, it should be emphasized that the equilibrium constant of protein G with certain animal species IgG's, which are non-reactive with protein A, is nevertheless extremely low (Åckerström *et al.*, 1985). To this end it was not surprising that in recent comparative immunocytochemical investigations using protein G–gold and protein A–gold, a high degree of variability in the results was encountered with both probes (Taatjes *et al.*, 1987b; Bendayan and Garzon, 1988).

1. Incubation Protocol for Semithin Lowicryl K4M Sections

We perform the pAg technique as follows:

1. Cover sections with 0.5% ovalbumin or 2–4% defatted milk powder dissolved in PBS for 5–10 minutes at room temperature to saturate protein-reactive tissue sites.

2. Drain away the blocking solution and cover the sections with appropriately diluted antibody for 2 hours at room temperature or overnight at 4°C. The appropriate antibody dilution has to be determined by titration. We dilute the antibody routinely with PBS. If background staining cannot be suppressed to reasonably low levels, we dilute the antibodies either with PBS containing 0.5–1% BSA, 0.05–0.1% Triton X-100, and 0.05–0.1% Tween 20, or with 1–2% defatted milk powder in PBS.

3. Rinse the sections in a staining jar with two to four changes of PBS for a total of 10 minutes at room temperature.

4. Cover the sections with pAg solution for 1 hour at room temperature. We use routinely pAg prepared from 8-nm gold particles. Of particular importance is the use of appropriately diluted pAg. Overly concentrated pAg will produce background staining that cannot be removed in most cases by prolonged washing with buffer. For work with Lowicryl K4M (as well as paraffin or frozen sections), we use the following standard dilutions:

   OD_{525} = 0.44 for pAg prepared with 15- or 20-nm gold particles
   OD_{525} = 0.06 for pAg prepared with 6–10 nm gold particles

   Of practical importance for the choice of the buffer solution for diluting pAg is the chemical means by which the colloidal gold was prepared. We use either plain PBS or PBS containing 0.02 mg/ml Carbowax 20 M (Fluka) if citrate (Frens, 1973) or ascorbic acid (Stathis and Fabrikanos, 1958) reduction was used for colloidal gold manufacture. However, pAg made from colloidal gold prepared with tannic acid–trisodium citrate (Slot and Geuze, 1985) exhibits a high tendency for nonspecific sticking to Lowicryl K4M that can only be overcome by adding 0.05–0.1% Triton X-100 and Tween 20 to the PBS.

5. Wash the sections in a staining jar with two to four changes of PBS for a total of 10 minutes at room temperature.

6. Cover the sections with 1% glutaraldehyde in PBS for 20 minutes at room temperature. This step is necessary to prevent low pH-induced release of antibody and pAg during photochemical silver amplification.

7. Rinse quickly with PBS followed by washing of the sections in a

staining jar with several changes of double-distilled water for a total of 5 minutes at room temperature and air-drying. The air-dried sections can be stored for at least 1 year or immediately processed by step 8.

8. This is the photochemical silver reaction for signal amplification. The procedure is carried out in a darkroom equipped with a photographic safe light. Due care should be exercised to use double-distilled water to prepare the solutions. They should be made up immediately before use outside the darkroom with the exception of the silver lactate solution, which is prepared in the darkroom. Furthermore, the glassware used for the photochemical silver reaction needs to be scrupulously cleaned and rinsed with double-distilled water. Two volumes of citrate buffer (0.5 *M*, pH 3.5–4.0) were mixed with 3 volumes of hydroquinone (Merck; 0.85 g per 15 ml of double-distilled water) and 12 volumes of double-distilled water. The slides are first placed for ~5 minutes in this solution followed by 2–4 minutes in the same solution containing 3 volumes of silver lactate (Fluka; 0.11 g per 15 ml double-distilled water). Afterwards, the slides were rinsed briefly in distilled water, then placed for 2 minutes in a photographic fixative. We use Superfix (Tetenal Photowerk, Norderstedt, FRG) diluted with 9 volumes of distilled water. Then, the slides are rinsed three times for 5 minutes each in distilled water. All positive structures in the sections appear dark brown to black by bright-field transmitted-light illumination.

9. Dehydrate sections through graded ethanol (30%, 50%, 70%, 90% for 2 minutes and 100% twice for 2 minutes), clear them with two changes of xylene, and mount them with synthetic medium such as DepeX (BDH Chemicals) or Tentellan (Merck).

If a protein A-nonreactive antibody is used in step 2, then after step 3 incubation with the corresponding affinity-purified rabbit anti-species antibody (20 μg/ml diluted with one of the buffers mentioned in step 2) is performed for 1 hour at room temperature. This is followed by rinses as described in step 3, and steps 4–9.

During the entire procedure sections should never become dry. All incubations are performed in a moist chamber.

2. Incubation Protocol for Lowicryl K4M Thin Sections

The incubation conditions (i.e., antibody and pAg dilutions and time of incubation) applied to thin sections for electron microscopy are basically identical to those just described for semithin light microscopy sections (Section V,A,1), with the exception that the glutaraldehyde fixation step after the pAg incubation step is omitted and no photochemical silver reaction is performed.

The grids with the sections facing downward are floated on the droplets (5–15 μl) of blocking buffer, antibody solution, and pAg during incubations. Rinsing between the individual incubation steps is done as follows: The grids hold at their outermost periphery with a pair of nonmagnetic tweezers are first washed by a *mild* spray of PBS from a plastic spray bottle for ~10 seconds and then immersed in PBS for ~2 minutes. This procedure is performed a second time. Before transferring them to the next droplet of incubation solution, the grid face without the sections is blotted dry with filter paper (Whatman no. 1, qualitative). Never dry the sections at any time during the procedure. After the pAg incubation, the grids are rinsed twice with PBS as just described, and then briefly with distilled water to remove salt and blotted dry with filter paper.

Double labeling can be performed with the use of pAg prepared from two different sizes of gold particles. Based on the mode of interaction between protein A and IgG, primary antibodies raised in the same animal species can be used as discussed in detail in Roth (1982b). Double labeling is performed principally in the sequence: first primary antibody–smaller pAg–second primary antibody–larger pAg (for specific details, see Geuze *et al.*, 1981; Bendayan, 1982; Roth, 1982b).

Some examples for the application of the pAg technique are shown in Figs. 1–4.

3. Controls

The demonstration of the specificity of the labeling is of utmost importance and has to be verified by several types of control experiments.

1. Incubate the tissue sections only with the pAg complex, which will reveal the degree and distribution of nonspecifically adsorbed pAg.
2. Preincubate the antibody with excess of purified antigen followed by pAg to verify the specificity of the antigen–antibody interaction.

Fig. 1. Thin section from low-temperature Lowicryl K4M-embedded rat kidney. Immunolocalization of the vitamin D-dependent 28-kDa calbindin with a specific antiserum and 14-nm protein A–gold. Gold particles demonstrating cytosolic localization of the antigen are found throughout the cytoplasm of a principal cell from cortical collecting duct. Note the absence of immunolabel over an adjacent intercalated cell. Low degree of nonspecific interaction with the Lowicryl K4M resin is indicated by the presence of a few gold particles over the tubular lumen (L). (For details, see Roth *et al.*, 1981b, 1982a.) ×23,000; bar = 1 μm.

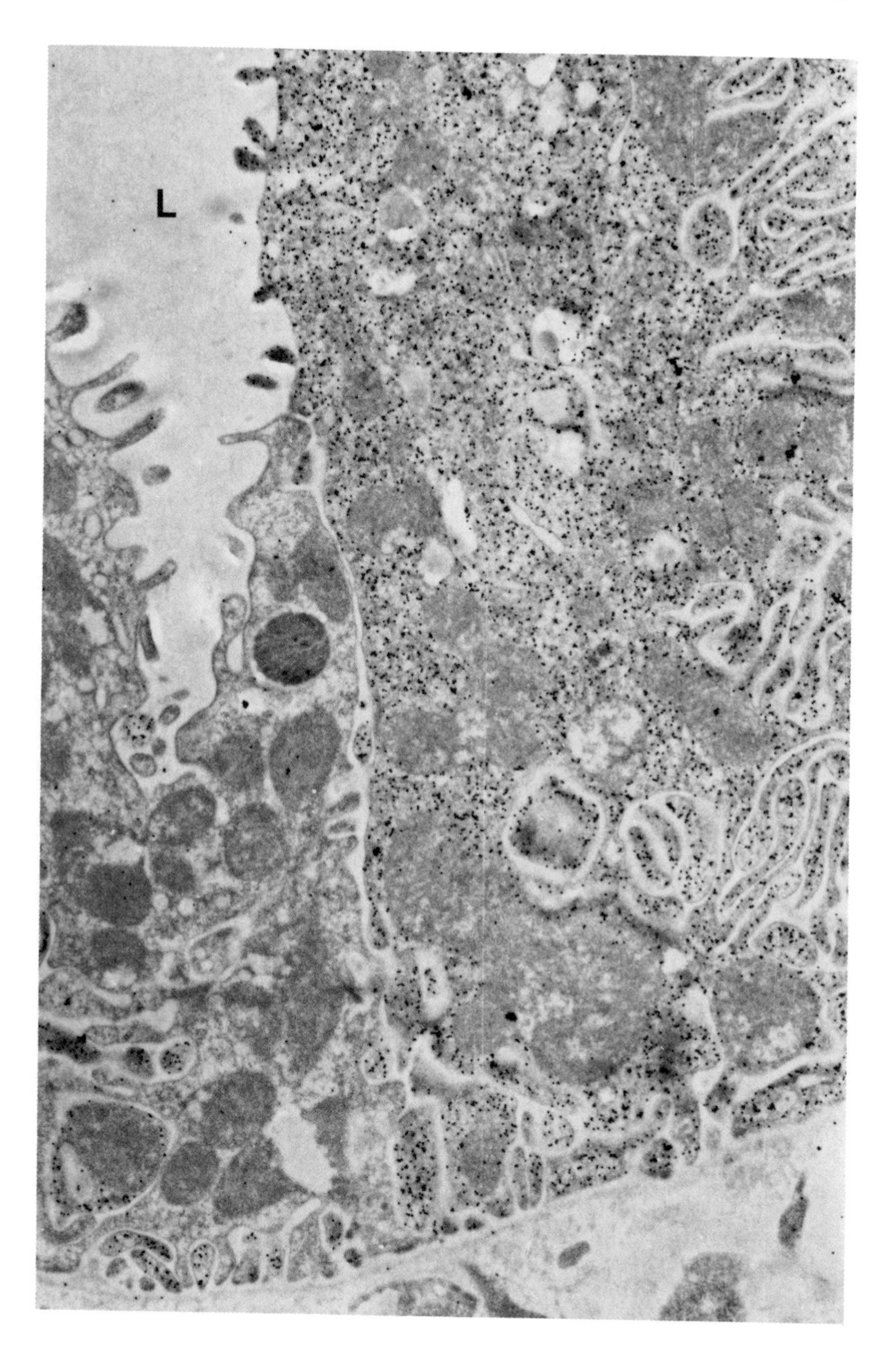
L

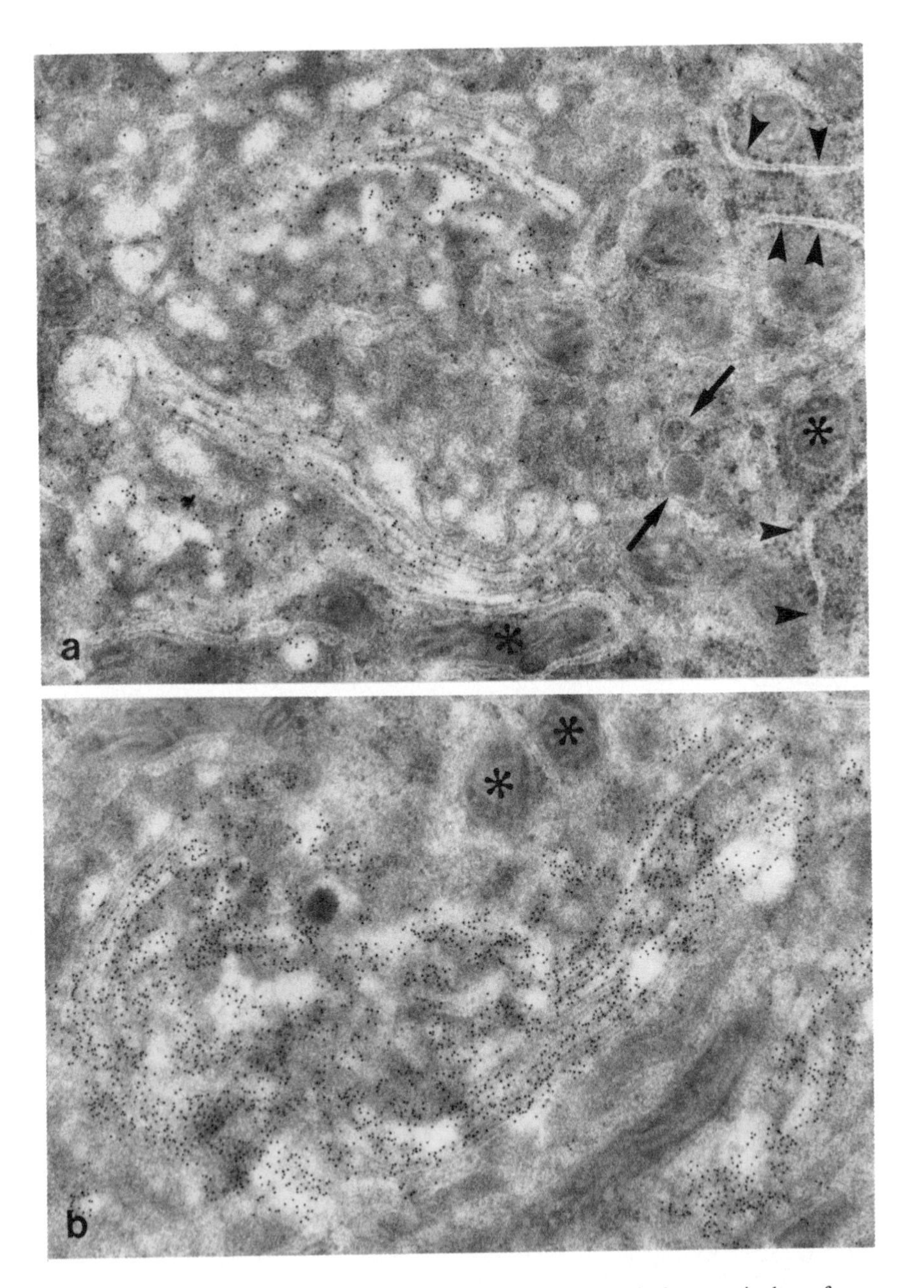

FIG. 2. Immunolocalization of blood group A α-1,3-*N*-acetylgalactosaminyltransferase and blood group A substance in the Golgi apparatus of human duodenal absorptive enterocytes, Lowicryl K4M, protein A–gold (8-nm) technique. The gold particle label indicating glycosyltransferase immunoreactivity (a) is found throughout the cisternal stacks and the trans-tubular network of the Golgi apparatus. Rough endoplasmic reticulum (arrowheads), mitochondria (asterisks), and small lysosomal bodies (arrows) are free of gold particle label. Immunoreactivity for blood group A substance (b) as revealed with a monoclonal antibody shows the same Golgi apparatus distribution as the blood group A transferase, namely diffuse throughout the cisternal stacks and the trans-tubular network. Mitochondria (M) are devoid of gold particle label. ×40,000 (a); ×37,000 (b); bar = 0.5 μm.

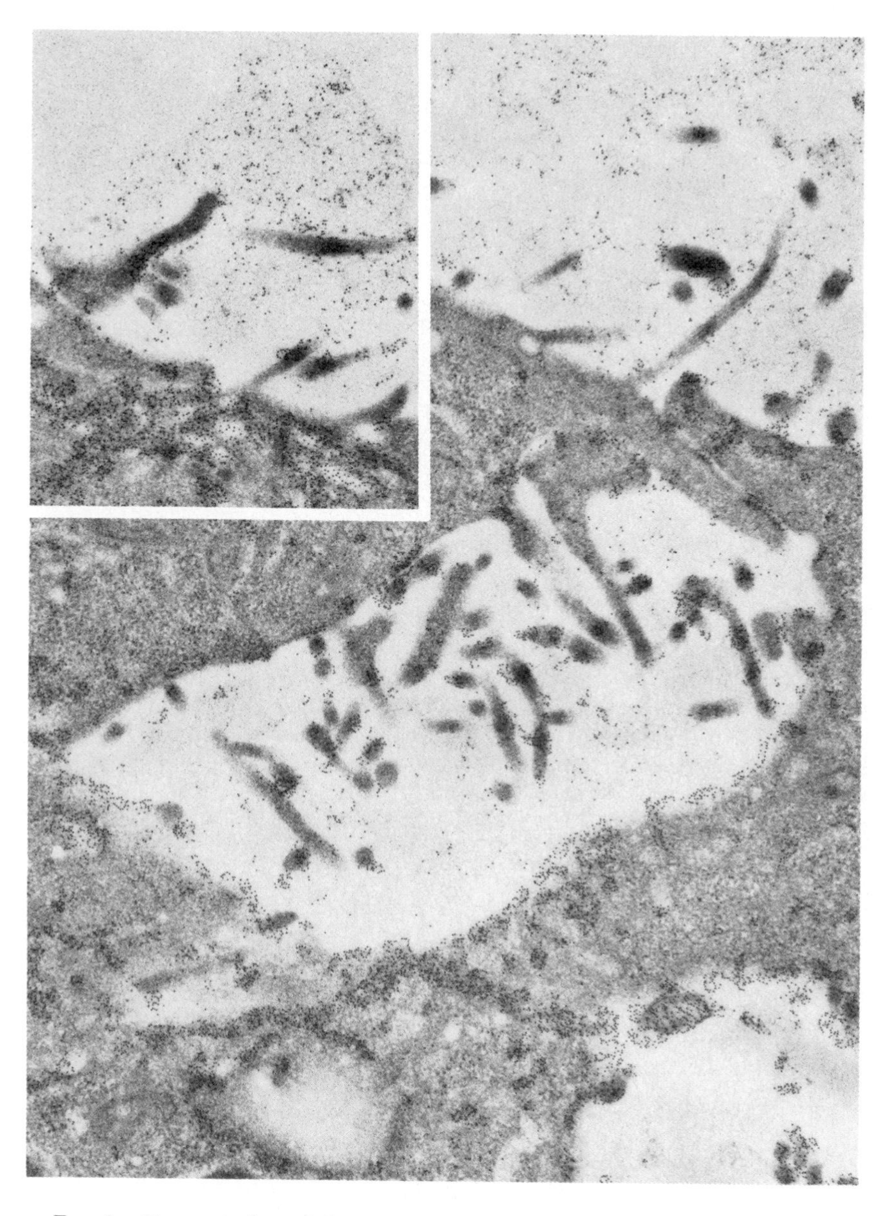

FIG. 3. Demonstration of the stage-specific embryonic antigen 1 (SSEA-1) with a monoclonal antibody and the protein A–gold (8-nm) technique in Lowicryl K4M thin sections from monolayer cultures of F9 cells. In addition to a well-developed, intensely labeled cell surface coat, tubulovesicular intracellular structures exhibit strong gold particle label. From unpublished work in collaboration with T. Feizi. ×26,500; inset ×44,000; bar = 0.5 μm.

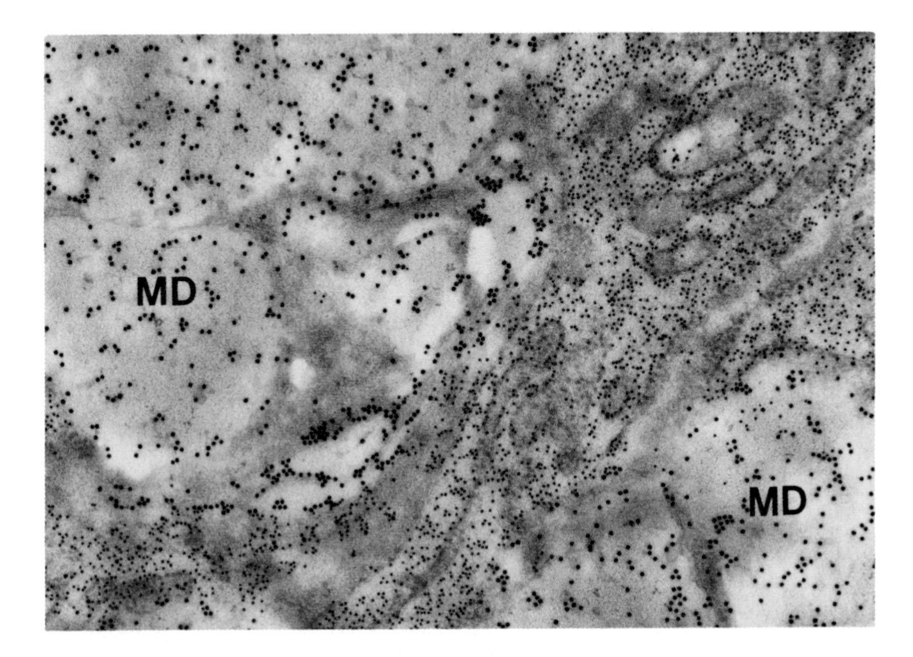

FIG. 4. Double labeling on Lowicryl K4M thin sections from pig submandibular gland mucus cells. Apomucin, as visualized with a specific antibody and 8-nm protein A–gold, is detectable over the cisternae of the rough endoplasmic reticulum but absent over the Golgi apparatus and mucus droplets (MD). Terminal *N*-acetylgalactosamine residues, as revealed with 14-nm *Helix pomatia* lectin–gold, are undetectable over the rough endoplasmic reticulum but found in all Golgi apparatus cisternae and the mucus droplets. (For details, see Deschuyteneer *et al.*, 1988.) ×39,000; bar = 0.5 μm.

Further means to demonstrate antibody specificity are Western blots, ELISA tests, and enzyme activity inhibition tests, in the case of antienzyme antibodies. Antibodies, even if affinity-purified, should be tested for reactivity with carbohydrates (for a detailed discussion, see Childs *et al.*, 1986; Feizi *et al.*, 1987; Taatjes *et al.*, 1988a).

3. Incubate with nonlabeled protein A (200 μg/ml for 1 hour) between the antibody and pAg incubation step to verify specificity of the IgG–protein A interaction.
4. Incubate with preimmune serum or unrelated polyclonal or monoclonal antibody followed by pAg to control the degree of nonspecific interaction between IgG and the resin.

B. The Lectin–Gold Techniques

Lectins have been extensively used in light and electron microscopy for the localization of various hexoses, hexosamines, and the sialic acids present in the oligosaccharide side chains of cellular glycoconjugates (for review, see Roth, 1978, 1987a,b). Most commonly, lectins are grouped in families according to their nominal sugar specificity (Goldstein and Poretz, 1986). Although it was initially believed that the specificity of a lectin could be sufficiently described in terms of the most potent monosaccharide in hemagglutination inhibition tests, it became soon clear that this was an oversimplification. The sugar-combining site of only some lectins seems to be complementary to a single glycosyl unit. Most have been found to possess extended binding sites accommodating two to five or six sugar residues (for reviews, see Kornfeld and Kornfeld, 1978; Goldstein and Poretz, 1986).

Most lectins can be used for complex formation with particles of colloidal gold and subsequently applied in a one-step labeling technique (Roth, 1983b,c; Lucocq and Roth, 1984; Taatjes *et al.*, 1987a). Some lectins, however, are problematic for complexing with gold because of their high isoelectric point and are applied in two-step affinity techniques in conjunction with an appropriate glycoprotein–gold complex (Geoghegan and Ackerman, 1977; Roth *et al.*, 1984). In addition to lectins, carbohydrate-specific monoclonal antibodies may be used in conjunction with the pAg technique (Roth *et al.*, 1986). Double labeling with different sizes of gold particles is possible for electron microscopy. But it should be kept in mind that most lectins are glycoproteins, and possible lectin–lectin cross-reactions need to be excluded.

1. One-Step Lectin Techniques

The incubation conditions with respect to lectin–gold complex dilution and incubation time are identical for semithin and ultrathin sections. However, the former are additionally subjected to glutaraldehyde fixation and photochemical silver reaction after the lectin–gold incubation as described earlier (Section V,A,1; steps 6–8). As for protein antigens, no etching of the sections is performed prior to incubation. The incubation conditions are as follows:

1. Semithin sections on glass slides are covered with PBS, or thin sections on grids are floated on droplets of PBS for 5 minutes at room temperature.

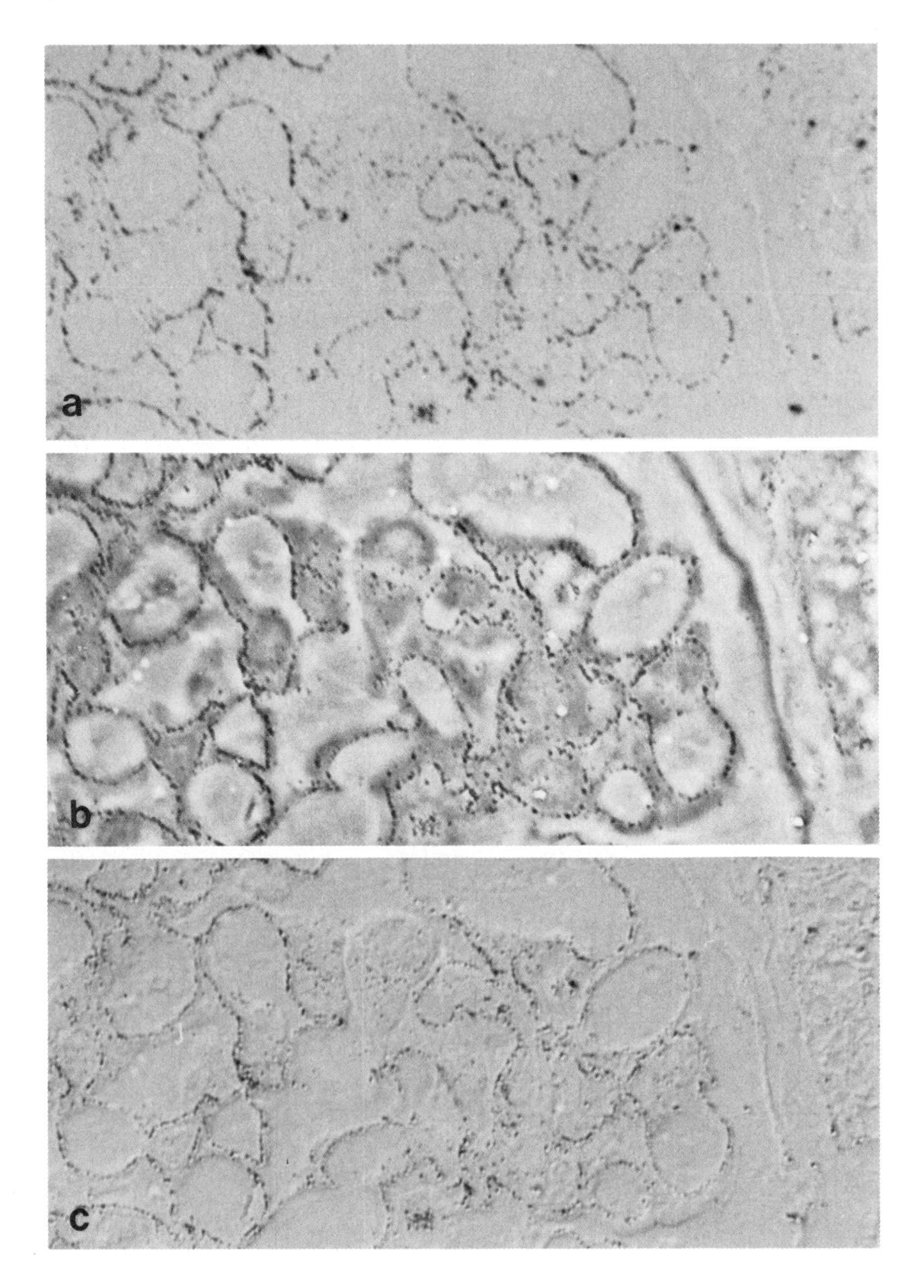

FIG. 5. Semithin (1-μm) section from low-temperature Lowicryl K4M-embedded rat kidney. The *Helix pomatia* lectin–gold was applied at an $OD_{525} = 0.7$, which corresponds to ≤ 5 μg lectin per milliliter, followed by photochemical silver reaction to visualize terminal,

2. Incubate with the lectin–gold complex for 30–45 minutes at room temperature. As before, semithin sections are covered with the solution, whereas the grids are floated on droplets. Depending on the particular lectin–gold complex, its working dilution may vary and needs to be determined by titration. Usually, we work with dilutions ranging between $OD_{525} = 0.05$ and 1.4 depending on the lectin. Lectin–gold stock solutions are usually diluted with PBS. But those prepared with tannic acid–citrate gold are diluted with PBS containing 0.5–1.0% BSA and 0.05–0.1% Triton X-100 and Tween 20 to prevent nonspecific interactions.

3. Wash the semithin sections with PBS in a staining jar two times for 5 minutes each, and continue with glutaraldehyde fixation and photochemical silver reaction (Section V,A,1). Grids with the thin sections are washed by a *mild* spray of PBS from a plastic spray bottle for ~10 seconds and then immersed in PBS for ~2 minutes. This procedure is performed a second time and followed by a short rinse with distilled water to remove the salt. Finally, the sections are blotted dry by filter paper.

The localization of tissue-binding sites with a lectin–gold complex in a semithin section is presented in Fig. 5.

2. Two-Step Lectin Techniques

The incubation conditions with respect to lectin concentration, glycoprotein–gold complex dilution, and incubation time are identical for semithin and ultrathin sections. However, the former are additionally subjected to glutaraldehyde fixation and photochemical silver reaction after the incubations as described earlier (Section V,A,1; steps 6–8). Two-step techniques have been worked out for concanavalin A (Con A), wheat germ agglutinin (WGA) *Limax flavus* lectin, and *Datura stramonium* lectin. The incubations are performed as follows:

1. Semithin sections on glass slides are covered with PBS, or thin sections on grids are floated on droplets of PBS for 5 minutes at room temperature.

---

nonreducing *N*-acetylgalactosamine residues. Specific staining is found at the level between the unlabeled podocytes and endothelial cells corresponding to labeling at the podocytes' foot process bases when investigated by electron microscopy (for details, see Roth *et al.*, 1982b). The label appears as black dots by bright-field transmitted light illumination (a), phase contrast (b), and Nomarski differential interference contrast (c). The latter two modes of imaging provide additional information about the tissue structure in the sections not counterstained. ×1100; bar = 10 μm.

2. Incubate with the appropriate lectin for 30–45 mintues at room temperature. We use the following lectin concentrations: Con A, 10–20 μg/ml; WGA, 10–20 μg/ml; *L. flavus* lectin, 100 μg/ml; *D. stramonium* lectin, 75 μg/ml.
3. Wash the semithin sections with PBS in a staining jar two times for 5 minutes each. The grids with the thin sections are washed by a *mild* spray of PBS from a plastic spray bottle for ~10 seconds and then immersed in PBS for ~2 minutes. This procedure is performed a second time.
4. Incubate with the glycoprotein–gold complex for 30 minutes at room temperature. The following complexes are used:
   a. Horseradish peroxidase (HRP)–gold (14-nm gold) diluted to give an OD_{525} = 1.0, together with Con A
   b. Ovomucoid–gold complex diluted to give an OD_{525} = 0.05 (5-nm gold), 0.2 (8, 10, and 15 nm gold), together with WGA and *D. stramonium* lectin
   c. Fetuin–gold complex (8-nm gold) diluted to give an OD_{525} = 0.35, together with the *L. flavus* lectin.
5. Wash the semithin sections with PBS in a staining jar two times for 5 minutes each, and continue with glutaraldehyde fixation and photochemical silver reaction (Section V,A,1). Grids with the thin sections are washed by a *mild* spray of PBS from a plastic spray bottle for ~10 seconds and then immersed in PBS for ~2 minutes. This procedure is performed a second time and followed by a short rinse with distilled water to remove the salt. Finally, the sections are blotted dry by filter paper.

Examples for the light and electron microscopic detection of sugar moieties by the two-step technique are shown in Figs. 6–9.

3. Controls

Specificity of lectin–gold labeling needs to be controlled by two types of experiments:

1. Lectins are preincubated for 30 minutes at room temperature with varying concentrations of as many as possible inhibitory *and* noninhibitory sugars or oligosaccharides. In addition, isolated defined glycopeptides or glycoproteins should be used whenever available.
2. Sections should be pretreated with exoglycosidases and/or endoglycosidases to remove sugar moieties (see Section VI,A).

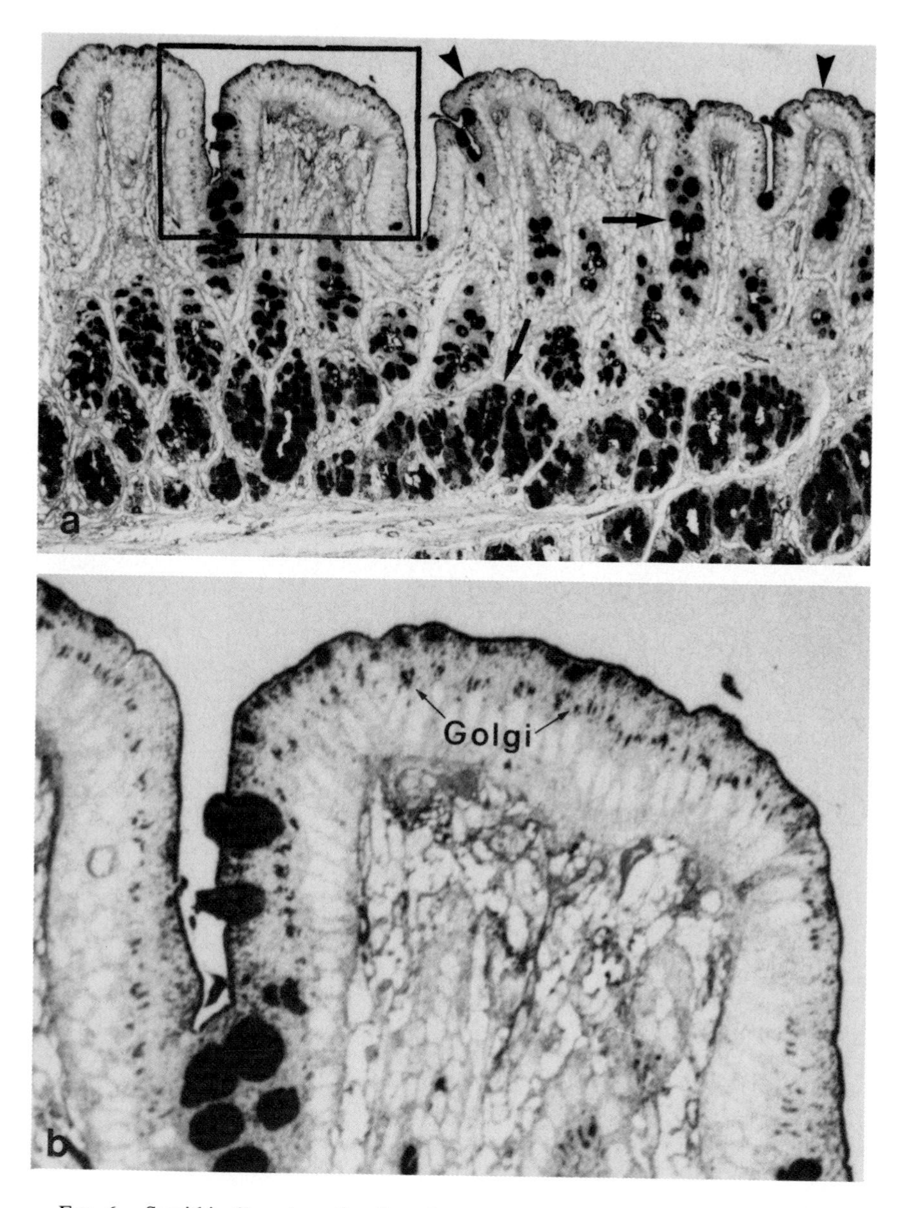

FIG. 6. Semithin (1-μm) section from low-temperature Lowicryl K4M-embedded rat colon. Detection of sialic acid residues with the *Limax flavus* lectin/fetuin–gold technique followed by photochemical silver reaction. The entire depth of the mucosa with adjacent lamina propria is shown at low magnification (a). The mucus of the goblet cells (arrows) as well as the brush border and the Golgi apparatus of the absorptive enterocytes (arrowheads) exhibit intense staining. In addition, elements of the lamina propria are positive. (b) The field marked in (a) at a higher magnification. See also Fig. 7 for the detection of sialic acid.

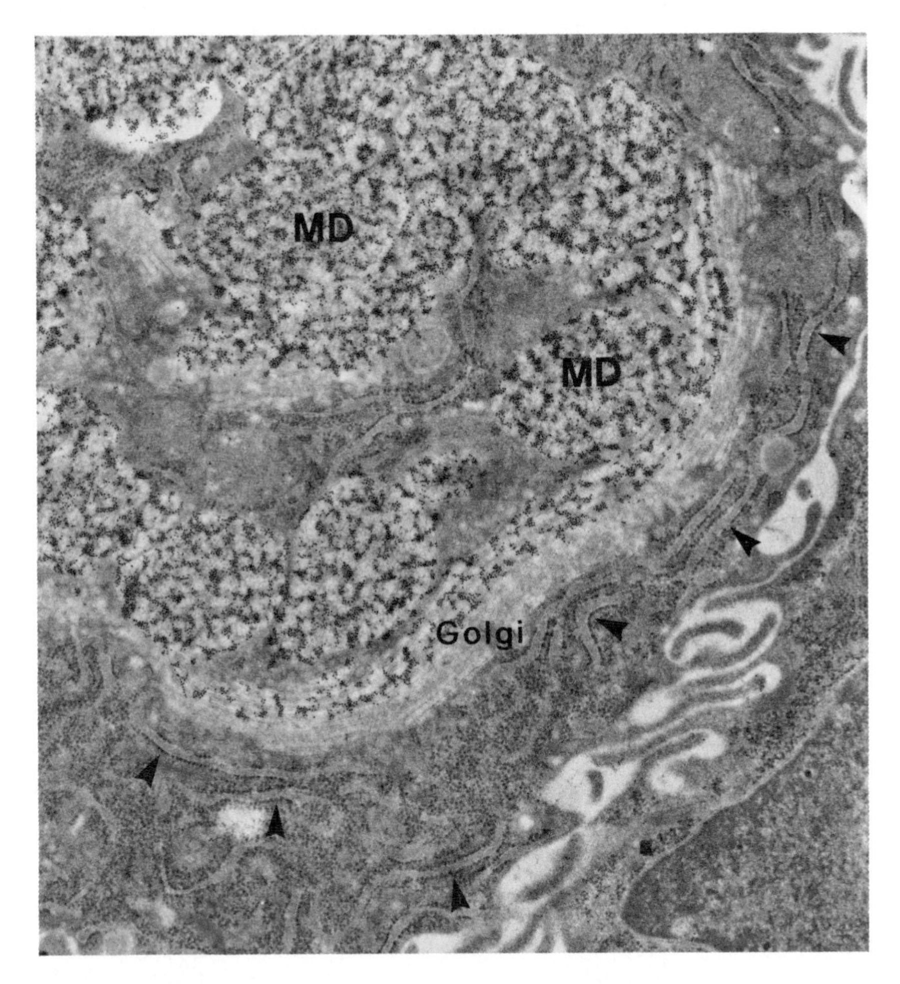

FIG. 7. Localization of sialic acid residues with the *Limax flavus* lectin/fetuin–gold technique in rat colonic goblet cells. The particles of colloidal gold (8 nm), which indicate the localization of sialic acid residues, are present over trans-Golgi apparatus cisternae and mucus droplets (MD) at different stages of formation. Note the absence of gold particles over middle and cis-Golgi apparatus cisternae, and the cisternae of the rough endoplasmic reticulum (arrowheads). ×20,000; bar = 1 μm.

Specificity of labeling obtained in the two-step technique needs to be verified by additional controls:

1. Incubate with excess amount of the respective glycoprotein between the lectin and the glycoprotein–gold incubation step to verify specificity of lectin/glycoprotein–gold interaction.

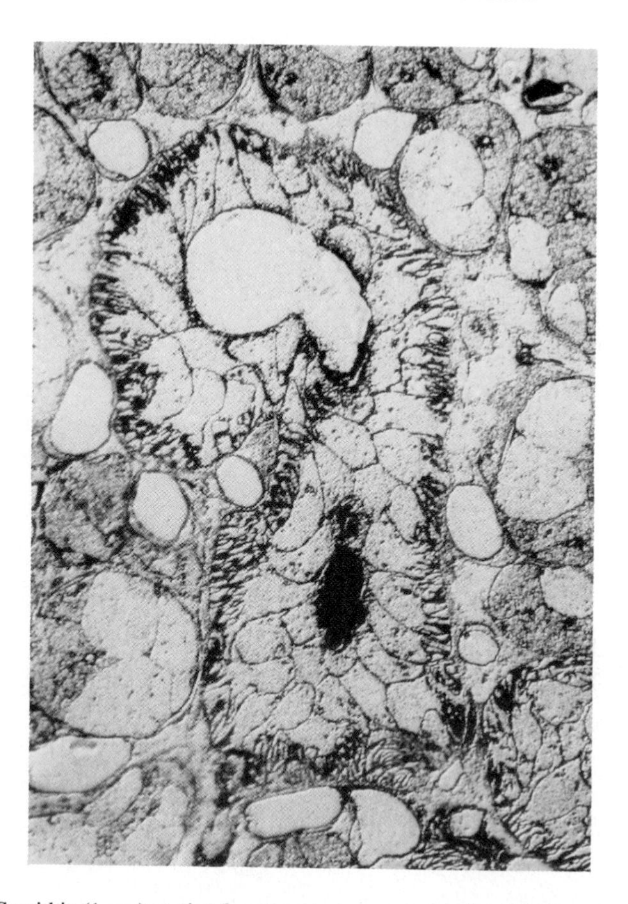

FIG. 8. Semithin (1-μm) section from low-temperature Lowicryl K4M-embedded sheep submandibular gland. The *Datura stramonium* lectin/ovomucoid–gold technique followed by photochemical silver reaction was applied to demonstrate galactose β-1,4-*N*-acetylglucosamine disaccharide units. Staining is present in serous but absent from mucous cells. A striated duct exhibits prominent plasma membrane staining (see Fig. 9 for the localization at the EM level) and luminal content staining. Note the resemblance of the duct to an archbishop's crook, the symbol of the Kanton Basel-Stadt. Nomarski interference contrast. ×600; bar = 25 μm.

2. Incubate only with the glycoprotein–gold complex to evaluate potential nonspecific adsorption.

C. Procedures for Contrasting Lowicryl K4M Thin Sections

We use two different procedures to obtain contrast in our specimens. It should be emphasized that in our experience, lead citrate is not very

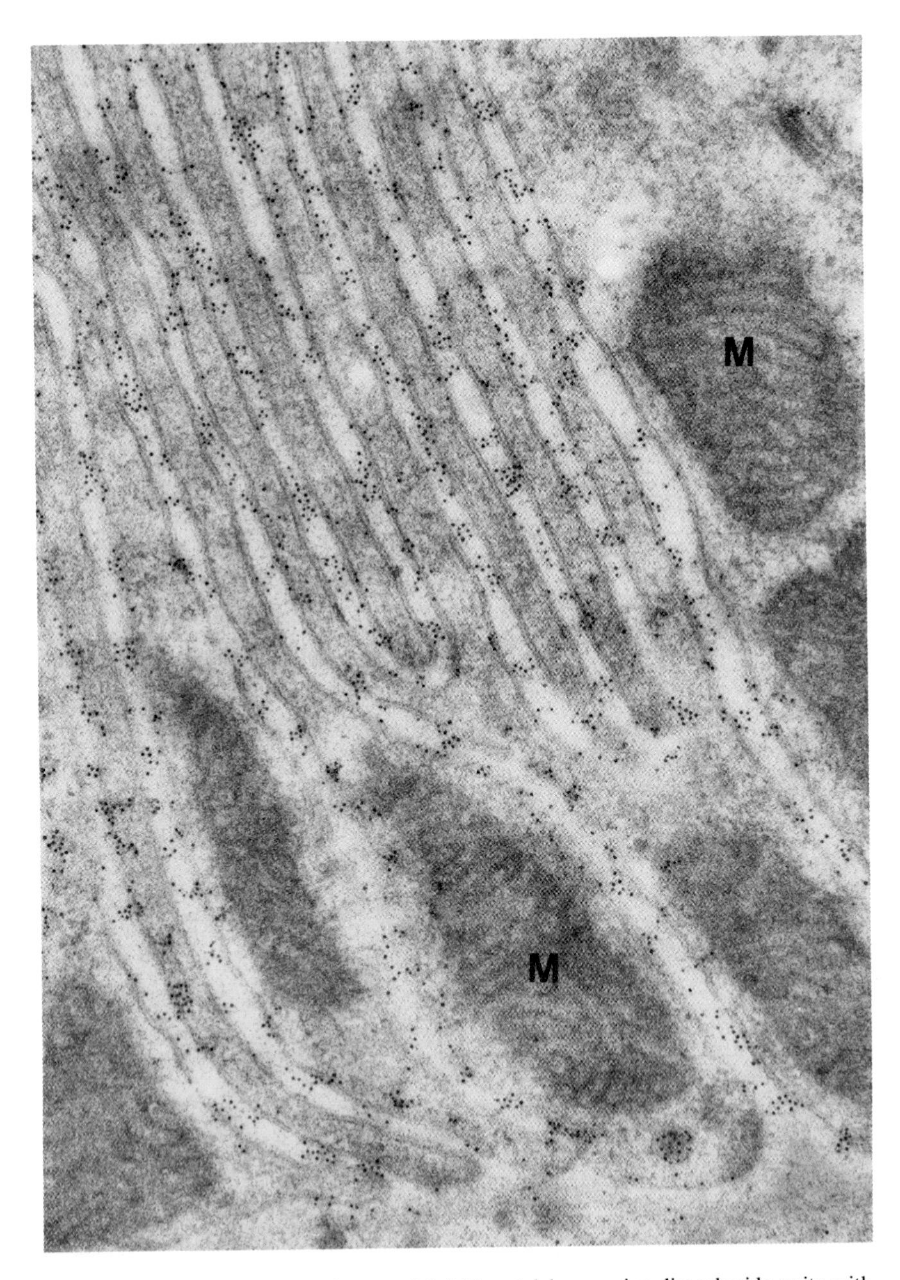

FIG. 9. Detection of the galactose β-1,4-*N*-acetylglucosamine disaccharide units with the *Datura stramonium* lectin/ovomucoid–gold technique in thin sections from sheep submandibular gland. Basal plasma membrane labyrinth of a striated duct cell exhibits strong labeling. Note the trilamellar appearance of the plasma membrane following uranyl acetate–lead acetate contrasting. M, Mitochondrium. ×69,000; bar = 0.25 μm.

sutiable, because it produces a relatively coarse and excessively intense staining. Before we start the contrasting step we let the grids air-dry for 30 minutes following the blot-drying with filter paper.

1. Uranyl Acetate–Lead Acetate Procedure

In this procedure the grids with the attached sections are floated on droplets of 3% aqueous uranyl acetate for 4–5 minutes followed by a quick rinse (~10 seconds) with distilled water and are dried by blotting with filter paper. This is followed by floating the grids for ~45 seconds on Millonigs lead acetate (Millonig, 1961). This step is performed in a nitrogen gas atmosphere to prevent precipitate formation and followed by a quick rinse with distilled water (~10 seconds). Grids are dried by blotting with filter paper.

This procedure provides sufficiently high contrast for the nucleoplasm and cytoplasmic ribosomes. However, the contrast of the usually positively stained cellular membranes is rather low in most cases, making the identification of smaller vesicular structures difficult (Fig. 10a). With the exception of collagen fibers, the extracellular matrix (ECM) components including the lamina densa of the basal lamina exhibit only faint contrast.

2. Uranyl Acetate–Methylcellulose Procedure

This procedure was worked out in collaboration with T. K. Tokuyasu and D. J. Taatjes and will be published in detail elsewhere. A 2% solution of 25 cP methylcellulose (Methocell, Fluka) is prepared by adding the powder under stirring to water preheated to 95°C. The solution is then placed on ice and mixed for additional 8 hours before being centrifuged at 55,000 rpm in a Beckman 60Ti rotor for 1 hour at 4°C. The centrifuge tubes are then stored in a refrigerator, and for staining purposes small amounts of the methylcellulose solution are pipeted carefully from the top. Furthermore, a 2% solution of uranyl acetate is made with distilled water. For contrasting, a 1.8% uranyl acetate–0.2% methylcellulose mixture is prepared and used as follows: Grids with the attached sections facing downward are floated on droplets of the mixture for 5 minutes at room temperature. Afterward each grid is picked up on a copper loop of 2.5–3 mm i.d., and after removing the excess of the stain, dried on air. The excess stain is removed to such an extent that the dried film of the staining mixture showed a silver-gold to gold-blue color.

This procedure gives a high degree of variation on the same grid. However, in suitable fields, cellular membranes, with the exception of endoplasmic reticulum membranes as well as those of the cis and middle-Golgi apparatus cisternae, exhibit intense positive staining that

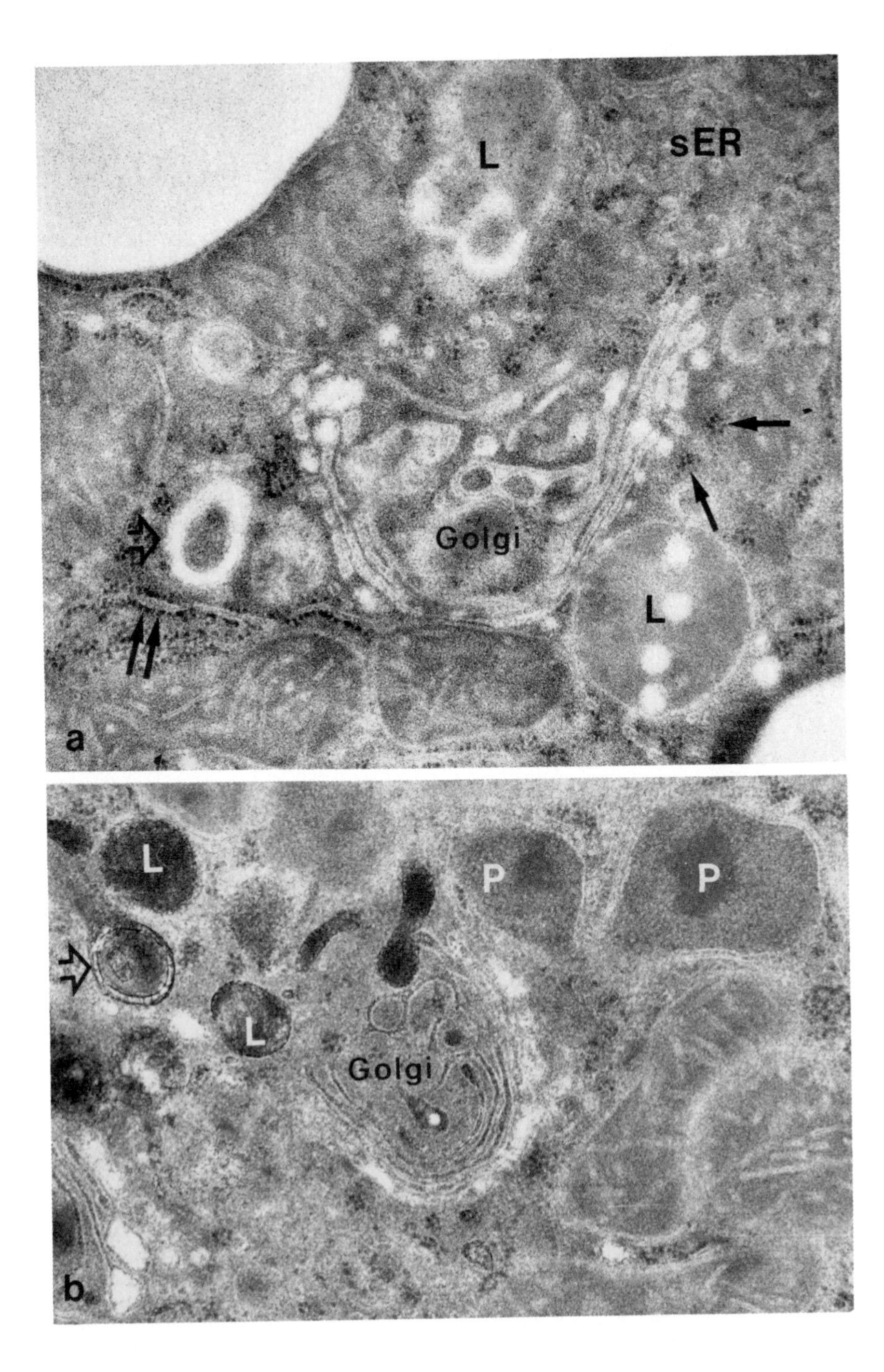
L
sER
Golgi
L
a
L
P
P
L
Golgi
b

makes even small vesiculotubular structures easily identifiable (Figs. 10 b and 11). The ECM components and basal lamina are intensely contrasted as well (Fig. 12).

VI. Enzymatic and Chemical Modifications on Lowicryl K4M Sections

Because glycosyl units are exposed on the section surface for interaction with carbohydrate-reactive reagents, we assumed they may also be accessible for enzymatic or chemical removal. Such procedures were useful in that they provided further control reactions for labeling obtained with lectins and monoclonal anticarbohydrate antibodies. In other applications they assisted in the control of antibody specificity (Deschuyteneer *et al.*, 1988). A very promising approach is the direct application of biochemical *in vitro*-glycosylation reactions to thin sections and their evaluation with electron-microscopic lectin techniques (Lucocq *et al.*, 1987).

A. Enzymatic Deglycosylation Reactions

The extent and rate of enzymatic deglycosylation of glycoconjugates on the surface of Lowicryl K4M sections depends, as in fluid-phase experiments with isolated glycoconjugates, to a high degree on the nature of the glycoprotein. Therefore, no general instructions with regard to the incubation conditions can be given. We usually apply conditions reported in the biochemical literature or follow the recommendations given by the supplier to incubate the thin sections attached to grids. In this way, we

---

FIG. 10. Lowicryl K4M thin sections from rat liver with a portion of the cytoplasm containing the Golgi apparatus. Thin sections were contrasted either with the uranyl acetate–lead acetate procedure (a), or the uranyl acetate–methyl cellulose procedure (b). In (a), all cisternal elements and the trans-tubular network of the Golgi apparatus as well as the smooth cisternal elements and the trans-tubular network of the Golgi apparatus as well as the smooth endoplasmic reticulum (sER) exhibit the same weak positive staining. Lysosomal bodies (L) and endosomal structures (open arrow) exhibit a similar quality of contrast. Rough endoplasmic reticulum with associated ribosomes (double arrow) and free ribosomes (arrow) are well contrasted. In (b), a similar field as shown in (a) is depicted. Here, trans cisternae and trans-tubular network as well as lyosomal bodies (L) and endosomal structures (open arrow) are intensely contrasted. P, Peroxisomes. ×46,000 (a); ×41,000 (b) bars = 0.25 μm.

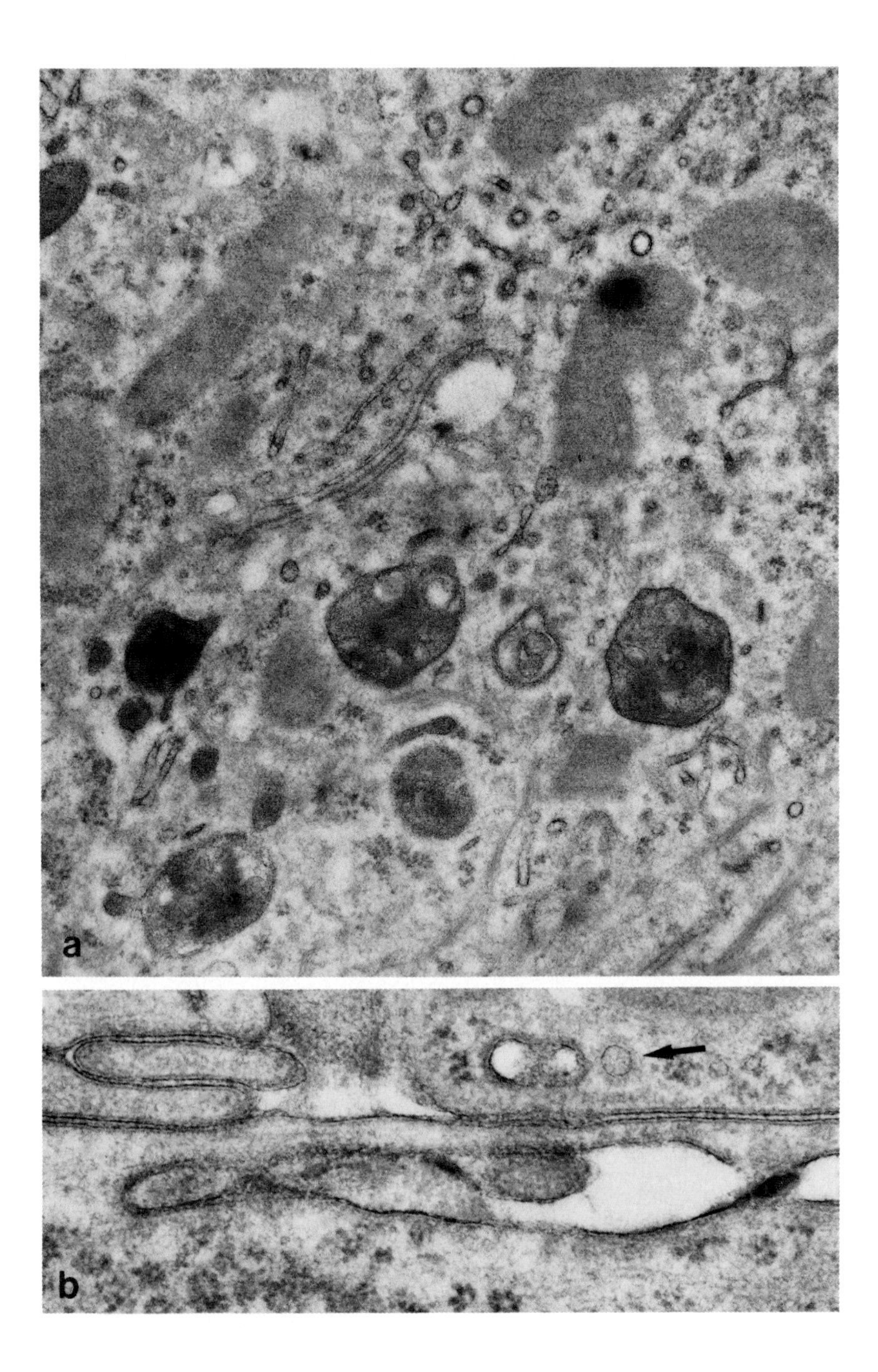
a
b

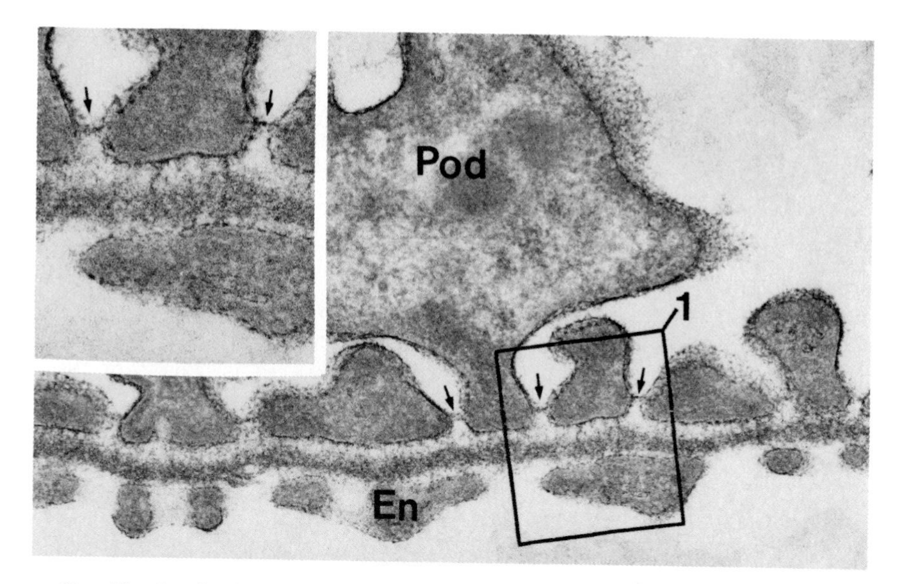

FIG. 12. Lowicryl K4M thin section from rat kidney showing a portion of a glomerular capillary loop. Staining with the uranyl acetate–methylcellulose procedure yields positively contrasted lamina densa of the glomerular basement membrane with fine fibrillar components visible in both laminae rarae. The slit diaphragm between the podocyte foot processes (arrows) and the plasma membrane of the podocytes (Pod) and endothelial cells (En) exhibit high contrast (see also inset enlargment of the region 1). All these structures are barely visible following the uranyl acetate–lead acetate procedure. ×50,000; ×100,000 (inset); bar = 0.1 μm.

have been able to use successfully a number of exoglycosidases either singly to remove a terminal sugar residue (Roth *et al.*, 1984) or as a mixture for stepwise deglycosylation (Deschuyteneer *et al.*, 1988). It is obviously also possible to remove oligosaccharidic structures with endoglycosidases. Endo-β-*N*-acetylglucosaminidase (endoglycosidase H) could be used for removal of high-mannose type *N*-glycans as evidenced by lack of labeling with Con A, and endoglucosaminidase F-peptide

---

FIG. 11. Lowicryl K4M thin section from rat colonic surface enterocytes contrasted with the uranyl acetate-methylcellulose procedure. In (a), a Golgi field with positively stained trans-Golgi apparatus cisternae and adjacent tubulovesicular structures as well as variously shaped endosomal-lysosomal elements is seen. In (b), the interdigitating lateral plasma membrane of two enterocytes is shown, which exhibits an intensely positively stained trilamellar structure. Note the coated vesicle (arrow). ×26,000 (a); ×53,000 (b) bars = 0.25 μm.

N-glycosidase F for removal of complex type *N*-glycans (Lucocq *et al.*, 1987).

B. Alkaline Cleavage of Protein–O-Glycan Linkage

Oligosaccharides can be released from mucus and other O-glycan-containing materials by treatment with alkaline solutions (Carlson, 1968). This procedure could be adapted to Lowicryl K4M sections (G. Egea and J. Roth, unpublished). For this purpose, free sections were floated on 0.1–0.2 *N* NaOH for 12–36 hours at 37°C in a moist chamber, and afterward rinsed four times 5 minutes each with double-distilled water. The extent of deglycosylation was checked with *Helix pomatia* lectin–gold.

C. Galactosylation of Terminal N-Linked *N*-Acetylglucosamine Residues

Purified human milk β-*N*-acetylglucosaminide β-1,4-galactosyltransferase was used to transfer galactose residues from UDP-galactose to terminal nonreducing *N*-acetylglucosamine residues present on thin sections of Lowicryl K4M-embedded liver. Thin sections were floated for 5–10 minutes on 1% BSA in 0.1 *M* cacodylate buffer (pH 7.4), quickly rinsed with BSA-free buffer, and transferred to the galactosyltransferase-containing solution for 2 hours at room temperature. The solution consisted of purified galactosyltransferase, 250 m*M* cacodylate buffer (pH 7.35), 25 m*M* $MnCl_2$, 0.5 m*M* UDP-galactose, and 0.25 mg/ml BSA. Substrate inhibition controls were performed by including 50–100 m*M* *N*-acetylglucosamine, D-glucose, D-galactose, *N*-acetylgalactosamine, or 2.5 m*M* ovalbumin in the galactosyltransferase solution 30 minutes before use. To release galactosyltransferase possibly bound to *N*-acetylglucosamine residues in the sections, 1.5 *M* urea, 0.005 *M* *N*-acetylglucosamine, 0.025 *M* EDTA, and 0.005 *M* 2-mercaptoethanol, or 0.5% SDS and 1% 2-mercaptoethanol, or 0.1 *M* *N*-acetylglucosamine were included in the washing buffer.

The galactosylated transferase product was detected with a *Ricinus communis* lectin I–gold complex. By comparing the distribution of the lectin–gold label for endogenous galactose residues with the distribution of the lectin–gold label after galactosylation, a shift of labeling for galactose residues in the Golgi apparatus could be revealed (Lucocq *et al.*, 1987), indicative of effective galactosylation. Further characterization of the galactosylated product (i.e., its presence on N- or O-glycans) was possible by endoglycosidase pretreatment of the thin sections.

VII. Prevention of Artifacts

Specificity of postembedding immunolabeling depends on the careful control of several factors. This includes not only the use of optimally diluted primary antibodies or other primary labeling reagents, but also the appropriate processing of the tissue after fixation as well as of the thin sections before commencing the labeling procedures and the correct use of the second-step reagent.

The basic observation that permitted the development of the pAg technique was that the pAg complex showed very little tendency to bind nonspecifically to thin sections of Epon or Lowicryl K4M-embedded tissues, and this appeared to be a general property of protein–gold complexes. Later observations, however, provided evidence that pAg and other protein–gold complexes tend to bind selectively, but nonspecifically to certain structures in cell cultures and in thin sections of glutaraldehyde-fixed and Epon-embedded tissues (Jessen and Behnke, 1986a,b; Asada-Kubota, 1988). Detailed analysis of their published data indicated that the observed phenomena were due to inappropriate use of published protocols. Selective nonspecific binding of pAg does not occur when the following conditions are fulfilled:

1. *Quenching of Unreacted Aldehyde Groups after Fixation.* After fixation of tissues with aldehydes, quenching of free reactive aldehyde groups with amino group-containing substances such as ammonium chloride or amino acids is an obligatory step. Otherwise nonspecific interactions between aldehyde groups and amino groups present on proteins such as immunoglobulins and protein A may occur.
2. *Blocking of Nonspecific Binding Sites by Reacting the Sections with Inert Proteins.* Nonspecific binding of the labeling reagents to so-called reactive groups in the tissue sections is prevented by conditioning the sections before commencing the labeling procedure. Most commonly, sections are incubated with 0.5% ovalbumin in PBS. Highly effective and inexpensive is the use of 2–4% defatted milk powder in PBS, which we call in the lab Swiss PBS.
3. *Use of Appropriate Concentrations of pAg Solutions.* Excessively high concentrations of pAg or lectin–gold may produce nonspecific binding. The proper dilution is also a function of the gold particle size.

From our experience we recommend the use of the following standard dilutions of pAg:

OD_{525} = 0.44 for pAg made with 15-nm gold particles

OD_{525} = 0.06 for pAg made with 6, 8, or 10 nm gold particles

These are standard dilutions that are used independent of the dilution of the primary antibody and the tissue or cell type under investigation. Corresponding information for lectin techniques are given in Section V,B,1 and 2.

We observed for pAg, lectin–gold, or glycoprotein–gold complexes made with tannic acid–citrate gold (Slot and Geuze, 1985) nonspecific binding to Lowicryl K4M sections even if properly diluted. This could only be overcome by diluting the complexes with PBS containing 0.5–1% BSA, 0.05–0.1% Triton X-100, and Tween 20. If in certain cases detergents cannot be used (Taatjes *et al.*, 1988b), ascorbic acid–gold (Stathis and Fabrikanos, 1958) or other gold preparations (phosphorus–gold, borohydride–gold; details of the procedures are given in Roth, 1983a) are good alternatives.

For light-microscopic studies on semithin sections, glutaraldehyde fixation after the incubation steps and before silver enhancement is obligatory (see Section V,A,1) to prevent low pH-induced removal of the label. In certain situations, glutaraldehyde fixation (1%, 10–20 minutes) of thin sections before contrasting has given some improvement of labeling intensity (Kellenberger *et al.*, 1987).

Some more general recommendations to prevent background staining or other troublesome effects are as follows. The sections should never become dry during the entire incubation procedure. Grids with the attached thin sections should be always floated on the droplets of reagents. Immersing them will result in additional heavy nonspecific sticking of all labeling reagents to the supporting film, resulting in superimposition with the specific label. If detergent-containing buffers are used, the rinsing step should be very carefully performed to prevent detachment of the thin sections and the supporting film from the grids. Colloidal gold, even when coated with macromolecules, has a slow tendency to aggregate. Larger aggregates can be removed by centrifugation in a bench centrifuge. The most effective means, however, is density gradient centrifugation with a continuous 10–30% glycerol or sucrose gradient (Geuze *et al.*, 1981). Even small aggregates have a high tendency for nonspecific sticking and cannot be washed off. Nonspecific binding to not well-preserved cellular structures is often observed, due to suboptimal fixation or cell death, dirt particle-contaminated sections, or imperfections of thin sections such as chatters and scratches.

ACKNOWLEDGMENTS

The work of the author and his co-workers described here was supported by grants from the Swiss National Science Foundation and the Kanton Basel-Stadt. I thank Werner Villiger for discussion.

REFERENCES

Acetarin, J.-D., and Carlemalm, E. (1982). *In* "Lowicryl Letters" (E. Carlemalm, W. Villiger, and E. Kellenberger, eds.), No. 1. Chemische Werke Lowi GmbH, Waldkraiburg, F. R. G.

Acetarin, J.-D., Carlemalm, E., and Villiger, W. (1986). *J. Microsc. (Oxford)* **143,** 81–88.

Acetarin, J.-D., Carlemalm, E., Kellenberger, E., and Villiger, W. (1987). *J. Electron Microsc. Tech.* **6,** 63–79.

Åckerström, B., and Björck, L. (1986). *J. Biol. Chem.* **261,** 10240–10247.

Åckerström, B., Brodin, T., Reis, K., and Björck, L. (1985). *J. Immunol.* **135,** 2589–2592.

Altman, L. G., Schneider, B. G., and Papermaster, D. S. (1984). *J. Histochem. Cytochem.* **32,** 1217–1223.

Armbruster, B., Garavito, R. M., and Kellenberger, E. (1983). *J. Histochem. Cytochem.* **31,** 1380–1384.

Asada-Kubota, M. (1988). *J. Ultrastruct. Mol. Struct. Res.* **98,** 147–157.

Bayer, M. E., Carlemalm, E., and Kellenberger, E. (1985). *J. Bacteriol.* **162,** 985–991.

Bendayan, M. (1982). *J. Histochem. Cytochem.* **30,** 81–85.

Bendayan, M. (1984a). *J. Electron Microsc. Tech.* **1,** 243–270.

Bendayan, M. (1984b). *J. Electron Microsc. Tech.* **1,** 349–360.

Bendayan, M., and Garzon, S. (1988). *J. Histochem. Cytochem.* **36,** 597–607.

Bendayan, M., Nanci, A., and Kan, F. W. K. (1987). *J. Histochem. Cytochem.* **35,** 983–996.

Binder, M., Tourmente, S., Roth, J., Renaud, M., and Gehring, W. J. (1986). *J. Cell Biol.* **102,** 1646–1653.

Björck, L., Kastern, W., Lindahl, G., and Widebäck, K. (1987). *Mol. Immunol.* **24,** 1113–1122.

Carlemalm, E., Garavito, M., and Villiger, W. (1982). *J. Microsc. (Oxford)* **126,** 123–143.

Carlemalm, E., Villiger, W., Hobot, J. A., Acetarin, J.-D., and Kellenberger, E. (1985a). *J. Microsc. (Oxford)* **140,** 55–63.

Carlemalm, E., Colliex, C., and Kellenberger, E. (1985b). *Adv. Electron. Electron. Phys.* **63,** 269–334.

Carlson, D. M. (1968). *J. Biol. Chem.* **243,** 616–626.

Childs, R. A., Berger, E. G., Thorpe, S. J., Aegerter, E., and Feizi, T. (1986). *Biochem. J.* **238,** 605–611.

Danscher, G. (1981). *Histochemistry* **71,** 81–88.

Danscher, G., and Rytter Nörgaard, J. O. (1983). *J. Histochem. Cytochem.* **31,** 1394–1398.

De Mey, J. (1986). *In* "Immunocytochemistry" (J. M. Polak and S. van Noordan, eds.), 2nd ed., pp. 115–145. Wright, Bristol, England.

Deschuyteneer, M., Eckhardt, A. E., Roth, J., and Hill, R. L. (1988). *J. Biol. Chem.* **263,** 2452–2459.

Faulk, W. P., and Taylor, G. M. (1971). *Immunochemistry* **8,** 1081–1083.

Feizi, T., Thorpe, S. J., and Childs, R. A. (1987). *Biochem. Soc. Trans.* **15,** 614–617.

Frens, G. (1973). *Nature (London) Phys. Sci.* **241,** 20–30.

Geoghegan, W. D., and Ackerman, G. A. (1977). *J. Histochem. Cytochem.* **11,** 1187–1200.
Geuze, H. J., Slot, J. W., Scheffer, R. C. T., and van der Ley, P. A. (1981). *J. Cell Biol.* **89,** 653–665.
Goding, J. W. (1978). *J. Immunol. Methods* **20,** 241–253.
Goldstein, I. J., and Poretz, R. D. (1986). *In* "The Lectins: Properties, Functions, and Applications in Biology and Medicine" (I. E. Liener, N. Sharon, and I. J. Goldstein, eds.), pp. 33–247. Academic Press, Orlando, Florida.
Guss, B., Eliasson, M., Olsson, A., Uhlen, M., Frej, H., Jörnvall, H., Flock, J.-I., and Linberg, M. (1986). *EMBO J.* **5,** 1567–1575.
Hobot, J. A., Carlemalm, E., Villiger, W., and Kellenberger, E. (1984). *J. Bacteriol.* **160,** 143–152.
Hobot, J. A., Villiger, W., Escaig, J., Maeder, M., Ryter, A., and Kellenberger, E. (1985). *J. Bacteriol.* **162,** 960–971.
Holgate, C. S., Jackson, J. P., Cowen, P. N., and Bird, C. (1983). *J. Histochem. Cytochem.* **31,** 939–944.
Humbel, B., and Müller, M. (1985). *In* "Science of Biological Specimen Preparation" (M. Müller, ed.), pp. 175–183. SEM, Inc., AMF O'Hare, Chicago, Illinois.
Humbel, B., Marti, T., and Müller, M. (1983). *Beitr. Elektronenmikrosk. Direktabb. Oberfl.* **16,** 585–594.
Hunziker, E. B., and Hermann, W. (1987). *J. Histochem. Cytochem.* **35,** 647–655.
Hunziker, E. B., and Schenk, R. K. (1984). *J. Cell Biol.* **98,** 277–282.
Hunziker, E. B., Hermann, W., Schenk, R. K., Müller, M., and Moor, H. (1984). *J. Cell Biol.* **98,** 267–276.
Jessen, H., and Behnke, O. (1986a). *Eur. J. Cell Biol.* **41,** 326–338.
Jessen, H., and Behnke, O. (1986b). *J. Invest. Dermatol.* **87,** 737–740.
Kellenberger, E., Schwab, W., and Ryter, A. (1956). *Experientia* **12,** 421–422.
Kellenberger, E., Carlemalm, E., and Villiger, W. (1985). *In* "Science of Biological Specimen Preparation" (M. Müller, ed.), pp. 1–22. SEM, Inc., AMF O'Hare, Chicago, Illinois.
Kellenberger, E., Villiger, W., and Carlemalm, E. (1986). *Micron Microsc. Acta* **17,** 331–348.
Kellenberger, E., Dürrenberger, M., Villiger, W., Carlemalm, E., and Wurtz, M. (1987). *J. Histochem. Cytochem.* **35,** 959–969.
Kornfeld, S., and Kornfeld, R. (1978). *In* "The Glycoconjugates" (M. I. Horowitz and W. Pigman, eds.), Vol. 2, pp. 437–449. Academic Press, New York.
Langone, J. J. (1980). *In* "Methods in Enzymology" (H. Van Vunakis and J. J. Langone, eds.), Vol. 70, pp. 356–375. Academic Press, New York.
Langone, J. J. (1982). *Adv. Immunol.* **32,** 157–252.
Lucocq, J. M., and Roth, J. (1984). *J. Histochem. Cytochem.* **32,** 1075–1083.
Lucocq, J. M., and Roth, J. (1985). *In* "Techniques in Immunocytochemistry" (G. R. Bullock and P. Petrusz, eds.), Vol. 3, pp. 203–236. Academic Press, Orlando, Florida.
Lucocq, J. M., Berger, E. G., and Roth, J. (1987). *J. Histochem. Cytochem.* **35,** 67–74.
Millonig, G. (1961). *J. Biophys. Biochem. Cytol.* **11,** 736–739.
Puvion-Dutilleul, F., Pichard, E., Laithier, M., and Leduc, E. H. (1987). *J. Histochem. Cytochem.* **35,** 635–645.
Richman, D. D., Cleveland, P. H., Oxman, M. N., and Johnson, K. M. (1982). *J. Immunol.* **128,** 2300–2305.
Roth, J. (1978). *Exp. Pathol., Suppl.* **3,** 1–186.
Roth, J. (1982a). *In* "Techniques in Immunocytochemistry" (G. R. Bullock and P. Petrusz, eds.), Vol. 1, pp. 107–133. Academic Press, London.

Roth, J. (1982b). *Histochem. J.* **14,** 791–801.
Roth, J. (1983a). *In* "Techniques in Immunocytochemistry" (G. R. Bullock and P. Petrusz, eds.), Vol. 2, pp. 217–284. Academic Press, London.
Roth, J. (1983b). *J. Histochem. Cytochem.* **31,** 547–552.
Roth, J. (1983c). *J. Histochem. Cytochem.* **31,** 987–999.
Roth, J. (1984). *In* "Masson Monographs in Diagnostic Pathology" (R. A. de Lellis, ed.), Vol. 7, pp. 43–65. Masson, New York.
Roth, J. (1986). *J. Microsc.* (*Oxford*) **143,** 125–137.
Roth, J. (1987a). *Scanning Microsc.* **1,** 695–704.
Roth, J. (1987b). *Biochim. Biophys. Acta* **906,** 405–436.
Roth, J., Bendayan, M., and Orci, L. (1978). *J. Histochem. Cytochem.* **26,** 1074–1081.
Roth, J., Bendayan, M., Carlemalm, E., Villiger, W., and Garavito, M. (1981a). *J. Histochem. Cytochem.* **29,** 663–671.
Roth, J., Thorens, B., Hunziker, W., Norman, A. W., and Orci, L. (1981b). *Science* **214,** 197–200.
Roth, J., Brown, D., Norman, A. W., and Orci, L. (1982a). *Am. J. Physiol.* **243,** F243–F252.
Roth, J., Brown, D., and Orci, L. (1982b). *J. Cell Biol.* **96,** 1189–1196.
Roth, J., Lucocq, J. M., and Charest, P. M. (1984). *J. Histochem. Cytochem.* **32,** 1167–1176.
Roth, J., Taatjes, D. J., Weinstein, J., Paulson, J. C., Greenwell, P., and Watkins, W. (1986). *J. Biol. Chem.* **261,** 14307–14312.
Rungger-Brändle, E., Englert, U., and Leuenberger, P. M. (1987). *Invest. Ophthalmol. Visual Sci.* **28,** 2026–2037.
Slot, J. W., and Geuze, H. J. (1985). *Eur. J. Cell Biol.* **38,** 87–93.
Springall, D. R., Hacker, G. W., Grimelius, L., and Polak, J. M. (1984). *Histochemistry* **81,** 603–608.
Stathis, F. C., and Fabrikanos, A. (1958). *Chem Ind.* (*London*) **27,** 860–861.
Taatjes, D. J., Schaub, U., and Roth, J. (1987a). *Histochem. J.* **19,** 235–245.
Taatjes, D. J., Chen, T.-H., Åckerström, B., Björck, L., Carlemalm, B., and Roth, J. (1987b). *Eur. J. Cell Biol.* **45,** 151–159.
Taatjes, D. J., Roth, J., Weinstein, J., and Paulson, J. C. (1988a). *J. Biol. Chem.* **263,** 6302–6309.
Taatjes, D. J., Roth, J., Peumans, W., and Goldstein, I. J. (1988b). *Histochem. J.* **20,** 478–490.
Van Tuinen, E., and Riezman, H. (1987). *J. Histochem. Cytochem.* **35,** 327–333.
Wroblewski, J., and Wroblewski, R. (1984). *Histochemistry* **81,** 469–475.
Wroblewski, J., and Wroblewski, R. (1985). *J. Microsc.* (*Oxford*) **142,** 351–362.

Note Added in Proof. We have tested a modification of the photochemical silver reaction for signal amplification described in Section V,A,1, which provides significant advantages. In particular, the entire procedure is performed in daylight. This is carried out as follows: Solution A, 100 mg of silver acetate (Fluka) dissolved in 50 ml double-distilled water using a magnetic stirrer. Silver acetate dissolves slowly. Solution B, 250 mg hydroquinone (Fluka) dissolved in 50 ml citrate buffer (0.5 *M*, pH 3.8). For signal amplification, place sections for 5 minutes in 50 ml of solution B to which 50 ml double-distilled water has been added. Afterwards, transfer sections to a solution which consists of equal volumes of solution A and solution B. Developing time is 15–18 minutes, and the reaction can be controlled under the microscope. Rinse the sections quickly in distilled water, fix with Superfix for 2 to 5 minutes, rinse with tap water, counterstain if required, dehydrate, and mount.

Chapter 25

Immunoperoxidase Methods for the Localization of Antigens in Cultured Cells and Tissue Sections by Electron Microscopy

WILLIAM J. BROWN

Section of Biochemistry, Molecular, and Cell Biology
Cornell University
Ithaca, New York 14853

MARILYN G. FARQUHAR

Department of Cell Biology
Yale University School of Medicine
New Haven, Connecticut 06510

METHODS IN CELL BIOLOGY, VOL. 31

I. Introduction

In recent years, the complexity of membrane traffic in cells has been established, in part, by the application of immunocytochemical techniques to visualize specific membrane components. Immunoelectron-microscopic methods have been especially useful for documenting the intracellular itineraries of membrane receptors such as those for low-density lipoproteins (LDL) (Pathak *et al.*, 1988), epidermal growth factor (EGF) (Beguinot *et al.*, 1984; Dunn *et al.*, 1986; Carpentier *et al.*, 1987), asialoglycoproteins (Geuze *et al.*, 1982, 1983; Hubbard *et al.*, 1985), and lysosomal enzymes bearing mannose 6-phosphate (Man6P) residues (Willingham *et al.*, 1983; Brown and Farquhar, 1984,a,b; Brown *et al.*, 1984, 1986; Geuze *et al.*, 1984, 1985). Two general methods have been developed for localizing antigens at the electron microscopic (EM) level: (1) the "diffusion" or "preembedding" methods, which rely on the permeation of cells or tissues with primary antibodies followed, in most cases, by an appropriate second antibody coupled to the enzyme horseradish peroxidase (HRP), with immune complexes then being rendered visible by the diaminobenzidine (DAB) reaction (i.e., immunoperoxidase), or antibodies coupled to ferritin particles (i.e., the ferritin-bridge method); and (2) the "surface-labeling" or "postembedding" methods, which utilize the direct application of primary antibodies to cellular compartments that have been exposed by one of several sectioning techniques, followed by a second antibody coupled to an electron-dense particle such as ferritin or colloidal gold.

There are many variations of these general methods, each with advantages and disadvantages (for descriptions of some of these methods, see Willingham, 1980; Tougard *et al.*, 1980; Roth *et al.*, 1981, and this issue; Geuze *et al.*, 1981; Armbruster *et al.*, 1982; Altman *et al.*, 1982; Griffiths *et al.*, 1983; Brown and Farquhar, 1984a,b; Keller *et al.*, 1984; Kerjaschki *et al.*, 1986; Tokuyasu, 1986, to name but a few). Rather than review the field of immunocytochemistry, it is the intent of this article to present our detailed methods for the visualization of antigens at the EM level by immunoperoxidase (IP) cytochemistry in cultured cells and tissue sections. Moreover, we hope that the methods given are of sufficient detail that they can be taken directly to the bench for use by a new investigator. It is worth mentioning, however, that IP does offer the following advantages: (1) It is highly sensitive because the immunocytochemical signal can be enzymatically amplified by the DAB reaction and therefore is very useful for the detection of sparse antigens. (2) The preservation and presentation of cellular structures approaches that of conventional transmission EM. (3) No special equipment is needed beyond that routinely used for transmission EM. The major disadvantages of the IP method are that it is not a quantitative method, and the DAB reaction product can diffuse from the site of its generation and bind nonspecifically to inappropriate structures. However, given the methods currently available, IP cytochemistry can be used to obtain reliable, qualitative information on the distribution of many interesting molecules at the EM level, much in the same way that immunofluorescence (IF) does at the light-microscopic (LM) level.

In our laboratories, we have used IP methods to determine the distribution of a variety of membrane-bound molecules. Some of these molecules include the cation-independent Man6P receptor for lysosomal enzymes (Brown and Farquhar, 1984a,b; Brown *et al.*, 1984), the pathogenic antigen of Heyman's nephritis, gp330 (Kerjaschki and Farquhar, 1983), tranferrin receptors (Woods *et al.*, 1986), cell surface proteoglycans (Stow *et al.*, 1985), and a 58-kDa resident protein of cis-Golgi cisternae (Saraste *et al.*, 1987). The same or similar IP methods have been used by others to visualize a variety of membrane molecules such as endogenous markers for the endoplasmic reticulum and Golgi complex (Louvard *et al.*, 1982; Chicheportiche *et al.*, 1984; Dunphy *et al.*, 1985; Yuan *et al.*, 1987), lysosomes (Reggio *et al.*, 1984; Lewis *et al.*, 1985; Barriocanal *et al.*, 1986; Lippincott-Schwartz and Fambrough, 1986), specific plasma membrane domains of hepatocytes (Hubbard *et al.*, 1985), endogenous markers of the endocytic and secretory pathways (Chicheportiche and Tartakoff, 1987), viral membrane glycoproteins (Saraste and Hedman, 1983; Saraste and Kuismanen, 1984; Rodriquez-Boulan *et al.*, 1984; Copeland *et al.*, 1986), and membrane receptors (Dunn *et al.*, 1986; Pathak *et al.*, 1988).

II. Localization of Antigens within Cultured Cells

The methods detailed in Sections II and III were modified from those developed by Feldman (1974), Ohtsuki *et al.* (1978), Maurice *et al.* (1979), Courtoy *et al.* (1980), Tougard *et al.* (1980), and Louvard *et al.* (1982).

A. Cell Attachment

The most convenient way to perform the IP procedure is on adherent, cultured cells grown on 35-mm plastic culture dishes, which yields a reasonable number of cells, and solutions can simply be pipeted or poured on and off the culture dish. Larger dishes require more antibody, which may be in short supply. Culture plates having 6 or 12 wells can also be used. Cells grown in suspension or those that do not adhere to standard tissue culture plastic surfaces can be made more adhesive in a variety of ways, most commonly by coating the substratum with poly-L-lysine (Mazia *et al.*, 1975). We have found that treating otherwise nonadherent tissue culture plastic with poly-L-lysine is useful for attachment of myeloma cells (Woods *et al.*, 1986) and HL-60 promyelocytic leukemia cells (unpublished results). If conditions cannot be found to attach cells to a surface, then cells can be carried through the procedure as a suspension requiring centrifugation at each step. In our experience, this is very time-consuming and, more importantly, damages the cells, so a few preliminary experiments to find attachment conditions are probably worth the effort. All of the subsequent incubations on cultured cells are done on a reciprocating platform shaker; rotary shakers tend to cause the center of the dish to dry out, especially if small volumes of fluid are used (as may be present during antibody incubations). Also, care should be taken to prevent the cells from drying out when changing solutions on the dishes, as this has deleterious effects on the ultrastructural preservation.

For most cell types, IP should not be done soon after trypsinization and replating, because many seem to be sensitive to oxidative damage during the DAB reaction. It is better to plate the cells out at lower density and wait 2–4 days for them to repopulate the dish. In most cases, cells are grown to ~75% confluency because it is easier to monitor the DAB reaction on these, rather than fully confluent dishes (see Section II,E). However, sometimes it is desirable to allow cells to form a confluent monolayer, for example, to study the biology of polarized epithelial cells in cultures (Vega-Salas *et al.*, 1987). Because oxidized DAB yields a visible (brown-black) reaction product, it is very convenient to perform many preliminary experiments by IP at the LM level. This can be useful for rapidly determining potential fixatives, titrating antibodies and detergents, and determining approximate DAB reaction times. For these preliminary experiments, cells are grown on microscope slide coverslips as commonly done for IF, and after the DAB reaction, coverslips are mounted on slides and viewed by LM.

B. Fixation

Many fixatives have been used for immunocytochemistry; however, the best fixative for a particular antigen and antibody can only be determined empirically. The best morphological preservation is usually achieved by a combination of formaldehyde (2%) and low concentrations of glutaraldehyde (0.05–0.1%). However, the antigenicity of many proteins is severely reduced in glutaraldehyde, and fixation in formaldehyde alone does not adequately preserve cellular structure. For example, on our hands the antigenicity of the Man6P receptor is completely lost by fixation in 0.01% glutaraldehyde for just 15 minutes. To avoid these problems, McLean and Nakane (1974) developed a combination fixative of periodate–lysine–paraformaldehyde (PLP), which was designed to crosslink carbohydrate residues rather than polypeptides and thus retain antigenic sites on proteins. Since membranes are particularly rich in glycoproteins and glycolipids, this fixative is especially useful for studies of membrane components. We have used this fixative extensively and have also found that it is very useful for preserving antigenicity and morphology. However, any amount of glutaraldehyde that can be tolerated by the antigen should be used, as better morphological preservation will result. Again, it is worth the effort to conduct preliminary experiments at the LM level to determine the fixative that is most likely to result in immunoreactivity as well as good ultrastructural preservation. The PLP fixative is prepared from stock solutions of paraformaldehyde and a lysine–sodium phosphate buffer as given in Section III,D. Cells are fixed by replacing the media with PLP fixative followed by incubation for 2–3 hours at room temperature (RT). Fixation for longer times results in better morphology, but there can be a significant loss of antigenicity for some molecules (McLean and Nakane, 1974; our unpublished results). In our experience, fixation at RT gives better preservation than at 4° or 37°C. This and all subsequent steps are done at RT, with the exception of osmium tetroxide postfixation and staining (see Section II,F). After fixation, cells are washed three times (10 minutes each) with 0.15 *M* NaCl, 0.01 *M* sodium phosphate (PBS), pH 7.4. Fewer cells will be lost and damaged if the buffer is gently poured on, rather than squirted with a pipet. With the PLP fixative, we have found that it is not necessary to quench free aldehydes after fixation; washing in PBS is sufficient (Brown and Farquhar, 1984b).

C. Permeabilization

The most satisfying method devised so far for permeabilizing cells and maintaining ultrastructural integrity involves the use of saponin (Ohtsuki *et al.*, 1978). Usually we permeabilize cells with 0.05–0.005% saponin in PBS for 5–10 minutes. The optimal concentration can vary depending on the source and batch of the detergent and the type of cell being investigated (Louvard *et al.*, 1982). Saponin permeabilizes cells by forming complexes with cholesterol; therefore, the ability of saponin to allow passage of antibodies through the plasma membrane (and other intracellular membranes) depends on the cholesterol concentration (Lucy and Glaubert, 1964). Membranes with lower cholesterol content will be less sensitive to saponin treatment. It is important to titrate the saponin to establish the absolute minimum concentration required to allow antibody penetration. This can easily be done by IF or IP at the LM level. Equally important, saponin at low concentrations probably results in the formation of reversible membrane pores (~8 nm in diameter) (Lucy and Glaubert, 1964; Seeman *et al.,* 1973); therefore, membranes can be permeabilized with saponin to allow antibody passage and then resealed (by rinsing in saponin-free buffers) to prevent oxidized-DAB diffusion. Establishing a minimum saponin concentration will ultimately result in less background staining during the DAB reaction. Whatever this concentration turns out to be, it should be used in all of the subsequent antibody incubations and washes to allow diffusion of antibodies through membranes (Ohtsuki *et al.,* 1978). Culture supernatants from hybridomas can be used as the primary antibody source without significant dilution by adding an appropriate volume from a 0.5% saponin stock solution. We have found Sigma (S-1252) saponin to be satisafctory for these experiments.

D. Antibody Incubation

1. Incubate cells with the first antibody diluted in PBS + 0.005% saponin (buffer A) for 1–2 hours. The appropriate antibody concentration can only be determined empirically; therefore, when first starting out it will be necessary to titrate the antibody. This can easily be done by IP at the LM level, as the antibody concentrations that give a good LM signal are generally the same as used for EM. Also, it is best to use affinity-purified antibodies or at least IgG fractions to reduce the background staining given by most antisera. Background can be decreased by including 0.1% ovalbumin, BSA, or other "inert" proteins in the antibody mixture. For 35-mm dishes, ~0.5 ml is the minimal volume (with shaking) needed to cover and keep the cells from drying out. After incubation, cells are washed three times (5–15 minutes each) with buffer A.

2. Incubate cells with a second antibody–HRP conjugate diluted in buffer A for 1–2 hours. We have found that the quality of the second antibody can have a significant effect on the end result. Most investigators use a goat or sheep anti-rabbit IgG (for rabbit polyclonals) or anti-mouse Ig (for mouse monoclonals) coupled to HRP. For some antibodies, it is also possible to use protein A–HRP as a detecting probe. We routinely

use Fab fragments of sheep anti-rabbit IgG coupled to HRP (Biosys, Compiegne, France), as the small conjugate size and low amount of unconjugated HRP result in good penetration and low background staining. Again, these reagents should be titrated to obtain the best results. After incubation, cells are washed three times (5–15 minutes each) with buffer A.

3. Wash cells three times over a period of 5–10 minutes with PBS alone to rinse out the saponin. By removing the saponin, the diffusion of oxidized DAB reaction product can be significantly reduced.

4. Fix with 1.5% glutaraldehyde in 0.1 *M* sodium cacodylate (pH 7.4) containing 5% sucrose for 30–60 minutes. This step is essential to prevent further deteriorization of morphology during subsequent processing and to fix the antigen–antibody complexes in place. The fixative and subsequent washes are made slightly hypertonic to shrink membrane-limited compartments and thereby inhibit the diffusion of oxidized, but not yet polymerized, DAB generated in membrane-enclosed compartments. After fixation, the cells are washed three times (10 minutes each) with 0.1 *M* cacodylate (pH 7.4), containing 7.5% sucrose.

5. Rinse cells three times with 50 m*M* Tris (pH 7.4) containing 7.5% sucrose (buffer B).

E. The Diaminobenzidine (DAB) Reaction

There are many variations of this reaction, which was originally described by Graham and Karnovsky (1966). Some of the important pitfalls in DAB cytochemistry have been discussed at length elsewhere and will not be reiterated here (Novikoff *et al.*, 1972; Novikoff, 1980; Courtoy *et al.*, 1983). The specific, currently employed steps followed by us are given here.

1. We have found that for best results, care should be taken when preparing the DAB solution, which is made just before use (e.g., while the cells are washing in cacodylate buffer). The solution consists of 0.2% 3,3′-diaminobenzidine tetrahydrochloride (Sigma) in buffer B. Store in the dark until use. Other concentrations of DAB can be used, but 0.2% is a good starting point. To make the DAB solution, add an appropriate amount of DAB to ~90% of the final buffer B volume, and mix vigorously with a magnetic stirring bar in the dark (put a box over the stirring plate). The DAB should dissolve in several minutes to yield a colorless or slightly brownish solution. At this stage the pH will drop to ~2.5. Readjust the pH of the solution with NaOH until it is ~7.0–7.4. (For 25 ml of DAB, one drop of 10 *N* NaOH plus several drops of 1 *N* NaOH is usually enough,

but this should be checked by continuous monitoring with a pH meter.) Bring up to the final desired volume with buffer B. Filter the DAB by passing through a 0.2-μm filter (we use a disposable Millex filter). The filter may turn brown, and the DAB solution should be transparent and colorless or very light brown. Keep the DAB in the dark until use. Use disposable gloves, as DAB is carcinogenic.

2. Add 1 ml of DAB to the cells; allow to incubate for 1 minute.

3. Initiate the DAB reaction by adding H_2O_2 to the DAB solution (final H_2O_2 concentration 0.005–0.01%). We add the H_2O_2 in small aliquots from a 3% stock solution. The length of reaction time is variable depending on how quickly the reaction proceeds. It is best not to have an explosive reaction, because this usually leads to diffusion of DAB away from the reaction site. It is better to have the reaction proceed slowly at first; in some cases, it is even desirable to initiate the reaction at 4°C to slow down the initial burst. Generally, our reactions proceed for 5–15 minutes at RT, and we add H_2O_2 in two or three aliquots over a period of ~5 minutes to reach the final concentration previously specified. During this time, the cells will probably turn brown or black. Again, this is somewhat variable depending on the antigen and antibody. Cells incubated with affinity-purified antibodies give less background and therefore will often not turn as dark as those incubated with serum, but the preciseness of the localization to specific cell structures will be better. The reactions are routinely monitored by regular bright-field LM (conventional or inverted microscopes with objective lenses of about ×20). In most cases, we initially localize antigens by IF in order to provide information as to the general cellular distribution of the antigen in question. This is very helpful, as the DAB reaction can be monitored to see if a similar pattern to that seen by IF emerges as expected. Many negative experiments can thus be terminated at this step rather than proceeding on with the time-consuming EM procedures. The reaction is stopped when the cells have turned "dark enough." In our experience, optimal EM staining is usually obtained when the reaction product looks slightly faint by LM observation. By the time an IP reaction reaches the point of being optimal by LM observation, it has usually gone too far for optimal EM visualization because DAB diffusion has occurred. In the beginning, however, it is better to err on the side of excess so that at least some signal will be seen. Given that starting point, adjustments of the antibody concentrations, reaction times, etc., can be made in subsequent experiments. As a matter of routine, we generally have enough samples to carry out a long and short DAB reaction for each experimental condition, in the hope that one or the other will be optimal. The reaction is stopped by rinsing the cells three times with buffer B.

In the cases where an antigen cannot be detected by these methods, further amplification can be achieved by applying the peroxidase–antiperoxidase techniques of Sternberger (1979) or by the use of biotinylated-IgG and streptavidin–HRP conjugates (available from many commercial suppliers).

F. Staining with Reduced Osmium Tetroxide

Postfix and stain cells in reduced osmium tetroxide [1% OsO_4, 1% potassium ferrocyanide ($K_4Fe(CN)_6$), in 0.1 *M* sodium cacodylate (pH 7.4)] for 1 hour at 4°C. Reduced OsO_4 gives significantly better staining of membranes and oxidized DAB deposits (Karnovsky, 1971; Courtoy *et al.*, 1982). We have found that adding solid $K_4Fe(CN)_6$ to the buffered OsO_4 solution [rather than dissolving $K_4Fe(CN)_6$ in buffer first, then adding OsO_4] results in the best staining. The osmium solution should be dark-brown to black after reduction. Note that this step is carried out at 4°C. The length of incubation time at this step can also significantly affect the amount of electron-dense reaction product formed; in most of our studies on relatively sparse membrane molecules, cells are incubated no less than 45 minutes in reduced osmium tetroxide. After osmication, rinse cells three times (or until no further osmium is removed) with cold 0.1 *M* sodium cacodylate, pH 7.4.

G. Dehydration and Embedding

1. Dehydrate cells by pouring on and off a graded series of ethanol in the usual manner (e.g., 70%, 95%, 100%). Dehydration of monolayer cells is very rapid, so a few seconds in each step of ethanol is sufficient. Rinse three times with 100% ethanol.

2. Cells can be embedded on the dish or removed to make a pellet, depending on the goals of the study. In some cases, it is desirable to obtain views of cells cut perpendicular to the plane of the substratum (e.g., monolayers of polarized epithelial cells). To accomplish this, subject the monolayer to several changes of the resin of choice (e.g., Epox, Spurr's), then incubate cells overnight in 100% resin at RT. Replace resin with a thin layer (~1–2 mm) of fresh resin and allow to harden. The embedded cells can then be pulled off of the dish by peeling away the polymerized-plastic sheet. Rather than section a single monolayer of cells, attach two sheets of cells together on their substrate surfaces with a small amount of resin and reembed in a flat mold. Be sure that no air bubbles are left between the attached sheets. To obtain a sampling of randomly oriented cells, a cell pellet is prepared. To do this,

add 100% propylene oxide to the dish, quickly score the dish with the end of a Pasteur pipet, and then gently remove cells from the dish by repeated pipetings (Bodel *et al.*, 1977). Propylene oxide rapidly solubilizes the plastic underneath the cells, so the trick is to harvest the cells before large amounts of plastic also come off. Microfuge the cells to a pellet, and wash several times in propylene oxide by gentle resuspension to remove any residual plastic, which may interfere with thin sectioning. Alternatively, amyl acetate (100%) can be used to remove cells, which has the advantage that the plastic is not so quickly solubilized. However, when amyl acetate is used, cells should then be centrifuged and washed several times in propylene oxide before embedding. Place the cells in a solution of 50 : 50 propylene oxide–plastic resin for 15–30 minutes, pellet the cells at maximum speed in a Microfuge, and then replace the 50 : 50 mixture with 100% unpolymerized resin overnight (in an uncapped tube). If the cell pellet floats in the plastic or tends to come apart, place the sample in a 60°C oven for 5–10 minutes (to reduce the viscosity of the plastic), and centrifuge at the highest speed with a Microfuge until a pellet is re-formed (5–10 minutes are usually sufficient). Polymerize the resin at the appropriate temperature.

3. Thin sections are cut with an ultramicrotome in the usual manner; however, to improve the electron-dense DAB–OsO_4 signal, it is usually desirable to stain the sections with lead citrate alone (omitting uranyl acetate). Contrast of specimens can then be enhanced by viewing at lower accelerating voltages (e.g., 60 kV versus 80 kV) in the electron microscope using a small-objective aperture (i.e., 20–30 μm).

III. Special Considerations for the Localization of Antigens within Cells of Tissue Sections

It is often desirable to obtain information on the distribution of membrane or matrix components *in situ*; however, performing IP on tissues presents many problems not encountered with cultured cells. Principal among these is providing a means for the antibodies to gain access to antigens located within intracellular compartments. To overcome this problem, methods have been developed that allow access to antigens inside cells, which usually involve the freezing of tissues followed by sectioning of fixed tissues prior to antibody incubation (McLean and Nakane, 1974; Courtoy *et al.*, 1980). We have modified procedures that utilize frozen cryostat tissue sections for the permeation of antibodies into intracellular compartments (Feldman, 1974; Maurice *et al.*, 1979; Courtoy *et al.*, 1980). The basic steps involve fixation of the tissue or organ, cryoprotection in dimethyl sulfoxide (DMSO), freezing of the tissue in isopentane, cutting of relatively thick frozen sections, and then incubation with antibodies. This and similar IP methods have been used to detect a variety of membrane receptors and other molecules (Brown and Farquhar, 1984a,b; Novikoff *et al.*, 1983; Hubbard *et al.*, 1985; Kerjaschki and Farquhar, 1983). Many of the important considerations for successfully localizing antigens in tissue sections are the same as described for cultured cells, so a review of Section II is suggested.

A. Fixation

Fix animal tissues by perfusion (if possible) with McLean and Nakane's PLP fixative for 5–10 minutes. Fixation by perfusion generally gives better ultrastructural preservation than immersion alone. Remove the tissues, carefully slice into small pieces of workable size (~2 mm), and continue fixing the tissue pieces by immersion in PLP fixative for 4–6 hours at RT. Glutaraldehyde-containing fixatives may be used as discussed before. After fixation, wash tissue three times (15 minutes each) in PBS, pH 7.4.

B. Cryoprotection and Cryosectioning

1. Cryoprotect the tissue by placing in 10% DMSO in PBS for 1 hour at 4°C.

2. Freeze by immersing tissues in isopentane (2-methylbutane, Kodak). To do this, fill a small metal beaker or cup with isopentane and place in a larger Styrofoam box or container. Fill the box with liquid nitrogen (N_2) up to the level of the isopentane. Wait for the isopentane to freeze (~20 minutes). With a small metal probe (e.g., spatula blade), melt a small well in the frozen isopentane until large enough to put in a piece of tissue. Wait a few seconds for the isopentane to cool (but not freeze); during this time, place a piece of tissue on the end of a wooden stick. Rapidly immerse the stick with tissue into the well of liquid isopentane. Within seconds the tissue will turn white and be frozen throughout. Leave in the isopentane for 5–10 seconds; do not leave the tissue in longer or the isopentane will freeze with the tissue in it. Remove the stick and place in the liquid N_2 bath. The tissue can be left here until all of the pieces have been frozen in a similar fashion. It may be necessary to occasionally remelt the frozen isopentane, if several pieces of tissue are being processed. With a scalpel equilibrated to liquid-N_2 temperature, cut the

tissue off the stick while still immersed in liquid N_2. With similarly chilled forceps, place the tissues in a cryovial for storage in liquid N_2 (punch several small holes in the vial so liquid N_2 will bathe the tissue). Tissue frozen in this manner can be stored in liquid N_2 for long periods of time (i.e., several months). This is a handy stopping point.

3. Relatively thick sections are cut from the frozen tissues with a cryostat microtome. The cryostat should be prechilled to the appropriate sectioning temperature. Retrieve the frozen tissue and keep in a liquid-N_2 bath until sectioned. Remove a piece of tissue and place in OTC mounting medium. Equilibrate the tissue to about $-18°$ to $-20°C$ in the cryostat and then cut sections 15–20 μm thick. The best sectioning temperature may vary with the tissue. After the sections are cut, place them into small tubes (e.g., 10 × 75 mm are convenient) containing PBS (pH 7.4) and 1% ovalbumin (all subsequent incubations will be done in these tubes). For immunoreactions, three to five sections are optimal. In principle, use of thinner sections improves antibody penetration, but they are difficult to handle and tend to curl. For most tissues, 15- to 20-μm sections are optimal. In some cases, thicker sections are required; for example, glomeruli tend to fall out of kidney sections, so we cut them ~30 μm. Sections tend to be rather fragile, so solutions should be gently added to the tubes.

4. It has been shown that Vibratome sections can be used as an alternative to cryostat sectioning (Novikoff *et al.*, 1983). The advantage of these sections is that the tissues do not have to be frozen, but in our experience antibodies do not penetrate these sections as effectively as those frozen and cut on a cryostat.

C. Antibody Incubation

1. Incubate in the first antibody diluted in PBS (pH 7.4) containing 1% ovalbumin and 0.02% NaN_3 overnight at RT. The longer incubation times for this procedure are suggested for convenience and because antibodies do not permeate the multilayered tissue sections as rapidly as they do a detergent-permeabilized cultured-cell monolayer. It is possible to saponin-permeabilize the thawed tissue sections prior to antibody incubations as done for cultured cells; however, for most antigens and antibodies this has not proved to be necessary or desirable. To change solutions, allow the sections to sink to the bottom of the tube, and aspirate the remaining fluid, being careful not to remove any sections. For 10 × 75-mm tubes, 200–300 μl is a sufficient volume to cover the sectons. For this and subsequent steps, we allow the tubes to rotate gently on a slowly moving Ferris wheel-type mixer. Care must be taken to ensure that

all sections remain immersed in the fluid and do not dry out. Seal the tubes with parafilm. All solutions containing ovalbumin should be filtered with Whatman no. 1 paper before use (i.e., before adding antibodies). As discussed before, it is necessary first to titrate the antibodies. This can easily be done by IF or IP on crystat sections attached to albumin-coated microscope slides (Courtoy *et al.*, 1980). It has been our experience with this technique that a good IP signal can be achieved at an antibody concentration that is about twice that which gives a good IF signal on cryostat sections. However, there is great variability depending on the nature of the antibody and antigen, but this should give a reasonable starting point. After incubation, wash the sections three times (15 mintues each) with PBS (pH 7.4) containing 0.1% ovalbumin (buffer C) at RT. Gently add several milliliters of wash buffer.

2. Incubate in the second antibody–HRP conjugate diluted in PBS containing 1% ovalbumin for 2–4 hours at RT. Again, reagents used at this step should be titrated to obtain best results (see Section II,D. for discussion). After incubation, wash the sections three times (15 minutes each) with buffer C at RT.

3. Fix the sections in 1.5% glutaraldehyde in 0.1 *M* sodium cacodylate (pH 7.4), containing 5% sucrose for 1 hour at RT. After fixation, wash sections three times (10 minutes each) with 0.1 *M* sodium cacodylate (pH 7.4) containing 7.5% sucrose at RT. Rinse sections three times with buffer B.

D. The DAB Reaction

1. The DAB reaction is carried out essentially as described for cultured cells. It is advisable to monitor these reactions with a dissecting microscope to determine the level of reaction product formed. Unfortunately, it is often difficult to make conclusions about the level of intracellular staining depending on the distribution of the antigen; nevertheless, it is useful to monitor reactions at the LM level when first using these methods so that comparisons with subsequent experiments can be made. Stop the reaction by rinsing sections three times in buffer B.

2. Postfix and stain sections in reduced osmium tetroxide, dehydrate, and embed in plastic resins as described for cultured cells. Embed sections in flat molds. Several sections can be stacked in one mold for a better sampling.

3. Thin-section, stain with lead citrate, and view by EM as described for cultured cells. Reaction product can also usually be visualized by LM in semithin (0.5-μm) sections.

4. The suggested sequence for investigators not familiar with these methods is as follows:

Day 1: Fix and freeze tissue, store until ready for sectioning.
Day 2: In the afternoon, cut the cryostat sections, and apply the first antibody for overnight incubation.
Day 3: Complete the procedure (i.e., second antibody, DAB reaction, OsO_4 staining, embedding).

With experience, days 1 and 2 can be done in one 12-hour day.

5. McLean and Nakane (1974) PLP fixative can be made in the following way:

Solution A: 0.1 *M* lysine HCl–sodium phosphate buffer
Add 1.83 g lysine HCl to 50 ml distilled water (dH_2O).
Add 0.1 *M* Na_2HPO_4 until pH 7.4 (~5 ml).
Bring up to 100 ml with 0.1 *M* $NaPO_4$, pH 7.4 buffer.
Solution B: 8% paraformaldehyde
Add 8 g paraformaldehyde (Fisher Scientific) to ~95 ml of dH_2O.
Heat to 60°C with stirring.
Add 1 *N* NaOH dropwise until solution clears.
Bring up to 100 ml with dH_2O. Filter through Whatman no. 1 paper.

Solutions A and B can be stored for several weeks at 4°C; however, we routinely use them freshly prepared. Mix three parts solution A with one part solution B, and add sodium periodate ($NaIO_4$) to a final concentration of 0.01 *M* (i.e., 2.13 mg $NaIO_4$/ml of A + B mixture). Use complete fixative within 4–5 hours at RT. Final concentrations: 2% paraformaldehyde, 0.01 *M* periodate, 0.075 *M* lysine, and 0.075 *M* phosphate buffer. It is also possible to use a final concentration of 4% paraformaldehyde for even better ultrastructural preservation.

IV. Summary

We have presented our detailed methods for localizing antigens in cultured cells and tissue sections by IP at the EM level. Immunoperoxidase cytochemistry is particularly well suited for the study of sparse antigens as a result of the enzymatic amplification afforded by the method, and of molecules confined within a membrane-enclosed compartment wherein the DAB reaction produce can accumulate. Although

IP is commonly used to localize membrane-compartmentalized molecules, reliable qualitative information can also be obtained on cytoplasmic antigens as well (Anderson *et al.*, 1978; Merisko *et al.*, 1986; Rodman *et al.*, 1984). For these and other reasons, it is likely that IP cytochemistry will continue to be an important tool for the cell biologist especially in the study of membrane traffic. Other inventive combinations of immunocytochemical methods will likely be forthcoming, for example, combining IP localization with postembedding labeling by colloidal-gold conjugates to provide triple EM labeling.

References

Altman, L. G., Schneider, B. G., and Papermaster, D. S. (1982). *J. Histochem. Cytochem.* **32,** 1217–1220.

Anderson, R. G. W., Vasile, E., Mello, R. J., Brown, M. S., and Goldstein, J. L. (1978). *Cell (Cambridge, Mass.)* **15,** 919–933.

Armbruster, B. L., Carlemalm, E., Chiovetti, R., Gravioto, R. M., Hobot, J. A., Kellenberger, E., and Villiger, W. (1982). *J. Microsc. (Oxford)* **126,** 77–92.

Barriocanal, J. G., Bonifacino, J. S., Yuan, L., and Sandoval, I. V. (1986). *J. Biol. Chem.* **261,** 16755–16763.

Beguinot, L., Lyall, R. M., Willingham, M. C., and Pastan, I. (1984). *Proc. Natl. Acad. Sci. U.S.A.* **81,** 2384–2388.

Bodel, P. T., Nichol, B. A., and Bainton, D. F. (1977). *J. Exp. Med.* **145,** 264–274.

Brown, W. J., and Farquhar, M. G. (1984a). *Cell (Cambridge, Mass.)* **36,** 295–307.

Brown, W. J., and Farquhar, M. G. (1984b). *Proc. Natl. Acad. Sci. U.S.A.* **81,** 5135–5139.

Brown, W. J., Constantinescu, E., and Farquhar, M. G. (1984). *J. Cell Biol.* **99,** 320–326.

Brown, W. J., Goodhouse, J., and Farquhar, M. G. (1986). *J. Cell Biol.* **103,** 1235–1247.

Carpentier, J.-L., White, M. F., Orci, L., and Kahn, R. C. (1987). *J. Cell Biol.* **105,** 2751–2762.

Chicheportiche, Y., and Tartakoff, A. L. (1987). *Eur. J. Cell Biol.* **44,** 135–143.

Chicheportiche, Y., Vassalli, P., and Tartakoff, A. M. (1984). *J. Cell Biol.* **99,** 2200–2210.

Copeland, C. S., Doms, R. W., Bolzau, E. M., Webster, R. G., and Helenius, A. (1986). *J. Cell Biol.* **103,** 1179–1191.

Courtoy, P. J., Kanwar, Y. S., Hynes, R. O., and Farquhar, M. G. (1980). *J. Cell Biol.* **87,** 691–696.

Courtoy, P. J., Timpl, R., and Farquhar, M. G. (1982). *J. Histochem. Cytochem.* **30,** 874–886.

Courtoy, P. J., Picton, D. H., and Farquhar, M. G. (1983). *J. Histochem. Cytochem.* **31,** 945–951.

Dunn, W. A., Connolly, T. P., and Hubbard, A. L. (1986). *J. Cell Biol.* **102,** 24–36.

Dunphy, W. G., Brands, R., and Rothman, J. E. (1985). *Cell (Cambridge, Mass.)* **40,** 463–472.

Feldman, G. (1974). *J. Microsc. (Oxford)* **21,** 293–300.

Geuze, H. J., Slot, J. W., Van Der Ley, P. A., and Scheffer, R. C. T. (1981). *J. Cell Biol.* **89,** 653–665.

Geuze, H. J., Slot, J. W., Strous, G. J. A. M., Lodish, H. F., and Schwartz, A. L. (1982). *J. Cell Biol.* **92,** 865–870.
Geuze, H. J., Slot, J. W., Strous, G. J. A. M., Lodish, H. F., and Schwartz, A. L. (1983). *Cell (Cambridge, Mmass.)* **32,** 277–287.
Geuze, H. J., Slot, J. W., Strous, G. J. A. M., Hasilik, A., and von Figura, K. (1984). *J. Cell Biol.* **98,** 2047–2054.
Geuze, H. J., Slot, J. W., Strous, G. J. A. M., Hasilik, A., and von Figura, K. (1985). *J. Cell Biol.* **101,** 2253–2262.
Graham, R. C., and Karnovsky, M. J. (1966). *J. Histochem. Cytochem.* **14,** 291–302.
Griffiths, G., Simon, K., Warren, G., and Tokuyasu, K. T. (1983). *In* "Methods in Enzymology" (S. Fleischer and B. Fleischer, eds.), Vol. 96, pp. 466–485. Academic Press, New York.
Hubbard, A. L., Bartles, J. R., and Braiterman, L. T. (1985). *J. Cell Biol.* **100,** 1115–1125.
Karnovsky, M. J. (1971). *Abstr. 11th Annu. Meet., Am. Soc. Cell Biol.* p. 146.
Keller, G. A., Tokuyasu, K. T., Dutton, A. H., and Singer, S. J. (1984). *Proc. Natl. Acad. Sci. U.S.A.* **81,** 5744–5749.
Kerjaschki, D., and Farquhar, M. G. (1983). *J. Exp. Med.* **157,** 667–686.
Kerjaschki, D., Sawada, H., and Farquhar, M. G. (1986). *Kidney Int.* **30,** 229–245.
Lewis, V., Green, S. A., Marsh, M., Vihko, P., helenius, A., and Mellman, I. (1985). *J. Cell Biol.* **100,** 1839–1847.
Lippincott-Schwartz, J., and Fambrough, D. M. (1986). *J. Cell Biol.* **102,** 1593–1605.
Louvard, D., Reggio, H., and Warren, G. (1982). *J. Cell Biol.* **92,** 92–107.
Lucy, J. A., and Glaubert, A. M. (1964). *J. Mol. Biol.* **8,** 727–748.
Maurice, M., Feldman, G., Druet, P., Laliberte, F., and Bouige, D. (1979). *Lab. Invest.* **40,** 39–48.
Mazia, D., Schatten, G., and Sale, W. (1975). *J. Cell Biol.* **66,** 198–200.
McLean, I. W., and Nakane, P. K. (1974). *J. Histochem. Cytochem.* **22,** 1077–1083.
Merisko, E. C., Farquhar, M. G., and Palade, G. E. (1986). *Pancreas* **1,** 95–109.
Novikoff, A. B. (1980). *J. Histochem. Cytochem.* **28,** 1036–1038.
Novikoff, A. B., Novikoff, P. M., Quintana, N., and Davis, C. (1972). *J. Histochem. Cytochem.* **20,** 745–749.
Novikoff, P. M., Tulsiani, D. R. P., Touster, O., Yam, A., and Novikoff, A. B. (1983). *Proc. Natl. Acad. Sci. U.S.A.* **81,** 4364–4368.
Ohtsuki, I., Manzi, R. M., Palade, G. E., and Jamieson, J. D. (1978). *Biol. Cell.* **31,** 119–126.
Pathak, R. K., Merkle, R. K., Cummings, R. D., Goldstein, M. L., Brown, M. S., and Anderson, R. G. W. (1988). *J. Cell Biol.* **106,** 1831–1841.
Reggio, H., Bainton, D., Harms, E., Coudrier, H., and Louvard, D. (1984). *J. Cell Biol.* **99,** 1511–1526.
Rodman, J. S. D., Kerjaschki, D., Merisko, E. C., and Farquhar, M. G. (1984). *J. Cell Biol.* **98,** 1630–1636.
Rodriquez-Boulan, E., Paskiet, K. T., Salas, P. J. I., and Bard, E. (1984). *J. Cell Biol.* **98,** 308–319.
Roth, J., Bendayan, M., Carlemalm, E., Villiger, W., and Garavito, M. (1981). *J. Histochem. Cytochem.* **29,** 663–671.
Saraste, J., and Hedman, K. (1983). *EMBO J.* **2,** 2001–2006.
Saraste, J., and Kuismanen, E. (1984). *Cell (Cambridge, Mass.)* **38,** 535–549.
Saraste, J., Palade, G. E., and Farquhar, M. G. (1987). *J. Cell Biol.* **105,** 2021–2029.
Seeman, P., Cheng, D., and Iles, G. H. (1973). *J. Cell Biol.* **56,** 519–527.
Sternberger, L. A. (1979). "Immunocytochemistry." Wiley, New York.

Stow, J. L., Kjellen, L., Unger, E., Hook, M., and Farquhar, M. G. (1985). *J. Cell Biol.* **100,** 975–980.
Tokuyasu, K. T. (1986). *J. Microsc. (Oxford)* **143,** 139–149.
Tougard, C., Picart, R., and Tixier-Vidal, A. (1980). *Am. J. Anat.* **158,** 471–490.
Vega-Salas, D. E., Salas, P. J. I., and Rodriquez-Boulan, E. (1987). *J. Cell Biol.* **104,** 1249–1259.
Willingham, M. C. (1980). *Histochem. J.* **12,** 419–434.
Willingham, M. C., Pastan, I. H., and Sahagian, G. G. (1983). *J. Histochem. Cytochem.* **31,** 1–11.
Woods, J. W., Doriaux, M., and Farquhar, M. G. (1986). *J. Cell Biol.* **103,** 277–286.
Yuan, L., Barriocanal, J. G., Bonifacino, J. S., and Sandoval, I. V. (1987). *J. Cell Biol.* **105,** 215–227.

INDEX

L

N

O

P

U

V

W

X

Y

Z

CONTENTS OF RECENT VOLUMES

Volume 26

Prenatal Diagnosis: Cell Biological Approaches

Volume 27

Echinoderm Gametes and Embryos

Volume 28

Dictyostelium Discoideum: Molecular Approaches To Cell Biology

Volume 29

Fluorescence Microscopy of Living Cells in Culture

Part A. *Fluorescent Analogs, Labeling Cells, and Basic Micropscopy*

Volume 30

Fluorescence Microscopy of Living Cells in Culture

Part B. *Quantitative Fluorescence Microscopy—Imaging and Spectroscopy*